90 0698624 3

Charles Seale-Hayne Library
University of Plymouth
(01752) 588 588
LibraryandITenquiries@plymouth.ac.uk

Developments in Water Science, 48

HYDROLOGICAL DROUGHT

Processes and Estimation Methods for Streamflow and Groundwater

OTHER TITLES AVAILABLE IN *DEVELOPMENTS IN WATER SCIENCE:*

Text design for Volume 48 by MARTIN MORAWIETZ (University of Oslo, Oslo, Norway)

Developments in Water Science, 48

HYDROLOGICAL DROUGHT

Processes and Estimation Methods for Streamflow and Groundwater

Edited by

LENA M. TALLAKSEN
University of Oslo, Oslo, Norway

HENNY A.J. VAN LANEN
Wageningen University, Wageningen-UR
The Netherlands

2004

ELSEVIER

Amsterdam – Boston – Heidelberg – London – New York – Oxford
Paris – San Diego – San Francisco – Singapore – Sydney – Tokyo

ELSEVIER B.V.
Sara Burgerhartstraat 25
P.O. Box 211, 1000 AE Amsterdam
The Netherlands

ELSEVIER Inc.
525 B Street, Suite 1900
San Diego, CA 92101-4495
USA

ELSEVIER Ltd
The Boulevard, Langford Lane
Kidlington, Oxford OX5 1GB
UK

ELSEVIER Ltd
84 Theobalds Road
London WC1X 8RR
UK

© 2004 Elsevier B.V. All rights reserved.

This work is protected under copyright by Elsevier B.V., and the following terms and conditions apply to its use:

Photocopying
Single photocopies of single chapters may be made for personal use as allowed by national copyright laws. Permission of the Publisher and payment of a fee is required for all other photocopying, including multiple or systematic copying, copying for advertising or promotional purposes, resale, and all forms of document delivery. Special rates are available for educational institutions that wish to make photocopies for non-profit educational classroom use.

Permissions may be sought directly from Elsevier's Rights Department in Oxford, UK: phone (+44) 1865 843830, fax (+44) 1865 853333, e-mail: permissions@elsevier.com. Requests may also be completed on-line via the Elsevier homepage (http://www.elsevier.com/locate/permissions).

In the USA, users may clear permissions and make payments through the Copyright Clearance Center, Inc., 222 Rosewood Drive, Danvers, MA 01923, USA; phone: (+1) (978) 7508400, fax: (+1) (978) 7504744, and in the UK through the Copyright Licensing Agency Rapid Clearance Service (CLARCS), 90 Tottenham Court Road, London W1P 0LP, UK; phone: (+44) 20 7631 5555; fax: (+44) 20 7631 5500. Other countries may have a local reprographic rights agency for payments.

Derivative Works
Tables of contents may be reproduced for internal circulation, but permission of the Publisher is required for external resale or distribution of such material. Permission of the Publisher is required for all other derivative works, including compilations and translations.

Electronic Storage or Usage
Permission of the Publisher is required to store or use electronically any material contained in this work, including any chapter or part of a chapter.

Except as outlined above, no part of this work may be reproduced, stored in a retrieval system or transmitted in any form or by any means, electronic, mechanical, photocopying, recording or otherwise, without prior written permission of the Publisher. Address permissions requests to: Elsevier's Rights Department, at the fax and e-mail addresses noted above.

Notice
No responsibility is assumed by the Publisher for any injury and/or damage to persons or property as a matter of products liability, negligence or otherwise, or from any use or operation of any methods, products, instructions or ideas contained in the material herein. Because of rapid advances in the medical sciences, in particular, independent verification of diagnoses and drug dosages should be made.

First edition 2004

UNIVERSITY OF PLYMOUTH

9006986243

ISBN: 0-444-51688-3 (Hardbound)
ISBN: 0-444-51767-7 (Paperback)
Accompanying CD: ISBN: 0-444-51768-5 (for copyright of CD – see CD Readme file)

ISSN: 0167-5648

♾ The paper used in this publication meets the requirements of ANSI/NISO Z39.48-1992 (Permanence of Paper).
Printed in The Netherlands.

Working together to grow
libraries in developing countries

www.elsevier.com | www.bookaid.org | www.sabre.org

ELSEVIER   BOOK AID
            International   Sabre Foundation

PREFACE

Hydrological drought – processes and estimation methods for streamflow and groundwater is a textbook for university students, practising hydrologists, lecturers and researchers who are engaged in the analysis of hydrological drought. The aim is to provide knowledge and understanding of the drought phenomenon and give the reader a comprehensive overview of methods and tools to estimate and manage hydrological drought. It includes a qualitative conceptual understanding of drought features and processes, a detailed presentation of estimation methods and tools, practical examples and key aspects of operational practice. Basic requirements are introductory courses in hydrology and statistics.

In contrast to floods, the literature available on hydrological drought is limited, in particular material that deals with quantitative methods for drought analyses. However, a large amount of knowledge in various fields of drought analysis has accumulated in scientific journals, reports and technical manuals. Clearly, there was a need to collate, review and supplement this material and subsequently present a synthesis of the current knowledge on hydrological drought in one volume. The main objective of the textbook is to make the material available to a larger audience and to present it to the reader in a concise and easily accessible form. The material presented ranges from well-established knowledge and analysing methods to recent developments in drought research. Its nature varies accordingly, from a more traditional textbook with its clear statements to that of a research paper, introducing new approaches and methodologies for drought analysis.

Drought is a worldwide phenomenon that originates from a prolonged deficiency in precipitation over an extended region. The deficiency may cause a hydrological drought to develop and give rise to below normal levels of streamflow, lakes, soil moisture and groundwater. The focus is here on streamflow and groundwater as these are useful as indicators of the available water resources in a region or at a site. The water levels of lakes and reservoirs are also commented on as these may, in some cases, particular in semi-arid regions, provide more relevant information. Precipitation deficit, soil water deficit, water quality and socio-economic aspects are topics only briefly commented upon due to space limitations. However, efforts have been made to briefly introduce these topics and to provide key references for the reader who seeks further information.

The book is divided into three parts. The drought phenomenon, its main features, regional diversity and controlling processes are discussed in general terms in Part I (Drought as a natural hazard). The chapters are descriptive and of an introductory nature and are therefore suited for undergraduate level. Part II (Estimation methods) presents contemporary approaches to drought estimation, and methods and calculation details are demonstrated using the sample data sets that are provided on the accompanying CD. Emphasis is given to estimation methods based on time series of hydrological drought characteristics and estimation of drought indices at the gauged and ungauged site. Alongside this, recommendations and possible limitations for application are given. The material in Part II is more advanced and aims at the graduate level. Part III (Living with drought) deals with possible human impacts on drought, ecological effects of drought on stream biota and key aspects of operational low flow hydrology. The chapters focus on hydrological drought and its interaction with man and ecology and are of an interdisciplinary character. All chapters conclude with a short summary which highlights the main issues and recommendations given in each chapter, whereas an overall outlook to future needs for knowledge on hydrological drought is presented in a final chapter.

The main text is supported with text boxes, worked examples, case studies, self-guided tours, appendices and supporting documents. An overview is given in the list of content. A *text box* presents details on a specific topic as an addition to the information given in the main text. *Worked examples* are sample methods that are being demonstrated in a stepwise manner using data and tools from the accompanying CD[1]. A *case study* presents the application of a method to a specific area or site and includes a discussion of the results. *Self-guided tours* are demonstrations of advanced methodologies that involve several calculation steps and are given as PowerPoint presentations on the CD. Some worked examples and case studies are presented on the CD as well as in the text. *Appendices and supporting documents* provide additional details of a method or modelling tool. The supporting documents are only available on the

---

[1] The authors and the organizations mentioned in the acknowledgements make no representation or warranties regarding the estimation methods (i.e. worked examples, case studies, self-guided tours and software), the accuracy of the data, the use to which the estimation methods and data may be put or the results obtained from the use of the estimation methods or data. Accordingly, the authors and the organizations mentioned accept no liability for any loss or damage (whether direct or indirect) incurred by any person through the use of the estimation methods or data.

CD. In addition the CD contains hydrological data, Excel procedures and software, which enable the reader to repeat the calculations, either with their own data or by using the time series included. The material on the CD is organized by category (e.g. data, worked example) and chapter, and the user is guided by a navigation procedure. A list of contents for the CD is provided in the textbook.

Three data sets are included on the CD; a global, a regional and a local data set. These data constitute the basis for the majority of the examples presented in the textbook. The drought phenomenon and its diversity across the world are illustrated using a global streamflow data set, whereas regional data and local aspects of drought are studied using a combination of hydrological time series and catchment information. The textbook uses the SI unit system.

The international FRIEND (Flow Regimes from International Experimental and Network Data) project, a contribution to UNESCO's International Hydrological Programme (IHP), has, since its start in Europe in 1985, encouraged cooperation and exchange of data across national boundaries. Its main focus has been on the use of international flow databases for regional studies, in particular hydrological extremes. The idea of compiling a textbook on hydrological drought first emerged in the Northern European FRIEND Low Flow Group as a result of many years of close cooperation in drought research. As editors we greatly acknowledge the spirit of this group and would like to thank all its members who in some way have made their contribution to this textbook, above all the authors who have contributed the chapters. They are all experts in their respective fields, and we would encourage the reader, whenever they may reference material from a particular chapter, to cite the authors of the chapter rather than the book as a whole.

The FRIEND initiatives are currently expanding into other regions of the world, and we trust that the textbook, despite its bias toward European experience and conditions, can benefit both professionals and students all over the world with increased knowledge and understanding of the drought phenomenon. We also hope that the textbook will contribute to an increased awareness of one of our main natural hazards, and thereby increase the preparedness and resilience of society to drought.

May, 2004   Lena M. Tallaksen & Henny A.J. van Lanen
                             University of Oslo   *Wageningen University (WUR)*
                             Norway   *The Netherlands*

Acknowledgements

Numerous individuals and organizations have been involved either directly or indirectly in the preparation of this textbook.

We thank Donald A. Wilhite and Kimberley D. Klemenz (National Drought Mitigation Center), University of Nebraska, USA, for contributing with 'The US Drought Monitor' in Chapter 2.

The authors of Chapter 3 are particularly grateful to the following people and organizations: Hervé Colleuille (Norwegian Water Resources and Energy Directorate), Norway for giving permission to include the illustration on soil frost; Istvan Tatrai (Balaton Limnological Research Institute), Hungary for providing the long time series of water levels from Lake Balaton, which were collected by the Middle-Transdanubian Water Authority, Lake Balaton Station, Siófok, Hungary; Centro de Estudios y Experimentación de Obras Públicas (CEDEX), Spain for providing recharge data from the Upper Guadiana; and the Netherlands Royal Meteorological Institute for meteorological data from the station Beek.

We gratefully acknowledge the contributions of the following ministries, authorities and organizations to the *Global Data Set* presented on the CD and described in Chapter 4: National Environmental Research Institute (NERI), Denmark; Ministry of Environment of Spain, for data provided by the Centre of Hydrographic Studies (Centro de Estudios Hidrograficos) of CEDEX; Department of Water Affairs of the Ministry of Agriculture, Water and Rural Development, Namibia; His Majesty's Government of Nepal, Department of Hydrology and Meteorology; Ministry of Transport, Public Works and Water Management, Directorate-General for Public Works and Water Management (Rijkswaterstaat), Programme Basic Information (Basisinformatie), RIZA, Lelystad and RIKZ, the Hague, the Netherlands; National Institute of Water and Atmospheric Research (NIWA), New Zealand; Glommen's and Laagen's Water Management Association (GLB), Norway; Norwegian Water Resources and Energy Directorate (NVE), Norway; Department of Water Affairs and Forestry (DWAF), Hydrological Service, South Africa; State Hydrological Institute, St. Petersburg, Russia; Environment Agency of England and Wales, for data provided by the National Water Archive at the Centre for Ecology and

Hydrology (CEH), Wallingford, UK; the US Geological Survey; and the Planificación Hidrológica y Servicio de Recursos Hidrocos, Ministerio de Medio Ambiente, Secretaria de Estado de AGUAS y COSTAS, Dirección General de Obras Hidráulicas y Calidad de los Aguas, Madrid.

We also thank Helmut Straub, Environmental Agency, Regional Office, State of Baden-Württemberg, Germany, for provision of flow data from Baden-Württemberg (*Regional Data Set*) and permission to store the data on the CD. The T.G. Masaryk Water Research Institute, Czech Republic supplied meteorological, groundwater and streamflow data for the Upper Metuje catchment (*Local Data Set*) and kindly gave permission to store the data on the CD.

We are indebted to Helge Brugaard (Norwegian Water Resources and Energy Directorate), Norway for his contribution to Worked Example 5.6 'Ranks and correlation coefficients' on the CD (Chapter 5); Anne Fleig at the Institute of Hydrology, University of Freiburg, Germany for the calculation of low flow indices in Tables 5.3 and 5.6, and Martin Morawietz at the Department of Geosciences, University of Oslo, Norway for his preparation of the BFI program on the CD. Wojciech Jakubowski (Agricultural University of Wroclaw), Poland kindly revised and provided the Nizowka program and gave permission to store it on the CD.

We would like to acknowledge Elisabeth Peters, Wageningen University, the Netherlands and Wendy S. McPherson at the US Geological Survey for their contribution to 'Analysing methods for groundwater' in Chapter 6.

The support from Mikhail Bolgov (Water Problems Institute), Russia in the initiation phase of Chapter 7 was greatly appreciated. We also acknowledge GLB and NVE, Norway for providing monthly flow data for the River Narsjø.

We would like to thank Gregor Laaha, University of Vienna, for his example on regionalization in Austria (Chapter 8) and Kerstin Stahl for providing the flow regimes of Baden-Württemberg. We appreciate the efforts of the following persons of the Institute of Hydrology, University of Freiburg, Germany who contributed to Chapter 8: Falk Scissek for his technical skills and his continuous engagement during the compilation of the Self-guided Tour; students for close cooperation and discussions about the chapter; Volker Abraham for providing the cartographic skills, maps and the layout of the front page of the Self-guided Tour; and Jürgen Strub for drawing the figures.

The Slovak Hydrometeorological Institute, Bratislava kindly provided hydrological data to show the effect of groundwater abstraction (Chapter 9). The T.G. Masaryk Water Research Institute, Czech Republic, revised the water balance model BILAN and gave permission to store it on the CD. The Czech Hydrometeorological Institute (CHMI), Czech Republic kindly provided meteorological and streamflow data for the Svitava and Bílina catchments. We thank Marjolein Mens (Wageningen University) for development of the Self-guided Tour on the assessment of the impact of human influence on drought.

We would like to thank Ian Jowett (NIWA) for making the RHYHABSIM (River HYdraulics and HABitat SIMulation) program available on the CD (Chapter 10). The program is very advanced and user-friendly, and the creator has spent many hours on its development. We also acknowledge the efforts of Mike Dunbar (CEH, Wallinford) for making helpful suggestions and comments to the 'User's Guide RHYHABSIM', and of David Hannah, Juan Felipe Blanco-Libreros and Tenna Riis for commenting on earlier drafts of Chapter 10. Thanks to Thomas Wilding (NIWA) and John McIntosh (EBOP) for making habitat data and results from the Whatakao Stream (New Zealand) available.

The following people and organizations kindly supplied information for Chapter 11: Rob Grew, Environment Agency of England and Wales (Section 11.2.9); Alternate Hydro Energy Centre (AHEC), India (Section 11.4.3); Himachel Pradesh Energy Development Agency, HIMURJA, India (Section 11.4.3); Ladislav Kašpárek, T.G. Masaryk Water Research Institute (Section 11.4.4); Slovak Hydrometeorological Institute and Miriam Fendeková, Comenius University, Slovakia (Section 11.4.4); The Czech Hydrometeorological Institute (Section 11.4.4); Elzbieta Kupczyk, University of Warsaw, Poland (Section 11.4.4); Mikhail Bolgov, Russia (Section 11.4.4); and Aleksandra Jaskula-Joustra, Rijkswaterstraat, Directie Limburg, the Netherlands (Section 11.4.5).

We are grateful to Mike Dunbar (CEH, Wallingford) and Wojciech Jakubowski (Agricultural University of Wroclaw) for teaching on the ASTHyDA International Study Course in Wageningen where the second draft of the textbook was discussed, and Marjolein Mens and Lisette van den Bos (Wageningen University) for compilation of the first and second draft of the textbook, respectively. We would especially like to thank Martin Morawietz (University of Oslo) for his detailed technical editing of the text, contribution to the compilation of the CD and for taking care of the flow of material between the authors, the editors and Elsevier Science Publisher in the final six months, and Kerstin Stahl (University of Oslo) for her valuable contribution in the early phase of the project, including the design of the ASTHyDA website and brochure and the compilation of the CD for the textbook. We also thank Anne Flcig (University of Oslo) for her careful revision of the reference list and subject index and Hilary Arnell (UK) for English proofreading of the text.

The textbook very much benefited from comments made by 13 European water managers and 26 researchers from the Mediterranean region during two workshops in La Grande Motte (France, March–April 2003) and the feedback received from 27 students participating in the International Study Course in Wageningen (September–October 2003). The accomplishment of these activities, involving 25 authors from 10 countries, reviews from water managers, researchers from the Mediterranean region, students and additional technical reviewers, involved a considerable degree of organization. This would not have been possible without the support from the European Union, through the Accompanying Measure: Analysis, Synthesis and Transfer of Knowledge and Tools on Hydrological Droughts Assessment through a European Network (ASTHyDA[1]). The Accompanying Measure (EVK1-2001-00166) was part of the EC Energy, Environment and Sustainable Development (EESD) program.

Finally, we would like to thank the Water Division of UNESCO (Paris) for supporting some of the authors' attendance at technical meetings, and in particular for their continuing support of the FRIEND project within the framework of the International Hydrological Program.

---

[1] ASTHyDA (http://drought.uio.no)

Contributors

Barry J.F. Biggs, National Institute of Water and Atmospheric Research,
P.O.Box 8602, Christchurch, New Zealand
Email: b.biggs@niwa.co.nz

Bente Clausen, Department of Civil Engineering, University of Canterbury,
Private Bag 4800, Christchurch, New Zealand
Now
National Environmental Research Institute, P.O.Box 314, 8600 Sillkeborg, Denmark
Email: bcl@dmu.dk

Siegfried Demuth, Institute of Hydrology, Fahnenbergplatz, 79098 Freiburg, Germany
Now
German Secretariat of UNESCO-IHP and WMO-HWRP programmes,
Federal Institute of Hydrology, Mainzer Tor 1, 56068 Koblenz, Germany
Email: demuth@bafg.de

Miriam Fendeková, Faculty of Natural Sciences, Comenius University,
Mlynská dolina, Pavilon G, 842 15 Bratislava, Slovak Republic
Email: fendekova@fns.uniba.sk

Lars Gottschalk, Department of Geosciences, University of Oslo, P.O.Box 1047
Blindern, NO-0316 Oslo, Norway
Email: lars.gottschalk@geo.uio.no

Alan Gustard, Centre for Ecology and Hydrology, Crowmarsh Gifford, Wallingford,
Oxfordshire OX10 8BB, United Kingdom
Email: agu@ceh.ac.uk

Hege Hisdal, Norwegian Water Resources and Energy Directorate, P.O.Box 5091 Maj.,
NO-0301 Oslo, Norway
Email: hhi@nve.no

Matthew G.R. Holmes, Centre for Ecology and Hydrology, Crowmarsh Gifford,
Wallingford, Oxfordshire OX10 8BB, United Kingdom
Email: mgrh@ceh.ac.uk

Ian G. Jowett, National Institute of Water and Atmospheric Research, P.O.Box 11115,
Hamilton, New Zealand
Email: i.jowett@niwa.co.nz

Ladislav Kašpárek, T.G. Masaryk Water Research Institute, Podbabska 30,
CZ-16062 Prague 6, Czech Republic
Email: ladislav_kasparek@vuv.cz

Artur Kasprzyk, Jan Kochanowski University, 25-406 Kielce, Poland
Email: artur.kasprzyk@pu.kielce.pl

Elżbieta Kupczyk, Jan Kochanowski University, 25-406 Kielce, Poland
Email: kupczyk@acn.waw.pl

Henny A.J. van Lanen, Subdepartment of Water Resources, Wageningen University,
Wageningen-UR, Nieuwe Kanaal 11, 6709 PA Wageningen, the Netherlands
Email: henny.vanlanen@wur.nl

Henrik Madsen, DHI Water & Environment, Agern Allé 11, DK-2970 Hørsholm,
Denmark
Email: hem@dhi.dk

Terry J. Marsh, Centre for Ecology and Hydrology, Crowmarsh Gifford, Wallingford,
Oxfordshire OX10 8BB, United Kingdom
Email: tm@ceh.ac.uk

Bjarne Moeslund, Bio/consult A/S, Johannes Ewalds Vej 42-44, 8230 Åbyhøj,
Denmark
Email: bm@bioconsult.dk

Oldřich Novický, Ministry of Environment, Department of Water Protection,
Vrsovická 65, CZ-100 10 Prague 10, Czech Republic
Email: oldrich.novicky@env.cz

Elisabeth Peters, Subdepartment of Water Resources, Wageningen University,
Wageningen-UR, Nieuwe Kanaal 11, 6709 PA Wageningen, the Netherlands
Email: elisabeth.peters@wur.nl

Wojciech Pokojski , Warsaw University, Faculty of Geography and Regional Studies,
Krakowskie Przedmiescie 30, 00-927 Warszawa, Poland
Email: wpokojski@uw.edu.pl

Erik P. Querner, Alterra, Green World Research, Wageningen-UR,
Droevendaalsesteeg 3, 6700 AA Wageningen, the Netherlands
Email: erik.querner@wur.nl

Gwyn Rees, Centre for Ecology and Hydrology, Crowmarsh Gifford, Wallingford,
Oxfordshire OX10 8BB, United Kingdom
Email: hgrees@ceh.ac.uk

Lars Roald, Norwegian Water Resources and Energy Directorate, P.O.Box 5091 Maj.,
NO-0301 Oslo, Norway
email: lar@nve.no

Kerstin Stahl, Department of Geosciences, University of Oslo, P.O.Box 1047 Blindern,
NO-0316 Oslo, Norway
Now
Department of Geography, University of British Columbia, 1984 West Mall,
Vancouver B.C. V6T 1Z2, Canada
Email: kerstin@2hydros.de

Lena M. Tallaksen, Department of Geosciences, University of Oslo,
P.O.Box 1047 Blindern, NO-0316 Oslo, Norway
Email: lena.tallaksen@geo.uio.no

Andrew R. Young, Centre for Ecology and Hydrology, Crowmarsh Gifford,
Wallingford, Oxfordshire OX10 8BB, United Kingdom
Email: ary@ceh.ac.uk

Brief Contents

Part I: Drought as a Natural Hazard

Part II: Estimation methods

Part III: Human Influences, Ecological and Operational Aspects

Contents

Part I: Drought as a Natural Hazard

List of Worked Examples, Self-guided Tours and Case Studies

Worked Examples

Self-guided Tours

Case Studies

Contents CD

File types: xls: Excel
                      pdf: Portable Document Format
                      pps: Power Point Show
                      exe+: Executables and additional files

Worked Examples   (file type)

Chapter 4	• Worked Example 4.1: Filling gaps in daily flow series	(xls)
Chapter 5	• Worked Example 5.1: Flow duration curve	(xls)
	• Worked Example 5.2: Mean annual minimum n-day flow	(xls)
	• Worked Example 5.5: Sequent peak algorithm	(xls)

Self-guided Tours

Chapter 6	• Regional frequency analysis (Section 6.7.5)	(pps)
Chapter 8	• Estimation of low flow indices at the ungauged site (Section 8.3.3)	(pps)
Chapter 9	• Groundwater abstraction in the Poelsbeek and Bolscherbeek catchments (Section 9.5.3)	(pps)

Case studies

Chapter 5	• Groundwater drought in the Metuje catchment	(xls)

Data

| **Global Data Set** | • Overview Global Data Set | (pdf) |
| | • 18 data files | (xls) |

Regional Data Set	• Overview Regional Data Set	(pdf)
	• Calibration data: 56 data files	(xls)
	• Validation data: 27 data files	(xls)

| **Local Data Set** | • Metuje Catchment | (xls) |

| **Monthly data series (Chapter 7)** | • River Göta | (xls) |
| | • River Narsjø | (xls) |

Software

| **BFI** | • User's Guide BFI | (xls) |
| | • BFI program | (pdf) |

BILAN	• Readme BILAN	(pdf)
	• User's Guide BILAN	(pdf)
	• Background Information BILAN	(pdf)
	• BILAN program (including five sample files)	(exe+)

NIZOWKA	• Readme NIZOWKA	(pdf)
	• Background Information NIZOWKA	(pdf)
	• Software Manual NIZOWKA	(pdf)
	• NIZOWKA program	(exe+)

RHYHABSIM	• Readme RHYHABSIM	(pdf)
	• User's Guide RHYHABSIM	(pdf)
	• Software Manual RHYHABSIM	(pdf)
	• RHYHABSIM program (including sample files)	(exe+)

Supporting Documents

Abbreviations, Symbols and Catchment Descriptors

Abbreviations

ADCP	Acoustic Doppler Current Profiler
AIC	Akaike Information Criterion
AM	Annual Minimum
AMS	Annual Maximum/Minimum Series
ARIDE	Assessment of the Regional Impact of Droughts in Europe
ARMA	AutoRegressive Moving Average
AR(1)	AutoRegressive first order model
ASCE	American Society of Civil Engineers
ASCII	American Standard Code for Information Interchange
BBM	Building Block Method
BFI	Base Flow Index
BILAN	A water balance model
CAMS	Catchment Abstraction Management Strategy
CASIMIR	Computer Aided SIMulation system for Instream flow Requirements
CEH	Centre for Ecology and Hydrology (UK)
cdf	Cumulative distribution function
CI	Confidence Interval
CMI	Crop Moisture Index
CPC	Climate Prediction Center (US)
CRED	Centre for Research on the Epidemiology of Disasters
ČSN	Czech Standard
CV	Coefficient of Variation
DTM	Digital Terrain Model
DVWK	German Association for Water, Wastewater and Waste (Deutsche Vereinigung für Wasserwirtschaft, Abwasser und Abfall)
ECA	European Climate Assessment
ECMWF	European Centre for Medium-range Weather Forecasts
EEA	European Environment Agency
ENSO	El Niño-Southern Oscillation
EOF	Empirical Orthogonal Functions
EOS	Earth Observation Systems
ES	Spain
EU	European Union
EV	Extreme Value

EVHA	EValuation of Habitat model
EWA	European Water Archive
EXP	EXPonential
FDC	Flow Duration Curve
FRE	FRÉchet
FREND	Flow Regimes from Experimental and Network Data
FRIEND	Flow Regimes from International Experimental and Network Data
GAM	GAMma
GCM	General Circulation Model
GEV	Generalized Extreme Value
GIS	Geographical Information System
GL	Generalized Logistic
GLS	Generalized Least Squares
GP	Generalized Pareto
GUM	GUMbel
IAHS	International Association of Hydrological Sciences
IC	Inter-event time and/or volume Criterion
IFIM	Instream Flow Incremental Methodology
IHA	Indicators of Hydrologic Alteration
IHP	International Hydrological Programme
iid	Independent, identically distributed
IMGW	Institute of Meteorology and Water Management (Instytut Meteorologii i Gospodarki Wodnej, PL)
IPCC	International Panel on Climate Change
ISO	International Organization for Standardization
ITCZ	InterTropical Convergence Zone
L-CV	L-Coefficient of Variation
LN	Log-Normal
LP3	Log-Pearson type 3
MA	Moving Average
MA(n-day)	Moving Average n-day
MAM	Mean Annual Minimum
MAM(n-day)	Mean Annual Minimum n-day
m.a.s.l.	Meter above sea level
ML	Maximum Likelihood
MODFLOW	MODular finite-difference groundwater FLOW model
MSLP	Mean Sea Level Pressure
NAOI	North Atlantic Oscillation Index
NAP	Normaal Amsterdam Peil, Normal Amsterdam Level, Dutch ordnance datum
NDMC	National Drought Mitigation Center (US)
NDVI	Normalized Difference Vegetation Index
NE FRIEND	Northern European FRIEND
NERC	Natural Environment Research Council (UK)
NIZOWKA	Program for estimating extreme drought characteristics

NL	The Netherlands
NOAA	National Oceanic and Atmospheric Administration (US)
NZ	New Zealand
OFDA	Office of US Foreign Disaster Assistance
P3	Pearson type 3
PC	Principle Component
PCA	Principal Components Analysis
pdf	Probability density function
PDS	Partial Duration Series
PDSI	Palmer Drought Severity Index
PHABSIM	Physical HABitat SIMulation model
PL	Poland
POT	Peak Over Threshold
PUB	Prediction in Ungauged Basins
PWM	Probability Weighted Moments
RBMP	River Basin Management Plan (EU)
RFD	Residual Flow Diagram
RHABSIM	River HABitat SIMulation model
RHYHABSIM	River HYdraulic and HABitat SIMulation model
ROI	Region Of Influence
RSS	River Simulation System
RVA	Range of Variability Approach
SDP	Stream Depletion Factor
SE	Standard Error
SHE	Systéme Hydrologique Européen modelling system
SIMGRO	SIMulation of GROundwater and surface water levels model
SOI	Southern Oscillation Index
SPA	Sequent Peak Algorithm
SST	Sea Surface Temperature
SUTN	Slovak Institute for Standardization (Slovenský Ústav Technickej Normalizácie)
SWSI	Soil Water Supply Index
TC	Technical Committee (of the ISO)
TIN	Triangulated Irregular Network method
UK	United Kingdom
UNEP	United Nations Environment Programme
UNESCO	United Nations Educational, Scientific and Cultural Organization
UNFCCC	United Nations Framework Convention on Climate Change
US	United States
USA	United States of America
WEI	WEIbull
WFD	Water Framework Directive
WMO	World Meteorological Organisation
WSP	Water Surface Profile
WUA	Weighted Usable Area

WUR	Wageningen University and Research Centre (NL)
WWW	World Wide Web
ZINX	Monthly Moisture Anomaly Index

Symbols

$\wedge$	Hat ($\wedge$) above a parameter indicates an estimated value
A	Area [L^2]
A_c	Catchment area [L^2]
A_{def}	Deficit area [L^2]
A_{ik}	Catchment descriptor k for catchment i
A_s	Surface water area [L^2]
$AM(n\text{-day})$	Annual minimum n-day flow [L^3T^{-1}]
$AM(n\text{-day})_T$	T-year annual minimum n-day flow [L^3T^{-1}]
AP	Soil moisture storage at anaerobe (excess water) point [L]
b	Drought criterion *(Chapter 5) or*
	regression coefficient (or parameter) *(Chapter 8)*
B	Half of aquifer width [L]
c	Hydraulic resistance of a semi-permeable layer [T]
C	Recession parameter [T]
$Corr\{\ldots\}$	Correlation of $\{\ldots\}$
$Cov\{\ldots\}$	Covariance of $\{\ldots\}$
CP	Soil moisture storage at critical point [L]
CV	Coefficient of variation
$Cv\text{-}R^2$	Cross validated coefficient of determination
d	Julian day number [1 … 365]
d_i	Duration of drought event i [T]
d_{min}	Minor drought duration criterion [T]
de	Weighted Euclidean distance
D	Duration [T]
D_a	Aquifer thickness [L]
e	Base of the natural logarithm
E_i	Evaporation from intercepted water [LT^{-1}]
E_s	Evaporation from surface water bodies [LT^{-1}]
$EF_X(x)$	Exceeeance frequency of x
$EP_X(x)$	Exceedance probability of x
ET	Actual evapotranspiration [LT^{-1}]
$E\{.\}$	Expectation operation
$E\{X\}$	Expected (or mean) value of X
$f_X(x)$	Probability density function (pdf)
FC	Soil moisture storage at field capacity [L]
$F_X(x)$	Cumulative distribution function (cdf)
h	Pressure head (negative) in the unsaturated zone [L]

H	Groundwater hydraulic head or water table [L] *(Chapter 3, 9) or*
	River water level [L] *(Chapter 4) or*
	Heterogeneity statistic based on L-moments *(Chapter 6)*
H'	Shannon Index
$\overline{H}$	Average groundwater hydraulic head [L]
H_0	River water level at zero flow [L]
H_b	Drainage base of groundwater discharge to surface water [L]
H_{de}	Groundwater hydraulic head of deep aquifer [L]
H_{pe}	Perched water table [L]
H_s	Groundwater hydraulic head of shallow aquifer [L]
H_{st}	Initial groundwater hydraulic head [L]
H_{su}	Stream water level [L]
H_x	Groundwater hydraulic head which is exceeded in $x\%$ of the time [L]
i	Rank
i_d	Rank on day d
I	Indicator function [0, 1] *(Chapter 5) or*
	Groundwater recharge [LT^{-1}] *(Chapter 3, 5, 9)*
I_r	Groundwater recharge which is exceeded in $x\%$ of the time [LT^{-1}]
j	Number of recession parameters
k	Recession parameter [-]
k_h	Hydraulic conductivity [LT^{-1}]
kD	Transmissivity [$L^2 T^{-1}$]
L	Channel length [L]
n, N	Number of observations, e.g. days, grid cells
n_e	Effective porosity [-]
n-day	Moving average interval in days [T]
m_i	Intensity of event i [$L^3 T^{-1}$]
m_{max}	Maximum deficit intensity [$L^3 T^{-1}$]
M	Number of stations in a region
$MAM(n$-day$)$	Mean annual minimum n-day flow [$L^3 T^{-1}$]
p	Non-exceedance probability
p_0	Probability that an observation is zero
p_i	Plotting position of the ith smallest event
P	Precipitation [LT^{-1}]
P_{gr}	Gross precipitation [LT^{-1}]
P_n	Net precipitation [LT^{-1}]
P_s	Surface water precipitation [LT^{-1}]
PE	Potential evapotranspiration [LT^{-1}]
Pr	Probability
q_{dr}	Aquifer discharge per unit area [LT^{-1}] or per unit length [$L^2 T^{-1}$]
q_{gr}	Groundwater discharge [LT^{-1}]
q_{if}	Throughflow or interflow [LT^{-1}]
q_{of}	Overland flow [LT^{-1}]
q_s	Infiltration across the soil surface [LT^{-1}]
q_r	Flux across the bottom of the root zone (percolation) [LT^{-1}]

q_x	Q_x divided by catchment area $[\mathrm{L\,T^{-1}}]$
Q	Discharge $[\mathrm{L^3\,T^{-1}}]$
$\overline{Q}$	Mean discharge $[\mathrm{L^3\,T^{-1}}]$
Q_0	Threshold discharge $[\mathrm{L^3\,T^{-1}}]$
Q_b	Base flow $[\mathrm{L^3\,T^{-1}}]$
Q_D	Drought threshold discharge $[\mathrm{L^3\,T^{-1}}]$
Q_{gr}	Groundwater discharge $[\mathrm{L^3\,T^{-1}}]$
Q_{if}	Throughflow or interflow $[\mathrm{L^3\,T^{-1}}]$
Q_{min}	Minimum discharge $[\mathrm{L^3\,T^{-1}}]$
Q_{MS}	Initial discharge $[\mathrm{L^3\,T^{-1}}]$
Q_{of}	Overland flow $[\mathrm{L^3\,T^{-1}}]$
Q_{rated}	Design flow for turbine $[\mathrm{L^3\,T^{-1}}]$
$Q_{residual}$	By-pass flow to meet water needs downstream of diversion $[\mathrm{L^3\,T^{-1}}]$
Q_t	Discharge at time t $[\mathrm{L^3\,T^{-1}}]$
Q_x	Flow which is exceeded $x\%$ of the time $[\mathrm{L^3\,T^{-1}}]$
QE_d	Discharge exceedance on day d $(\%)$
r_P	Pearson correlation coefficient
r_S	Spearman rank correlation coefficient
$r(k)$	Sample autocorrelation function (or coefficient) of lag k
R^2	Coefficient of determination
s	Sample standard deviation
s^2	Sample variance
s_x	Standard deviation of X
S	Storage (or stored) water $[\mathrm{L}]$ or $[\mathrm{L^3}]$
S_{gr}	Stored groundwater $[\mathrm{L}]$
S_s	Stored surface water $[\mathrm{L}]$
S_{so}	Stored soil water $[\mathrm{L}]$
S_{sor}	Stored soil water in the root zone $[\mathrm{L}]$
S_{sos}	Stored soil water in the subsoil $[\mathrm{L}]$
S_{su}	Stored water at the surface $[\mathrm{L}]$
S_t	Storage at time t $[\mathrm{L}]$ or $[\mathrm{L^3}]$
S_{veg}	Stored water on the vegetation $[\mathrm{L}]$
S_y	Storage coefficient $[\text{-}]$
t	Time $[\mathrm{T}]$
Δt	Time step or interval $[\mathrm{T}]$
t_i	Inter-event time $[\mathrm{T}]$
t_{min}	Inter-event time criterion $[\mathrm{T}]$
T	Return period
T_a	Annual return period (T-year)
T_D	Time it takes for the discharge to decline from Q_{MS} to Q_D
T_p	Average return period for level x_p in a PDS
u	Upper limit above which extreme events are selected
v_i	Volume of drought event i $[\mathrm{L^3}]$
V	Volume $[\mathrm{L^3}]$ *or*

V *(cont.)*	Record-length-weighted standard deviation of L-CV estimates *(Chapter 6) or*
	Regional L-moment variability measure *(Chapter 6)*
V_{area}	Total area deficit volume [L^3]
$Var\{\ldots\}$	Variance of $\{\ldots\}$
w_i	Weight assigned to site (or item) i
WP	Soil moisture storage at wilting point [L]
x	Real number
$\overline{x}$	Sample mean or average
x_i	The ith value in a series
x_p	Quantile of a distribution function
x_T	T-year event
X	Random variable
X_{pi}	Catchment descriptor p for catchment i
$X_{(i)}$	The value of order (or rank) i
$X_{(m:n)}$	The value of order m in a sample of size n ranked in ascending order
y	Reduced variate
z_i	Inter-event volume [L^3]
z_T	Normalized regional T-year event
Z	Goodness-of-fit statistic based on L-moments
α	Fraction of maximum drought deficit volume *(Chapter 5) or*
	Recession parameter (constant, coefficient) [T^{-1}] *(Chapter 5, 8) or*
	Scale parameter of a distribution *(Chapter 6) or*
	Significance level *(Chapter 6)*
β	Amplitude function *(Chapter 6) or*
	Regression parameter (or coefficient) *(Chapter 6)*
γ_3	Skewness
γ_4	Kurtosis
δ	Residual model error
ε	Random error
θ	Soil moisture content [-] *(Chapter 3) or*
	Model parameters *(Chapter 6)*
κ	Shape parameter of a distribution
λ	Number of events per time interval
λ_1	First order L-moment
λ_2	Second order L-moment
λ_3	Third order L-moment
λ_4	Fourth order L-moment
Λ	Covariance matrix
μ	Mean, average or expected value
ξ	Location parameter of a distribution
ρ	Intersite correlation coefficient
σ	Standard deviation
σ^2	Variance

τ	Kendall's Tau
τ_2	L-coefficient of variation
τ_3	L-skewness
τ_4	L-kurtosis
τ_i	Time of occurrence of event i [date]
τ_r	r^{th}-order L-moment
Φ	Standard Normal distribution function
Φ_0	Threshold level (in general)

Catchment descriptors

AAR	Average annual rainfall (mm)
ALPHA	The recession parameter (constant, coefficient) α used as a catchment descriptor [T^{-1}]
AREA	Catchment area (km^2)
BASE	Mean base flow (m^3 s^{-1})
BE	Catchment width (km)
BFI	Base Flow Index [0, 1]
DD	Drainage density (km km^{-2})
FALAKE	Percentage of lake (%); ratio between the area of lakes and the catchment area
FOREST	Percentage of forest (%)
GEO	Geological index [-]
GEOHCMEAN	Weighted mean hydraulic conductivity (m s^{-1})
GEOVLHC	Percentage of rock formations with a very low hydraulic conductivity (%)
HMAX	Maximum elevation in the catchment (m.a.s.l.)
HMEAN	Mean catchment elevation (m.a.s.l.)
HMIN	Minimum elevation in the catchment (m.a.s.l.)
HOST	Hydrology of Soil Types
LAKE	Lake parameter (%); sum of the areas of the catchment draining through a lake divided by the catchment area
LE	Catchment length (km)
LFHG	Low Flow Host Groups
LG	Length of the river network
MOUNT	Percentage of area above the tree line (%)
MSL	Main stream length [L]
NEIG	Mean slope of the catchment [-]
PAST	Percentage of pasture (%)
RATIO	Runoff ratio (Q_{20}/Q_{90}) (%)
RB	Bifurcation index [-]
RELIEF	Relief ratio
ROOTSMEAN	Mean water-holding capacity in the effective root zone (mm)

ROOTSHIGH	Percentage of soils with high water-holding capacity in the effective root zone (%)
RR	Relative gradient
SL	Stream slope (m km^{-1})
SLOPEMAX	Maximum slope (%)
SLOPEMEAN	Mean slope (%)
SLOPEMIN	Minimum slope (%)
SLxxXX	Slope of the river channel between xx and XX percent of the river stretch (m km^{-1})
SOIL	Soil index [-]
SOILH	Percentage of soils with high infiltration capacity (%)
SOILHCMEAN	Mean hydraulic conductivity of the soils (cm day^{-1})
SOILM	Percentage of soils with medium infiltration capacity (%)
SOILL	Percentage of soils with low infiltration capacity (%)
SOILLHC	Percentage of soils with low hydraulic conductivity (%)
SOILVL	Percentage of soils with very low infiltration capacity (%)
URBAN	Percentage of urbanisation (%)
VGS	Volume of trees (m^3 ha^{-1})
WSEA	Shortest distance to the western coast line (km)
WPLAKE	Weighted lake percentage (%); 100 [$\Sigma(A_i \, a_i)$]/A^2 where a_i is area of lake i, A_i is drainage area of lake i and A is total catchment area

Part I

Drought as a Natural Hazard

1

Introduction

Lena M. Tallaksen, Henny A.J. van Lanen

Water – the most fundamental and indispensable resource of the world. Drought means lack of water; water that normally would be available in a region and to which nature and mankind have adapted over centuries (Figure 1.1) History has shown how vulnerable regions all over the world are to severe and prolonged droughts, which have caused major social, economic and environmental problems. Increasing demand for water, following a growing global population and extensive use of water for irrigation and industry, has raised the awareness of our vulnerability to drought. Any deficit or limitation in water supply will be most critical in drought periods, and competing water needs may be the cause of

Figure 1.1 River l'Eygues (France) during the drought in August 2003 (photo by H.A.J. van Lanen).

conflicts. Current global change scenarios suggest that the magnitude, frequency and impacts of extreme drought events could increase due to mankind, through climate and large-scale changes in vegetation. A prerequisite for sound water management is a thorough understanding of drought, considered by many to be the least understood of all major natural hazards (Wilhite, 2000a).

Drought is a sustained and regionally extensive occurrence of below average natural water availability, and can thus be characterized as a deviation from normal conditions of variables such as precipitation, soil moisture, groundwater and streamflow. It is a reoccurring and worldwide phenomenon, with spatial and temporal characteristics that vary significantly from one region to another. The most severe social consequences of drought are found in arid or semi-arid regions where the availability of water is already low under normal conditions. Drought should not be confused with aridity, which is a permanent feature of a dry climate.

1.1   Scope

Throughout most of this textbook, drought is considered a natural phenomenon caused by a meteorological anomaly and modified by the physical properties of a catchment. Doubtless, today's hydrology is often far from natural. Wherever water resources are utilized, the hydrological cycle is affected. In addition, emissions are changing the world's climate. Air pollution and shifts in vegetation, which particularly affects evapotranspiration, further alter the regional or local climate. The predicted impact of climate change on water resources varies regionally, largely following the projected changes in precipitation and affects in particular the volume and timing of streamflow and groundwater recharge. As there are large uncertainties in the predictions and even in the direction of change in some parts of the world, climate change adds foremost additional uncertainty to future predictions, in particular for extremes events. Water resource management needs to adapt to these potential changes in the hydrological system. Still, non-climatic changes may have a greater impact on the natural system than climate change. Dams, diversions and abstraction of surface water, change in land use, extensive use of irrigation, but also industrial discharge to rivers, can greatly modify the quantity and quality of streamflow. Groundwater abstraction lowers the water table and reduces groundwater discharge to rivers. Yet despite, or rather because of, all those possible influences to the hydrological cycle, the understanding of the natural drought phenomenon and its meteorological and hydroclimatological causes is essential for a considerate management of our limited water resources now and in the future.

The textbook describes the many factors that give rise to drought in different hydroclimatological regions of the world, how to quantify and analyse hydrological drought and how to estimate the impact of environmental change. It focuses on hydrological drought conditions defined by the available water resources in a catchment or an aquifer. The main scope is to provide the reader with a comprehensive understanding of processes and estimation methods for streamflow and groundwater drought. It is accompanied by computational details, general recommendations and possible limitation for application of a particular methodology. Drought diversity is illustrated using a global set of daily streamflow series. The regional aspects of drought are studied using daily streamflow series and catchment information for several sites in a region. Streamflow and groundwater drought processes are studied at the local scale using hydrological and climatological time series combined with detailed catchment information. The book concludes with ecological issues and examples of procedures used for designing and operating water resources schemes which are sensitive to droughts. A majority of the examples are taken from regions where the rivers run most of the year, not only occasionally. The hydrology of arid and semi-arid areas is known to be substantial different from that in more humid regions (Simmers, 2003), and is only briefly introduced.

Although the basic physical principles are well understood and the estimation methods presented are powerful tools, the degree of knowledge and the level of precision that can be obtained in our estimates will always depend on the information contained in the observed data. Often the observations are limited in space and time, they may contain gaps and one might question the quality of the data. The data sets included with the book are real observations, and no interpolation or corrections of the series have been undertaken. The user will therefore encounter real-life problems when working with the data series and should always keep these limitations in mind. In general, the purpose of the study, the region under investigation and the data available should guide the choice of drought characteristics, estimation methods and conclusions that can be drawn.

1.2   Hydrological drought

The primary cause of a drought is the lack of precipitation over a large area and for an extensive period of time, called a *meteorological drought*. This water deficit propagates through the hydrological cycle and gives rise to different types of droughts. Combined with high evaporation rates a soil water deficiency might cause a *soil moisture drought* to develop. The term *agricultural drought* is used when soil moisture is insufficient to support crops. Subsequently groundwater recharge and streamflow will be reduced and a *hydrological drought* may develop. A reduced recharge leads to lower groundwater heads and storage. The relationship between the different types of droughts is illustrated in Figure 1.2.

Within the hydrological cycle, groundwater is normally the last to react to a drought situation, unless surface water is mainly fed by groundwater. In deep aquifers the slow reaction of groundwater implies that only major meteorological droughts will finally show up as *groundwater droughts*. The lag between a meteorological and a groundwater drought may amount to months or even years, whereas the lag between a meteorological and a *streamflow drought* varies from days in a flashy catchment to months in a groundwater-fed catchment.

This disciplinary perspective of drought classification also encompasses *ecological drought*, which can be perceived as a shortage of water causing stress on ecosystems, adversely affecting the life of plants and animals, and *socio-economic drought* which is primarily concerned with the impacts of drought and drought mitigation strategies on economy and society. The economic impact of a drought will be most comprehensive in areas with large populations and extensive industrial development, whereas human society suffers most in developing countries that lack the ability to cope with drought. Excessive water use and human activity may intensify and even cause a drought to develop. Whereas one scientist might prefer a definition of drought that is objective, quantitative and applicable over a large region, others might not label a water deficit a drought unless it has an adverse effect on society, referring to degraded terrestrial and aquatic ecosystems or socio-economic impacts.

The general definition of drought as "a sustained and regional extensive occurrence of below average natural water availability" implies that both the time and spatial aspect of the drought are considered. The definition is relative in the sense that the concept of a drought refers to a certain threshold that distinguishes a drought event from a non-drought situation, and the event has thus a beginning and an end. Key aspects of a drought include its duration, severity, time of occurrence and spatial extent. Streamflow drought characteristics are obtained from time series of discharge, observed or simulated, and encompass both low flow and deficit characteristics (Figure 5.1). A time series of low flow characteristics can, for instance, be the lowest observed streamflow each year, i.e. the annual minimum series. Low flow characteristics are particularly suitable for characterizing the hydrological regime, i.e. the seasonal variation in streamflow, but consider only one feature of the event, the drought severity. A method that simultaneously characterizes

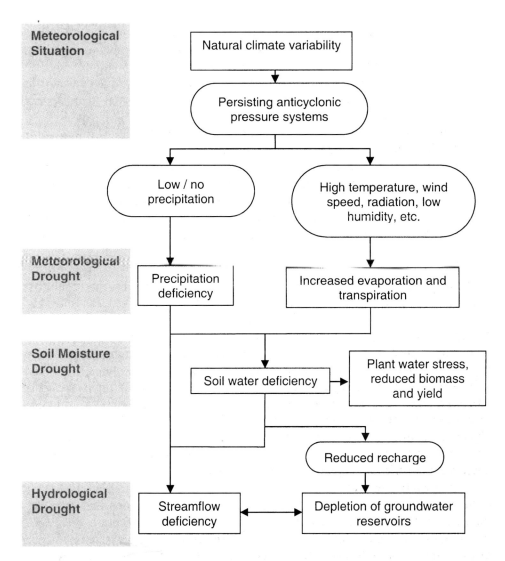

Figure 1.2 Propagation of drought through the hydrological cycle (modified from Stahl, 2001).

streamflow drought in terms of severity and duration is the threshold level method, which defines drought as a period during which the flow is below a certain threshold level. Time series of observed groundwater level and derived time series of groundwater discharge and recharge are used for characterizing groundwater drought.

Based on time series of hydrological drought characteristics, corresponding indices (single values) can be derived, for example the *mean* annual minimum flow or the *mean* annual deficit duration. As droughts are regional in nature and critical drought conditions occur when there is an extreme shortage of water for long durations over large areas, a drought study often includes the spatial extent of the drought as a measure of the severity of the drought. The purpose of the study guides the choice of time step; daily or monthly data for within year or seasonal droughts, and annual or monthly time step for multiyear droughts. Further details on the definition of hydrological drought and the derivation of drought characteristics, including time series and indices as well as at-site and regional characteristics, are given in Chapter 5.

1.3   The drought hazard

Drought differs from other natural hazards in several ways and is rarely the first hazard that comes to mind for most people. The sudden effect of floods, tornadoes and earthquakes is often more striking and thus better known. One reason for the smaller public awareness of drought is that droughts develop slowly and imperceptibly and may thus remain unnoticed for a long time. When obvious, they often cover and affect large areas making mitigation and aid programs difficult to organize. Not many people realize the full financial impact of drought, and how destructive it may be. For instance, in the USA drought is costlier than any other natural disaster.

The impact of drought may vary considerably in the different climatological and hydrological regimes found across the world. Figure 1.3 presents some of the most recent and severe droughts in the world and their impacts as registered in the international disaster database EM-DAT (OFDA/CRED, 2002). Chapter 2 provides more information on the EM-DAT database.

In 2003 Europe was hit by a severe drought (Box 2.3). The drought had enormous adverse effects, such as the destruction of large areas of forest by fire, substantial agricultural losses, power cuts and transport problems. Central and southern Europe suffered the most. In late September 2003, the damage to agriculture and forestry was estimated at 13 billion € for the member and candidate states of the European Union (COPA-COGECA, 2003). The largest damage occurred in France. With about 186 million tonnes (MT), the EU arable sector showed a fall in production of more than 23 MT, i.e. approximately 11%. The forest fires mainly affected Portugal, Spain, France and Italy. Some 650 000 ha forest were destroyed, of which 60% were in Portugal and 20% in Spain. The direct damage as a result of the forest fires in Portugal has been

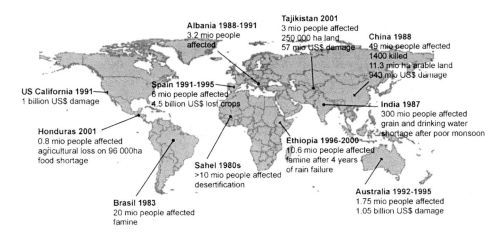

Figure 1.3 Selected severe droughts (data from OFDA/CRED, 2002).

estimated at 1.03 billion €. Europe was not the only region to suffer from drought this year. For instance, in Central Vietnam about 10% of the total area experienced extreme drought conditions and approximately 67% of the population was short of water (UNDP/MARD, 2003). Also Argentina went through a severe drought in 2003 and serious impacts were reported in many provinces. Mendoza had the harshest drought in 40 years, whereas in the north-eastern region the drought was declared the worst in history. Losses in wheat crops represented a value close to 200 million €. Drought was also reported in some African countries (e.g. Ethiopia) and some states in the USA (e.g. Missouri, Colorado) this year. As demonstrated by these examples drought occurs in all regions of the word and has a wide range of severe impacts.

Bradford (2000) and Wilhite (2000b) give extensive reviews of the wide and sometimes far-reaching range of impacts related to drought that have been reported. The impacts can be classified as environmental (ecological), social and economic, and often a distinction is made between direct and indirect effects. Environmental effects are both direct and indirect effects, whereas socio-economic consequences are principally indirect effects (Table 1.1). For example, a low lake level is a direct effect of drought, whereas the associated effects on energy production or recreational activities are indirect effects. The most important economic losses of drought are related to a reduction in crop production, in particular in areas that depend on irrigation, dairy and livestock production and industrial and hydropower production. In addition, drought affects the economy through the navigability of rivers and recreation activities.

Table 1.1 Examples of impacts of drought

Aspect	Effects[1]	
	Direct	**Indirect**
Environmental	Soil moisture	Water quality
	Groundwater levels	Plant growth
	Groundwater discharge	Habitats
	Spring yield	Endangered species
	Streamflow	Dust storms
	Lake levels	Forest fires
	Reservoir levels	
	Water velocity	
	Water depth	
Economic	Groundwater abstraction	Irrigation water
	Surface water abstraction	Domestic water
	Reservoir outflow	Crop failure
	Crop yield	Farm animal death
		Navigability
		Hydropower
		Higher food and fodder prices
		Economic growth
Social	Drinking water	Conflicts water users
		Unemployment
		Famine
		Poverty
		Health (heat stress and respiratory)
		Migration
		Deaths

[1] drought has a negative effect meaning that magnitude of variable is lower, smaller or reduced, except for the indirect social effects

In general plant growth is the first to be affected by a drought because of the small water store in most soils. This may impose stress on drought-sensitive, natural plant species as well as agricultural crops, and subsequently lead to loss of wildlife habitat and agricultural production. Natural vegetation and riparian areas might dry up, and wild animals may suffer. In response to a regional extensive and long lasting drought, groundwater level will drop, spring yield decrease, streamflow reduce and lake levels go down. The rate and impact of the decline depends on weather conditions and properties of the river basin. Drought does not only affect water quantity, but also quality. Concentration of most components increases, leading to stress for aquatic communities and restrictions on water use.

Drought usually enhances the demand for water leading to a higher pressure on groundwater and surface water resources. The hazard can be short-lived, and society and ecosystems will quickly recover. Drought may also have prolonged impacts that will be intensified if aquifers or lakes are heavily overexploited. This may lead to below average outflow for a long recovery period of up to years. Damage might even be permanent if an endangered plant or animal species does not survive the drought and becomes extinct.

Drought combined with human activities can also induce land degradation, such as *desertification* in dry environments. Desertification implies a degeneration of productive ecosystems into desert. This irreversible chain reaction starts with a drought that rigorously affects plant growth. The vegetation cover is seriously affected, and the process is often enhanced by too high cattle grazing intensity. As a result the topsoil becomes compact and severe soil erosion (water and wind), including loss of soil fertility may result. This may prevent the recovery of the vegetation and sometimes bring about permanent changes in the ecosystems. In the absence of human pressures, the ecosystem will slowly recover from drought as the dryland plants are adapted to cope with dry conditions (Dregne, 2000). Human activity combined with poor management of soil, crop and herds may thus cause an increase in the impact of drought, which in turn may tend to extend or intensify drought and the development towards desertification.

Drought as a hazard must be viewed in the context of *risk* (Chapter 6). Risk is the chance that the natural or man-altered hydrological system in a region fails, i.e. it cannot meet the demand for water. Risk is a function of the probability of the event to happen and the value of the consequence, which is also referred to as the vulnerability of the system. Vulnerability has environmental, social and economic aspects. Regions with a high probability of drought but with a low vulnerability show a low risk, and the opposite occurs if both the probability and the vulnerability are high. The impact of drought depends on the complex interaction between the physical environment and the socio-economic system.

Countries in development in arid or semi-arid regions where the availability of water is already low under normal conditions are particularly prone to suffer adverse effects as they often lack the resources to mitigate droughts. Droughts in these regions are likely to have a huge economic and social impact, i.e. the vulnerability is high and drought has caused crop failures, extensive cattle death, and subsequently famine and deaths of millions of people. In regions with fossil groundwater this non-renewable resource may be over-exploited. In more financially well-off regions, like Europe or the USA, a temporary shortage of water has adverse effects primarily on ecology and economy. Economic vulnerability is here commonly related to large investments in the agricultural sector and hydropower production through increase in energy prices and restrictions on use. However, socio-economic resilience is sufficient for recovery. Social impacts like increase in poverty, social unrest and starvation that take place in developing countries do not happen.

Human protection from drought has traditionally been through structural measures such as storage reservoirs for water supply and water transfer schemes. Non-structural mitigation measures like water use restrictions and water demand management schemes are increasingly being required (Kundzewicz *et al.*, 1993). The potential conflicts between different users can be accounted for in the decision-making process by the involvement of users and affected bodies. Modern tools, like decision support systems, focus on the interaction between the users, the environment and models designed for analysis and impact assessment. A national drought watch system is recommended by Wilhite *et al.* (2000b) as a necessary component of drought assessment and management.

1.4   International low flow and drought studies

There is an extensive literature on the different processes that operate during a low flow or drought period. In particular catchment response and the low flow regime of a river or a region have been described. Less material is available on methods for low flow and drought assessment, including prediction, forecasting and estimation at the ungauged site. The latter was already pointed out in a survey on literature relevant to low flow prediction for ungauged locations (Institute of Hydrology, 1980), and still holds. The situation is even more striking when compared to the vast literature on flood and flood estimation. A selection of international drought studies and reviews are briefly presented below.

McMahon (1976) offers a useful overview of computational procedures and an annotated bibliography for low flow analyses. The international bibliography goes back to 1939 and gives a short summary of a total of 148 studies. An ASCE Task Committee (Riggs *et al.*, 1980) reviewed the types of low flow information and corresponding low flow characteristics and data needed in planning and design of water supplies, setting instream flow requirements and modelling stream water quality. It is concluded that improved methods for estimating low flow indices at the ungauged site are needed along with measures of the accuracy of various estimation methods. In McMahon & Arenas (1982) different methods for low flow computation are presented and illustrated with case studies from a number of countries with different low flow conditions and amounts of available data. Theoretical aspects of natural and man-induced factors affecting low flows are discussed, as well as data issues related to low flow analysis.

A comprehensive review of low flow hydrology is given in Smakhtin (2001). Low flow hydrology is defined as a discipline which deals with minimum flow in a river during the dry periods of the year. The review covers a range of topics; low flow generating mechanisms, methods for low flow estimation, relationships between indices, techniques for estimation at the ungauged site, environmental flow management and recent research initiatives. Literature on drought analysis and management is only marginally referred to, however, and droughts are seen as distinguished from low flows and connected with implications for resource availability. This view is supported by Beran & Rodier (1985) who do not regard drought as a definable entity in itself but as a prime mover which has attributes and consequences. They classify hydrological drought based upon variation in duration, severity and season, and various aspects of drought are discussed in terms of their implication for different sectors such as agriculture and hydropower. Yevjevich *et al.* (1983) similar define droughts as social phenomena that can be conceived as only human-related disasters which have by definition an impact on society. The authors focus on several themes related to efforts needed to cope with or manage drought. An overview of socio-economic, political and environmental impacts of drought is given, problems related to mitigation measures and drought control strategies are raised, and several case studies from around the world are presented.

Hisdal (2002) discusses the interpretation of terms frequently applied in low flow and drought studies, and provides an overview of different approaches to the definition of drought; both at-site and regional methods are presented. Drought in this study is considered a relative measure that is not linked to the impact of drought. Other reviews of drought definitions are found in Dracup *et al.* (1980), Wilhite & Glantz (1985), Rodda (2000), Tate & Gustard (2000) and Panu & Sharma (2002). The various interpretations of the term drought are frequently referred to as one of the main obstacles to the investigation of drought. It is generally recognized that as drought affects many sectors in society it is not considered appropriate to define one unique measure of drought. Nevertheless, the lack of well-defined criteria for drought identification is still considered to be a major obstacle to permitting a suitable response in the management system in time of drought (EEA, 2001).

International cooperation has enabled communication and exchange of data across national boundaries and thus encouraged joint research activities, particular in regional analysis. The FRIEND (Flow Regime from Experimental, International and Network Data) project, a contribution to UNESCO's International Hydrological Programme (IHP) aims to develop better understanding of hydrological variability and similarity across time and space. A central feature of the FRIEND project is the sharing of data and the establishment of a common data base for each regional FRIEND project. The first project, established amongst northern European countries (Gustard *et al.*, 1989), explored the relationship between low flow and catchment characteristics with special focus on the role of soil and geology in determining catchment response. Later the project was extended into other regions of Europe as well as into other parts of the world. This is reflected in subsequent FRIEND reports (Gustard, 1993; Oberlin & Desbos, 1997; Gustard & Cole, 2002) where a variety of analysing techniques applied both at-site and to regional data sets of hydrological drought characteristics are presented.

In response to a growing concern about the drought hazard and its impact on European economies, a workshop was organised by the European Commission in 1999 to analyse the state-of-the-art in drought research, drought planning and drought mitigation strategies (Vogt & Somma, 2000). The collection of workshop papers covers aspects of drought definition, risk assessment, drought monitoring and mitigation and gives recommendations for future actions. Bradford (2000) reports on recent drought events in Europe, with emphasis on hydrological droughts over the past decade. The EU supported ARIDE project investigated the nature of European droughts; their duration, magnitude, extent and sensitivity to environmental change (Demuth & Stahl, 2001). The study was based on data from the European Water Archive (EWA), containing nearly 5000 daily streamflow series. This allowed studies at the pan-European scale, focusing on the temporal and spatial variability in drought behaviour. Additional groundwater, meteorological and catchment data constituted the basis for detailed process studies at the catchment scale. On behalf of the European Environment Agency an overview of the main natural and artificial causes and impacts of extreme hydrological events in European countries has been presented (EEA, 2001). The report provides information on policy responses and national strategies in case of flood and drought.

"Floods and Droughts: the New Zealand experience" (Mosley & Pearson, 1997) combines basic knowledge about hydrological extremes, including controlling processes, estimation methods, effect of land use and management practise, with examples and case studies from New Zealand. It gives equally weight to the two extremes and provides an overview of the many aspects related to understanding and coping with flood and drought that goes beyond the interest of water professionals in the region. A detailed analysis of streamflow drought in Southern Africa is presented by Tate *et al.* (2000). The study aims to provide improved capabilities for the assessment of drought by developing tools to analyse and monitor current droughts. Daily time series from 15 streamflow stations from eight countries, mainly perennial rivers, were analysed. It is concluded that no single approach to the assessment of streamflow drought is satisfactory across the whole Southern African region, a region that suffers from a high degree of water scarcity, and a variety of methods therefore have to be considered.

A global assessment of drought, it causes, predictability, monitoring, forecasting, impacts and approaches for drought management, is presented by Wilhite (2000b). The two volumes include an extensive range of case studies and discuss methodologies and mitigation strategies from recent drought experiences worldwide, including monitoring and early warning techniques as well as integrated drought management schemes. The diversity and complexity of drought, its causes, characteristics and impacts, are demonstrated through the many examples provided and it is stressed that addressing the range of problems associated with drought requires a multidisciplinary approach.

1.5 Outline

The book is separated into three main parts. Part I (Drought as a natural hazard) provides basic knowledge on the drought phenomenon as a prerequisite for the analysis of hydrological drought. *Chapter 1* introduced the drought terminology and presented the aspects to be dealt with in the book. The drought hazard and the impact of drought were discussed followed by an overview of major international drought studies and reviews. *Chapter 2* looks at droughts at the global scale and focuses on the need to study droughts within their climatological context. Following a brief presentation of the major hydroclimatological regions of the world, spatial and temporal characteristics of drought are discussed including some important historical events for illustration purposes. It concludes with a section on drought forecasting and a general description of climate change impacts on drought. *Chapter 3* presents the natural processes of hydrological drought. It discusses how the local and regional climate governs the occurrence of a drought, i.e. how a drought develops within a specific hydroclimatological region. The chapter studies the propagation of a meteorological drought through the hydrological system, possibly leading to groundwater and streamflow drought.

Part II (Estimation methods) presents contemporary approaches to hydrological drought estimation, ranging from simple procedures with low data requirement to data demanding, advanced techniques involving several calculation steps. All chapters in Part II starts with simpler tools and then gradually increase the complexity of the methods presented. *Chapter 4* introduces the three data sets included on the accompanying CD; the Global Data Set consisting of 18 daily streamflow series selected with reference to the hydroclimatological regions defined in Chapter 2; the Regional Data Set with 83 daily streamflow series from southern Germany; and the Local Data Set that comprises meteorological, catchment and hydrological data from an experimental catchment in the Czech Republic. The chapter provides an overview of the hydrological data that are typically available and the procedures that ought to be applied to ensure data of good quality for drought analysis. *Chapter 5* gives an overview of the most common hydrological drought characteristics, both in terms of time series of low flow and drought deficit characteristics and indices. Methodology of their derivation is presented in detail, and demonstrated using time series of streamflow, groundwater recharge and discharge, and groundwater level. *Chapter 6* presents the different steps involved in frequency analysis and special attention is given to extreme value analysis of time series of low flow and deficit characteristics. Both at-site and regional frequency analysis are covered, and the procedures are demonstrated using the accompanying data sets. *Chapter 7* discusses time series modelling of drought by the use of random models. Different approaches to simulate annual, monthly and daily time series are presented. The distributions of drought characteristics are compared for observed and simulated time series, and the chapter ends by describing analytical calculations of drought characteristics. *Chapter 8* is devoted to the fundamental principle of estimating low flow indices in ungauged catchments. Different approaches are introduced focusing on simple estimation methods (empirical methods and flow correlation procedures) as well as multivariate techniques like statistical regression methods. It also introduces advanced estimation procedures, such as hydrological mapping methods, river network approaches and a region of influence approach. Examples are presented and demonstrated using the regional data set, and the applicability of the procedures for specific purposes is discussed.

Part III (Living with drought) covers human and ecological aspects of hydrological drought, in particular the interaction between drought, stream ecology and human activities. *Chapter 9* looks at possible impacts of human activities on hydrological drought. International studies, observations and physically-based models are used to analyse different environmental impacts of human activities on hydrological drought characteristics, including land use change, climate change, groundwater abstraction, land drainage, urbanization and water transfer. *Chapter 10* deals with the relationship between ecology and streamflow, in particular the effect of streamflow drought on stream biota. It includes an introduction to stream ecology for readers unfamiliar with the topic. The chapter also describes practical aspects of flow management for ecological protection, especially during drought, and contains practical examples as well as a program for modelling physical habitat. *Chapter 11* deals with the operational aspects of drought. It demonstrates how some of the methodologies presented in previous chapters are applied in the process of water resources management. The chapter includes a section on the user requirements for low flow information and development of national design procedures in Europe. Operational aspects are illustrated with practical cases. *Chapter 12* looks to the future need for advances in drought research, operational techniques and management strategies to better cope with the drought hazard.

2

Hydroclimatology

Kerstin Stahl, Hege Hisdal

2.1   Introduction

Hydroclimatology is the study of hydrological events within their climatological context (Hirschboeck, 1988). Hydrological droughts are events that develop from meteorological droughts (Section 1.2). They have different causes, characteristics and a variety of consequences in a range of climates and environments: a particularly dry spring or early summer may cause crop failures in the agricultural plains of North America, Europe or China; several dry winters leave reservoir levels low and lead to water restrictions in the Mediterranean countries or in California; decades of low precipitation expand desertification in the Sahel zone. Though diverse, these examples have one thing in common; they result from a meteorological *anomaly* – i.e. a deviation from the normal weather patterns that determine the long-term average climatic behaviour with its typical variability in time and space. Meteorological drought is therefore perceived in a relative way: as a departure from the normal hydroclimatic features of a region. Knowledge of a region's climate is therefore crucial to understand the meteorological causes of drought and ultimately to manage hydrological drought. This chapter aims to explain the climatic context in which droughts occur around the world, and illustrates the causes, characteristics and consequences hydrologists therefore face in their work to prepare for and manage hydrological droughts. The entire chapter deals with hydrological drought in a general way. Chapter 3 explains more specifically the processes of how a meteorological drought propagates through the hydrological system. Specific hydrological drought characteristics are introduced in Chapter 5.

After a brief introduction to the global atmospheric circulation, Section 2.2 gives an overview of the world's climate zones and discusses hydroclimatic characteristics such as the seasonality of precipitation, temperature, evapotranspiration, and the resulting hydrologic regimes. Examples illustrate how anomalous atmospheric circulation and weather patterns produce the combination of hydrometeorologic conditions that may cause a drought in a particular climatic region. The spatio-temporal scales of the drought causing meteorological features strongly influence the typical scale of hydrological droughts and their variability in space and time. These properties are discussed in Section 2.3. Section 2.4 presents examples of recent drought events in the context of the discussed causes and characteristics. Knowledge from all these areas: atmospheric causes, spatio-temporal characteristics of drought, and recent events are important for sound water resources management including drought monitoring and forecasting. Current forecasting practice is briefly described in Section 2.5. Section 2.6 looks into the long-term future and summarizes the current knowledge on the influence of climate change on drought around the world. The last section (Section 2.7) summarizes the important points of the chapter.

2.2   Drought in different climates

2.2.1   Global atmospheric circulation

The world's climate provides a challenge for hydrologists. Precipitation, temperature, evapotranspiration, and runoff not only vary in space in terms of their annual sums, but also vary differently throughout the year in different locations. These seasonal variations are predominantly a result of the global circulation, which is a consequence of global energy redistribution. Generally, the global insolation causes a temperature difference between the equator and the poles. This difference invokes flows of airmasses to maintain a global energy balance. Figure 2.1 shows a simplified global circulation system with the prevailing wind directions. The equator has a belt of low-pressure centres (*equatorial trough*). To compensate for the pressure difference, airmasses flow equatorwards. Due to the earth's rotation, the Coriolis effect deflects airmasses from their north-south direction towards the right (Northern Hemisphere) or left (Southern Hemisphere). The tropics in the low latitudes from the equator to about 30° Latitude (N and S) are therefore governed by NE *trade winds* (SE in the Southern Hemisphere). The airmasses from the trades converge in the Inter Tropical Convergence Zone (ITCZ). Under convection, the air rises to the upper atmosphere, where a compensating poleward flow forms the tropical *Hadley cell*. Sinking and diverging airmasses from the Hadley cell form a zone of large permanent high-pressure centers (anticyclones) called the *Subtropical High-pressure Belt* at about 20°–40° Latitude. Poleward of this boundary, in the mid-latitudes between about 60° and 80° Latitude, *westerly winds* or *Westerlies*

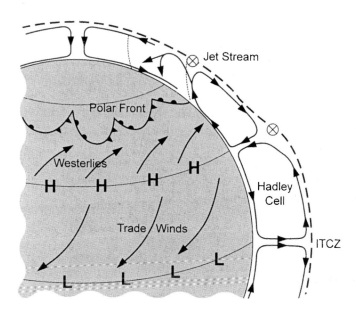

Figure 2.1 Simplified model of the global circulation.

prevail. The boundary to the *Polar Easterlies* in the high latitudes above 60° Latitude is known as the *Polar Front*. In this zone, cold polar air mixes with warm tropical air, turning the upper-air westerly flow into a strong meandering *Jetstream*. The location of the Jetstream greatly affects the extra-tropical mid-latitude weather. With the seasonal change in global radiation, the ITCZ migrates north and south of the equator also shifting the other circulation features (e.g. subtropical high pressure belt and jetstream) north and south with the seasons.

These general circulation characteristics create the world's major climate zones. Surface characteristics such as topography and land-water distribution may strongly modify the regional climate characteristics. The oceans, with their large heat storage capacity, also play a major role and their currents transport energy over great distances. They also provide moisture for evaporation. Airmasses carried from the oceans onto the continents often provide precipitation as they cool and condensate due to orographic rise effects. Within the mid-latitude westwind zone, the result is the typical west coast climate with abundant precipitation. Warm currents, such as the Gulf Stream, cause warm and moist climatic conditions in western and northern Europe. The role of the ocean also becomes obvious during temporary reversals of ocean currents and associated changes in the sea surface temperatures, for example during El Niño

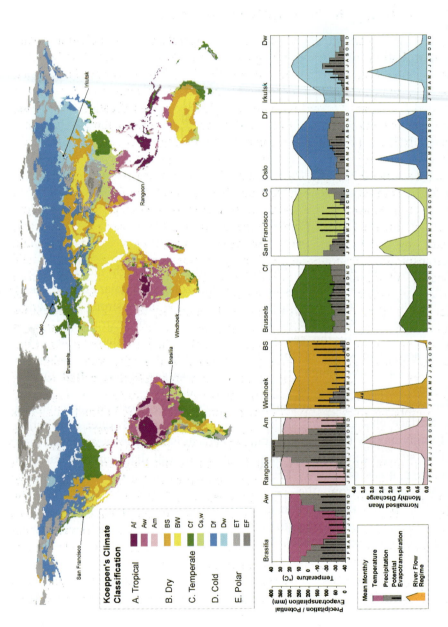

Figure 2.2 Generalized Köppen's Climate Classification (data from FAO, 1997), selected climate diagrams (data from www.klimadiagramme.de and Ahn & Taishi, 1994), and typical river flow regimes in the region.

events. Such anomalies have a strong impact on the weather in the region. In fact, they may cause anomalies in climatic features at great distances, a phenomenon called 'teleconnections', and have been linked to hydrological extremes such as floods and droughts (Section 2.2.3).

2.2.2   Hydrology in the world's climate zones

Hydroclimatic variables, such as temperature, precipitation and evapotranspiration, together with vegetation provide a basis for classifications of the world's climates. Mapped, climate classifications illustrate the regional climatology and can help to understand the causes of droughts, to estimate their effect, and to prepare for and mitigate drought. Basic classifications divide the world's climate into arid, semi-arid and humid referring to the annual water balance. UNESCO (1997), for example, bases the degree of *aridity* on the ratio of the mean annual precipitation to the mean annual potential evaporation. For a ratio smaller than 0.2, the climate is classified as arid, for a ratio between 0.21 and 0.5 it is semi-arid, between 0.51 and 0.75 it is sub-humid and greater than 0.75 is humid. However, to understand the temporal characteristics of drought in a certain climatic region, the existence and timing of a seasonal *water deficit*, defined by the difference between precipitation and potential evapotranspiration, is important. A variety of classifications exist that include seasonality. Wladimir Köppen (1918) introduced a still widely applied climate classification, which uses the mean annual and mean monthly precipitation and temperature to derive a letter symbol classification (Appendix 2.1). It classifies the dimensions of the *water availability* and the average *seasonality* of precipitation and temperature, information usually shown in *climate diagrams*. When potential evapotranspiration exceeds precipitation, the water balance is negative causing a water deficit. A seasonal water deficit is called an *arid season*, and a seasonal precipitation excess is called a *humid season*. Figure 2.2 shows a global map of Köppen's climate zones with selected climate diagrams and typical river flow regimes from the region. River flow *regimes* show the variability of stream flow over the year. They are often illustrated by the long-term mean monthly stream flows, normalized by the long-term mean annual flow. Due to the influence of catchment properties and a lack of data, global classifications and maps of river flow regimes are difficult to derive. A study by Dettinger & Diaz (2000) recently analysed and discussed global characteristics of streamflow seasonality and variability. The following paragraphs discuss how atmospheric circulation features cause the major seasonal and inter-annual hydroclimatic variations found in the different climates.

The tropical climates (A-climates) dominate the lower latitudes between approximately 30°N and 30°S. The average temperature of the coldest month is warmer than 18°C and convective precipitation of the ITCZ dominates the hydroclimatology. For locations close to the equator, the ITCZ's influence is effective all year round (Robinson & Henderson-Sellers, 1999). In these *Af-climates*, despite high temperatures, the atmospheric water balance is always positive. *Perennial rivers* flow all year and vegetation is abundant with mostly rainforest. Towards the north and south, however, the annual displacement of the ITCZ determines the amount of precipitation and the timing of the rainy seasons. These *Aw-climates* have a distinct dry season (Figure 2.2, Brasilia). The general rule in the tropics is, the further away from the equator, the longer are the dry seasons and the shorter are the rainy seasons. Köppen's definitions for a dry season in the different climates are given in Appendix 2.1. A particular phenomenon of the tropics is the monsoon. This *Am-climate* is pronounced in Asia due to the unique setting of land and water with the Himalayan Mountains to the north. A reversal of the wind system brings warm and moist airmasses from the Indian Ocean onto the continent. Therefore, monsoonal rainy periods are more intense and bring higher amounts of precipitation (Figure 2.2, Rangoon) than the ITCZ convection in an Aw climate's rainy season. Precipitation and potential evapotranspiration in both Aw- and Am-climates show a distinct seasonality with a dry season.

   The dry climates (B-climates) are arid, and annual evapotranspiration exceeds annual precipitation. They are found mainly in the regions north and south of the tropics, where the ITCZ's influence becomes minimal, and dry diverging air dominates in the form of a permanent subtropical belt of *anticyclones* (high-pressure systems). Under certain local circumstances (e.g. continental interiors sheltered by high mountains), however, dry climates may also occur at higher latitudes. The two main subclasses refer to the dominant vegetation types. The *BS-climates* have a wet season (Figure 2.2, Windhoek). However, mean annual and mean monthly precipitation (and streamflow) amounts are often misleading, because the inter-annual variability of precipitation in these regions tends to be high and the amounts may result from rainfall events that occur only every few years. Rivers are often *ephemeral* and only flow periodically after a storm. While the semi-desert grassland environment may be able to cope with rainless years, longer periods with only minimal rainfall may become hazardous to the flora, fauna, and people living in these regions. An example is the Sahel region in west Africa, where long and recurrent drought periods seem to have provoked feedback mechanisms that cause desertification (Section 1.3). A *BW-climate* rarely experiences rain. The water deficit is permanent. Vegetation is largely absent, and people rely on scarce water resources. Some fortunate regions receive water from *exogenic rivers* such as the Nile in Egypt, which originates in a different (humid) climate zone (Box 3.1). While locations with access to groundwater or along an exogenic river used to be the only habitable places in the desert over the last century, human development has increasingly brought water into desert environments via pipelines or canals from more humid regions or distant surface or groundwater resources, including fossil groundwater. Dry climates can further be subdivided by temperature into hot and cold subtypes.

The temperate climates (C-climates) are characterized by an average temperature of the coldest month between −3°C and 18°C, and an average temperature of the warmest month above 10°C. They dominate in the mid-latitudes, but may also occur at higher elevations in lower latitudes. Temperate climates are subdivided by the timing and intensity of the dry season. *Cf-climates*, which have no distinct dry season, occur along the west coasts of the continents that are exposed all year round to meso-scale *cyclones* (low pressure systems) and *anticyclones* (high pressure systems) moving west-to-east. Summer precipitation may be slightly lower than winter precipitation, because wintry systems tend to be stronger. Cf-climates also occur in the continental interiors, where the amount of convective summer precipitation is equal to that of frontal winter precipitation. The water balance is generally positive, but can be slightly negative during the summer months due to high temperatures and associated evapotranspiration (Figure 2.2, Brussels). Consequently, rivers are predominantly perennial with a stream flow minimum during the summer months. The recovery from the summer low flow will lag slightly behind the precipitation and evapotranspiration cycle, as soils are rewetted. The *Cs-climate*, which has a dry summer season, is characterized by at least three times as much rain in the wettest month of the winter as in the driest month of the summer, the latter having less than 30 mm of precipitation. Often called 'Mediterranean climate', Cs-climates are a result of the permanent anticyclones of the subtropical high pressure belt creating stable dry and hot weather conditions with a pronounced seasonal water deficit in summer (Figure 2.2, San Francisco). During the winter months, however, the subtropical high-pressure belt is usually displaced to lower latitudes as a result of the higher energy difference between the equator and the pole. The westerly circulation then influences these regions with cyclones producing a rainy season with a positive water balance. The winter months are therefore critical to vegetation and water supply. Depending on local conditions, convection or orographic condensation may also provide some precipitation in summer. In Cs-climates, the flow regime follows the precipitation cycle. Amplified by high temperatures and evapotranspiration, summer flows are generally low. Particularly smaller rivers are *intermittent,* i.e they regularly dry up during the summer months. The seasonal differences are reversed in the *Cw-climates* (dry winter), which often occur in high altitudes of subtropical regions (e.g. northern India and Nepal).

The cold climates (D-climates), which are characterized by an average temperature of the warmest month above 10°C and that of coldest month below −3°C, are typical in high latitudes, continental interiors, and high elevations, where winter precipitation usually falls as snow. Similar to the classification for the temperate climates, *Df-climates* have no distinct dry season. Since winter precipitation falling as snow is stored until the temperatures rise in spring, the river flow regimes show a distinct snowmelt peak. A second streamflow peak occurs in the autumn slightly lagging behind the increasing precipitation. The subtype also occurs mainly along west coasts due to the year-round influence of cyclones (Figure 2.2, Oslo). The *Dw-climate* has a dry winter season. In the continental interiors and high mountains, low temperatures during a long winter cause surface water to freeze. The combination of the precipitation minimum and freezing conditions may cause a strong reduction or even interruption of streamflow during winter. In the continental plains, snowmelt in spring is often sudden, while it is more gradual in mountain rivers due to the contribution of different elevation levels. High temperatures in summer can cause a negative water balance despite a precipitation maximum due to convective events in continental regions (Figure 2.2, Irkutsk).

Polar climates (E-climates) are characterized by very low temperatures. As snow and ice dominate the land cover, vegetation is rare and population density is low. E-climates occur in the Arctic and Antarctic and in high altitude regions. E-climates are mainly relevant to drought where they are part of a catchment, i.e. in mountainous regions where glacial meltwater feeds a stream that flows down into another climate zone (exogenic rivers).

The hydroclimate shows great differences around the world. Depending on the climate zone and local physical properties of the catchments, the annual cycle of precipitation, temperature and evapotranspiration determines groundwater recharge and river flow regimes. Knowing the annual cycle of the hydroclimatic input and the lag time between a precipitation minimum and the typical streamflow response will be helpful in choosing analysis methods (Chapter 5) or preparing for hydrological drought (Section 2.5).

2.2.3   Atmospheric circulation and drought

While the general atmospheric circulation creates the average climate characteristics of a region, its characteristics vary from year to year. Occasional strong variations cause anomalies in the hydrometeorological characteristics. In recent years, therefore, researchers have increasingly investigated the relationship between atmospheric circulation and drought in order to better understand drought occurrence, to assess forecasting possibilities and to evaluate the effect of climate change on drought. Studies on this topic from several countries around the world covering different climate zones have recently been assembled in Wilhite (2000b).

Regional atmospheric causes

In the Tropics (A-climates), the distinct rainy seasons depend on the ITCZ and its related convective rainfall or monsoon storms, which occur over the continents when the ITCZ shifts poleward in summer. High inter-annual variability in the strength and location of the ITCZ and the monsoon troughs provides a great risk for the rain-fed agriculture in, for example, south Asia and sub-Saharan Africa, where the crop cycle is strongly affected by a delayed onset or failing of the rainy season.

Barry & Chorley (1992) suggest that regions with marginal climates that are alternately influenced by differing climatic mechanisms are particularly susceptible to drought. Their theory applies to the transitional zones from the tropics to the dry climates and from the temperate climates to the dry climates. For some parts of the year these subtropical regions are under the influence of the anticyclones of the subtropical high-pressure belt, at other times they are influenced by tropical or mid-latitude circulation. An extension of the usual dry season, because of the persistence of the subtropical anticyclones into the humid season, may cause droughts. For example, an abnormally southern position of the subtropical anticyclones during the summer may prevent rain from the ITCZ from moving into outer tropical regions such as the Sahel region in west Africa. Located too far north during winter, subtropical anticyclones may prevent Atlantic cyclones from bringing the typical winter rain into the Mediterranean countries which rely on the recharge of impounding reservoirs and groundwater during winter.

With constantly changing atmospheric conditions and precipitation expected all year around, the mid-latitude temperate climates (C-climates) are vulnerable to extended dry weather periods at any time of the year. During summer, however, they may become critical to plants and water supply schemes if temperatures and associated evapotranspiration are high and surface water reservoirs, which are often filled during autumn and winter, are already depleted. *Blocking action* is the major atmospheric anomaly to cause extended dry weather periods in western and northern Europe and in North America. Strong high-pressure centres develop and persist over an extended region and only move westward very slowly if at all, diverting storm tracks either to lower or higher latitudes. During the years of 1975 and 1976, western Europe, particularly the UK and north-western France, experienced one of their most severe droughts when blocking high-pressure centres persisted over the region in the summer periods (Figure 2.3).

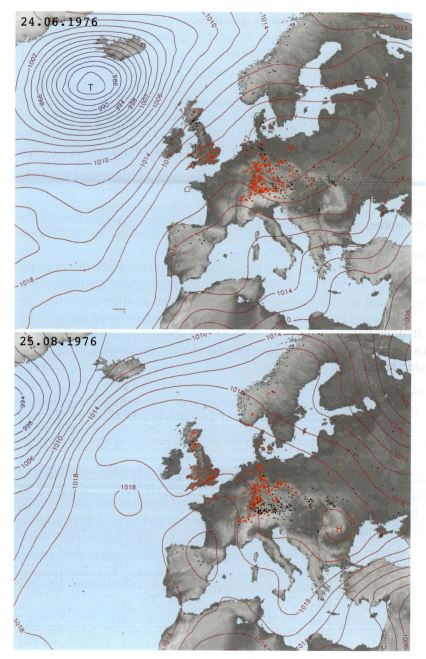

Figure 2.3 Mean sea level pressure distribution (in hPa) and low flows during the 1976 drought in Europe. Red dots indicate rivers with very low streamflow; black dots indicate rivers with normal flows (mapping according to Stahl, 2001).

The typical circulation and weather patterns associated with drought have been analysed by correlating mean sea level pressure (MSLP) or circulation indices and patterns derived from it, with a hydroclimatic or hydrological variable of interest. Adler *et al.* (1999), for example, used canonical correlation to identify periods with lower than normal southwesterly Mediterranean circulation in winter leading to low precipitation periods in Romania. Synoptic situations responsible for dry periods in the Mediterranean basin have been identified by principle component analysis (Section 5.5.2) of MSLP data during anomalously dry months (Maheras, 2000). Analysing the link between routinely classified weather patterns from the German Weather Service and streamflow deficiency showed the strong influence of persistent high-pressure patterns in various regions across Europe (Stahl, 2001). Generally, regional drought-causing atmospheric situations are characterized by an anomalous:

a) timing of a seasonal phenomenon;

b) location of pressure centres and the track of cyclones;

c) persistence or persistent recurrence of dry weather patterns.

Extreme hydrological droughts are likely to be caused by a combination of anomalous circulation patterns and already low antecedent soil moisture, groundwater, and lake storage.

Large-scale atmospheric causes

Some of the regional drought-causing atmospheric phenomena are linked to larger scale ocean-atmosphere interactions and fluctuations that occur with a certain periodicity. As described earlier for the Asian monsoon, seasonal changes in the prevailing wind direction may also cause atmospheric and hence, hydroclimatic changes. Well known is the less regular El Niño-Southern Oscillation (ENSO). The Southern Oscillation Index (SOI), measured by the pressure difference between Tahiti and Darwin (Australia), is used to monitor the seesaw of atmospheric pressure that exists in that part of the world. Every two to seven years, an *El Niño* event occurs. Characterized by high negative values of the SOI, it is a strong reversal of the prevailing pressure situation. This reversal also invokes changes in the Sea Surface Temperatures (SST) in the Pacific: warm water replaces the usually cold currents off the coast of Peru. This 'warm phase' usually peaks around Christmas along the coast of Peru and Chile, where direct influence on rainfall is greatest. Local fisherman therefore called the event El Niño, the Christ Child. The opposite 'cold phase' with high positive SOI is often called *La Niña*. Teleconnections with the two phases have been noticed around the world and suggest particularly strong linkages to drought. Ropelewski & Halpert (1987) analysed El Niño-associated rainfall

December - February

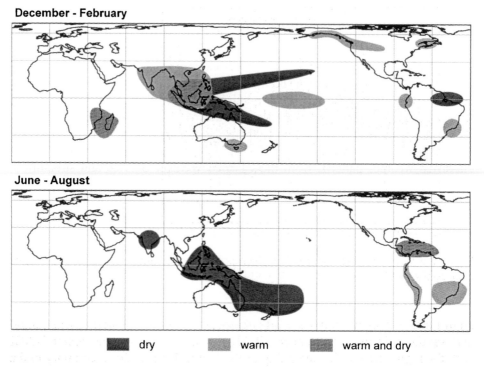

June - August

| | dry | | warm | | warm and dry |

Figure 2.4 Global climatic impact of El Niño (modified and redrawn from CPC, 2002).

from 1700 stations worldwide and for different lag-times. They found a coherent relationship with less than normal rainfall in Indonesia and the southwest Pacific, large parts of Australia, southeast Africa and a monsoon rainfall deficiency in central-northern India. Similar continental or regional studies confirming these patterns have been carried out for rainfall and also for streamflow. A worldwide analysis of ENSO-streamflow teleconnections was published by Chiew & McMahon (2002). Figure 2.4 summarizes the climatic impacts of El Niño around the world and their relevance to drought.

During the 1990 to 1995 drought in Spain and Portugal, winter precipitation was considerably below the long-term average from November to March in the years 1991–1992, 1992–1993 and 1994–1995 (Peral Garcia *et al.*, 2001). During these winters, storm tracks from the Atlantic Ocean into Europe were shifted northwards and spared the south. This behaviour was related to a certain phase of the fluctuating North Atlantic Oscillation Index (NAOI), which can be measured by the pressure difference between Iceland and either the Azores, Gibraltar or Lisbon. The NAOI indicates the strength of the westerlies.

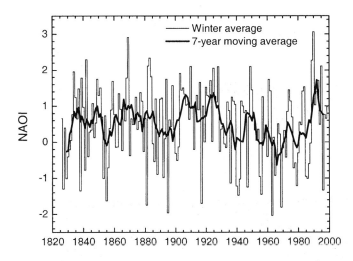

Figure 2.5 Winter average (December–February) North Atlantic Oscillation Index (data from Jones, 1997).

High index values in winter are often associated with dry conditions in central and southern Europe (Greatbach, 2000). Figure 2.5 shows the winter NAOI with the high phase in the beginning of the 1990s. Despite the uncertainty in the periodicity of these atmospheric phenomena, the link to hydroclimatic variables promises potential for drought forecasting and prediction (Section 2.5).

2.3 Space-time variability

2.3.1 Scales in Hydrology

Hydrological phenomena occur at a wide range of spatial and temporal scales. Figure 2.6 places drought in the space-time domain in relation to floods and the major components of the hydrological cycle, as well as the meteorological processes generating hydrological response. It can be seen that drought scales differ from the much shorter and smaller scales of flood. Droughts tend to persist spatially and temporally. Floods are generally produced by a distinct storm or a series of storms, convective or frontal, by snow melt or a combination of both. Droughts, in contrast, develop as the result of an accumulated water deficit. While within-year persistence is an important characteristic of drought everywhere, inter-annual persistence seems to be prevalent in climatic regions such as the outer tropics and subtropics where

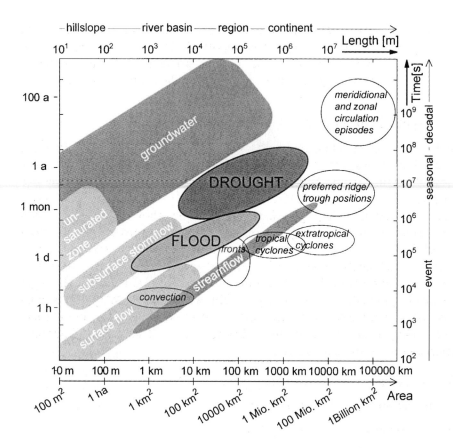

Figure 2.6. Space-time domain of flood and drought in hydrology and meteorology (hydrological scales as defined by Blöschl & Sivapalan, 1995a; meteorological scales as defined by Hirschboek, 1988).

replenishment of the water resources depends on a distinct rainy season (Section 2.2.3). Climatic feedback mechanisms may enhance drought due to factors that are also closely linked to desertification, such as increased albedo of bare and dry soil and increased aerosol concentrations.

The spatial and temporal persistence of droughts raises important questions for their analysis. At what scale should the data be analysed? Are there sufficient data available to characterize drought in the area concerned? Are the time series long enough? At the local scale, meteorological and hydrological records usually have a resolution of days or months. Larger scale studies often use time-aggregated seasonal or annual values. The spatial and temporal characteristics of drought typical in a certain region depend on the climate and the seasonal hydrological cycle as well as other factors such as vegetation, soil, and geology (Chapter 3). Although it is impossible to completely separate the space and time domain when discussing drought, principal spatial and temporal characteristics can be identified and are described in the following sections.

2.3.2 Spatial variability

Drought is a regional phenomenon and the area covered by a drought influences the perception of drought as a hazard (Section 1.3). However, spatial studies have revealed that despite the large spatial extent, there might be considerable heterogeneity within drought-affected areas. The 1976 drought in Europe discussed in Section 2.2.3 (Figure 2.3), for example, shows some rivers with average discharge among the dry rivers. Local thunderstorms in the UK and a particularly severe one near Paris, provided local alleviation (Beran & Rodier, 1985). Also, catchments with extensive aquifers usually have higher streamflow than those without a well-developed groundwater system (Chapter 3). Drought often spreads gradually across a region. Knowledge of such typical spatial patterns and a continuous monitoring may help to better plan for drought and improve early-warning systems (Section 2.5). Therefore, important characteristics to be studied include:

a) the spatial extent of individual events;

b) the variability within the affected area;

c) the dynamics of the spatial extent;

d) recurrent patterns in space.

Obtaining information about the spatial extent of a drought is both important and challenging. Hydroclimatic variables of precipitation or streamflow provide the basis for characterization of drought in quantitative terms (Chapter 5). Most variables, however, are measured at a limited number of stations. To obtain information at the regional scale, the spatial pattern of at-site drought characteristics can be analysed. However, for the derivation of regional drought properties such as the area covered by drought, assumptions about the behaviour between stations must be made. The derivation of regional drought characteristics is described in Chapter 5.

Information about the spatial patterns of droughts is often presented in the form of a map. The simplest way is to *directly display at-site results* on maps. This is illustrated in Figure 2.3. Rossi *et al.* (1992) give various examples of the most common map presentations. Maps with *isolines or contours* of the average dry spells of a specific drought require interpolation. Unfortunately, available data are often variable and irregular when looking at the regional, national or even continental to global scale. Therefore, the original data available should always be kept in mind when using interpolated spatial data. Another option for map presentation is to *shade areas* with equal drought characteristics. Map presentations can be also be used to study the real time development of a drought in space by making snapshots of the drought situation at individual points in time as demonstrated in Figure 2.3. Grid based data presentation and GIS features provide a variety of computational opportunities to map and study spatial relations. Recently, remote sensing data have provided some new opportunities to analyse and monitor drought in space and time (Section 2.5). A different description of the extent of a historical drought can be given by presenting graphs relating the area covered to the drought characteristic. These presentations, however, do not reveal the spatial pattern as a dynamic feature.

2.3.3   Temporal variability

As a result of natural climate variability, hydrological behaviour also fluctuates at different timescales. Understanding patterns of climate variability assists in understanding the large variation in hydrologic characteristics over time. For drought, it is necessary to analyse both persistence and recurrence in time (Chapters 6 and 7). Such knowledge, however, strongly depends on the available data (period and resolution) to be analysed.

The analysis of *annual* or *multi-year fluctuations* generally relies on instrumental meteorological and hydrological records. Time series of about 30 years are commonly applied to define a particular climate or hydrological regime, or to look for a trend (e.g. the WMO standard periods are 1931–1960 and 1961–90). Hence, such records are also often used in drought studies. Figure 2.7 illustrates how a drought characteristic derived from a streamflow record fluctuates from year to year. The example also shows fluctuations at a multi-year scale due to a clustering of wet and dry years. The lower figure shows trends that were calculated for the same drought characteristic for consecutive 30-year sub-periods. Depending on the chosen sub-period, both positive trends towards more and negative trends towards less severe droughts have been calculated. Consequently, adding or subtracting a few years of data in the time series may lead to a change in trend. This is important to bear in mind when comparing results of different trend studies and for drought analysis in general. A related question in drought studies is whether a constant hydrologic 'baseline' exists and what record length is required to define it. Long instrumental records extend to about 200 years. Despite frequent inconsistencies due to changing the observation equipment or other human influences, these rare records are long enough to document droughts that vary at multi-year

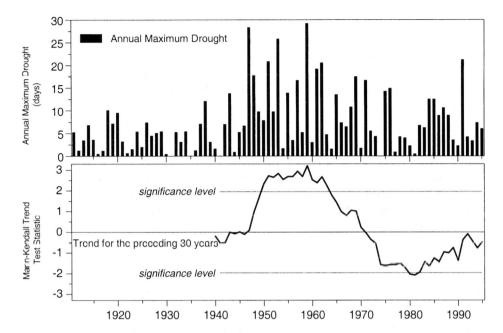

Figure 2.7 Series of annual maximum drought (normalized deficit volume in days) for River Vils at station Pfronten Ried in Germany (upper) and a trend test statistic calculated for consecutive 30-year periods (lower) (modified from Hisdal *et al.*, 2001; reproduced by permission of John Wiley & Sons Limited).

timescales. Series of this length can also be used to detect relationships with NAO and ENSO fluctuations, which can be used for forecasting purposes (Sections 2.2.3 and 2.5).

On the scale of *decadal* to *centennial variability*, the knowledge of climatic fluctuations and the occurrence of dry periods depends largely on paleoclimatic reconstruction from analyses of proxy-data such as tree rings, fossils, corals, and ice cores, spanning the last 500 to more than 2000 years. Yu & Ito (1999) studied a sediment core from Rice Lake in the northern Great Plains of the USA, and their study indicates that century-scale drought frequency is related to solar-oscillation periods. Fluctuations in lake salinity of Moon Lake, North Dakota (USA) indicate that droughts more intense than the 1930s Dust Bowl (Box 2.1) were more frequent prior to AD 1200 (Laird *et al.*, 1998). An abrupt change was found to coincide with the end of the Medieval Warm Period and the onset of the Little Ice Age (AD 1300–1850). Using tree ring records, Hughes & Brown (1991) found that drought frequency in the Sierra Nevada, California (USA) from 1850 to 1950 was lowest in any one hundred year period in the record (101 BC to AD 1988). They also found that the twentieth century up to 1988 had a below average frequency of extreme droughts.

Evidence from combined archaeological and paleoclimatic records indicates that multi-decadal to multi-century droughts have been the cause for population dislocations, urban abandonment and state collapse. In a review, deMenocal (2001) emphasizes that the climate perturbations throughout the late Holocene have been extreme in their duration and intensity. Multi-decadal to multi-century droughts are rare, but nonetheless are integral components of natural climate variability. Droughts played an important role in the collapse of the Akkadian (~2200 BC, Mesopotamia), Maya (AD ~800, Yucatán peninsula), Mochica (AD ~500, coastal Peru) and Tiwanaku (AD ~1000, Bolivian-Peruvian altiplano) civilizations. Records of paleoclimatic data sources also show that an unusually severe drought persisted in what is now called the Middle East in the thirteenth century BC. The drought is described in the Bible as having devastating effects with people migrating searching for food and water. Cities were abandoned and whole empires vanished (Stiebing, 1989). This latter episode is an example of the clustering of dry years, which in hydrology is referred to as the 'Joseph effect' (Mandelbrot & Wallis, 1968).

2.4   Recent drought events

The preceding sections have given an overview of the different hydroclimatological regions of the world and causes and characteristic spatial and temporal patterns of drought in these widely different environments. This section discusses some major recent drought events. In general, two types of historic records are available: continuous time series of climatological and hydrological data, and reports on natural disasters that received international public attention because of their impact. To date there is no database holding and combining the two, but several attempts have been made to gather large samples of both types of records. Major drought events with serious economic, social and environmental impacts are recorded in the international disaster database EM-DAT, which is kept by the Office of US Foreign Disaster Assistance (OFDA) and the Centre of Research on the Epidemiology of Disasters (CRED). EM-DAT contains essential core data on the occurrence and effects of over 12 000 disasters in the world from 1900 to the present (Figure 1.3). The database is compiled from various sources such as UN agencies, non-governmental organizations, insurance companies, research institutes and press agencies (OFDA/CRED, 2002). It does not, therefore, include a common definition of drought, but rather relies on what the individual data sources have classified as a drought disaster. Africa clearly dominates the

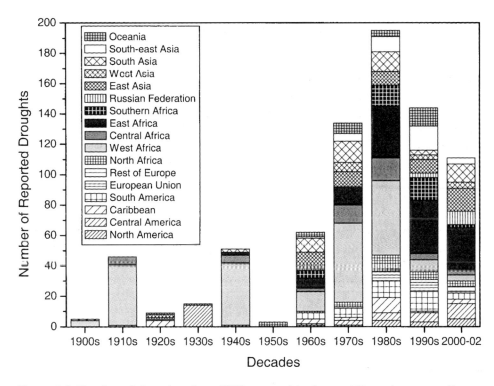

Figure 2.8 Number of droughts since 1900 reported in the world's regions according to EM-DAT (data from OFDA/CRED, 2002).

records. The regional distribution and temporal development shown in Figure 2.8, however, may be biased by the increasing news coverage in the twentieth century. Droughts seem to be registered mostly in developing countries that are located within marginal climatic conditions and lack the institutional capacity to compensate for the hydrometeorological variability. Drought consequences in industrialized countries are different and characterized more by financial damage than by famine (Section 1.3). Before reinsurance data were available droughts might not have been recorded to the same degree in developed countries. Hayes (2002) concludes that increasing economic losses worldwide are more due to an increased vulnerability than to an increased number of events. Despite these caveats, the database highlights the droughts that caused public attention and the EM-DAT records provide a general chronology and illustrate that drought is a hazard around the world.

Box 2.1

The Dust Bowl

Much attention was caused by the so-called 'Dust Bowl' in the United States and Canada in the 1930s. During the drought farmers kept plowing and planting though nothing would grow. As the ground cover that held the soil in place was gone, the plains winds whipped across the fields raising clouds of dust to the sky. In some places the dust would drift like snow, covering farmsteads (Figure 2.9). Svoboda (2002) summarizes: "In the 1930s, drought covered virtually the entire Plains for almost a decade (Warrick, 1980). The drought's direct effect is most often remembered as agricultural. Many crops were damaged by deficient rainfall, high temperatures, and high winds, as well as insect infestations and dust storms that accompanied these conditions. The resulting agricultural depression contributed to the Great Depression's bank closures, business losses, increased unemployment, and other physical and emotional hardships. Although records focus on other problems, the lack of precipitation would also have affected wildlife and plant life, and would have created water shortages for domestic needs." In fact, hydro-meteorological records show that there were at least four distinct drought events: 1930–1931, 1934, 1936, and 1939–1940 (Riebsame *et al.*, 1991).

Figure 2.9 The Dust Bowl (http://www.usd.edu/anth/epa/dust.html, photo credit: US National Archives).

North America was affected by the so-called Dust Bowl in the 1930s (Box 2.1) and again by severe droughts during the 1980s and 1990s. Crop damage was noted in the south and west of the USA. Regions with a strongly rising population and increasing water demand, such as southern California, experienced water supply shortages. The threat seems to continue at the beginning of the twenty-first century: in 2001 and 2002 water shortages once more troubled most of the country. In March 2002 New York City's water supply was so depleted that the area was declared a drought emergency region.

In Africa, the west African Sahel zone suffered from two severe droughts in the first half of the twentieth century. The first drought lasted from 1910 to 1914, the second one occurred 30 years later and lasted from 1940 to 1944. Since the 1970s west Africa has more or less constantly been in a critical condition with rainfall deficiencies and depleted streamflow (Box 2.2).

Box 2.2

The Sahel Drought

The Sahelian zone on the southern edge of the Sahara desert in north Africa stretches from Senegal through Mauritania, Mali, Burkina Faso, Niger, and Chad into Sudan. Located on the northern edge of the tropics it has one rainy season in June–August, when the SW monsoon brings moisture into the region. However, the ITCZ and the associated monsoon rainfall do not always reach the Sahel, making the dry savannah environment particularly prone to drought. While the long-term average rainfall amount is between 100 and 500 mm, typically several years of below average rainfall alternate with several years of above average rainfall. Though not uniform across the entire Sahel, since the late 1960s the region has endured a series of extensive and severe droughts with serious consequences for the population, including famine and desertification. Scientists are now investigating whether the situation is a phase of natural variability or a permanent change. Statistical analyses in several studies have identified a discontinuity in the rainfall series around 1970 (e.g. Hubert, 1998). Recently, a study by L'Hôte *et al.* (2002) concluded that this drought still continues. Nicholson (2000) illustrates that despite the diverse factors producing the mean climate, general rainfall fluctuations occur synchronously over the African continent (Figure 2.10). However, a strong inter-annual persistence is only visible in the Northern Hemispheric part, suggesting land-atmosphere feedback mechanisms (Nicholson, 2000). Increasing population, agriculture and overgrazing are often blamed for the desertification in the Sahel. The extent of a positive feedback of removal of vegetation raising the albedo, reducing convection, reducing rainfall and ultimately more reduction of plant growth, however, is still uncertain. Nevertheless, the Sahelian region has a high economic vulnerability to drought and currently remains on the agenda of aid organizations and development politics. (→ Box continued on next page)

Box 2.2 (continued)

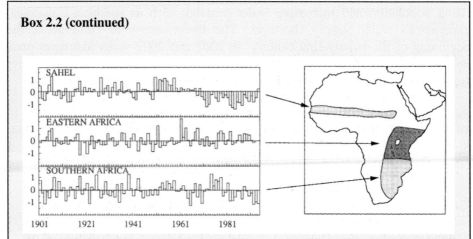

Figure 2.10 Rainfall fluctuations in the Sahel, eastern and southern Africa from 1901 to 1994, expressed as regionally averaged standardized departure. Regions represented are shown in the small map (from Nicholson, 2000; reproduced by permission of Elsevier Science Publisher).

Oceania and southeast Asia's strong response to El Niño events (Section 2.2) are visible in Figure 2.8. In particular the events in 1983–1984 and 1997–1998 caused drought over much of the region. Australia has suffered during many El Niño events, and was again affected by the most recent event in 2002–2003.

Droughts in Europe (as discussed in Sections 2.2 and 2.3) tend to be less extreme and uniform in scale due to the complex and dynamic scale circulation features. However, some events, particularly since the 1970s, have affected wide areas. Bradford (2000) summarizes the major European droughts in the twentieth century and notes large-scale events for north-western Europe in 1973, 1976 and 1988–1992. Spain and Portugal recently had to cope with droughts from 1990 to 1995 (Section 2.2.3). In the summer of 2003 most of Europe suffered from drought (not included in Figure 2.8) (Box 2.3).

2.5   Drought monitoring and forecasting

Caused by climatological phenomena, hydrological drought is a natural hazard and can therefore not be prevented (Section 1.3). However, the impacts of droughts on economic, environmental and social sectors can be reduced through mitigation and preparedness (Wilhite *et al.*, 2000a). Constant monitoring of the

---

Box 2.3

The 2003 drought in Europe

During the summer of 2003, the European Continent experienced extremely high temperatures and below average rainfall. In August, the EU-MEDIN (2003) portal reported: "the heat was caused by an anticyclone which has anchored itself over the west European land mass, holding off rain-bearing depressions over the Atlantic and funnelling hot air north from Africa". Occurring after an already unusual series of high-pressure weather types in spring, this circulation pattern broke many long-term records. For a composite of Swiss stations, Schär *et al.* (2004) demonstrated statistically how extreme the summer 2003 was, with an average June to August temperature of 5.4 standard deviations above the mean, it had the characteristics of an outlier within the climatic variability of the twentieth century (Section 2.6). The heat wave and forest fires in many Mediterranean countries drew most of the attention. However, considerable problems also occurred due to hydrological drought. Newspapers reported low river stages that put navigation on hold in several European rivers. In Berlin, Germany, the River Spree began flowing in the opposite direction. Croatia reported the worst drought for 50 years; the River Sava was at its lowest level in 160 years and low flow in the Danube revealed relics from World War II (EU-MEDIN, 2003).

In the Netherlands the average summer temperature was 18.6°C, which is the second highest since 1706. Summer rainfall was 74 mm (normally 195 mm); only 1887 had lower rainfall (59 mm). The Rhine at Lobith reached the lowest level ever recorded (6.91 m above NAP, Figure 2.11). The surface water temperature

(→ Box continued on next page)

Figure 2.11 Stage meter at the River Rhine at Lobith during the 2003 drought (NAP: Normaal Amsterdam Peil, Normal Amsterdam Level, Dutch ordnance datum).

Box 2.3 (continued)

reached 28.2°C in mid-August. As in other European countries, streamflow temperatures endangered the life of fish. Power plants using river water for cooling had to be shut down because permitted water temperature levels were exceeded. Some power plants delivered only 10% of normal production. Initially the Dutch government decided to let brackish water into some polders in the western Netherlands to prevent surface and groundwater levels falling below a critical level. Low water levels damage the wooden piles of monumental buildings and the dikes. However, high chloride concentrations are detrimental to the aquatic ecosystem, agriculture and horticulture. Subsequently, in the next phase of the drought, it was decided to stop the inlet of brackish water and to pump fresh water from Lake IJssel into River Amstel, which implied a reversal of the natural flow direction. The water supply was not able to prevent a peat dike in the village of Winis from sliding down, which led to the flooding of part of the village in the low-lying polder. The high temperatures had caused shrinkage and oxidation of the peat, making these dikes unstable.

---

hydrological system, the identification of drought characteristics, the forecasting of drought occurrence, and the prediction of drought characteristics and their probability of recurrence, preferably built into an early warning system, are key services that can help to be prepared. *Forecasting* is the estimation of conditions at a specific future time, or during a specific time interval. It is distinguished from *prediction,* which is the estimation of future conditions, without reference to a specific time (Part II).

   Monitoring is important at two levels: long-term monitoring and real-time monitoring. As shown in Section 2.3.3, the length of the observed time series of hydrologic variables greatly affects the knowledge and understanding of the temporal variability of drought. Long time series increase the reliability in the statistical prediction of drought indices (Chapter 6). Real-time monitoring provides information on the current status of the hydrological system and is therefore crucial for drought preparedness. Monitoring of drought together with a good knowledge of a region's hydrology may provide an early warning indicator of the impact on, for example, agriculture, navigation and ecology. Which parameters to monitor depends on a region's climate, vegetation, and hydrology, and finally on the specific purpose of the monitoring. The weekly-published *US Drought Monitor* uses a blend of several current and past indicators routinely monitored by the Climate Prediction Center and agricultural authorities (NDMC, 2003) (Box 2.4). The monitoring of rainfall amount or snowpack accumulation during winter may be highly important for regions

Box 2.4

The US Drought Monitor

Donald A. Wilhite

Monitoring drought presents numerous challenges because of its slow onset characteristic, the lack of a universal definition, the duration of the event, and the fact that the different types of drought require multiple indicators and climatic indices to best illustrate severity. An excellent tool recently developed in the United States to portray the spatial extent and severity of drought conditions is the US Drought Monitor. The US Drought Monitor (DM) was developed in 1999 and represents an integrated approach to drought monitoring involving several US federal agencies and the National Drought Mitigation Center (NDMC) at the University of Nebraska. The DM is available on the NDMC's website at drought.unl.cdu/dm. An example is shown in Figure 2.12. The website, which also contains a variety of climate and hydrologic information and forecasts, receives over 6 million hits/year. In addition, the map is widely disseminated through the print and electronic media. (→ Box continued on next page)

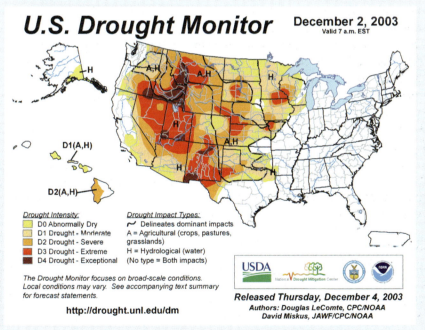

Figure 2.12 The US Drought Monitor, an example from 2 December 2003 (NDMC, 2003).

Box 2.4 (continued)

The US Drought Monitor is not a forecast. Instead, it is a snapshot of current conditions. The DM characterizes the intensity, spatial extent and resultant impacts associated with drought conditions across the United States. The product is updated weekly and is based on a thorough analysis of multiple climatic and hydrologic indicators, including precipitation and temperature, streamflow, snow pack, soil moisture, reservoir levels, and others. It also incorporates expert input from over 150 people across the country. This input from local experts helps to verify the results of the data analysis and calibrate these findings to impacts occurring at the regional and local scale.

The US Drought Monitor classifies drought into four categories (D1–D4), with a fifth category (D0) indicating an abnormally dry area (possible emerging drought conditions or an area that is recovering from drought but may still be seeing lingering impacts). The D1–D4 categories reflect increasing drought intensity levels, with D1 representing areas experiencing moderate drought and D4 depicting a region experiencing an exceptional drought event (the 'drought of record'). The Drought Monitor uses a percentile approach with a D0 equal to the 30th percentile, D1 the 20th, D2 the 10th, D3 the 5th and D4 the 2nd percentile.

The US Drought Monitor has been evolving since it was first introduced in 1999 as new technology, tools, and indicators become available. User feedback is solicited on a routine basis. The map has become the most widely used product in the United States for assessing drought conditions and is now used by the US Department of Agriculture for policy decisions associated with the declaration of areas eligible for some types of drought assistance.

with a strong seasonality such as in Dw- or Cs-climates, while continuous monitoring of precipitation surplus and deficit may be more effective in regions with a more equal annual rainfall distribution such as Cf-climates. Where rainfall is scarce and strongly linked to atmospheric circulation, as in a tropical monsoon climate, the monitoring of the location of the rain-bringing atmospheric features is important. The US Climate Prediction Center (CPC), for example, closely monitors the location and movement of the ITCZ for Africa. The Famine Early Warning Systems Network (CPC, 2003) uses this information to monitor and forecast drought hazards across the continent. Furthermore, recent drought monitoring efforts focus on the use of remote sensing data. Green vegetation reflects little radiation in the visible part of the solar spectrum and much in the near-infrared part. During drought this difference becomes less, an observation that has led to the construction of the

Normalised Difference Vegetation Index (NDVI) (e.g. Kogan, 1995). Combined with hydrologic parameters, this index is now used for drought monitoring.

While real-time monitoring can indicate and provide information on the extent of current drought conditions, it cannot forecast whether these will continue to exacerbate or whether alleviation is underway. Many countries in the world have regional flood forecasting centres. Operational *drought forecasting* is not as common, but the topic has gained more attention after recent economic losses due to drought and climate change studies predicting more extreme hydrological conditions. Hydrological forecasting through the use of models is based on the interactions between the oceans and the atmosphere, and between the land surface and the atmosphere. Atmospheric and hydrologic systems can be coupled dynamically using physically-based hydrological models at the subgrid scale nested into a General Circulation Model (GCM), or by statistical downscaling (Murphy, 2000). The latter uses a statistical relationship between synoptic meteorological situations and hydroclimatic parameters derived from (lag-) correlation and regression analyses of historic series. Drought causing processes are difficult to capture in a physically-based model due to their slow development and large regional extent. Compared to the flood processes, models cannot be confined to one river basin and a short time span. Nevertheless, medium-range weather forecasts with a lead-time of six to ten days are available and useful for short-term hydrological forecasting purposes. Knowledge of the typical recession behaviour of a river together with hydrologic monitoring can allow a forecasting of low flows, which may be useful for river navigation.

For substantial decisions in water resources management and planning, however, longer term forecasting would be helpful. The European Centre for Medium Range Weather Forecast (ECMWF) provides seasonal forecasts (1–3 months lead-time) of mean sea level pressure, precipitation and temperature for the tropics including the uncertainty range from the ensemble runs of the coupled ocean-atmosphere model. Such information, together with known statistical relationships between hydrologic indices and the status of soil and groundwater storages, can produce helpful seasonal drought forecasts as well. However, uncertainties are usually high. The link between spatial and temporal patterns of drought occurrence and atmospheric circulation or sea surface temperatures also holds potential for long-term forecasting. Ropelewski & Folland (2001) recently summarized the prospects of predictability of meteorological, and consequently also hydrological, drought along these lines.

2.6   Impact of climate change

There is a strong interrelation between climate and the hydrological system. A change in one of the systems will therefore induce a change in the other (Kundzewicz, 2002). Understanding the link between past and current droughts and climate and atmospheric circulation is therefore not only a prerequisite to understanding past drought characteristics and predicting the next drought, it is similarly important for the assessment of the impact of climate change on the characteristics of drought. This task ranks high on the political agenda. Have droughts become more frequent and severe? Will the characteristics of hydrological events change if global warming continues? Which parts of the world are and will be affected? These are important questions, because changes in the magnitude and frequency of drought may increase the uncertainty in water resources planning, which is often based on assumptions of *stationary conditions*. Most standard methods in hydrology, including approaches described in Part II and III of this book, assume stationary data. Awareness and identification of non-stationarity is therefore important when assessing the result of hydrological analyses.

Climate change is controversially defined in the literature. As a result of a growing awareness and concern that increased emission of greenhouse gases would raise the global temperature the Intergovernmental Panel on Climate Change (IPCC) was jointly established by the World Meteorological Organisation (WMO) and the United Nations Environment Programme (UNEP) in 1988. In IPCC (2001a), the panel defines climate change as "a statistically significant variation in either the mean state of the climate or in its variability, persisting for an extended period (typically decades or longer). Climate change may be due to natural internal processes or external forcing, or to persistent anthropogenic changes in the composition of the atmosphere or in land use." Climate variability is defined as "variations in the mean state and other statistics (such as standard deviations or the occurrence of extremes) of the climate on all temporal and spatial scales beyond that of individual weather events. Variability may be due to natural internal processes within the climate system (internal variability), or to variations in natural or anthropogenic external forcing (external variability)." In contrast, in Article 1 of the United Nation Framework Convention on Climate Change (UNFCCC, 1992) climate change is "attributed directly or indirectly to human activity that alters the composition of the global atmosphere" and furthermore "is in addition to natural climate variability observed over comparable time periods". In this textbook we refer to *climate variability* as a result of natural causes and to *climate change* as a consequence of human influences.

Conclusion

The impact of climate change on hydrology is difficult to determine. Hydrological records are often short and human influences may also directly affect streamflow, soil moisture and groundwater through interference in the basin itself. Therefore, the various possible causes of hydrological change, or of a certain mode of variability, are often difficult to separate. Generally two approaches can assess climate change and its impact:

a) the analysis of observed data for changes and trends;

b) scenario calculations using physically-based models.

Kundzewicz & Robson (2000) summarize methods for the detection of trends and other changes in hydrological data. Scenario calculations are usually performed with physically-based models (Chapter 9). In the following two sections we briefly summarize observations of trends and scenarios of greenhouse-gas induced climate change that influence the hydrological cycle. Possible consequences for drought are summarized based on recent reports of the IPCC. Though somewhat selective, the IPCC publications to date contain the most complete assessment of the available scientific information on climate change and its impacts.

2.6.1 Observed trends

Observed trends and changes in climatic variables can be summarized as follows (IPCC, 2001a): an increase of 0.6°C in the average global temperature over the last 100 years has been observed and most of the observed warming during the last 50 years is likely to be due to an increase in the greenhouse gas concentration. The 1990s was very likely the warmest decade. The summer of 2003 broke many records (e.g. the all time maximum temperature record in the UK was broken on August 10 with 38.1°C (EU-MEDIN, 2003). A newsletter on the NOAA website reported that the average temperature for the Northern Hemisphere during June–August 2003 was the second warmest on record (since 1880) with 0.55°C above the mean. IPCC (2001a) also notes that the temperature increase has reduced the global snow and ice cover since the late 1960s, whereas precipitation amounts and atmospheric moisture have generally increased over the 20th century. More extreme and heavier precipitation events are observed in many areas. Furthermore, IPCC (2001a) concludes that the global ocean heat content has increased and the global average sea level has risen during the twentieth century. Warm phase ENSO episodes have been more frequent since the mid 1970s compared with the previous 100 years and the winter NAOI has been in a mostly positive phase since the 1970s. Both ENSO and NAOI have been linked to floods and droughts in different regions of the world (Section 2.2.3). However, the results of statistical analyses of the El Niño variability are controversial, for example a recent test on an air pressure and sea surface temperature record of the last 125 years shows no significant change (Solow & Huppert, 2003).

Global climate change effects on hydrology and water resources show trends in annual, seasonal and extreme streamflow, floods as well as droughts in many regions (IPCC, 2001b). However, the observed global warming and precipitation increase is not uniform, neither in time nor in space. For some areas of the globe no change or changes in the opposite direction have been detected, and calculated trends often depend on the period analysed. Examples of regional studies of trends in observed low flow and drought are Lins & Slack (1999) for the United States and Hisdal *et al.* (2001) for Europe. Confidence that the trends detected are solely due to global warming is mostly low because hydrological records are often short, natural hydrological variability over time is high (Section 2.3.3), and human interference in the basin is common. Clearly associated with the observed increasing winter temperatures is the widespread glacier retreat and a shift in the timing of snowmelt from spring to winter which in some regions affects water resources management.

In addition to global climate change, regional anthropogenic influences may have increased the drought risk by changing climatic features. In southern Asia, for example, streamflow has decreased considerable over the past years. Air pollution over the densely populated areas created the so-called 'Asian Brown Cloud'. This brown haze extends over south, southeast and east Asia and reduces the solar radiation reaching the earth. Generally, the aerosol effect on temperature and precipitation (cooling and drying) is the opposite of that of greenhouse gases (warming and increased rainfall). The long-term effects on the hydrological cycle are not yet fully studied. However, they are expected to include a reduction in rainfall and a perturbation of the wintertime rainfall pattern, which will influence water availability and agricultural production (UNEP, 2002). This is in agreement with already observed trends.

2.6.2 Scenarios

General Circulation Models (GCMs), large-scale physically-based models of the atmosphere, are used to simulate the climate under scenarios such as the doubling or more of the pre-industrial carbon dioxide concentration by 2100. The outputs of different GCMs to scenarios such as this $2xCO_2$ scenario vary as they cannot yet adequately take into consideration features that influence the regional climate, such as local effects of mountains, coastline, lakes, vegetation boundaries and heterogeneous soils (Hayes, 2002). According to IPCC (2001a), however, the model runs agree that the increasing greenhouse gas concentration will further raise the average global surface temperature; the ocean water will get warmer and hence the sea level will rise. There are strong indications that global precipitation will further increase. Though less certain, scenarios also predict an evapotranspiration increase. A study by Gregory *et al.* (1997) shows that climate change is associated with reduced soil moisture in Northern Hemisphere mid-latitude summers. A modelling experiment by Oerlemans *et al.* (1998) shows how glaciers around the world would react to different climate change scenarios: while for 0.04°C per year without increase in precipitation, few glaciers would survive until 2100, for 0.01°C per year with an increase in precipitation of 10% per degree warming, the overall loss would only be 10 to 20% of the 1990 volumes.

While in large parts of the world the initial changes predicted due to global warming are small compared to changes caused by multi-decadal variability, many scientists still expect that the hydrologic cycle will intensify causing hydrological extremes of floods and droughts to become more common. Furthermore, the oscillations in the ocean-atmosphere system such as ENSO and the NAO might be altered, which is again likely to have an impact on the frequency of floods and droughts (Section 2.2.3). Hydrological scenarios have been obtained by numerous studies, usually by using climate scenario output from a GCM as input to a hydrological model (Chapter 9). New approaches directly couple hydrological models and regional climate models. Although increasingly sophisticated, in addition to the uncertainty from the climate change scenarios and GCMs, estimates of the impact of climate change on groundwater recharge and streamflow, for example, are affected by the uncertainty of the hydrological model.

According to IPCC (2001b), the effect of climate change on streamflow and groundwater varies regionally and between different scenarios, largely following projected changes in precipitation. In general, with 33–67% confidence, an increase in annual mean streamflow in high latitudes and southeast Asia and a decrease in central Asia, in the area around the Mediterranean, in southern Africa, and in Australia is expected. Examples of impacts related to droughts resulting from scenario calculations are (IPCC, 2001b):

a) "increased summer drying over most mid-latitude interiors and associated risk of drought" (66–90% chance);

b) "intensified droughts and floods associated with El Niño events in many different regions" (66–90% chance);

c) "increased Asian summer monsoon precipitation variability (66–90% chance) and associated increased flood and drought magnitude and damages in temperate and tropical Asia".

The amount of change varies between the scenarios, and the hydrological effect of one specific scenario will vary between basins. Gellens & Roulin (1998) showed how the same scenario leads to different changes in low flow in different basins depending on geological conditions. Small headwater streams might become especially affected by changes. Low flow is expected to decrease in many areas due to higher evapotranspiration. However, the expected changes in precipitation may strengthen or weaken the effects of increased evaporation.

The frequency of drought is affected by the combination of changes in the seasonal and inter-annual variability in all hydroclimatic variables. With the motivation to better understand European summer heat waves as the one in 2003, Schär *et al.* (2004) simulated possible future European climate with a regional climate model. The simulations show an increase of the temperature variability and of the drought frequency particularly in central Europe. The study also shows that drought conditions that develop due to anticyclonic forcing may nonlinearly amplify local temperature anomalies through the suppression of evapotranspiration due to the lack of soil moisture. In a regime of increased variability, extreme summers as in 2003 would be more common.

The number of people living in countries that are water-stressed is projected to increase due to population growth. Vörösmarty (2000) found that rising water demands greatly outweigh the effect of climate change in defining the state of global water systems in 2025. The projected climate change could further decrease groundwater recharge and streamflow in many of these countries, e.g. central Asia, southern Africa and countries around the Mediterranean Sea, but streamflow may also increase in some water-stressed countries. To increase resilience to drought, a sound water management, particularly an integrated water resources management, ought to be initiated to adapt to increased hydrologic variability and additional uncertainty due to climate change.

2.7   Summary

This chapter has discussed droughts in their climatological context. Around the world droughts have different characteristics, which are closely linked to the seasonality of a region's climate. Knowing and understanding a region's hydroclimate is therefore an important prerequisite for any drought study.

The global atmospheric circulation controls the average pattern of rainfall, temperature and associated evapotranspiration in the different climate zones. An atmospheric circulation anomaly may cause drought. Usually, regional weather patterns cause drought due to their unusual timing, location, or persistence. In the outer tropics, drought is often related to the delay or failure of the annual rainfall, which follows the movement of the ITCZ. In the mid-latitudes persisting high-pressure systems associated with dry weather may lead to drought during summer. Strong regional anomalies are often linked to (sometimes distant) oscillations of large-scale ocean-atmosphere interactions. Strong teleconnections exist with the El Niño Southern Oscillation, which has been related to drought occurrence, for example, in Australia and India. Weaker teleconnections are found in the Northern Hemisphere.

The impact of large-scale atmospheric circulation together with feedback mechanisms results in large spatial and long temporal persistence of hydrologic drought compared to the more instantaneous and local occurrence of floods. However, there is also considerable spatial and temporal variability in drought, which provides a challenge to the estimation of future drought risk. A good and long record of historic events is necessary in order to understand drought characteristics. Together with the monitoring of hydroclimatic variables this is an important prerequisite for early detection and warning of drought. Known links to atmospheric circulation patterns also allow short-term and seasonal forecasting of drought, though still with high uncertainties. For regions with strong teleconnections to ENSO, long-term forecasting is possible.

The strong relationship between climate and the hydrological system finally raises the question of the impact of climate change on drought. Although trends from observed historic series and scenario predictions from coupled GCMs provide a mixed picture, scenario calculations indicate that drought risk is likely to increase for some regions, e.g. in central Asia, the Mediterranean, southern Africa and Australia. Continued research will shed more light on the question of the impact of climate change in the future.

3

Flow Generating Processes

Henny A.J. van Lanen, Miriam Fendeková, Elżbieta Kupczyk, Artur Kasprzyk,
Wojciech Pokojski

3.1 Introduction

It is well-known that low precipitation over prolonged periods triggers the development of hydrological drought. This is often combined with high temperature and evapotranspiration. An intriguing question is why should lack of precipitation cause hydrological drought in some catchments, but not in others in the same climatic region. A thorough understanding of catchment properties (characteristics) and hydrological processes under drought conditions can help to explain differences between catchment response.

Under natural conditions, hydrological drought usually develops from a water deficit at the surface that is larger than normal. The *climatic water deficit* is the difference between potential evapotranspiration and precipitation. The deficit affects catchment processes that are linked to precipitation, such as infiltration into the soil or overland flow. For instance, in sloping catchments where overland flow might regularly take place a high climatic water deficit causes no contribution of overland flow to the stream during the dry period. This is typical for catchments that respond quickly to precipitation. In other catchments low infiltration causes reduction in soil moisture content, possibly leading to reduced recharge that feeds the saturated groundwater system. A drop in the water table is consequently the first sign of a developing groundwater drought. Under these circumstances borehole yield may decrease and shallow wells may dry up. Groundwater storage properties strongly control how fast water levels fall. In extended groundwater systems the decline of water levels is slow, indicating that these systems are not sensitive to short periods with below-normal precipitation. Low water tables reduce groundwater flow towards the stream implying that the stream receives less water. The decrease of groundwater flow, which is associated with falling water tables, determines if a streamflow drought develops. Catchments with large groundwater systems show a very gradual flow decrease indicating that the response to precipitation is slow (delayed and attenuated response).

Basin properties and processes determine how fast groundwater and streamflow respond to high climatic water deficit (Section 2.2.2). Topography, land use, soil type, hydrogeological conditions, lakes and stream network are important properties controlling the response. For example, in a certain climatic region an agricultural catchment with deep permeable soils on top of a large groundwater system shows a completely different response than a catchment covered with forest on steep slopes with a shallow soil overlying impermeable hard rock. This chapter seeks to provide an overview of how hydrological processes affect the development of hydrological drought for catchments with different properties. This will form the physical background for subsequent chapters dealing with hydrological data, estimation tools, human impacts, ecology and operational management. The emphasis is on natural hydrological processes; the impact of artificial influences is considered in Part III. The chapter does not intend to present a general and comprehensive description of hydrological processes in a catchment as given in, for example, Davis & De Wiest (1966); Freeze & Cherry (1979); Chow *et al.* (1988); WMO (1994); Kendall & McDonnell (1998); Ward & Robinson (2000); Dingman (2002); Davie (2003), but it focuses on the main processes and catchment properties relevant for drought development.

The chapter starts with an overview of how climatic water deficits affect hydrological processes in different type of catchments (Section 3.2). Hydrological processes are linked to the flow systems in the different catchments. It then continues with a more comprehensive description of drought-relevant processes. Two catchments in climatologically contrasting regions are used for illustrative purposes. The description starts with the role of the subsurface (unsaturated and saturated zone), including the effect of a climatic water deficit on key water transfers in a catchment, such as actual evapotranspiration, groundwater recharge and groundwater discharge (Sections 3.3 and 3.4). The description concludes with the surface water system by focusing on the effect of a climatic water deficit on streamflow generation processes, including the role of lakes and wetlands (Section 3.5). Finally, a summary of the chapter is given in Section 3.6.

3.2   Flow systems in different types of catchments and drought-related processes

The effect of climatic water deficits that may induce hydrological drought is different for *quickly-responding* and *slowly-responding catchments*. Water may

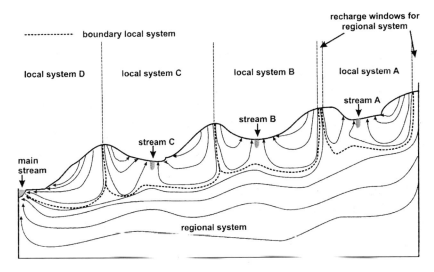

Figure 3.1 Water flow paths for a hypothetical hydrological system of three small valleys and a main valley.

feed the stream quickly as a response to rainfall or melting snow (*quick flow*), or it may recharge an extended groundwater system leading to a slow response to precipitation. In this context it is relevant to investigate the path that water follows in the various catchments. This implies that not only water deficits have to be considered, but also water surpluses to understand the development of drought. The *water surplus* is defined as the difference between precipitation and actual evaporation during a certain period of time.

Catchment properties (e.g. climate, topography, geological conditions) control the *flow paths* of water in catchments leading to water flow at different scales, e.g. local and regional flow systems. In *local flow systems* (e.g. local systems A and B, Figure 3.1) water surplus flows to an adjacent discharge area. Note that the water table is the upper boundary of the hydrological system in Figure 3.1. The water table marks the top of the saturated zone over which the pressure is equal to the atmospheric pressure. *Discharge areas* receive water, e.g. overland flow or groundwater. Riparian areas along streams, spring lines or wetlands are typical discharge areas. Discharge areas of *regional flow systems* receive groundwater from remote recharge areas; part of the water flows underneath a number of local systems. In this context *recharge areas* are defined as regions where water surplus flows downwards to feed the saturated groundwater system. Some recharge of a regional system takes place in an area between two local systems, which is called a *recharge window* (Figure 3.1). The

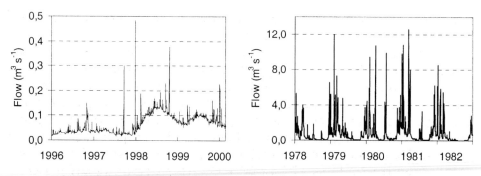

Figure 3.2 Daily discharge in a humid temperate climate: slowly-responding catchment (left), and quickly-responding catchment (right).

catchment area of local systems is much smaller than that of regional systems. Local systems can be found in areas with a pronounced topography mostly with impermeable material at shallow depth. Regional flow systems favour deeply buried geological layers with a high hydraulic conductivity and streams at different elevations.

In general, water stores of regional flow systems are substantially larger than those of local systems. Therefore the main stream of a regional system collects more base flow from beyond the adjacent local system (local system D, Figure 3.1) than quick flow. *Base flow* is defined as the part of discharge which enters a stream channel from groundwater and other stores, such as large lakes and glaciers. Most catchments do not have lakes or glaciers and depend solely on base flow from groundwater. This base flow is slowly released from groundwater storage maintaining streamflow in dry periods. As a consequence of the high portion of base flow in a catchment with a regional system, hydrological droughts do not readily develop. Typical for streams in these catchments is the still relatively high streamflow during dry periods per unit of catchment area (specific low flow). Figure 3.2 (left) presents an example of a streamflow hydrograph for a slowly responding catchment. Overland flow from a local system in this catchment is superimposed on base flow that slowly changes. The wet seasons cannot be recognized in the hydrograph.

On the contrary, streams of local systems may receive different portions of quick flow and base flow depending on catchment properties. In drainage basins with a high portion of quick flow, hydrological droughts rapidly show up, but may cease directly after the first major rainfall event. An example of quickly-responding catchment is given in Figure 3.2 (right). Wet seasons can clearly be recognized and in some dry periods a significant response to precipitation can be observed (e.g. 1980). Some catchments with a local flow system still have substantial base flow. These catchments take an intermediate position in the development of a hydrological drought.

In conclusion, drainage basins with a regional flow system respond slowly to climatic water deficits implying that hydrological drought does not develop quickly. However, in catchments with only a local system a drought may start quickly as the stream is predominantly fed by quick flow. It is important to note that large river basins mostly consist of a large number of local and regional systems and that base flow may come from groundwater, peat bogs, lakes and glaciers. These basins show a complex, composite response to climatic water deficits. Sections 3.2.1 and 3.2.2 describe the general processes and properties of catchments either dominated by quick flow or by base flow.

3.2.1   Systems dominated by quick flow processes

Quick flow processes are not only important for flood generation, but also for the susceptibility of a catchment to hydrological drought. If a large portion of the water surplus flows fast to a stream, then the precipitation does not feed the groundwater system, and hence the base flow is low leading to drought developing quickly as outlined in the previous section. The following section briefly describes some quick flow processes relevant for drought development. The description includes:

a)   overland flow and throughflow;

b)   groundwater flow in sloping areas with shallow impermeable hard rock;

c)   groundwater flow in delta areas.

Quick flow from overland flow and throughflow

Quick flow is characteristic for basins with high rainfall intensity (Box 3.1), steep slopes, slowly-conductive topsoils covered with sparse or no vegetation. *Overland flow* or *surface runoff* can easily be generated under such conditions (q_{of}, Figure 3.3). Overland flow is defined as water flowing over the ground before it enters a definite channel. In regions with climate type D (Section 2.2.2) conductivity can be temporarily low in the case of frozen soil. When the topsoil allows infiltration of rainfall or snow-melt water, it may still contribute to quick flow. In many areas the hillslopes are built by stratified deposits with a sequence of permeable and slowly-permeable layers. Percolation (q_r, Figure 3.3) often exceeds vertical hydraulic conductivity of the slowly-permeable layers leading to a temporary saturated zone overlying unsaturated layers. Downslope flow in these thin saturated layers, also called *throughflow*, interflow, shallow lateral flow or rapid subsurface flow (q_{if}, Figure 3.3), may

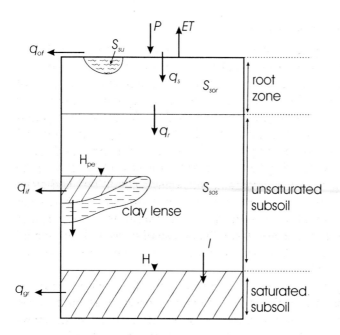

Figure 3.3 Components of the soil water balance.

contribute to downhill flow into a stream. Overland flow and throughflow, which is usually fast, results in flashy streamflow hydrographs (Figure 3.2, right).

In local systems with a high portion of quick flow, lack of rainfall or delayed snow melt shows up almost directly in the observed streamflow. Hence, streamflow droughts can develop quickly, but they are typically short-lived depending on the precipitation pattern. Groundwater systems are not extensive in these local systems and groundwater droughts are rarely reported from such regions. In semi-arid regions with local systems dominated by quick flow, which are very common due to high rainfall intensities, streamflow drought may last longer if the irregular rainfall holds off.

In Figure 3.3, P is precipitation [LT^{-1}], ET is actual evapotranspiration [LT^{-1}], q_{of} is overland flow [LT^{-1}], q_s is infiltration [LT^{-1}], q_r is percolation [LT^{-1}], q_{if} is throughflow [LT^{-1}], I is recharge [LT^{-1}], q_{gr} is groundwater flow [LT^{-1}], S_{sor} is soil water stored in the root zone [L], S_{sos} is soil water stored in the unsaturated subsoil [L], H_{pe} is perched water table [L] and H is water table [L].

Quick flow from saturated groundwater flow

Quick flow in local flow systems may also come from saturated groundwater flow through an aquifer under some conditions. An *aquifer* is defined as a permeable water-bearing layer capable of yielding exploitable quantities of water. Vegetated hillslopes with permeable soils overlying impermeable hard rock at some depth generate saturated groundwater flow during the wet period in humid climate types (e.g. Cf-climate, Section 2.2.2). Overland flow and throughflow are rare due to the vegetation cover and the high infiltration capacity of the soils and weathering layer. The water surplus infiltrates into the soil and feeds a sloping aquifer. Favourable geological conditions and steep groundwater gradients in the wet season cause a large downhill groundwater flow (q_{gr}) contributing to quick flow in the stream. Streamflow hydrographs in regions with a sloping aquifer are less flashy than those of areas with substantial overland flow and throughflow. Lack of rainfall does not immediately show up in the streamflow hydrographs.

In snow-affected humid climates (D-climate), quick flow due to saturated groundwater flow cannot occur if the soil is frozen. Extensive overland flow can then be observed. These processes lead to high streamflow peaks in the snow-melt period and lower base flow during the dry season. In the D-climate not only precipitation controls streamflow generation, but also temperature plays an important role. In dry climate types (e.g. B-climate) quick flow as saturated groundwater flow is rare. Usually the soil is so dry that the whole water surplus is used to replenish the soil store.

Quick flow in delta areas

Flashy streamflow hydrographs are also observed in flat delta areas, e.g. Bangladesh and the Netherlands. The dense natural channel network and shallow groundwater tables cause a quick catchment response to precipitation. There is little chance for the water surplus to be stored, which flows to the streams as shallow saturated groundwater flow. As in similar areas with considerable quick flow, streamflow droughts are frequent and short-lived.

3.2.2 Systems dominated by base flow

Streams of both regional and local flow systems with a high portion of base flow have smooth groundwater and streamflow hydrographs (Section 3.4). Groundwater flow and streamflow do not immediately respond to individual rainfall events. Droughts are less frequent than in local systems with a high portion of quick flow, but if they develop they often last longer. Local flow systems with a high portion of base flow develop in areas with permeable layers to a depth largely exceeding the valley bottom. Typical examples are the chalk regions in northwest Europe (Downing *et al.*, 1993). A regional flow system can develop if elevations of streams in a region are very pronounced (Figure 3.1) and deep geological layers can transmit sufficient amounts of groundwater. *Transmissivity, kD*, which is the product of the hydraulic conductivity and the saturated thickness of the aquifer, is used to quantify the amount of water flowing through an aquifer. Note that regional systems do not occur, if (a) transmissivity of the deep layers is too small (then only local systems develop), and (b) transmissivity of the deep layers is too large leading to water tables deep below the bottom of the smaller valleys. Then only one large local system develops (streams A–C would be dry, Figure 3.1). The probability of having one large local system is higher in dry than in humid climates (Box 3.1), due to the lower recharge (deep water tables and no streams in small valleys).

Recharge areas in regions dominated by base flow typically have deep groundwater levels resulting in a strong attenuation of the percolation (q_r, Figure 3.3) into the thick unsaturated subsoil. The recharge hydrograph has a very smooth nature showing no relation with individual precipitation events. Only wet and dry seasons can be observed. This characteristic seasonal behaviour of recharge influences the groundwater hydrographs. These are smooth and in a series of consecutive dry years they may even be continuously declining (Figure 3.16). The very slow rise or decline of water tables causes the groundwater discharge to change only gradually. Water table fluctuations near the water divide can be large (up to more than 10 m). The sustained groundwater discharge towards a stream may therefore vary considerably between series of wet and dry years. The slow, but significant variation in groundwater discharge causes less frequent, but more prolonged groundwater and streamflow droughts.

In areas with a regional flow system, the main stream can be fed both by regional groundwater flow and water from a local system (Figure 3.2, left), leading to a mixed regime if the local system generates considerable quick flow. A regional flow system can be built by various geological deposits. It is likely that these deposits have different hydraulic conductivities. This results in layers with a relatively high conductivity and ample thickness in which horizontal flow dominates (aquifers) and layers of a rather impermeable nature (*aquitards*). An aquitard transmits water at a very slow rate compared with an aquifer, and vertical flow dominates (leakage, Figure 3.4). A regional system with interacting shallow and deep aquifers (*multiple aquifer system*) mostly results in even more smoothed groundwater and streamflow hydrographs than a *single aquifer system*. In series of wet years the increased recharge of the shallow aquifer leads to an enlarged downward leakage leading to a smaller groundwater discharge. *Leakage* is defined as flow of water from or into an

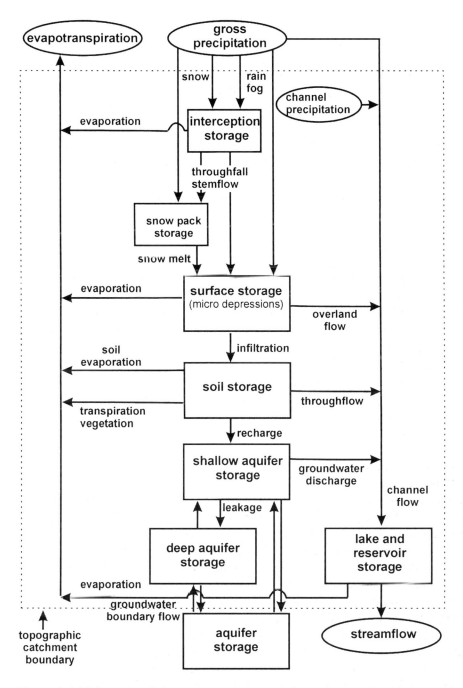

Figure 3.4 Major water balance components and fluxes in a surface water catchment with deep water tables and groundwater boundary flow.

Box 3.1

Arid and semi-arid climates

The low and highly variable rainfall regime in arid and semi-arid areas mostly leads to intermittent or ephemeral streams, if no persistent groundwater discharge takes place. These streams are called wadi or ouedd. Most often, persistent groundwater discharge is rare and only occurs if a regional flow system exists with recharge areas located in higher-rainfall areas (mostly mountains, Figure 3.5), so that groundwater is transmitted over long distances. In dry climates, part or even all streamflow generated in upstream reaches can recharge an aquifer downstream (indirect recharge) and finally leave the aquifer as groundwater discharge near the watershed outlet. Water from the aquifer may also evaporate in wetlands or salt lakes. Lack of precipitation in the upstream, usually mountainous reaches, leads to a reduced indirect recharge. This can induce groundwater drought. Occasionally, *fossil groundwater*, i.e. water infiltrated into an aquifer during an ancient geological period under climatic and morphological conditions different from the present, contributes to sustained groundwater discharge.

Intermittent streams flow only in direct response to quick flow or to the flow of an intermittent spring. Sustained groundwater discharge is uncommon in arid and semi-arid regions because rainfall is scarce and its high intensity easily leads to the generation of overland flow and throughflow. The remaining rainfall

Figure 3.5 Palm trees along a stream leaving the mountains in Chebika (Tunisian desert); the stream disappears in the plain in front of the mountains (photo by H.A.J. van Lanen).

replenishes the dry soil meaning that hardly any water is left to recharge the groundwater system. Intermittent streams that feed salt lakes or surface water reservoirs are widespread. A climatic water deficit causes a reduction in the discharge volume of the intermittent streams feeding the salt lakes and reservoirs. In contrast to humid regions, the presence of dry streams is not typical evidence of droughts in arid and semi-arid regions. If an aquifer is present drought shows up as declined water tables due to lower recharge. Kruseman (1997) and Simmers (2003) give more details of arid zone hydrology, especially concerning the interaction between intermittent streams and groundwater.

aquifer through an underlying or overlying semi-permeable layer. However, in series of dry years hydrological droughts develop slower in these regions because upward leakage towards the shallow aquifer prevents a strong decrease in the groundwater discharge. As a result, occurrence of hydrological drought decreases due to the leakage.

In areas with a regional flow system underlying local flow systems (Figure 3.1) topographic water divides usually do not coincide with groundwater divides. This implies that groundwater flows beneath the topographic water divide (*groundwater boundary flow*, Figure 3.4). The *topographic water divide* is a boundary line separating adjacent surface water basins, and a *groundwater divide* is a line on the water table surface on either side of which the groundwater flow diverges. If a high water deficit occurs in a catchment then the boundary flow from neighbouring catchments may contribute to the base flow of the catchment that is suffering.

The textbook does not address *water quality processes* (e.g. Appelo & Postma, 1993; Chapman, 1992), although drought certainly affects these (Section 1.3). Surface water quality issues often emerge during low streamflow. For example, more salt water may intrude in estuaries or lagoons, or effluent from sewage treatment plants and cooling water from power plants is diluted too little. Clearly, water supply and aquatic ecosystems suffer under these conditions, which may influence the groundwater and surface water quality. Prolonged droughts, especially occurring in arid and semi-arid regions, may cause salt water intrusion in coastal aquifers. In some regions shortage of recharge leads to insufficient dilution of otherwise chloride-rich groundwater in aquifers consisting of marine deposits.

3.2.3   Water balance

Water balance components identify the principal hydrological processes in a catchment that are relevant for drought development (Sections 3.2.1 and 3.2.2). A water balance component is the integrated value of one or more hydrological processes over a certain period of time (e.g. month, year). A change of any of the governing processes gives key information for drought development.

The water balance for the surface water system of a catchment, including the three major contributors, overland flow, throughflow and groundwater flow, can be written as follows:

$$Q = A_c \left(q_{of} + q_{if} + q_{gr} \right) + A_s \left(P_s - E_s + \frac{\Delta S_s}{\Delta t} \right) \tag{3.1}$$

where Q is discharge at the outlet [$L^3 T^{-1}$], A_c is catchment area excluding large surface water bodies [L^2], q_{of} is overland flow [$L T^{-1}$], q_{if} is throughflow [$L T^{-1}$], q_{gr} is groundwater flow to the surface water system [$L T^{-1}$], $\Delta S_s/\Delta t$ change in surface water storage [$L T^{-1}$], A_s is surface water area [L^2], E_s is evaporation of open water surface [$L T^{-1}$], and P_s is precipitation directly falling on the surface water system [$L T^{-1}$].

In most small and medium-sized river basins without large lakes and surface water reservoirs the second term on the right-hand side of Equation 3.1 is negligible. Consequently, the discharge equals the sum of the overland flow, the throughflow and the groundwater discharge. Sections 3.2.1 and 3.2.2 showed that catchment properties determine their relative contributions. The term $\Delta S_s/\Delta t$ can be neglected if an appropriate period, i.e. a *hydrological year*, is chosen for climates with marked seasonality. A hydrological year, or water year, is a continuous 12-month period selected in such a way that the overall changes in storage are minimal. The carryover is thus reduced to a minimum. Note that the start of the hydrological year predominantly depends on climate conditions. In some countries the hydrological year starts at the beginning of the dry period (April in the Netherlands and June in Denmark), in others at the start of the wet or winter season (October in USA, Spain and South Africa, or November in Germany and Czech Republic) and in cold or mountainous climates at the time of minimum snow cover (e.g. September in Norway and October in Switzerland). In other countries with a wide variety of climate types or no marked seasonal differences, e.g. New Zealand, the calendar year is chosen as the hydrological year. In Russia the selection of the start of the water year depends on the problem to be investigated. The time when the rivers freeze (October or November) is often used. Note that in many investigations the start of the hydrological year selected is dependent on the objective of the study.

Appendix 3.1 gives detailed water balance equations, including those of the soil and groundwater domains. These equations are used in Sections 3.3–3.5 to explain the impact of a high climatic water deficit on hydrological drought development, which is the propagation of a pulse through the hydrological system.

3.3   Unsaturated zone – Actual evapotranspiration and groundwater recharge

This section elaborates the effect of high water deficits on actual, or real evapotranspiration and groundwater recharge (flux across the upper and lower boundary of the unsaturated zone, Figures 3.3 and 3.4). Actual evapotranspiration together with precipitation determines the input (recharge) to the groundwater system. A deviation from normal water deficits might initiate hydrological drought.

Actual evapotranspiration is composed of *soil evaporation* through the soil surface and *transpiration* of soil water taken up by plant roots (Figure 3.4). If sufficient soil moisture is available actual evapotranspiration equals the potential evapotranspiration rate. In dry soils vegetation reduces transpiration to cope with lower soil water contents, implying that the actual evapotranspiration is lower than the potential rate. In extremely wet soils the reduction process may occur due to lack of oxygen (excess water) that prevents optimal water uptake. *Soil moisture storage*, which is water available to plants, is the key factor controlling actual evapotranspiration (Figure 3.6). The soil storage is

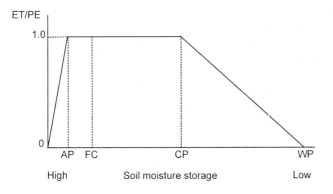

Figure 3.6 Relationship between the ratio of the actual, *ET*, and the potential evapotranspiration, *PE*, and the soil moisture storage (*AP*: excess water point, *FC*: field capacity, *CP*: critical point and *WP*: wilting point).

replenished by precipitation infiltrating across the soil surface (q_s, Figure 3.3). Shortage of precipitation often combined with high potential evapotranspiration leads to a low soil moisture content. This causes a reduction in actual evapotranspiration, a lower groundwater recharge, or both.

Direct *groundwater recharge*, or precipitation recharge, is the water flux across the water table feeding the shallow aquifer (Figures 3.3 and 3.4). Over a longer period of time the percolation, q_r, equals the recharge, *I*, if no throughflow takes place. In areas with deep water tables, recharge considerably differs from percolation for short periods (months, or even a couple of years). Thickness and the hydraulic properties of the thick unsaturated subsoil control the distribution of recharge over time. Lower recharge than normal may induce hydrological drought. Groundwater recharge is one of the key variables that are used to identify a groundwater drought (Chapter 5).

In the following the effect of low precipitation and high potential evapotranspiration on the water balance is elaborated. This is followed by a description of the influence of soils, land use and water-table depth on evapotranspiration and groundwater recharge. The description is illustrated with actual evapotranspiration and recharge for two climatologically contrasting catchments, i.e. the Noor and the central part of the Upper-Guadiana catchment (La Mancha Occidental region). The Noor is located on the Belgian-Dutch border and is representative of a temperate humid climate (Cf-climate), Section 2.2.2). The average precipitation in this region is 758 mm year^{-1}. La Mancha Occidental belongs to the driest part of southern Spain, representing a semi-arid climate (Cs-climate), having an average precipitation of not more than 425 mm year^{-1}. The great differences between the two climate types are reflected in the recorded annual precipitation for the second part of the 20th century. In the Noor the precipitation varies between 465 and 1074 mm year^{-1}; in 10% of the years it is lower than 606 mm year^{-1}. In the Upper Guadiana the range is 190–828 mm year^{-1} with 10 out of 100 years with a precipitation lower than 285 mm year^{-1}. In the Noor and La Mancha Occidental, respectively, the average annual potential evapotranspiration equals 574 and 958 mm. The inter-annual variation is lower than for precipitation. In the Noor the annual potential evapotranspiration varies between 490 and 717 mm.

3.3.1   Effects on the water balance

Low precipitation

The water balance equation of the soil (Appendix 3.1) illustrates that in principle a lower precipitation than normal, ΔP, can affect all other terms:

$$\Delta P = \Delta(ET + \frac{\Delta S_{so}}{\Delta t}) + \Delta(q_{if} + I) \tag{3.2}$$

where ΔS_{so} is change in stored soil water [L]. Section 3.2.1 explains the other symbols.

The first term of Equation 3.2 indicates soil and vegetation processes, whereas the second term includes subsurface flow processes. A low precipitation leads to a low soil moisture storage ΔS_{so}. Subsequently, two different flow situations can develop depending on the prevailing weather conditions, i.e. whether a dry or a wet period occurs.

Dry seasons are characterized by low or no precipitation often combined with a high potential evapotranspiration and probably hardly any downward soil moisture flow ($q_{if} + I \approx 0$). First the reduction in soil moisture storage causes low actual evapotranspiration. This applies to catchments with deep water tables where capillary rise does not occur. The smaller evapotranspiration results in low biomass production due to higher drought stress leading to the development of a soil water drought. In terrestrial ecosystems the drought stress affects the competition between plant species and might have detrimental effects on the abundance of rare species. Under extreme drought conditions these rare species may wilt and disappear for a while, or even permanently. These negative effects are indicators of an ecological drought (Section 1.2). Whether low precipitation during the dry season contributes to the development of a hydrological drought depends on the decrease in soil moisture. If the decrease is substantial then it has to be replenished in the subsequent wet season leading to lower throughflow and recharge. In lowlands with shallow water tables a high water deficit has a direct effect on groundwater and streamflow.

Wet seasons are considered to be periods when actual evapotranspiration is close to or equal to the potential rate. It is thus likely that a downward soil moisture movement occurs ($q_{if} + I > 0$). It is assumed that a reduction in precipitation does not result in decreased evapotranspiration as for dry seasons. Consequently, a reduction in soil moisture storage occurs, soon leading to a decrease in throughflow, q_{if} (if present) and groundwater recharge, I (Equation 3.2). This might induce the development of a hydrological drought.

To summarize, low precipitation in the wet season may lead to the development of a hydrological drought in the subsequent dry season, whereas for reduced precipitation in dry seasons the chance of drought in areas with deep water tables is reduced and depends on the increase in soil moisture deficit.

In catchments where precipitation partly falls as snow, not only precipitation, but also deviations from the normal temperature over the year determine if a hydrological drought develops. Temperature-related catchment descriptors (e.g. maximum temperature in certain months, mean annual air temperature, latitude) are therefore included in models for prediction of low streamflow at ungauged sites (Chapter 8). As discussed in Section 3.2.1 early snow melt on frozen topsoils may lead to extensive overland flow, which causes low soil moisture contents and probably reduced recharge. This might cause drought due to low base flow late in the hydrological year (*summer drought*). On the other hand, a late melt might lead to drought in the period preceding the melt (*winter drought*) because the groundwater has received no recharge, and surface water no groundwater discharge since the start of the winter. Winter droughts may also develop due to extreme low air temperatures resulting in complete freezing of streams. Reduction in groundwater inflow, which usually has a higher water temperature, contributes to freezing.

High evapotranspiration

The water balance of the soil (Appendix 3.1) shows that high potential evapotranspiration leads to high actual evapotranspiration and accordingly low soil moisture storage and decreased recharge and throughflow:

$$\Delta PE = \Delta(\frac{\Delta S_{so}}{\Delta t}) + \Delta(q_{if} + I) \tag{3.3}$$

where ΔPE is change of potential evapotranspiration [LT^{-1}]. Section 3.2.1 explains the other symbols.

High potential evapotranspiration results in a decrease in moisture stored in the soil (ΔS_{so}), as long as it is available. The climate determines which term on the right side of Equation 3.3 is most affected by high evapotranspiration. In Table 3.1 the climates introduced in Section 2.2.2 are grouped according to the effect of high evapotranspiration on the water balance.

Under cold conditions (Condition I, Table 3.1) evapotranspiration is usually at its potential rate, and it is low compared to most other water balance components, such as precipitation. Note that sublimation of intercepted snow in forested areas may be substantial. In cold regions, the effect of high potential evapotranspiration is often negligible and does not contribute significantly to drought development.

In humid regions with medium to high temperatures (Condition II, Table 3.1), high potential evapotranspiration results in increased depletion of the soil moisture storage. However, the soil moisture storage in the root zone never reaches the wilting point (Figure 3.6). Subsequent precipitation has to replenish a larger soil moisture deficit, which implies a reduction in recharge and throughflow. Low recharge may lead to hydrological drought.

Under dry conditions (Condition III, Table 3.1) the low soil moisture storage due to high evapotranspiration losses hardly influences the downward

Table 3.1 Grouping of climate types (Section 2.2.2) and description of the refinement of some climate types in conditions relevant for explanation of high evapotranspiration

Condition	Climate type and possible refinement
I	E
	C and D in winter period, with low temperatures
II	Af
	Aw and Am in moist periods
	C and D in moist summer periods
III	Aw and Am in dry periods
	B
	C and D in dry summer periods

movement of soil water, because this flux is insignificant. The decrease in soil moisture storage (Equation 3.3) due to increased potential evapotranspiration, however, is not infinite. It is bounded by the soil moisture capacity, which can be derived from the rooting depth of the vegetation and the soil moisture retention curve (pF-curve). Below a particular storage (*CP*, Figure 3.6), the evapotranspiration rate is reduced, which implies that the actual rate is lower than the potential evapotranspiration. Under these conditions ΔPE in Equation 3.3 has to be replaced by ΔET. Under dry conditions this situation continues until the moisture storage reaches the wilting point (*WP*). The vegetation then stops transpiring and the soil moisture storage does not change anymore providing that soil evaporation is negligible. The soil moisture storage completely depletes, irrespective of the potential evaporation rate. The only difference is that wilting is reached earlier than normal, affecting soil water and ecological droughts. Note that in case of negligible evapotranspiration (latent heat flux), incoming solar radiation is converted into sensible heat leading to higher air temperatures and possible heat waves (Box 2.3). Subsequent precipitation replenishes soil moisture, and then throughflow and recharge starts. In dry climate types, the second term on the right-part of Equation 3.3 is not affected, which means that hydrological drought does not primarily develop because of high potential evapotranspiration.

To summarize, high potential evapotranspiration only has a significant impact on drought development in regions with medium or high temperatures and abundant soil moisture (Condition II, Table 3.1). This applies to catchments with deep water tables. In lowlands with shallow water tables, high potential evapotranspiration has an impact on drought in nearly all types of catchments.

3.3.2   Illustration of inter-annual variability of evapotranspiration and groundwater recharge in contrasting climates

Actual evapotranspiration

Actual evapotranspiration, *ET*, shows inter-annual variability due to variation in precipitation, potential evapotranspiration and soil water content. In regions with a temperate, humid climate, soil water is regularly replenished by precipitation. This means that the rate of actual evapotranspiration is usually close to the potential rate. The *evapotranspiration deficit* (potential – actual evapotranspiration, *PE – ET*) illustrates this well. In the Noor catchment located in a temperate humid climate, the average deficit in the second part of last century was only 14% of the potential evapotranspiration (Figure 3.7). The evaporation deficit for a semi-arid climate is substantially higher, as demonstrated for the La Mancha Occidental region. Every year a deficit develops, and the average deficit equals 62% of the potential rate. Supporting Document 3.1 describes the methods used to calculate actual evapotranspiration. Note that *ET* does not include interception losses.

Both precipitation and potential evapotranspiration determine the inter-annual variability of the evaporation deficit. In dry climates, annual precipitation has greater control on the deficit than in wetter climates (Figure 3.7).

For the temperate humid climate the relationship between the evapotranspiration deficit and the annual precipitation is relatively weak,

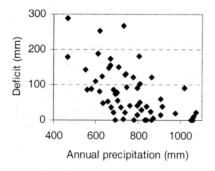
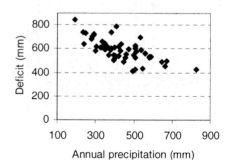

Figure 3.7 The relationship between annual evapotranspiration deficit and annual precipitation for a temperate, humid climate (the Noor in the period 1945–2001, left), and a semi-arid climate (the La Mancha Occidental in the period 1941–1996, right).

whereas it is slightly stronger for the semi-arid climate. In arid and semi-arid climate types, such as the La Mancha Occidental region, the variation in the potential evapotranspiration hardly affects the actual evapotranspiration. In almost all years soil moisture reaches the wilting point, implying that the potential rate has no effect (Condition III, Table 3.1). Thus actual evaporation is strongly controlled by precipitation and soil moisture capacity in these catchments. In the dry climates low summer rainfall generally initiates a soil water drought, but not necessarily a hydrological drought. In the temperate, humid climate of the Noor catchment, the soil seldom reaches the wilting point, which means that not only the precipitation, but also the potential evapotranspiration determines the actual evapotranspiration losses (Condition II, Table 3.1). Consequently, the relationship between the evapotranspiration deficit and the annual precipitation is weaker as compared to a semi-arid catchment (Figure 3.7).

Groundwater recharge

In a temperate humid climate the groundwater recharge exceeds the recharge in a semi-arid climate due to high precipitation and low potential evapotranspiration. In the example catchments the average recharge is 263 and 59 mm year^{-1} for the temperate humid climate and the semi-arid climate, respectively. Recharge in the semi-arid climate is not only lower in total but is also more variable.

The maximum deviation from the average recharge (Figure 3.8) in the semi-arid climate is strongly controlled by the rare occurrence of wet years. In the La Mancha Occidental region the years 1976–1978 and 1981–1985 are wet.

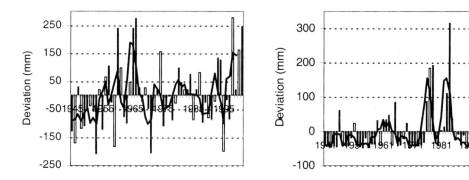

Figure 3.8 Deviation from average recharge for a temperate, humid climate (the Noor in the period 1945–2001, left), and a semi-arid climate (the La Mancha Occidental in the period 1941–1996, right). Annual deviation (bars), and 3-year moving average deviation (line).

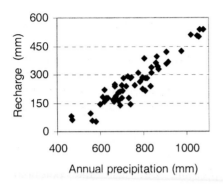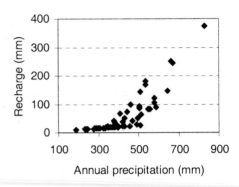

Figure 3.9 Relationship between the annual groundwater recharge and the annual precipitation for a temperate, humid climate (the Noor in the period 1945–2001, left), and a semi-arid climate (the La Mancha Occidental in the period 1941–1996, right).

A recharge below average may indicate a drought. In the temperate humid climate, only one period (i.e. in the period 1949–1954) has five or more subsequent years with recharge below average. In the drier climate four such periods occur.

The n-year moving average recharge appears to be a good way to identify groundwater drought. In the semi-arid region prolonged periods with below average recharge appear using 3-year moving average. In fact, no more than three periods are wet, i.e. having recharge above the average (the early 1960s, the late 1970s and the early 1980s). In the temperate humid climate a regular pattern of dry and wet periods can be observed with a limited number of prolonged dry periods, e.g. 1945–1955 for the Noor. Catchments in temperate humid climates in general show a stronger seasonality than drainage basins in semi-arid climates (Peters, 2003).

Potential evapotranspiration, soils and land use all affect groundwater recharge, but the most dominant controlling element is precipitation, as demonstrated in Figure 3.9.

In the temperate humid climate a more or less linear relationship exists between annual precipitation and recharge (Corr{x,y} = 0.94). In the semi-arid climate, the correlation is slightly weaker (Corr{x,y} = 0.85). A break in the plot is observed at P ~ 400 mm year^{-1}. The correlation slightly improves if the years with P < 400 mm year^{-1} are omitted (Corr{x,y} = 0.88).

The dominant role of precipitation in the generation of recharge and subsequent possible drought is confirmed by the many catchment descriptors for precipitation that are included in the regression methods for low flow estimation at ungauged sites (Chapter 8). Examples are: annual precipitation, mean precipitation in different seasons, the ratio between precipitation of groups of months and intensity of rainfall events.

Short and intensive rainfall in dry climates requires comprehensive methods and high-resolution meteorological data (e.g. daily values) to obtain reliable groundwater recharge estimates. However, these detailed data are often unavailable as in the case of the example catchment in Spain (Figure 3.8) for which only monthly data are available.

3.3.3   Role of soils, land use and water-table depth

Soils

Catchment characteristics, e.g. land use, soils and water-table depth, determine the actual evapotranspiration and consequently affect percolation (q_r, Figure 3.2) and recharge (I), and hence can lead to more frequent or more severe hydrological droughts.

To illustrate the effect of soil conditions, actual evapotranspiration from different soils in the two climatologically contrasting catchments is considered. The most important soil characteristic controlling actual evapotranspiration is the *available soil moisture*, or soil moisture capacity. This capacity consists of readily available soil moisture (stored between *FC* and *CP*, Figure 3.6) and maximum available soil moisture that can only be extracted at low rates (stored between *CP* and *WP*). Soil texture, structure, organic matter content and pH determine available soil moisture, which is reflected in rootable depth and soil moisture retention. The soil moisture retention curve specifies the relationship between soil moisture content and pressure head, *θ-h*, or moisture content and decimal logarithm of absolute pressure head (pF-curve). In regions with shallow water tables, the unsaturated hydraulic conductivity also needs to be specified to account for capillary rise. The unsaturated hydraulic conductivity curve gives the relationship between the conductivity and the moisture content (k_h-*θ*). In this section rootable depth and soil moisture retention are considered. The importance of capillary rise through the k_h-*θ* is explained in a following section.

In the temperate humid climate (the Noor catchment) silty clay soils occur that have developed in loess and chalk weathering deposits. For these soils moisture retention data are given in Table 3.2 for grassland with deep water tables. Soil moisture capacity varies from 100 to 137 mm. Actual evapotranspiration is calculated using these data. Clearly, actual evapotranspiration is positively correlated with soil moisture capacity. For the example catchment the average evapotranspiration increases by 28 mm year^{-1} with an increase in the capacity from 100 to 137 mm. In the semi-arid climate (La Mancha Occidental) fluvial deposits with loamy clay soils are observed.

Table 3.2 Rootable depth, soil moisture retention data and available soil moisture for some soils and crops in the Noor catchment and the La Mancha Occidental region

Catchment	Crop	Rootable Depth (m)	Moisture content (vol. %)			Soil moisture storage (mm)			Available Soil Moisture (mm)
			FC	CP	WP	FC	CP	WP	
Noor	grass	0.5	30.0	15.0	10.0	150	75	50	100
	grass	0.5	35.0	12.5	7.5	175	63	38	137
	maize	1.0	30.0	15.0	10.0	300	150	100	200
La Mancha Occidental	grapes/olives	0.9	31.3	27.0	15.0	282	243	135	147
	grapes/olives	2.0	31.3	27.0	15.0	626	540	300	326

Moisture retention data are presented in Table 3.2. In some of these soils, crop roots can reach deeper levels (e.g. grapes and olives), that is 2.0 m as compared to 0.9 m below the soil surface. This has a substantial effect on the soil moisture capacity, which increases by 179 mm (about 120%) for the example catchment. Average evapotranspiration, however, does not increase by more than 30 mm year^{-1}. In dry years with an actual evapotranspiration of not more than 200–300 mm year^{-1} a higher soil moisture capacity does not result in a noticeable increase in evapotranspiration. The low precipitation is the main reason for this. In dry years, including the winters, the soil moisture capacity is not replenished to the level of the critical point *CP* (i.e. 540 mm for the example catchment; Table 3.2). In 95% of the years, the actual evapotranspiration increases by not more than about 35–40 mm year^{-1} as a result of the increase in soil moisture capacity. Only in a few wet years does a high available soil moisture result in an increase in actual evapotranspiration that exceeds 100 mm year^{-1}.

To summarize, a low or high soil moisture capacity does not automatically lead to a corresponding decrease or increase of the actual evapotranspiration. In a temperate, humid climate, precipitation is often sufficient to prevent complete depletion of soil moisture storage. In dry years, soils with a high soil moisture capacity have a higher actual evapotranspiration. On the other hand, in semi-arid climates precipitation is insufficient to replenish the large soil moisture stores in most years, and the soil moisture capacity will not have a large impact on actual evapotranspiration in dry years.

Soils indirectly affect groundwater recharge through evapotranspiration losses and thus influence the development of hydrological drought. In Figure 3.10 the recharge for the example catchments is given. Clearly, soils

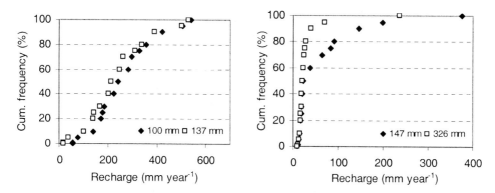

Figure 3.10 Effect of soils on the cumulative frequency distribution of groundwater recharge for a temperate, humid climate (the Noor in the period 1945–2001, left), and a semi-arid climate (the La Mancha Occidental in the period 1941–1996, right).

with a high soil moisture capacity have a low groundwater recharge. In the temperate humid catchment the recharge decreases by about 29 mm year^{-1} in response to an increase in soil moisture supply capacity from 100 to 137 mm. In the semi-arid region the average recharge reduces by about 29 mm year^{-1} as a result of an increase in soil moisture capacity from 147 to 326 mm. Over a longer time span the decrease in groundwater recharge will approach the increase in actual evapotranspiration, provided no throughflow takes place.

In the semi-arid region, soils have no major impact on groundwater recharge in dry years. In 50% of the years there is hardly any influence (Figure 3.10, right) due to nearly identical actual evapotranspiration in these years. In wet years the differences in recharge are significant (up to about 140 mm year^{-1}). As a result, areas with soils having a low moisture capacity might actually have a lower susceptibility to hydrological droughts, if wet years precede dry years. In areas with low moisture capacities the additional stored groundwater in the wet years leads to higher groundwater levels, which are beneficial for succeeding dry years. In contrast to the dry climates, areas with low moisture capacity in humid temperate climate also have a high groundwater recharge in dry years. The low capacity results in high recharge both in dry and wet years (Figure 3.10, right).

Soil characteristics may thus have an important impact on drought development due to the difference in actual evapotranspiration, percolation and subsequent recharge rates. This is reflected by including soil descriptors into methods for prediction of low streamflow and associated drought at ungauged sites (Chapter 8). These are, for instance, mean hydraulic conductivity, mean water holding capacity of the root zone and percentage of soils with high water holding capacity in the root zone.

Land use

Vegetation cover affects actual evapotranspiration through the net precipitation, potential evapotranspiration and the rooting depth. The potential evapotranspiration depends on vegetation type because of geometry (e.g. height), physiological properties (e.g. stomatal resistance) and land coverage (e.g. arable crops do not cover the whole soil surface in the initial growth phase). The rooting depth determines available soil moisture capacity. Clearly, the deeper the crop rooting system, the more soil moisture is available.

To illustrate the effect of land use, actual evapotranspiration of different crops in the humid temperate climate (the Noor) is considered. In this basin permanent grassland is converted to land with maize. This results in a low average annual potential evapotranspiration (about 50 mm year^{-1}), although in the middle of the growing season when the maize crop reaches its mature phase, the potential evapotranspiration for maize is 20% higher than for grass. The rooting depth for grass is 0.5 m, whereas for maize it is 1 m (Table 3.2). For a homogeneous soil, this implies that soil moisture capacity increases from 100 to 200 mm.

The average actual evapotranspiration for both crop types over the last 50 years is nearly identical at 496 mm year^{-1}. For low annual actual evapotranspiration rates, the evaporation of the maize crop is clearly higher than for grass (about 40 mm year^{-1}, Figure 3.11, left), whereas for the high rates the opposite occurs (about 20 mm year^{-1}). In dry years with low precipitation and high potential evapotranspiration, the actual evapotranspiration is lower than the potential rate. In these years, the higher soil moisture capacity for maize causes higher actual evapotranspiration losses despite the slightly lower potential evapotranspiration. In wet years, the actual evapotranspiration is close to the potential rate.

In conclusion, the effect of land use on actual evapotranspiration is not straightforward due to the change in the potential evapotranspiration, the distribution of the potential evapotranspiration over the year and soil moisture capacity. In the case discussed, conversion of grassland into maize leads to higher evapotranspiration and thus lower recharge and higher risk of hydrological drought. In the example catchment with a temperate humid climate, conversion of grassland into maize causes a lower groundwater recharge in the dry years, although the differences are relatively small (Figure 3.11, right). The cumulative frequency distribution shows that over a wide range of *ET* values (20–80%) the groundwater recharge is not really different, except for the tails. In the wet years more recharge takes place in the

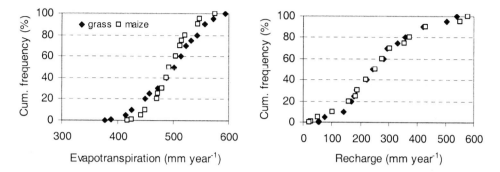

Figure 3.11 Effect of land use, i.e. grass versus maize in a temperate, humid climate (the Noor in the period 1945–2001); cumulative frequency distribution of actual evapotranspiration (left), and groundwater recharge (right).

areas with grass from which subsequent dry years might benefit, if large stores, such as aquifers or lakes, occur in the basin.

Land use generally affects groundwater recharge through actual evapotranspiration. Therefore, land use descriptors (for example, the percentage of area covered by forests, pasture land and urban area in catchments) are incorporated in models for prediction of low streamflow at ungauged sites (Chapter 8).

Land-use change will also influence interception losses (Figure 3.4) and thereby affects water surplus and drought development. The impact of land-use change is further discussed in Section 9.3.

Water-table depth

For the depth of the water table two different situations are distinguished. If water levels are close to the surface (within 1–2 m), capillary rise can supply the root zone if a soil water deficit develops. This leads to high actual evapotranspiration. When the water levels are deep below the soil surface, the impact of capillary rise is negligible. Furthermore, deep water levels influence the distribution of groundwater recharge over time so that it substantially differs from that of the percolation (I versus q_r, Figure 3.3), although the long-term averages are identical.

When the water table is close to the soil surface, water-table depth, thickness of the root zone and the unsaturated hydraulic conductivity determine the capillary rise. In silty clay soils, capillary rise is still sufficient to supply crops with sufficient water over a wide range of soil moisture contents provided that the water table does not drop deeper than 1.0–1.8 m below the root zone. This is due to high unsaturated hydraulic conductivities. In contrast, in coarse

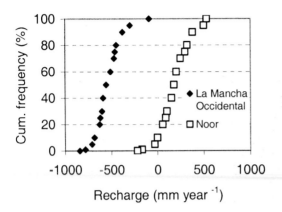

Figure 3.12 Cumulative frequency distribution of the groundwater recharge for areas with shallow water tables for a temperate, humid climate (the Noor in the period 1945–2001), and a semi-arid climate (the La Mancha Occidental in the period 1941–1996).

sands the distance between the water table and root zone should not exceed 0.3–0.7 m to maintain abundant capillary rise. In fact, two extreme situations are observed, namely (a) capillary rise is sufficient to avoid soil moisture stress, which means that the evapotranspiration is maintained at the potential rate, and (b) water tables are so deep, i.e. about 2–2.5 m below the bottom of the root zone, that the capillary flux is negligible. The latter situation was implicitly assumed in the previous paragraphs on soils and land use. Here the situation with evapotranspiration at the maximum rate is further elaborated. When the capillary rise is maximal, crops evaporate at the potential rate. Then percolation (q_r, Figure 3.3) equals precipitation minus potential evapotranspiration. In a humid temperate climate the annual percolation is positive on average (e.g. 184 mm year^{-1} for grassland in the Noor, although extreme years occur with negative percolation). In a semi-arid climate, like in the La Mancha Occidental, the annual percolation is negative in areas with shallow water tables (e.g. wetlands), which means that on average groundwater is depleted. In Figure 3.12 the recharge for areas with shallow water tables (e.g. the wetlands) in the two example catchments is shown.

Depletion of the aquifer appears in about 10% of the years for the Noor with a temperate humid climate. Under semi-arid conditions the recharge of a wetland is always negative. In the La Mancha Occidental more than 80% of the years have a recharge of less than −400 mm year^{-1}. In dry years, valley bottoms with shallow water tables naturally abstract vast amounts of groundwater for evaporation, which contributes to the development of hydrological drought.

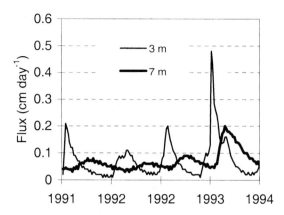

Figure 3.13 Simulated soil moisture flux at two depths illustrating the delay and attenuation in thick unsaturated subsoils.

In areas with deep water tables, low percolation (Δq_r) due to a high water deficit can affect the following water balance components (for the meaning of symbols see Figure 3.3):

$$\Delta q_r = \Delta(\frac{\Delta S_{so}}{\Delta t}) + \Delta(q_{if} + I) \tag{3.4}$$

If throughflow is neglected, then low percolation influences soil moisture storage in the thick unsaturated subsoil and recharge. Note that in a humid, temperate climate the percolation of 4–5 years (i.e. about 1000 mm) is stored in each 10 m of unsaturated subsoil. Hence, small changes in soil moisture content reflect significant amounts of water (first term in Equation 3.4). As a consequence the distribution of q_r and I (including Δq_r and ΔI) over time are different. The flux I is delayed and attenuated compared to q_r.

Comparison of the flux at different depths in the unsaturated subsoil clearly illustrates the delay and attenuation (Figure 3.13). The clear peaks in the years 1991 and 1992 at 3 m depth hardly occur at 7 m depth. However, the wet period of 1993–1994 can be observed at 7 m depth, but is delayed for about 3–4 months. The degree of delay and attenuation depends on the depth of water table and on the hydraulic properties of the unsaturated subsoil, i.e. the h-θ and k_h-θ relations. The effect of thick unsaturated subsoil on hydrological drought development is similar to that of aquifers with large groundwater storage, and it is discussed in the next section.

3.4   Saturated zone – Groundwater discharge and water tables

Hydrogeological conditions in combination with groundwater recharge govern natural groundwater flow in a catchment. Groundwater can follow a short and shallow flow path, and contribute as quick flow to a stream. It can also follow a longer path in a local flow system or regional system contributing to the base flow of a stream (Section 3.2). In this section the focus is on the latter (i.e. groundwater discharge as part of the base flow).

Hydrological drought may develop due to below-normal recharge. Key variables to identify groundwater droughts are (Section 5.6):

a)   groundwater recharge;

b)   water tables of unconfined aquifers or hydraulic heads of semi-confined and confined aquifers;

c)   groundwater discharge.

The recharge and water levels have a spatially-distributed nature, whereas the discharge is distributed along the longitudinal profile of the stream. In general, use of groundwater storage as a variable is avoided because it is difficult to obtain reliable estimates. For storage estimation spatially-distributed water levels, the bottom of the aquifer and the storage coefficient are required, which are not routinely observed. Low recharge leads to a decrease of the groundwater storage, which means that groundwater gradients decay meaning less groundwater flux to the streams. Hydrological drought can be enhanced by human activities, in response to drought (e.g. groundwater abstraction) or more general activities (e.g. land drainage). Human impact is discussed in Chapter 9. Properties such as aquifer hydraulic conductivity and storage coefficient determine how fast the groundwater head gradient and the associated flux drop. The influence of aquifer properties and groundwater recharge on water levels, groundwater discharge and possible drought development are elaborated in this section.

3.4.1   Illustration of inter-annual variability of groundwater discharge in contrasting climates

Groundwater recharge (Figures 3.10 and 3.12) controls groundwater discharge. Discharge is thus very different in the various climate types. Groundwater discharge is difficult to measure because it mostly feeds the stream in a diffuse way and is possibly mixed with overland flow and throughflow (Section 3.2).

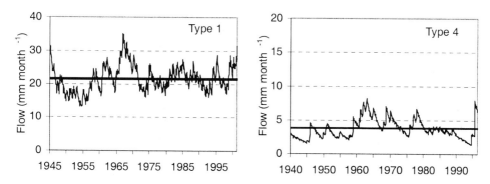

Figure 3.14 Groundwater discharge simulated for areas with a recharge from a temperate humid climate (left), and a semi-arid climate (right). Please note the different scale of y-axis. Table 3.3 defines Types 1 and 4. The bold horizontal line represents mean monthly discharge.

Therefore groundwater discharge is simulated in this chapter with a simple conceptual model (i.e. a linear reservoir). The simple model is adequate to describe the overall effect of a groundwater system on drought development (Peters *et al.*, 2003). In reality, of course, the hydrogeological setting is more complex and requires more comprehensive models (Chapter 9). To illustrate groundwater discharge in a temperate humid climate, the simple model uses groundwater recharge data from the Noor catchment (period 1945–2001). For the semi-arid climate, recharge data from the Upper Guadiana are used in an otherwise identical model. Supporting Document 3.1 describes the methods used to simulate groundwater discharge.

Figure 3.14 shows the widely different groundwater discharge regimes in the two climate types. Obviously the discharge is lower in the semi-arid region, but it also has a very different inter-annual variability as compared to the temperate humid climate. Groundwater discharge in dry climates has more prolonged dry periods, e.g. 1940–1960 and the early 1990s. In the example for the temperate humid climate dry periods are present in the 1950s, in the 1970s and 1990s. In the latter two periods the dry periods are separated by a wet period The event definition, which is defined in Chapter 5, determines whether the dry periods are classified as distinct or pooled hydrological drought events.

3.4.2   Influence of groundwater system properties on discharge

Important hydraulic properties controlling groundwater discharge are the saturated hydraulic conductivity (k_h), storativity and the storage coefficient (S_y) of the geological layers and the distance between streams. The groundwater

Table 3.3 Properties of groundwater systems

Type	Recharge	kD	S_y	Distance between streams	Area
1	temperate, humid climate[1]	high	low	large	large
2		high	high	large	large
3		low	low	small	small
4	semi-arid[2]	high	low	large	large
5		high	high	large	large
6		low	low	small	small

[1] Noor catchment, [2] La Mancha Occidental

storage in a aquifer is defined as $S_{gr} = S_y D_a$, where D_a is aquifer thickness. Groundwater systems may consist of a single aquifer or a complex of aquifers and aquitards (multiple aquifer system). They range from hilly or mountainous catchments dominated by shallow saturated subsurface flow with high portion of quick flow, to lowland basins characterized by extensive regional systems dominated by slowly-responding groundwater discharge (Section 3.2). In local or regional systems with a high portion of groundwater discharge, an interaction often exists between shallow and deep aquifers. The hydraulic resistance, c, characterizes the possibility of vertical water transfer through the aquitard, i.e. leakage (Figure 3.4).

Typical properties of a regional groundwater flow system are (a) medium to high transmissivity, (b) medium to large storage, (c) extended area, and (d) large distance between streams. The properties of the groundwater system have a distinct influence on the distribution of the groundwater discharge over time. Table 3.3 presents different types of groundwater systems. The groundwater discharge hydrographs for the temperate humid and semi-arid climate in Figure 3.14 belong to Type 1 and Type 4, respectively. In Figure 3.15 the groundwater discharge of four other types of groundwater systems is given.

Groundwater systems of Types 3 and 6, which are typical for local systems with shallow subsurface saturated groundwater flow, react very quickly to recharge resulting in a strongly fluctuating groundwater discharge. In the temperate, humid climate (Type 3) hardly any prolonged dry periods are found. In the semi-arid climate (Type 6), dry periods are present even for quickly responding groundwater systems, for instance the early 1940s and early 1990s in the example. In dry climates (B-climate, Section 2.2.2), Type 6 hydrographs

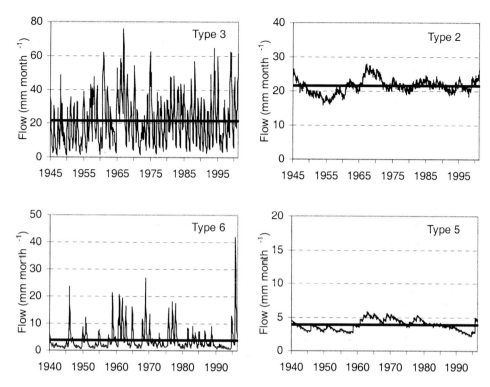

Figure 3.15 Groundwater discharge for a temperate humid climate and aquifer Type 3 (upper left), temperate humid climate and aquifer Type 2 (upper right), semi-arid climate and aquifer Type 6 (lower left), and semi-arid climate and aquifer Type 5 (lower right). Aquifer types are defined in Table 3.3. Please note the different scale of y-axis. The bold horizontal line represents mean monthly discharge.

are observed, however, with zero flow between the peaks. In catchments with precipitation falling as snow in the cold season (e.g. Dw-climate, Section 2.2.2), the groundwater discharge hydrograph is less flashy with a more pronounced peak in the early warm season than Type 3.

Groundwater systems of Types 1-2 and 4–5 are typical for systems with slowly-responding groundwater discharge that vary with climate type and storage capacity. If kD, S_y, distance between the streams and aquifer area increase, the fluctuation in the groundwater discharge decreases (Figure 3.15). Groundwater systems of Types 2 and 5 hardly show dry periods implying that drought development is less frequent, and when a drought develops it is long-lived but not intensive (e.g. 1950–1960, Figure 3.15, upper right). Multiple aquifer systems behave as Types 1–2 and 4–5 providing that water tables are not shallow.

A special type of groundwater system is karst. This occurs mainly in areas built by limestone and dolomite, and it is characterized by the occurrence of large underground passages or fractures due to rock solution, which enable underground movement of large quantities of water. In well-developed karstic systems, deep groundwater levels and high transmissivities often occur. Nevertheless, the response in streamflow to precipitation is very quick and does not resemble groundwater of Types 1 or 4 as might be expected. In terms of drought development, karst regions have much in common with groundwater systems of Types 3 or 6. However, the response of karstic systems is more difficult to predict because it depends on the extent of rock solution, actual groundwater storage, interconnection of underground stores, and the interaction between groundwater and surface water (e.g. presence of sinkholes, karstic springs).

In models to predict low streamflow at ungauged sites in a river basin (Chapter 8) the different groundwater systems are implicitly included in groundwater-related properties. Geological descriptors, which describe the transmissivity and storage capacity of a catchment, are commonly used. In addition, the relative area of low-permeability rocks and hydraulic conductivity are used. The effect of the hydrogeological configuration of a particular catchment on the generation of groundwater discharge is hard to assess. Therefore, recession curves derived from measured hydrographs are also used to characterize groundwater behaviour. Recession coefficients (Section 5.3.4) can be estimated at ungauged sites by relationships with catchment properties.

3.4.3   Water tables

Water tables are easy to measure, and when these are low a groundwater system suffers from drought. Water levels are key variables for drought assessment and reflect the state of groundwater storage at a particular location and time. Figure 3.16 gives the groundwater hydrograph for representative wells in a temperate humid climate region and in a semi-arid climate. Both wells are situated in aquifers with deep groundwater levels. The levels in the semi-arid climate are naturalized (Chapter 4) because of extensive groundwater abstraction (Chapter 9).

The hydrographs show large fluctuation (10–16 m) as a response to the inter-annual variability of recharge. Distinct dry periods can be observed in the time series, which might be classified as periods of groundwater drought (Section 5.6). The groundwater levels in these two wells resemble the time series of groundwater discharge for slowly responding groundwater systems,

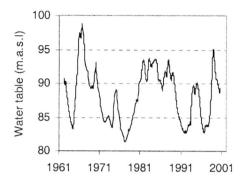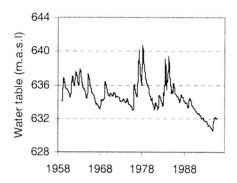

Figure 3.16 Water tables in a temperate humid climate, well WP99 (the Noor, left), and semi-arid climate, well 2030-30001 (the La Mancha Occidental, right).

e.g. Type 1 for well WP99 and Type 4 for well 2030-30001 (Figure 3.14). Thus, dry periods in the groundwater recharge, groundwater levels and the groundwater discharge reasonably coincide. In Section 5.6 the extent of coincidence is further elaborated. Snow and soil frost also affect recharge thereby having a distinct influence on the water tables as shown in Box 3.2.

Box 3.2

Groundwater levels in cold climates

In areas where precipitation falls as snow in the cold season groundwater hydrographs have a characteristic shape. Recharge is delayed for several months. At the beginning of the cold season frost starts penetrating the soil (Figure 3.17, left). Liquid soil water stored in the pores freezes leading to redistribution of soil water and extremely low hydraulic conductivities. This is characteristic for soil frost. Several thawing and freezing periods may take place in the early cold period dependent on the climate region and air temperature in a particular year. Soil frost penetration depth depends on soil moisture content, soil type and snow accumulation. Lower moisture contents (water has high heat capacity), thinner snow covers (snow has low thermal conductance) or coarser-grained soils (e.g. gravelly soils) lead to deeper penetration. After some time an impervious ice layer may form at the soil surface. Precipitation does not recharge the aquifer during the soil frost period implying falling groundwater levels. At the beginning of the warm period the soil frost disappears and the snow melts. Groundwater

(→ Box continued on next page)

Box 3.2 (continued)

levels show a sharp rise in response to the percolation and following recharge due to melting of the precipitation accumulated over the cold period.

Groundwater hydrographs in some snow-affected regions have two falling and rising limbs (Figure 3.17, right, hydrograph 1). The first rise comes from the snow melt and the second one from rainfall at the beginning of the cold season. In colder climates precipitation does not appear as rainfall but as snow in the early cold season leading to one long recession during the warm and the cold period with snow until the snow melt starts (hydrograph 3). As discussed in Section 3.3.1 early snow in the interior region (hydrograph 1) instead of rain leads to no water-table response in the early cold period. Winter droughts may develop then because in that particular year the hydrograph resembles that from the colder regions (hydrograph 3).

Groundwater levels in coastal areas (hydrograph 2) are typical for areas with a warm and cold season with regular precipitation falling as rain and not too deep groundwater levels.

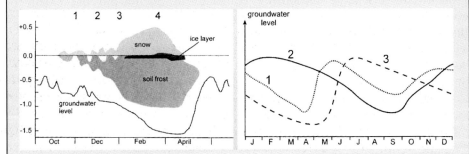

Figure 3.17 Cold regions; groundwater levels, snow accumulation and soil frost development (left). 1: slow penetration of frost, 2: thawing and freezing periods, 3: soil pores fill with ice, and 4: development of impermeable ice layer on top of the soil (modified from H. Colleuille, NVE-Norway), and typical groundwater hydrographs in Norway (right). 1: interior, 2: coastal regions, and 3: mountainous regions and northern (Kirkhusmo, 1986).

3.5   Surface water

The surface water system collects water from overland flow, throughflow and groundwater discharge (Section 3.2). Generation of groundwater discharge is described in Section 3.4. In this section the contribution from overland flow and throughflow is described. Furthermore, total surface water flow as a composite of the three flow components is discussed. The section concludes with the influence of lakes and wetlands on streamflow.

3.5.1 Streamflow generation

The contribution of the three components of streamflow varies over time. During and directly after a rainfall event, the contribution of overland flow and throughflow increases whereas during a dry spell the surface water system is completely fed by base flow. The contribution of the three components over time depends on catchment characteristics, precipitation form, precipitation intensity and antecedent soil moisture. In the following sections overland flow, throughflow and streamflow are elaborated. Note that occurrence of zero streamflow in dry environments is perceived differently than in humid climates (Box 3.1). Intermittent streams are there common, and hydrological drought is predominantly characterized by low reservoir and groundwater levels.

Overland flow and total streamflow

Overland flow takes place if the rainfall intensity or snow melt exceeds the infiltration rate of the soil (Section 3.2). Water that cannot infiltrate is stored in small depressions (e.g. plough furrows, Figure 3.3), if present, or in the organic litter layer in forests. Usually the surface storage capacity is limited resulting in water flow over the soil surface. Overland flow follows the slopes towards the stream. After a short while (quick process), it reaches the stream, if it does not infiltrate into the soil. Slope angle, slope length and slope direction are important catchment properties for the generation of overland flow. Another important property is land use, e.g. a forested area has a lower probability on overland flow than temporarily, partly covered arable land or sparsely-vegetated land in arid and semi-arid regions. Physical properties (e.g. k_h-θ and θ-h relations) of the topsoil also play a major role because they determine the highly dynamic infiltration capacity. Some swelling clay soils develop wide cracks that readily allow infiltration. Stability of the topsoil structure is important because it can prevent crust development that leads to an impermeable top layer. In addition rainfall amount and intensity govern overland flow. Overland flow is a complex process that takes place at detailed spatial (field) and temporal (minutes) scales.

Overland flow reduces groundwater recharge and discharge and contributes to flashier streamflow hydrographs. In general, catchments with a high portion of overland flow have low minimum flow, and enter quickly into a drought condition. To illustrate this, the effect of overland flow on (total) streamflow and associated streamflow droughts is roughly simulated by decreasing rainfall. Overland flow is assumed to be 5% of the rainfall in this example. Throughflow is expected not to take place. Overland flow is superimposed on groundwater discharge to achieve the total streamflow. Figure 3.18 illustrates this for the

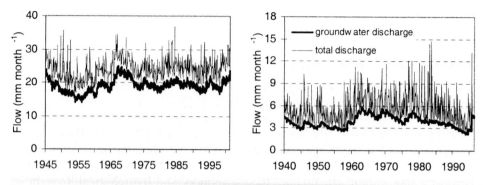

Figure 3.18 Overland flow, groundwater discharge (Types 2 and 5) and total streamflow for recharge typical for a temperate humid climate (the Noor, left), and recharge for a semi-arid climate (the La Mancha Occidental region, right). Overland flow is the difference between total streamflow and groundwater discharge.

temperate humid and semi-arid climates and a slowly-responding groundwater system (Types 2 and 5, Table 3.3).

Clearly, for both climate types the temporal variability of overland flow is higher than the variability of the groundwater discharge. Overland flow has a major influence on the streamflow hydrograph and occurrence of dry periods. The prolonged dry periods that could be distinguished in the groundwater discharge, for example the early 1990s for the semi-arid climate (Figure 3.15, lower right) are subdivided into many minor dry periods. Overland flow is more likely to occur in dry than in temperate humid climates due to the sparse vegetation, crusted soils and high rainfall intensities. Thus, the streamflow hydrograph for the semi-arid climate (Figure 3.18, right) is more likely to occur than that for the temperate humid climate (Figure 3.18, left).

In temperate humid catchments with a flashy response to recharge (groundwater system Types 3, Figure 3.15), overland flow has a smaller impact on the streamflow hydrograph, because long dry periods do not exist. In semi-arid catchments with a flashy response, overland flow can still subdivide dry periods, such as in the early 1990s for the example in Figure 3.15 (lower left).

The important role of overland flow is reflected by inclusion of physiographic catchment descriptors into models for prediction of hydrological drought at ungauged sites (Chapter 8). These descriptors comprise minimum, mean and maximum slope, and the percentage of soils with low, medium and high infiltration capacity. Therefore, streamflow, including overland flow, usually varies within a particular climate type due to these catchment properties as shown in Box 3.3.

Box 3.3

Spatial patterns in Poland

Temporal variability of streamflow does not only deviate among different climate types, but it also shows a spatial variability in a region with a particular climate type. For example, Poland is located in the zone with a continental climate type having precipitation in all seasons (type Df, Section 2.2.2). Nevertheless, the precipitation deficit (precipitation minus potential evapotranspiration, $P - PE$) and the streamflow deficit have a clear spatial structure during summer (Figures 3.19, middle and lower). The central part of Poland (52° latitude) is the driest part of the country (< -300 mm). The wetter areas occur in the north along the Baltic Sea and in the Carpathian Mountains in the south. The pattern illustrates the effect of land sea interactions and altitude. This is the reason why the distance

(→ Box continued on next page)

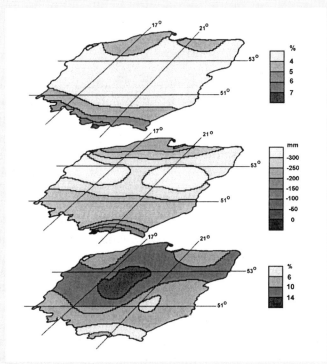

Figure 3.19 Spatial pattern in Poland of rainfall frequency: mean annual number of days per year (%) with daily precipitation exceeding 10 mm (upper), precipitation deficit of the summer (May–October, middle), and streamflow deficit volume in % of mean annual streamflow volume (lower).

Box 3.3 (continued)

to the coast and altitude-associated descriptors (e.g. minimum, mean and maximum elevation, portion of catchment above tree line, topography) are included in models for prediction of hydrological drought at ungauged sites (Chapter 8).

The streamflow deficit volume is a measure of the deviation of streamflow compared to normal flow and is further elaborated in Chapter 5. The highest deficit (> 14% of mean annual streamflow deficit volume), which occurs in the summer, can be observed in central Poland. The lowest deficits are in high-altitude areas of the south, in the central-east uplands, and in the northern area influenced by the Baltic Sea.

Throughflow

Throughflow, or interflow, is lateral flow over a slowly permeable layer or through natural pipes in the unsaturated zone (Section 3.2). It takes places if the percolation (q_r, Figure 3.3) is large and exceeds the saturated conductivity of semi-pervious layers in the unsaturated soil. If these slowly-permeable layers or pipes are continuous and reach the stream, then throughflow; q_{if}, feeds the stream, otherwise the lateral flow recharges groundwater. The response of throughflow on rainfall or snow melt takes an intermediate position between the instantaneous response of the overland flow and the slower response of groundwater discharge. In areas with typical groundwater systems of Types 1 and 4 (Figure 3.14) the response of throughflow can clearly be observed as peaks on smooth groundwater discharge curves. This implies that throughflow, like overland flow, subdivides long prolonged periods with low streamflow into several dry periods. The peaks caused by throughflow are less pronounced than a series originating from overland flow, meaning that the probability of subdividing a prolonged dry period is slightly lower. A lower precipitation in these areas might lead to less throughflow, which might affect drought development. In regions with quickly-responding groundwater systems (Section 3.2 and Figure 3.15, upper left and lower left), the effect of throughflow does not differ much from the influence of groundwater discharge. Both throughflow and groundwater discharge respond here very quickly to precipitation.

Streamflow network

The drainage network of a river influences the discharge at a particular point in the stream. It determines how fast the different flow components are conveyed to that point. The travel time (i.e. time elapsing between the passage of a water parcel or packet between a given point and another point downstream) in the open channel has hardly any influence on low flows and drought development, unless large surface water bodies occur (Section 3.5.2). Catchment properties, such as drainage density, stream length, bifurcation ratio and number of confluences, reflect the type of groundwater system (Section 3.4.2). For instance, dense drainage networks and high stream lengths indicate quickly responding groundwater systems (Type 3 and 6, Table 3.3). Catchments with dense drainage networks have low groundwater discharge. These stream network characteristics are thus included in models to predict hydrological droughts at ungauged sites (Chapter 8).

3.5.2 Lakes and wetlands

In some river basins lakes are present, which are fed by streams and groundwater discharge. Usually the water surplus of the lake flows into a river at the outlet. Lake levels indicate large variations in storage over the years. As an illustration, Figure 3.20 gives the observed levels of Lake Balaton (Hungary) for the period 1863–2002. Lake Balaton is central Europe's biggest freshwater lake and one of Hungary's main tourist attractions. The lake fluctuated more in the late 19th and early 20th century than it did in the last decades. However, the last four years were hot and rainfall was low. Evaporation exceeded precipitation, leading to a serious decline of the lake level (see the last part of the hydrograph, Figure 3.20). Large mudflats are now visible, forcing holidaymakers far out into the lake before they can swim. A potential ecological and economic catastrophe is emerging (Geoghegan, 2003). In the following passages the effect of low precipitation on the water balance of a lake is

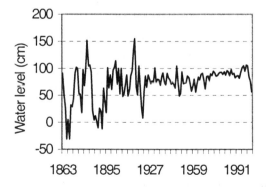

Figure 3.20 Mean annual level of Lake Balaton (Hungary) for the period 1863–2002.

elaborated. This is followed by a description of the role of lakes on the streamflow downstream of a lake in two contrasting climate types.

The effect of lower precipitation on the water balance of a lake

Many catchment properties and processes with different time constants determine if a streamflow drought develops downstream of a lake as a consequence of higher water deficit. The storage changes of the lake smooth fluctuations of the inflow. Lack of rainfall has a quick influence on the downstream flow through the absence of rainfall directly falling on the lake, reduced overland flow and throughflow from the upstream catchment. Reduced precipitation in the catchment can also influence lake inflow after months or years, as a result of slowly-responding groundwater systems. In the following the flow downstream of a lake in two contrasting climates is described.

The role of lakes on downstream flow in two contrasting climates

The streamflow downstream of a lake is calculated using Equation 3.1. The input is time series of precipitation and open water evaporation for a temperate humid region and a semi-arid region (Section 3.3). Other inputs are overland inflow (i.e. 5% of precipitation) and groundwater discharge from different types of groundwater systems (Section 3.4.2), lake area and catchment area upstream of the lake. Furthermore, the relationships between lake level and discharge (outlet rating curve), and lake level and storage (reservoir volume curve) are specified. These characteristics can vary widely from lake to lake. For instance, the physical environment (geological and geomorphological conditions) at the outlet controls the relationship between lake level and discharge. Some lakes have a natural threshold preventing outflow below this level. Other lakes have a wide valley with gentle slopes at the outlet, whereas still others have a narrow canyon. Supporting Document 3.1 describes the methods used to simulate lake outflow.

Figure 3.21 gives the simulated outflow from lakes in a temperate humid and semi-arid climate for an arbitrarily-defined lake covering 1% of the catchment area. Obviously, lake outflow is lower in the dryer climate because it is controlled by the low precipitation, the high open water evaporation and the low overland flow and groundwater discharge as demonstrated in the previous sections. The hydrographs of lake outflow have much in common in terms of shape with the groundwater hydrographs as discussed in Section 3.5.1, although the lake outflow hydrographs are more smoothed. This means that the streamflow droughts downstream of the lake occur more or less in the same years, but they are different in duration and severity. Naturally, the outflow is also strongly controlled by the type of groundwater system feeding the lake. Quickly responding groundwater systems (Types 3 and 6, Table 3.3) give a

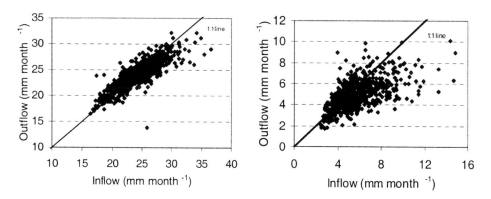

Figure 3.21 The relationship between lake inflow and outflow for catchments with slowly responding groundwater systems (Types 2 and 5) for recharge from a temperate humid climate (the Noor, left), and semi-arid climate (the La Mancha Occidental, right).

more fluctuating lake outflow than lakes fed by slowly responding groundwater systems. In Figure 3.21 results are presented for groundwater systems of Types 2 and 5.

For the lake in the temperate humid climate, the number of months in which the outflow exceeds inflow (i.e. 44% of the months) is about equal to the number of months that the inflow is higher than the outflow. More importantly, in dry periods (outflow < 18.5 mm month^{-1}) outflow is higher than inflow, meaning that there is less likelihood of streamflow drought downstream. In the semi-arid climate the outflow from the lake is often smaller than the inflow (i.e. 70% of the months) because of the high lake evaporation. In extremely dry periods (outflow < 3.0 mm month^{-1}) the outflow is smaller than the inflow leading to more severe streamflow drought.

Clearly, lake inflow has a strong effect on outflow, but lake properties may also have a considerable influence. For instance, the lake area in a dry climate clearly affects the lake outflow due to evaporation losses. Figure 3.22 shows the impact on the lake outflow of an increase in lake area from 1 to 5% (of the catchment area). The average outflow drops by more than 50% when the lake area increases. It is clear that streamflow drought will be significantly different for the basin with the larger lake area. The droughts in the period 1940–1960 grow together to become one major drought, and the drought in early 1990 becomes more severe.

The important role of lakes is underpinned by including some of the lake properties into models for prediction of hydrological drought at ungauged sites (Chapter 8). Usually the relative area and the location of lakes in a catchment is included.

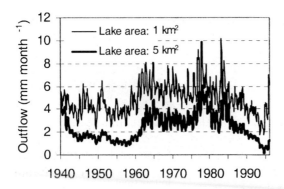

Figure 3.22 Streamflow downstream of lakes with different areas for a catchment with a slowly responding groundwater system (Type 5) and recharge of a semi-arid climate (the La Mancha Occidental).

Wetlands

There are many different types of wetlands. Riverine wetlands usually receive surface water from the upstream catchment, groundwater discharge and precipitation, and lose water to drainage at the outlet and through open water evaporation. Thus, wetlands have much in common with lakes. However, wetlands also differ from lakes. For instance, parts of the wetland consist of nearly saturated soil, meaning that a change in water storage leads to larger level fluctuations. Furthermore, evaporation losses consist of evaporation from open water surfaces and potential transpiration from the aquatic and wet terrestrial vegetation. Under dry conditions wetlands may save water by having smaller open water surfaces and reducing the potential transpiration.

3.6   Summary

This chapter has dealt with hydrological processes under drought. Its three main objectives were (a) to describe how processes in various types of catchments are affected by lack of precipitation, possibly in combination with high temperature and potential evapotranspiration, (b) to list an overview of catchment properties that control drought development, and (c) to illustrate the different processes for a humid-temperate and a semi-arid climate.

The development of hydrological drought depends on the relative contribution of overland flow, throughflow and base flow to streamflow over time. In catchments without lakes, glaciers and peat bogs, base flow consists entirely of groundwater discharge. Catchment properties (e.g. local climate, land use, topography, soils, geological conditions and drainage network) control their contribution. Local flow systems with high portions of overland flow and throughflow (quick flow) generate flashy streamflow hydrographs with frequent, but usually no long droughts. Examples are catchments with pronounced topography and impermeable rock at shallow depth, basins in flat delta areas and catchments with snow melt on top of a frozen soil in the early warm season. Large amounts of quick flow may also occur in sloping aquifers during the wet season. Quick flow is very common in arid and semi-arid regions due to the high rainfall intensity and percentage of bare soil. Intermittent or ephemeral streams are often observed and low lake or reservoir levels are preferred drought indicators rather than streamflow. In most regions with a high portion of quick flow, large aquifers are not present. Hence, long-lived groundwater drought does not develop.

In contrast to local systems with high proportions of quick flow, regional flow systems have generally smooth streamflow hydrographs due to groundwater discharge, which only respond slowly to long, wet or dry periods (e.g. months, years). Droughts do not readily develop, but if they do they are generally persistent. Basins with a regional groundwater flow system generally have deep water tables, permeable geological environment to great depths, marked topography and streams in the different valleys at different elevation. Local systems with permeable geological environment deep below the valley bottom show similar behaviour, although drought mostly develops earlier. Long-term recharge controls the development of groundwater drought in regional flow systems. Deviations from normal recharge over a long period of time cause low groundwater levels and reduced groundwater discharge. In dry regions well-developed regional flow systems are less common because groundwater recharge is low (a large proportion of quick flow and high evapotranspiration). If regional systems occur, they mostly stretch out with recharge areas far away in wetter mountainous reaches. Under such conditions sustained groundwater discharge is of vital importance, because communities rely on it during dry periods. Drought develops in these regions when a high water deficit builds up in the remote recharge areas.

The influence of a climatic water deficit on actual evapotranspiration and resulting groundwater recharge has been illustrated by discussing processes in the unsaturated zone. In climates with a wet and dry season, lack of precipitation in the wet period may lead to hydrological drought in the subsequent dry season. On the other hand, shortage in dry periods usually leads to a lower chance of drought as compared to lack of precipitation in the wet season. The effect of shortage in the dry period depends on the increase of the soil moisture deficit. A high potential evapotranspiration only has significant impact on drought development in regions with medium or high temperatures and abundant moisture. In cold climates, temperature deviation from the normal can lead to drought due to shift of the snow melt period or soil freezing affecting the portion of quick flow.

The effect of aquifer properties (transmissivity, storage coefficient) and drainage density on groundwater discharge to streamflow has been shown by explaining processes in the saturated zone. Dependent on the properties, flashy or smooth groundwater discharge hydrographs are generated. Water tables are strongly related to aquifer properties, although they contain more local information than groundwater discharge. Data from the Noor catchment (temperate-humid climate) and the La Mancha Occidental (semi-arid climate) are used to illustrate the effect of high climatic water deficit on processes in the unsaturated and saturated zone.

Finally, the effect of overland flow and throughflow (quick flow) on streamflow has been described. It was shown that catchments with quick flow have streamflow hydrographs with a flashy component on top of a slowly-responding groundwater discharge. The flashy component may subdivide periods with low streamflow into separate periods. Furthermore, it was shown that lakes and riverine wetlands considerably modify outflow. Usually streamflow hydrographs at the outlet are smoothed, generally leading to fewer, smaller droughts and more, low-frequency, long-lasting droughts. In some cases, for example in semi-arid climates, outflow can also be substantially lower due to high evaporation losses. Lake and wetland properties, together with groundwater discharge into the lake and features of the upstream catchment, determine whether a particular climatic water deficit leads to higher or lower streamflow drought at the outlet.

Part II

Estimation methods

4

Hydrological Data

Gwyn Rees, Terry J. Marsh, Lars Roald, Siegfried Demuth,
Henny A.J. van Lanen, Ladislav Kašpárek

4.1   Introduction

Hydrological data are available in many forms and are used for many purposes. They may be used in the *operational management* of *water resources*, to *monitor* and *control* abstractions from and discharges to rivers and the inflows to and outflows from reservoirs and lakes. In periods of low flow or drought, the data assist water resources managers to maintain essential water supplies, enable river navigation and safeguard the ecological quality of the river. Hydrological data are also used for the *forecasting* of drought and flood, providing local authorities, farmers, businesses and the general public with early warning of these extreme events and enabling the timely implementation of *mitigation* measures. Hydrologists, engineers and planners use hydrological data for the long-term *planning* of water resources or for the *design* of new schemes and hydraulic structures, relying on good quality data that are likely to be representative of site conditions in the foreseeable future. The data may also be used to assess the technical and economic feasibility of prospective water management projects, such as irrigation systems, hydropower schemes, river diversions and flood control schemes. With concerns growing over climate change and the effect human activity has on the environment, hydrological data are increasingly being employed for *environmental monitoring* and *impact assessment* purposes. Trends observed in long-term records provide evidence of changes in the general climate and may influence the *policy* of national and regional governments. At international level, hydrological data help to ensure compliance with *international obligations* on monitoring and water sharing. Good quality data are also vital for *hydrological research* and the objective of improving the understanding of the hydrological cycle and the interactions between the many physical processes it involves, which ultimately leads to the development of *tools* and *methods* to assist in the operational management of water resources.

Irrespective of purpose, a meaningful analysis of how groundwater and rivers behave in periods of low flows, or drought, requires good quality data that adequately represent the hydrological conditions of the catchment or region of interest. Hydrological data may include time series of hydrometeorological variables, such as river flows, groundwater levels, precipitation and temperature, and thematic data, such as maps of topography, soils and land use and other information that describe the physical properties of relevant catchments. The type of data required varies according to its application. For example, the hydrological study of a single catchment may be adequately served by the time series data from a nearby rain gauge or river gauging station and information from local maps. An assessment of regional water resources would, however, require long-term data from many gauges, to represent the spatial and temporal variability of the hydrological regimes of the region. Maps that consistently describe the distribution of key physical features in the entire study area would also be necessary.

The temporal extent of the data is an important factor too. One or two years' river flow data barely provides a snapshot of the hydrological behaviour of a catchment and, on its own, provides little indication of how the catchment has behaved, or is likely to behave, in the long-term. Generally, the longer the time series the better, although caution should be exercised when dealing with longer records, in case changes in the method of measurement, or changes in land use and water management, have affected the values obtained. For climate impact studies, standard period data are commonly used, where only data for a set 30-year period (e.g. 1961–1990) are considered (Section 2.3.3). Time series data should also be continuous, containing as few gaps, or missing data, as possible and providing data across the entire range of possible values, from the highest to the lowest. The approach to the measurement of the hydrometeorological variables will ultimately have a bearing on the quality of the data in this respect. Unfortunately, those conducting hydrological analyses are often several steps removed from the procedures of data collection and processing but, to correctly interpret and apply the results of their analyses, it is important that users are aware of these procedures and the limitations and applicability of the data.

The objective of this chapter is to provide an overview of the hydrological data that are typically available for the analysis of hydrological drought and the procedures that ought to be applied to ensure that this data is of good quality. In Section 4.2, the types of hydrological data are considered, classifying them generically as time series data, thematic data or metadata. Section 4.3 provides an introduction to the discipline of collecting hydrometric and groundwater data and describes procedures for ensuring the delivery of good quality data. Details are then provided in Section 4.4 of the variety of thematic data that are typically available on a global, regional and catchment scale. The penultimate section of the chapter, Section 4.5, presents the data sets that will be used for worked examples and case studies of other chapters. They include a global data set, representing a number of different hydrological regimes around the world, a regional data set, representative of a specific part of northern Europe and detailed data for a single catchment in central Europe (local data set). A brief summary of the chapter is given in Section 4.6

4.2   Types of data

As we have learnt from Chapter 3, the flow of water in a river and the behaviour of groundwater are essentially the result of the interaction between the natural processes and human activities upstream. Variations in river flow reflect changes, or events, occurring in the catchment. The key to hydrological analysis is to determine relationships between the observed river flows, the catchment processes (natural or otherwise) and events. While river flow data are vital for most hydrological analyses, data that describe the catchment processes and events are also very important. An introduction to some of these different types of data is provided below.

4.2.1   Time series data

For any quantity that varies in time, a sequence of readings, or measurements, generate a *time series* of data for that quantity. For instance, repeated measurements of rainfall at a single rain gauge generate a time series of rainfall data at the gauge. Other quantities that are regularly measured at specific locations to form time series include air temperature, solar-radiation, wind-speed, barometric pressure, humidity, soil moisture and groundwater levels. Repeated measurement of river levels and flows also produce time series data. However, in contrast to variables whose information refers to conditions at the location of the gauge, or instrument, only, river flow data represents the conditions of the entire catchment area upstream of the gauge. This, therefore, gives rise to the further definition of a *point measurement*, corresponding to conditions at a specific location (point) only, and an *areal measurement*, which represents the conditions of a particular area or region.

Once a time series has been established and a sufficient number of readings have been taken, the data can be manipulated to produce further time series of *aggregated* data. For example, a daily time series may be calculated as the mean, or sum, of 15-minute values during each day. As can be seen in Figure 4.1, the aggregation of daily data may produce further time series data;

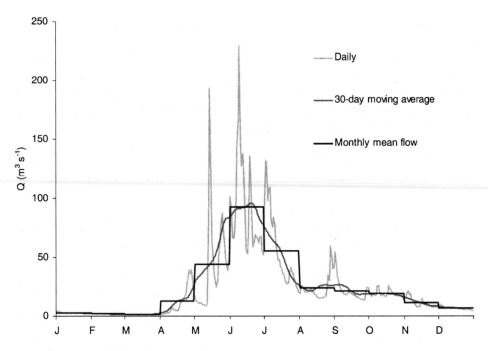

Figure 4.1 Aggregation of time series data.

monthly time series and 7-day, or 30-day, moving average data are often used in low flows analyses. A whole range of *statistics*, such as, means, maxima, minima, standard deviation, coefficient of variation, decile and percentile values, can also be derived for different periods covered by the time series to provide a useful summary of the data and form the basis of analysis.

4.2.2   Thematic data

Thematic data may be generically defined as information that can be used to describe the features and attributes of a certain entity. In hydrology, that entity would typically be a catchment or a region, or area, of interest. Thematic data that would describe such an entity may include name, location, area, shape, orientation, slope and elevation. Data describing the distribution of climatic variables (e.g. precipitation, evaporation, relative humidity) and other physiographical features (e.g. topography, soil types, geology, land use, vegetation) within the entity would also be classed as *thematic data*.

The classification of time-series and thematic data, however, is not necessarily exclusive. With climatological data, for instance, the areal distribution of several variables can be represented in map form by the interpolation of data from a time series of point measurements. The resulting map, in describing the distribution of the climatological variable within an entity (in this case, the entity is the area covered by the map), is, by definition, thematic data. Rainfall is one variable that is commonly represented in map form, using interpolation methods, such as the Thiessen (1911) polygon area averaging method, Inverse Distance Weighting (e.g. Barnes, 1973; Cressman, 1959; Sheppard, 1968), Thin-Plate Splines (Champion *et al.*, 1996) and Kriging (Dingman *et al.*, 1988). These methods can be used to derive map representations of instantaneous (e.g. air temperature at 9 a.m. today) or aggregated values (e.g. average annual precipitation) and, clearly, a sequence of such maps can result in a time series of thematic data.

In general, any data that can be represented on a map can be referred to as thematic data. However, it is sometimes convenient to distinguish the thematic data that can be represented spatially, in map form, by calling it *spatial data*. Topographical maps, showing the elevation contours, rivers and lakes and specialist maps, showing geological formations, the distribution of soils, vegetation and land use, are all examples of thematic data, commonly used in hydrological analyses, that are also referred to as spatial data.

Topographical data, featured as contours on a map, or represented by a digital elevation model, are particularly useful as they can be used to delineate the boundary of a catchment. The *catchment boundary* (Section 3.2.2) is one of the most important types of spatial data in hydrology, as it not only describes the areal extent of the catchment, giving the catchment area, but it can also be used to derive other pertinent *catchment characteristics* or *descriptors*. By overlaying the boundary onto other thematic data, many catchment descriptors, such as mean and maximum elevation, mean slope, aspect and the proportion of different soils, geological formations and land cover (e.g. forest, urban, lake), can be readily determined. By definition, catchment descriptors, as attributes of the entity, are also classed as thematic data.

Ideally, all spatial data would be available in digital form so that they could be conveniently analysed and manipulated in a Geographical Information System (GIS). Unfortunately, much of the information that a hydrologist would require is only available in hard-copy form on paper maps. Although paper-scanning technologies have improved vastly in recent years, manual digitizing remains the most effective way of preparing maps for use within a GIS.

A type of digital spatial data that is increasingly being used in hydrology is the data from remote sensing. Remote sensing images, obtained from Earth Observation Satellites and aerial photography, provide valuable data, often at a high resolution, on weather, climate and a variety of terrestrial and morphological features (e.g. topography, stream density, lake levels, land use, vegetation, snow cover, soil erosion and soil moisture). Orbiting the earth several times a day, satellites provide instantaneous snapshots of land features and processes, a sequence of which can be collated to provide a time series of thematic data.

4.2.3   Metadata

Another valuable form of thematic data is *metadata*. Essentially, metadata provides information, or data, on data (the entity). For time series data, metadata would include a definition of the variable being measured, the method of measurement, the type of instrumentation used, its age, when it was installed and calibrated. Details of the measuring authority, the individual responsible for the observation, telephone numbers, site directions, grid reference, and any other relevant information would make up the metadata for a single gauge or monitoring point. Similarly, information on how the time series data is processed, stored and disseminated would be considered to be metadata. For thematic data, metadata may comprise details of its derivation, such as source maps, digitizing methods, data collection methods, its ownership and copyright restrictions. International standards have been defined for the design and handling of metadata (e.g. ISO19115/TC211) and those developing hydrological metadata catalogues or databases are well advised to conform to these.

4.3   Hydrological time series data

In this section, information is provided on the acquisition and management of two main types of hydrological time series data, focusing first on methods of collecting river flow data, then those of groundwater data and, finally, on approaches to data quality control.

4.3.1   River flow data

Arguably, the most useful data in hydrological analysis are those which describe the flows (or streamflow) in a river. River flow "is the only phase of the hydrological cycle in which the water is confined in well-defined channels which permit accurate measurement to be made of the quantities involved" (Herschy, 1995). Ensuring that the observed flow data are an acceptably accurate and reliable representation of the actual river flow is essential for hydrological analysis.

The practice of measuring river levels and flows is called *hydrometry*. The point on a river where flows are regularly measured is called a *gauging station*. There are several types of gauging station, including velocity-area stations, weirs and flumes, ultrasonic stations and electromagnetic stations.

Velocity-area gauging stations

Velocity-area gauging stations are by far the most common as they are typically established at a natural section of the river and do not require extensive civil works for their construction. At a velocity-area gauging station, there is generally a unique relationship (the *stage-discharge relation*) between the observed water level, or *stage*, and the river flow, or *discharge*, which is determined by taking a series of instantaneous discharge measurements, or *spot gaugings*, and corresponding stage readings. The series of spot gaugings should be representative of the range of river flows that are likely to be encountered. The discharge values can be plotted against the corresponding stage and a smooth curve drawn between the points (Figure 4.2). The curve, which defines the stage-discharge relationship, is known as the *rating curve*. Once the rating

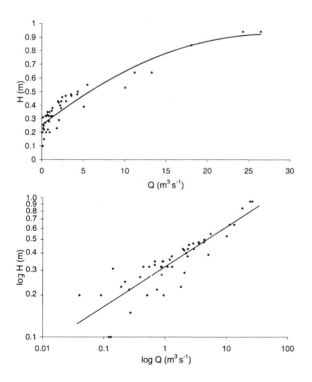

Figure 4.2 Rating curve; linear scale (upper) and logarithmic scale (lower).

curve has been obtained, continuous monitoring of stage at the gauging station provides a continuous record of discharge (river flow) data. The rating curve is represented with a relationship of the form:

$$Q = const \ (H + H_0)^b \tag{4.1}$$

This provides an estimate of the discharge, Q, for a simultaneous value of water level, H. The constant, *const*, the exponent, b, and the intercept (water level at zero flow), H_0, are determined from measurements. For convenience, the relationship is often linearized by taking the logarithm of Equation 4.1

$$\log Q = \log const + b \log(H + H_0) \tag{4.2}$$

The discharge measurements used in establishing a rating curve are most frequent around the average discharge but are often missing at extreme water levels. A single relationship does not necessarily apply to all conditions and may be significantly different during periods of low flows. The rating curve can, therefore, comprise several linear segments with each segment being valid for a particular range of water levels and flows.

To ensure the discharge measured at the gauging station is not affected by flow conditions further downstream, a *station control* is usually present at, or some short distance downstream of, the gauging station. Station controls can be natural or artificial. Waterfalls, rapids and a widening of a river are types of natural control, while a low dam is an example of an artificial control (WMO, 1994). Such controls generally provide a stable stage-discharge relation in periods of low flows.

The most common approach to spot gauging involves the use of a *propeller-type* (or *cup-type*) *current meter* to measure the stream velocity at certain depths and at appropriate intervals across a river. These mechanical types of current meters tend to be insensitive to variations at the lowest flows, as stream velocities approach zero. In such circumstances, the use of an *electromagnetic current meter* ought to be considered. For large rivers, the Acoustic Doppler Current Profiler (ADCP) may be used. The ADCP integrates the velocity profile across the river to obtain the total discharge for the cross-section. Another method of instantaneously measuring discharge, that is particularly applicable in shallow or turbulent streams, is *dilution gauging*. By adding a suitable tracer (chemical) to the river some distance upstream, its subsequent downstream concentration gives a measure of discharge.

Weirs and flumes

Weirs and flumes, by generating a critical flow control within their structure, provide a reliable stage-discharge relation, which can be determined theoretically. There are several different types of weirs and flumes. *Thin-plate weirs*, generally installed in smaller streams, include full-width weirs, rectangular-notch weirs, v-notch weirs and compound weirs, which comprise a combination of all three. On larger streams and rivers, *broad-crested weirs* or *flumes* may be seen. The three most common types of broad-crested weirs are the triangular-profile (or Crump) weir, the flat-v weir and the rectangular-profile weir. Flumes typically comprise an approach channel, a constriction (throat) and exit channel having a rectangular, trapezoidal or u-shaped profile. Despite the theoretical hydraulic behaviour of these structures, the accuracy of the rating curves should always be verified by flow measurements in the field and corrected as necessary.

Ultrasonic and electromagnetic gauging stations

Ultrasonic gauging stations and electromagnetic gauging stations are two further types of gauging stations that are increasingly being used. *Ultrasonic gauging stations* have sets of submerged transmitters and receivers set obliquely on either side of the river. The average velocity of flow is determined by recording the difference in time that it takes ultrasonic pulses, travelling in different directions, to cross the river. By measuring the corresponding depth of water and the cross-sectional area, the discharge can be calculated. An *electromagnetic gauging station* has a large electromagnetic coil installed over, or under, the river. As water passes through the coil's magnetic field, an electromagnetic force is induced (and measured), which is proportional to the discharge.

Stage measurements

Measurement of stage is required with all types of gauging stations. The accuracy of the stage measurement has a direct impact on the quality of the flow record. This is especially so during low flows, when a relatively modest error in the measurement of water depth results in substantial errors in the computed discharge (Marsh, 1999). The most basic method of measuring stage is by means of a staff, or reference, gauge. The *staff gauge* is usually a graduated metal measuring plate installed at the side of the river. The stage of the river is simply read off the staff gauge and recorded to the nearest centimeter.

At many gauging stations a *stilling well* provides a sheltered environment for the measurement of stage. Commonly there is a float attached to a graduated metal tape that passes over a pulley connected to a suitable recording device. This arrangement is called a *float-tape gauge*. The water level in the stilling well may also be measured by use of an *ultrasonic water level gauge* or a *pressure transducer*, either of which are generally more sensitive to variations in water levels than measurements from a staff gauge or float-tape gauge.

The stage readings at gauging stations are recorded by a variety of means, including *manual recording, autographic (paper) charts* and *digital data loggers*. Manual observations may be taken routinely once, twice or three times a day. Autographic recorders provide a continuous trace of stage onto a paper chart. Stage measurements are read from the chart at quarter-hour, hourly or three-hourly intervals either manually or by digitizing. Digital loggers usually store readings at 5 or 15-minute intervals, although some can be triggered to record at higher frequencies if a rapid change in stage is detected or if the stage exceeds a certain value. Digital recording can record to an accuracy of about a millimetre and is preferred where the monitoring of low flows is important. Some digital loggers are connected to telemetry (i.e. telephone, radio or satellite communications), which allows readings to be transmitted in real time, at prescribed times each day, or on interrogation.

Factors affecting the stage-discharge relation

In all types of gauging stations a number of factors can affect the stage-discharge relation. For instance, the erosion, or deposition, of *sediment* can alter the cross-section of the river and, hence, the relation. The build-up of aquatic *vegetation* (weed growth) reduces the cross-sectional area of the river, increases the flow resistance, and can cause an overestimation of discharge. This is a particular problem for low flows as, in many parts of the world, the season of lowest flows (summer) coincides with the period of greatest weed growth. If removal of the vegetation is not possible a different stage-discharge relation may be required for the different phases of growth, and reliability of the data can only be improved by increasing the frequency of spot gaugings.

The partial formation of *ice* has a similar effect to vegetation because it reduces the area of the cross-section. Again, this is an important consideration when analysing the low flows of catchment that experience winter minima (e.g. mountainous or high-latitude catchments). *Backwater* effects may further influence the relationship between water level and discharge. A frequent cause is vegetation or ice in the river channel, tidal effects from the sea or backwater from downstream tributaries.

When such factors render the normal stage-discharge relationship inapplicable, data infilling or revision of provisional flows is often required. Intensive current metering programmes may be required to establish a family of ratings or provide a guide to the stage adjustment necessary to allow the normal stage-discharge relation to be used. Under such circumstances a comparison with data from a similar catchment allows a more cost-effective first estimation of the low flow sequence to be made (Section 4.3.3). This employs the fact that runoff rates during drought normally display a considerable degree of spatial coherence.

Measurement at low flows

Over the last 30 years, major developments in flow measurement, water level sensing, data recording and data communication technologies have had a beneficial impact on the way river flow data during low flows are acquired, processed and archived. The increasing use of new measurement techniques, and a fuller understanding of the hydraulic behaviour of weirs and flumes, has enabled low flows to be determined with greater precision. However, many factors can combine to reduce the accuracy of processed or archived flow data. Generally, the factors of most significance to low flows are the uncertainties associated with the measurement of very limited water depths, the disturbance to low flow regimes caused by an increasing range of artificial influences, and the failure to adjust stage levels to account for backwater effects.

In order to reduce the uncertainty associated with computed flow values, especially in the low flow range, gauging stations are commonly sited where any significant change in discharge is accompanied by a substantial change in water level. Thus, by exploiting natural channel characteristics, or by the installation of a gauging structure, attempts are made to maximize the sensitivity of the measuring station. Even so, the limited water depth which normally accompanies low flows places a premium on reliable water level sensing and recording. Where hourly or 15-minute recording intervals have been adopted, random errors in computed daily mean flows tend to be very low (Herschy, 1995). By contrast, systematic bias in measured river levels, caused, for example, by algal growth on weir crests or datum errors, can be substantial and difficult to eliminate. For a notional 15 m wide rectangular channel, Figure 4.3 illustrates how a modest systematic error in the determination of

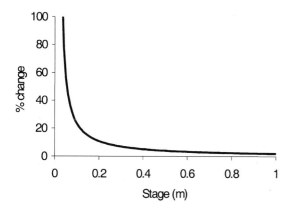

Figure 4.3 Change in flow (%) resulting from a 10 mm increase in stage (derived for a rating curve equal to $Q = 30H^2$).

water depth can result in a substantial error in computed flows. In small rivers, or those with exceptionally low dry season flows, systematic errors in stage measurement can dominate the overall uncertainty in processed river flow data. In the UK a systematic stage error of 5 mm at the Q_{95} flow translates into a flow error of at least 10% for more than a third of the gauging station network (Marsh, 2002). Hydrometric standards need to be maintained at a high level if confidence is to be placed in computed flow values, particularly those likely to be experienced during periods of drought.

4.3.2   Groundwater measurements

The behaviour of *groundwater* is often of critical importance for investigating the status of the subsurface in a catchment and for determining the low flow response of a river (Section 3.4). Data that describe variations in groundwater hydraulic head and groundwater recharge and discharge are, therefore, valuable for drought analysis. While hydraulic head and groundwater discharge can be readily measured, it is practically impossible to directly gauge groundwater recharge.

Data on groundwater hydraulic (elevation + pressure) head are obtained using piezometers. A *piezometer* is a non-pumping tube, or standpipe, in which the elevation of a water level can be measured (e.g. Freeze & Cherry, 1979; Schwartz & Zhang, 2003). The bottom tip of the piezometer is fitted with a perforated screen, 0.5 to 1.0 m long, to allow inflow of groundwater. Groundwater entering the piezometer rises up the pipe and reaches some position characterizing the hydraulic head. Piezometers are sometimes called observation wells. An *observation well* is a special piezometer that measures the position of the phreatic water table with a standpipe having a longer screen (e.g. 2 or 3 m). The bottom of the screen should be below the lowest level of the water table and the top should be above the highest level to allow free inflow of groundwater over the whole range of the water table fluctuation. With a diameter of between 0.1 and 0.2 m, an observation well is generally of larger diameter than a piezometer, which typically has a diameter of around 0.05 to 0.1 m. Piezometers are used to measure the hydraulic head of deep groundwater or the head of shallow groundwater that has a distinct vertical (head) gradient. Comparisons between individual conventional piezometers at different depths (*nested piezometers*) with a long-screened piezometer have shown that the long screen may act as a vertical conduit for water to flow, thereby neglecting the vertical gradients of the hydraulic head in an aquifer. Heads of shallow groundwater measured at long-screened piezometers tend to be too high in recharge areas and too low in discharge areas (Section 3.2). This effect is also seen in *boreholes* and *hand-dug wells* (i.e. wells of relatively large diameter having a long screen or no standpipe at all in consolidated aquifers, such as, chalk or sandstone). Nested piezometers provide reliable measurement of the hydraulic head distribution, especially in *anisotropic aquifer systems* (i.e. where the hydraulic conductivity and storage characteristics are not uniform in all directions) or in areas where significant vertical groundwater flow occurs. Rapid and accurate water table measurements are best made in small diameter observation wells. If the diameter is too large, the volume of water contained in the well may cause a time lag in water table changes, although this is not so critical during drought because of the slow decline of the groundwater level. Large-diameter wells show a delayed response of the water table to groundwater recharge and, consequently, tend to overestimate the duration of a drought. Both automated-recording and manual-operated instruments are available to measure the groundwater hydraulic head in piezometers and the water table in observations wells.

Automatic recording of the groundwater hydraulic head

Automatic recording of the hydraulic head is common in many countries because it is less laborious than manual monitoring. Over the last decade, methods based upon the *pressure transducer* have increasingly been used. With a pressure transducer the water pressure is converted to an electrical signal, which is stored by a *data logger*. Although pressure transducers are able to take high-frequency measurements, one measurement per day is generally sufficient for drought studies. Pressure transducers are typically accurate to 1 cm, or less, and have the capability to store data for several months. In some regions a float-tape gauge is used as an alternative to the pressure transducer. It is important to regularly check the friction of the pulley to prevent a staircase appearance to the resulting data. Furthermore, the length of the cable and counterweight pipe should be sufficient to avoid the counterweight reaching the pulley during drought. A well-maintained float-tape gauge can be as accurate as a pressure transducer.

Manual groundwater head monitoring

Regular manual monitoring of the groundwater hydraulic head, or water table, is still needed, even where data are collected automatically. Readings of a pressure transducer or mechanical float recorder should be compared with a manual observation at least once a month. In countries where labour is not too expensive, manual monitoring is often the most practical method of collecting groundwater hydraulic head levels during drought.

A popular manual method is the *dipper*. The dipper is a cylindrical probe with a hollow space at the end, which is connected to a plastic-coated tape or flexible steel cable. If the dipper reaches the water level an audible signal is produced. The length of tape or cable used equates to the groundwater head, or water table level.

A more sophisticated device is the *two-electrode device*, which consists of a portable reel having a plastic-coated tape, or cable, that is connected to a probe at one end. The probe has two adjacent electrodes such that, as soon as the probe reaches the water level, a circuit is made to activate a light bulb or buzzer at the reel. The depth of the water table is read from the length of tape or cable used. These instruments can also be used for artesian groundwater, where the hydraulic head rises above the soil surface, as long as the head is not more than 1.5 to 1.8 m above the surface. If the head of the artesian aquifer is more than 1.5 m above the surface a pressure transducer needs to be used. Otto (1998) and van Lanen (2003) review the maximum depth, accuracy, required skills and costs of each observation method, while Jousma & Roelofsen (2003) give an extensive bibliography of groundwater monitoring tools.

Monitoring groundwater discharge

Groundwater can be discharged either diffusely, in a brook, stream or river, or, more concentrated, in a spring. Consequently, the measurement of groundwater discharge is identical to conventional streamflow measurement, as described earlier in Section 4.3.1.

4.3.3   Quality control

Quality control should begin well before data is delivered for processing. It should be integrated into every step of data collection, as well as in the data processing. In this section the various aspects of data quality control are considered, primarily in respect of the processing of hydrometric data. Nevertheless, the same principles apply to the quality control of groundwater data and most other types of time series measurements.

The nature of errors

Even with the most thorough of quality control procedures, *errors* will always be present in data. An error is the difference between the measured quantity and its true value. With hydrometric data, errors are usually *random* (or stochastic), *systematic* or *spurious* (Herschy, 1995). A random error is the type of error that is unavoidable whenever any type of measurement is made. If a large number of measurements are taken, the random error will tend to be normally distributed around the true value of the quantity being measured. A systematic error is the type of error that appears consistently in a set of measurements, due to a problem in the method of measurement. For example, an incorrect stage-discharge relation due to weed growth, or a problem with the instrumentation, would introduce a systematic error in a flow record. A spurious error is one that is introduced accidentally into the data, most often by human error or equipment malfunction.

Little can be done to eliminate random errors, although the extent of uncertainty varies according to the method of measurement and the characteristics of the cross-section. Systematic and spurious errors, however, can be minimized with good practice. Where such errors are detected in the data, corrections should be applied and be properly recorded.

Good practice

Compliance with International Organization for Standardization (ISO) standards for data collection generally ensures good practice, thus helping to minimize systematic and spurious errors occurring in data. Some of the ISO standards commonly referred to include ISO 748 (Velocity-Area Methods), ISO 1100/1, (Establishment and Operation of a Gauging Station) and ISO 1100/2 (Determination of the Stage-Discharge Relation). In addition, adequate and proper maintenance of gauging structures and their instrumentation are essential to ensure that water levels are correctly recorded and are representative of flow conditions at the respective sites (Marsh, 1999).

As well as supplying raw data, it is important that those responsible for data collection in the field provide up-to-date information on methods of collection and report any deviations to those responsible for processing and storing the data. For every gauging station, a record should be kept of the *gauging methods* employed, changes in *instrumentation* and, for hydrometric data, *past spot gaugings* and the *history of rating curves*. Details should also be kept of the *factors that have affected the record*, such as changes to site conditions and activities in the upstream catchment. Quality assurance books, detailing how to undertake data collection and quality control, have been developed by many national data collection agencies.

Data validation

As soon as a data set is received for processing, its identity should be verified and linked to the correct gauging station. An audit-trail should be associated with the data set to record all actions that are subsequently performed. The 'raw' time series of stage data that is collected from the gauge is initially pre-processed into a usable format, as dates, times and levels, and stored digitally on a computer in text or binary files, in a spreadsheet or on a database. Data digitized from autographic charts (Section 4.3.1) should be corrected at this juncture for minor errors in time or levels.

Once in a usable format, the raw stage data should be checked to see if it falls within an expected range of values and whether the difference between two consecutive readings is acceptable. These limits of allowable *range* and *rate of change* are defined according to knowledge and experience of local site conditions and may vary during the year. *Visual checking* of graphical plots of stage data is another useful method of data validation that should be conducted prior to their conversion to flows.

Application of the stage data to the relevant stage-discharge relation (rating curve) results in a *time series* of *river flow data*. A record should be made of the rating curve that was applied and the raw stage data should be stored intact, in case it needs to be revisited later.

The derived flow data should be checked as soon as possible. One of the initial checks is to compare the new data with past records. A *hydrograph*, a graphical representation of the variation in flow against time (Figure 4.4), can be used to identify a *discontinuity*, or jump, between the end of the previous record and the start of the current. This may indicate a possible mistake in the *identity* of the *calibration* of the gauge, the data or the *units of measure* (e.g. $m^3 s^{-1}$, $1 s^{-1}$, $ft^3 s^{-1}$, mm). A hydrograph may also be used to identify *spurious peaks* and *troughs*. The flow data should also be compared to limits of allowable range and rate of change. A useful technique is to compare the data within an envelope of the corresponding long-term daily maximum and minimum flows (Figure 4.4); the envelope helps to direct attention to the most hydrologically significant low flows – those which approach, or exceed, the extreme recorded minima for the target station. Figure 4.4 shows some typical problems that may be observed using a hydrograph, such as:

a) a step change (shown here in January), which would normally be a problem of calibration but could also be due to a change in the measurement unit;

b) missing data entered as zero flows;

c) missing data (in April);

d) an isolated erroneous peak (in May);

e) an artificial increase in flow (in June/July) caused by weed growth, followed by a steep decrease after weed cutting;

f) an unrealistic steepening of the recession (in late July and August), normally due to unchecked extension to the rating curve, with the 'staircase' effect reflecting limited stage resolution at low flows;

g) truncated peak flows (in November and December), where the processing threshold is reached by one, or more, sub-daily levels.

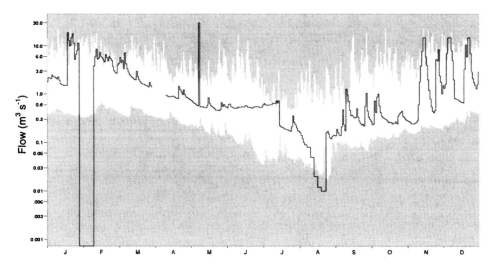

Figure 4.4 Data validation using a hydrograph.

Anomalies and errors that are detected should be examined carefully and, if necessary, removed or corrected. Any corrections, or adjustments, should always be made in a timely manner. They may be applied manually or according to an appropriate *infilling* method (see below). Whenever an individual data item is altered, a simple quality control flag should be associated with it to signify that it has been changed and, if possible, to describe the method of correction.

The accuracy of the recorded water levels is particularly important for the quality of low flow data. For example, the recession portion of a hydrograph may occasionally be shaped as a staircase. This is the effect of inadequate vertical resolution in the measurement of stage that, typically, occurs where the water level has been read to the nearest centimetre from a staff gauge.

Hydrological validation

The *comparison of data* from another nearby gauging station is a valuable method for validating a new data record hydrologically. A graphical plot (e.g. hydrograph) of the two data records should show similar, if not identical, behaviour over the same period. Any variances between the two records indicate possible errors. Additional information for the validation can be derived from time series of catchment precipitation (especially in the rainy season) and air temperature (in winter and during snow-melt season).

A *double-mass curve*, which plots the accumulated values of two time series against each other, is another useful validation check (Figure 4.5). The

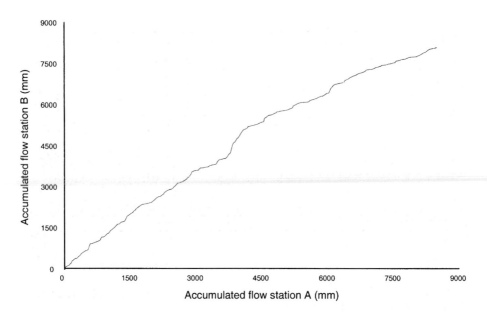

Figure 4.5 Double-mass curve.

double-mass curve constructed from the flow records of two gauging stations in the same vicinity should approximate a straight line. With this method, the two time series need not necessarily be of the same variable; data from a rain gauge within the catchment may also be used to check the flow record. A similar method that can be used is the *residual-mass curve*, which plots accumulated departures of the two time series from some datum (normally the mean).

Missing data

Missing data, or gaps, are a common problem that tend to cluster disproportionately in the extreme flow ranges (Marsh, 2002). Even a small proportion of missing data can greatly reduce the ability to derive meaningful summary statistics (e.g. annual 30-day minima). Data validation and infilling procedures should be applied vigorously to low flow data sequences. Judgement needs to be exercised when infilling missing low flows, to avoid archiving misleading flow estimates. It will not always be possible to derive realistic flows to infill lengthy sequences of missing data, but in many circumstances the inclusion of auditable and flagged estimates, rather than leaving a gap in the record, will produce significant benefits in relation to the overall utility of the time series.

    Experience demonstrates that hydrograph appraisal undertaken by personnel familiar with the behaviour of individual rivers is the most effective

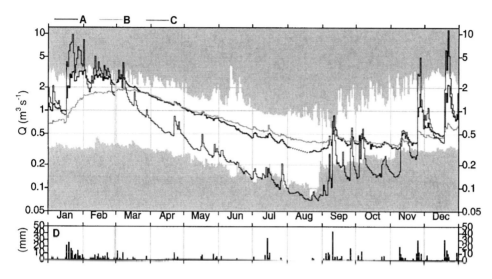

Figure 4.6 Infilling of missing data at target A with information from analogue stations (B, C) and a representative rain gauge (D).

means of infilling missing low flow data. Short gaps, when there is no indication of a flood event, may be filled *manually* (by 'eye'), by *linear interpolation* or by *polynomial fitting* with values from the time series itself. Longer gaps can be filled by use of data from an *analogue gauging station, or stations*, or by simulation with an appropriate *hydrological model*.

Comparison with one, or more, time series for nearby (or analogous) gauging stations, together with daily rainfall totals for a representative rain gauge within the catchment, provides the opportunity to derive missing flows estimates. In the example shown in Figure 4.6, a logarithmic scale has been used to help emphasize the low flow range. Flows have been plotted as daily runoff totals, making for more direct comparison between the gauging station flows. The limited rainfall over the period for which flows are missing allows estimated flows to be inserted with confidence. Where rainfall may be expected to influence the recession, or where artificial influences may be significant, the time series should be scanned for analogous flow sequences which would help to reconstruct the missing data.

An analogue gauging station would usually be upstream or downstream of the gauging station concerned (the 'target gauging station') or be in a nearby catchment. The choice of analogue stations should be determined according to the following criteria:

a) geographical proximity of the catchments, to ensure that the catchments are in climatologically similar areas;

b) similarity of runoff generation mechanisms, similar soil and hydrogeological conditions, approximately equal proportions of lake, forest, swamp or plough land area in the catchments, best indicated by the base flow index (BFI) of each (Chapter 5);

c) similarity of catchment area, elevation and topographical relief;

d) absence of artificial influences significantly affecting the natural flow regime (such as river regulation, sewage or industrial effluent, or intakes for irrigation or other needs).

As well as BFI, which is an index of catchment response, the coefficient of correlation (calculated from the flow series over an identical period of both target and prospective analogue stations) gives a good indication of the suitability of an analogue station.

One of the simplest methods of infilling missing data, using one or more analogue gauging stations, is to *visually compare* the hydrographs of both target and analogue stations and manually enter estimated values that are sympathetic with the observed behaviour. Alternatively, a simple relationship between the flows at the target station Q and the flows at the analogue station Q_a can be used:

$$Q = Q_a k \qquad\qquad\qquad (4.3)$$

This simple method is based on the assumption that the ratio of the flows, k (i.e. Q/Q_a), does not change suddenly in time. Values of k are calculated from observations both before and after the gap. Linear interpolation is then used to calculate k values of each missing interval within the gap. Finally, the missing flow values are calculated according to Equation 4.3. The method is illustrated in the following worked example for River Metuje in the Czech Republic (Section 4.5.3).

Worked Example 4.1: Filling gaps in daily flow series (CD)

Water stage records from stations located on the Upper Metuje are continually converted to daily flows. Artificial impacts on the flows are insignificant. In the 1981 water year (1 November to 31 October), the water stage recorder at the M VIII gauging station was out of order on three different occasions and observations are missing. To infill these gaps in the daily time series, a relationship between the flows at this station and another downstream (M XII) on the Metuje was used.

The basin areas of the Metuje are 41 km^2 and 73.6 km^2 at the M VIII and M XII stations respectively. The length of the river between the two stations is 6.4 km, and consequently the lag-time between the flows at the two stations could be neglected. It was also assumed that, for the individual days, the runoff from the incremental area between the two stations (the inter-basin) is proportional to that from the whole basin above the M XII station, so that:

$$Q_{M\,XII} - Q_{M\,VIII} = z\,Q_{M\,XII} \qquad (W4.1.1)$$

$$Q_{M\,VIII} = Q_{M\,XII} - z\,Q_{M\,XII} \qquad (W4.1.2)$$

$$Q_{M\,VIII} = k\,Q_{M\,XII} \qquad (W4.1.3)$$

where:

$$k = 1 - z = Q_{M\,VIII}\,/\,Q_{M\,XII} \qquad (W4.1.4)$$

The inflow from the inter-basin cannot be negative, therefore, $k < 1$. With respect to the fact that the missing series are relatively short, k can be calculated from flows immediately before and after the gap and the missing flows calculated using linear interpolation.

The individual steps of the calculation are as follows:

1. Values of k are calculated from the observed series by using Equation W4.1.4.

2. k values for the first day after the gap $k(f)$ and the last day before the gap $k(0)$ are used for calculating the difference (dk) between the two k values (if some of the k values are extremely high or low as compared to the surrounding values (outlier), the arithmetic mean of k values from several consecutive days can be used).

3. k values for individual days during the gap are calculated by linear interpolation:

   $$k(i) = k(0) + i\,dk1 \qquad (W4.1.5)$$

   where:

   $i = 1$ to $(f - 1)$ is the rank number of the days between day 0 (last day before gap) and day f (first day after gap),

   f is the number of days in the gap plus 1,

   $dk1 = dk/f$ is difference in k expressed proportionally for each day.

4. Equation W4.1.3 is used for calculation of the flows for each day $i = 1$ to $(f - 1)$, as shown in Table 4.1 below.

Table 4.1 Infilling data for the River Metuje, Czech Republic

Date	Daily flow (l s^{-1}) $Q_{M\,VIII}$	$Q_{M\,XII}$	Calculation step 1 k	2 dk	i	3 $dk1$	$k(i)$	Results 4 $Q_{M\,VIII}$
15/11/80	263	719	0.3654		0		0.3654	263
16/11/80		2368			1		0.4407	**1044**
17/11/80		2322			2		0.5160	**1198**
18/11/80		2681			3		0.5913	**1585**
19/11/80	1337	2005	0.6666	0.3011	$4 = f$	0.0753	0.6666	1337

An even simpler approach, which is often used, is to scale, by the ratio of respective catchment areas, the data from the analogue catchment to the target. An alternative, more complex yet effective method, involves calculating the flow percentile for each missing day from the flow record of an analogue catchment and then extracting, from the existing data record of the target gauging station, the flow values that correspond to the flow percentiles.

All of the methods described above assume synchronicity of flows between target and analogue catchments. It would be inappropriate, therefore, to apply them to sub-daily data (e.g. 15-minute or hourly flow data) or where one of the catchments is much further downstream of the other, without accounting for the lag-time between the gauging stations. The attenuation of the flow in the river channel between the stations has also to be taken into account when infilling flow data according to this method. The effect can be serious when infilling flood data, but has minimal consequences for low flows data, unless there is a lake between the stations. Further information on these empirical methods is available in Section 8.2.

For groundwater level data, equipment failure (i.e. failure of the pressure transducer or float-tape gauge) leads mostly to gaps in the observed record. During droughts, however, gaps can occur when observation wells run dry as a result of the base of the standpipe being shallower than the lowest level of the water table. Short gaps can readily be filled in by linear interpolation, assuming no recharge occurs to cause the water table to rise in the interim. Multiple regression is applied to infill longer gaps, or where recharge occurs, with missing values calculated using observations from adjacent piezometers.

Coping with artificial influences

The continuing development of water supply and sewerage systems, irrigation schemes, land drainage, land use change and hydro-electric power generation, all combine to disturb the pristine relation between rainfall and river flow. The net effect on low flow patterns is often profound and (unadjusted) gauged flows can be very unrepresentative of the natural response.

In highly developed catchments, the artificial component of flow may exceed the natural contribution, particularly at low flows. Information on the location, volume and timing of artificial influences is vital for water resources managers to allocate freshwater to users and preserve the ecological quality of the rivers during periods of low flows or drought. The *water user* may sometimes monitor major abstractions and discharges on a daily or sub-daily basis. Access to this data is useful for identifying the effect of the artificial influences on the natural flow regime and for the management of water resources. In some countries, water use is regulated within a legal framework, where licences, or consents, are issued to those who need to abstract water (from a freshwater body) or discharge effluent (into a freshwater body). Such licences, typically, prescribe limits on the volume of water (or effluent) to be abstracted (or returned) at a particular site on an annual or monthly basis. Users may be obliged to periodically submit information on the volume of water

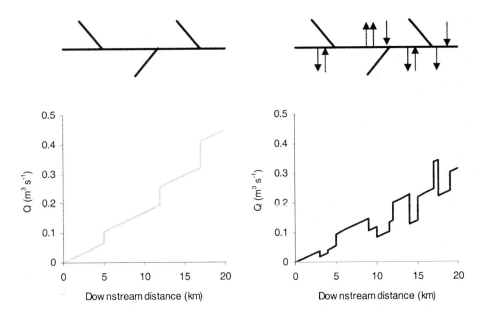

Figure 4.7 Flow accretion diagram (left) and residual flow diagram (right).

actually used to the regulatory authority. Where there is no direct monitoring, these data are estimated by the user. Similarly, reservoir operators are required by licence to follow a certain regime of regulated releases (compensation flows) to ensure that at least a minimum flow is retained in the river downstream. They, too, periodically submit data of the releases to the regulatory authority. The information provides a good indication of past and future water use within a catchment and helps the water manager to determine appropriate operational responses, or strategies, for maintaining supply.

Residual flow diagrams and flow accretion diagrams, as shown in Figure 4.7, provide a useful tool for identifying the effects of artificial influences within a catchment (see also Figure 8.2). The diagram shows how, for a given flow statistic (the mean flow in the example shown), the estimated flow increases, or decreases, along a river: first, when *no* artificial influences are considered (flow accretion diagram, left); and, second, when artificial influences *are* considered (residual flow diagram, right). Step-changes are seen where a tributary, abstraction or discharge occurs. Such diagrams are particularly useful for identifying reaches where augmentation or compensation flows are required to maintain the necessary flows in a river (Environment Agency, 2002a).

Flow naturalization

In some countries attempts are made to adjust gauged daily or monthly flows to account for the largest and most easily quantified artificial influences (e.g. non-returning abstractions above the gauging station). Normally, no attempt is made to account for the, often subtle, impacts of land-use change but the longest of such naturalized series can be exceptionally valuable – e.g. in relation to the detection of trends in low flows.

The process of distinguishing the natural flow component from the artificially influenced flow record is called *naturalization*. Resources are seldom available to undertake flow naturalization for a comprehensive network of gauging stations, and the process is usually only conducted where the number of good quality flow records is insufficient for an analysis to proceed without the naturalized record. Essentially, naturalization involves the adjustment of the observed (artificially influenced) flow record according to the known artificial influences upstream. The procedure requires a *systematic record* of the influences, including times, rates and durations of abstractions, discharges and compensation flows. The data should be in *compatible units* and an appropriate method to *temporally disaggregate* the data should be applied, as necessary (e.g. from monthly to daily). On any day, the natural flow, Q_n, at a gauging station can be approximated by simply adjusting the observed flow, Q, according to the net upstream artificial influences (compensation flows plus

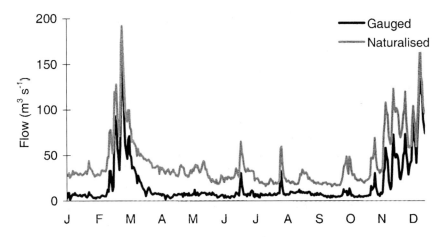

Figure 4.8 Disaggregation of river flows into artificial and natural components (River Thames at Kingston, 1997).

discharges minus abstractions). It should be noted that there could be a high degree of uncertainty to the approximation; while, typically, there may be an error of ±5% in the observed flow, the error associated with the artificial influence can be around ±40%, or higher. An example of a naturalized hydrograph for a heavily influenced catchment in the UK is shown in Figure 4.8. Further details of methods to naturalize river flow data are presented in Section 9.7.

Observed *groundwater* hydrographs may be naturalized to assess the impact of artificial influences using statistical tools or groundwater models. Assuming the affected area is not too large, multiple regression of water table elevation data from observation wells outside the affected area (the depression cone) can be used to calculate the water table without abstraction. Time series analysis can also be applied to detect a trend in the groundwater hydrographs due to human influences. Furthermore, various types of groundwater models have been developed to simulate time series without artificial influences. In Chapter 9, the application of such models is illustrated with respect to the effects of, for instance, land-use change and urbanization.

Network design

The gauging stations that comprise most existing hydrometric networks were originally installed for a variety of water resources management purposes, such as, to monitor water use, for flood forecasting and management, for long-term planning and design, or for environmental monitoring and impact assessment. Accordingly, networks tend to be heterogeneous in respect of the gauging

Table 4.2 Strategic categories of a national gauging station network (from Bradford & Marsh, 2003)

Category	Main objectives
Benchmark	Identify and interpret hydrological trends – principally climate-driven
Artificial impacts	Monitor heavily impacted catchments to establish the scale of disturbance to the natural flow regime. Demonstrate the effectiveness of water management strategies and remedial measures
Regionalization	Underpin the development of regionalization techniques and modelling procedures. (Not all 'regionalization' stations will be suitable for both high and low flow analyses)
Integrated monitoring	Provide a focus for the improved understanding of hydrological processes from the sub-catchment to the basin scale

methods, the instrumentation and the approach to data management (Rees *et al.*, 1996). With escalating costs and reducing budgets, hydrometric agencies in many countries are now being asked to strategically review the design of their networks and assess the value and utility of the data collected.

In any monitoring programme, data from individual gauging stations contribute unevenly to the understanding of low flow behaviour in individual countries. Those with the best hydrometric performance, least disturbance to their natural flow regimes and the longest records are usually the most strategically valuable. Formal identification of such catchments within a broad station classification programme allows low flow analysts to select data sets appropriate for particular projects. The classification programme should recognize that a large proportion of low flow estimates are required for ungauged sites (Chapter 8) and that the development of suitable models to address this need relies heavily on those datasets which best characterize the natural response of the likely range of catchment types to be found in particular countries or regions.

Table 4.2 outlines four categories which may be used to classify individual stations in a network. The categories are not mutually exclusive: many well-gauged catchments will qualify for selection in several categories. Of particular importance for drought hydrology are the *benchmark* and *regionalization* categories. Many countries maintain benchmark gauging station networks incorporating catchments carefully selected to include only those where the effect of man's activities on the flow regime is non-existent (or minimal). The maintenance of such networks depends on galvanising long-term support for essential monitoring programmes (Rodda, 1998) but such networks have assumed an enhanced status given the probability that climate change will significantly alter natural flow regimes. Further guidelines on the design of hydrometric networks are presented in the WMO *Guide to Hydrological Practices* (WMO, 1994).

Geostatistical approaches are used to design *groundwater networks* (WMO, 1994; van Lanen, 2003; Jousma & Roelofsen, 2003). Network density is often defined according to the required accuracy of the spatial interpolation, such as the standard deviation of the spatial interpolation error. A suitable spatial interpolation technique (e.g. kriging) is needed that provides not only estimates of the groundwater head, but also of the standard deviation of the estimation error. Stein (1998) demonstrates that, for an area with 0.8 well per km^2, uncertainty can even decrease if 10% of the wells are removed and 1% of new wells are introduced at strategic locations. Time series analysis is used to determine monitoring frequency, in which auto-correlation and non-stationarity are accounted for. Under particular conditions, there is a trade-off between sampling density and frequency. In some regions, a stochastic-deterministic approach is applied that combines Kalman filtering and a physically-based groundwater model. In regions sparse of groundwater data, background monitoring, based upon expert knowledge, should be first undertaken.

4.4 Thematic data at different spatial scales

In Section 4.2, an introduction was provided to the types of thematic data that may be used in low flows and drought analyses. Here, further information is given on the data that is typically available at global, regional and local scales. These represent the scales at which hydrological studies are often conducted.

4.4.1 Global scale data

With the recent rapid development of Geographical Information Systems (GIS) and Earth Observation Systems (EOS), spatial data that describe the physical features of the earth and its atmosphere are becoming more readily available to hydrologists. The type and volume of global spatial data that is available is vast, ranging in scale from degrees of longitude and latitude down to tens of metres on the Earth's surface.

Many *global data* sets are now freely available over the *world-wide web* (WWW). Global data that are of interest to a hydrologist can be broadly categorized as *climatological* or *physiographical*. Global climatological spatial data are generally derived either from interpolating ground-based meteorological point measurements or from satellite observations. Variables that are described globally include precipitation, air temperature, vapour pressure, sunshine hours, cloud cover and wind speed. Global physiographical data are derived from maps or remote sensing. There is a plethora of such data available at a variety of scales and resolution, including topographical data (e.g. digital elevation models), drainage patterns (rivers and lakes), land cover (or land use), soils, geology, snow and ice cover. Global-scale thematic data are prepared for use within a GIS and may be presented as *points*, as *grids* or in the form of *lines* or *polygons* (e.g. isotherms, contours).

4.4.2   Regional scale data

For regional scale hydrological studies, the drainage basin (catchment) is the most convenient unit to consider, since the derivation of catchment properties is related to that unit. There are several *catchment properties* that are relevant for low flow analyses, including *physiographic* (*geomorphologic, land cover, soil and geology*) and *climatic descriptors*, with each capable of having a number of different indices to describe their properties. For instance, there are several different ways of indexing precipitation: annual rainfall, seasonal rainfall or statistics of extreme rainfall. Therefore the selection of significant catchment properties in establishing models to estimate low flow indices usually depends on the scale of the problem, the data available and experience of the user.

Usually the problem under consideration suggests the temporal and spatial scale of required catchment descriptors. The catchment descriptors may be derived from several maps of different scale. For example, 1:500 000 and 1:1 000 000 scale maps are generally suitable for the presentation of rainfall data in regional analyses. For drought forecasting and modelling at a catchment scale, *physiographical properties* are derived from topographical, hydrogeological and soil maps at a scale of 1:50 000 or larger. *Stream channel* and *slope* should be extracted from maps of scale not smaller than 1:50 000, assuming a digital river network is not available. The *scale* at which most of the hydrological problems are investigated depends on the size of a catchment and the desired accuracy of the assessment. Catchment boundaries and land cover can generally be discerned from 1:250 000 scale maps for meso-scale catchments, and a scale of 1:1 000 000 and 1:500 000 is generally suitable for the presentation of soil and geology.

The physiographic descriptors that are used most frequently in low flow studies are *catchment area*, *shape* and *stream network characteristics*. The significance of the individual variable depends on the objective of the study and the time and areal scale requested for the analysis. The catchment area, which is usually defined as an area bounded by topographic water divide, does not always reflect the area contributing water to the stream channels (Section 3.2.2). In drought conditions the size and shape of the area that actually supplies water to the stream strongly depends upon areal extent of the groundwater aquifer, aquifer properties and groundwater head and slope. In many parts of the world the upstream part of a catchment becomes dry during a drought and does not contribute to the streamflow. An estimation of the fraction of the watershed fed by aquifers is often an essential parameter in low flow and drought modelling.

Among the *geomorphological* (or morphometric) descriptors of a watershed, the denivelation, or the *average basin slope*, is essential to understand the river system behaviour under drought conditions (Strahler, 1957). It is defined as the difference in elevation between the peak at the water divide (summit elevation) and catchment outlet. A further geomorphological index, generally used as an independent descriptor for low flow estimation at the ungauged site, is the *slope of the main stream*. Usually, slope is defined as the ratio of the difference in elevation and the distance between two points on the main stream (m km^{-1}). In the United Kingdom, standard points are used at 10 and 85% of the main-stream length, defined as SL1085 (Wilson, 1990). Besides the main-stream length and the length of the river network, the *drainage density* also plays an important role. The drainage density reflects the geological and petrographical conditions of a catchment and acts as an index representing the infiltration capacity and the transmissivity of the subsurface.

Land cover has an important influence on runoff generation and the low flow behaviour in a catchment. Interception, infiltration and evaporation, in particular, depend strongly on land cover descriptors (Section 3.3) such as the proportion of catchment area under urban development, the proportion of forest and the proportion of agricultural land. During a prolonged drought, a significant forest area can significantly affect low flows, as the roots of trees accelerate the drying process of a catchment. In Nordic countries, Canada and Siberia, the proportion of catchment occupied by bogs and wetlands is especially important. The effect of lakes, by providing storage, is similarly important for maintaining low flows (Section 3.5). Three different parameters, LAKE, FALAKE and WPLAKE, may be used to index lakes (Institute of Hydrology, 1980). The difference between the three definitions is based on the *weighting of the lake area*. FALAKE describes the ratio between the area of the lakes and the basin area, LAKE the sum of the areas of the catchment which drain through a lake divided by the catchment area and WPLAKE is the weighted lake percentage taking lake surface area into account together with contributing area.

In many catchments, the *geology* and *soils* control low flow response (Sections 3.3 and 3.4). However data that characterize these features remain difficult to establish and quantify. The difficulty is not only due to the lack of large-scale, extensive, geological or hydrogeological maps, but is also due to the problem of quantifying the impact of geology or soil on catchment runoff. Most attempts to characterize the hydrological response of rocks and soils have involved grouping catchments according to the geology or soils occurring within, and then comparing these groupings with the observed flow data, either by simple correlations or graphically. Such hydrological classification of geology and soil has been successfully applied in many countries including the state of Baden-Württemburg in Germany (Demuth, 1993) and across the United Kingdom (NERC, 1975; Boorman & Hollis, 1990).

In addition to physical catchment properties, *climatic characteristics*, especially rainfall, are of use in analysing low flows. Usually the characteristics are derived from maps at a scale of 1:1 000 000. Annual and monthly average *rainfall* data are particularly useful. Short-term rainfall measures, or indices of rainfall intensity, are usually used in flood studies and are not particularly relevant when dealing with low flows. For estimating hydrological drought severity the identification of *atmospheric features* which initiate drought is important (Chapter 2). The initial drought phase, a meteorological drought (Sections 3.3 and 3.4), is usually defined on the basis of rainfall deficiency. Rainfall volume may be examined at various intervals (seasonally, half-yearly or yearly) and related to the average total characteristic for the respective period. Hydrological drought during summer periods is the result of accumulated deficits between low rainfall and high evaporation rates. Evaporation losses are usually indexed by values of *potential evaporation*. Satisfactory results have been obtained in regression analyses aimed at predicting summer hydrological drought characteristics based on chosen catchment properties and climatic descriptors, expressed as the difference between rainfall amount and sum of potential evaporation in the summer half-year. The factor is widely used as the critical factor in soil water drought and usually named climatic water balance (Palmer, 1965) or water deficit (Chapters 2 and 3).

4.4.3   Local scale data

A wide range of thematic data is required for catchment, or local, scale studies. For detailed hydrological research, *maps* of a *high spatial resolution* are most often used. Data describing the distribution of *precipitation, potential evapotranspiration* and *recharge* within the catchment are particularly useful. First, a map with the location of meteorological stations is needed. Data collected from such stations may be used to calculate the potential evapotranspiration. Soil and land-use maps may be used to calculate the *actual evapotranspiration* and *groundwater recharge* (Section 3.3). In some catchments (e.g. those with arable land), *land use* should be mapped every year. *Remote sensing* can support such a yearly survey. In catchments with thick unsaturated subsoils, a map with the superficial geology (e.g. the lithology) should be available.

Geological maps (e.g. lithololology, depths of layers, and geological structure, such as faults, folding) are used to derive various hydrogeological maps (e.g. occurrence of aquifers). Meanwhile, *hydrogeological maps* provide valuable information on the distribution of *hydraulic parameters* (e.g. conductivity, storage coefficient, transmissivity, hydraulic resistance). Maps with *groundwater head contour lines* of aquifers are useful, particularly if they represent different hydrological conditions (e.g. summer and winter conditions or dry years versus wet years). In some catchments, a map with the location and depths of *observation wells* and *piezometers* will be available. If groundwater exploitation occurs, a map should be compiled with the location of the *wells*. Maps with the *elevation* of the soil surface are also useful to describe the water table depth below soil surface, which can be essential for computing possible capillary rise or recharge in thick unsaturated subsoils (Figure 3.12).

For comprehensive catchment studies, a map of the *drainage network* (rivers, brooks, springs, seepage areas, lakes, gauging stations) would need to be compiled. This map should also include properties of the drainage network, such as the slope, width, length of reaches, depth of bottom, average water depth and so on. *Digital elevation maps* with a high resolution (e.g. 5 m) are being increasingly used to generate the drainage network or to produce maps with the slope length, slope direction and slope angle.

4.5 Example data

The CD that accompanies this book contains a variety of data sets to enable the reader to work through the many worked examples and case studies of the subsequent chapters. The data are arranged in three groups: a global data set of river flow data from around the world; a regional data set from the state of Baden-Württemberg in Germany; and a local data set for the Upper Metuje catchment in the Czech Republic. A brief description of the data is given below.

4.5.1 Global Data Set

The Global Data Set comprises long-term daily river flow data from many gauging stations around the world. It is provided as a means of demonstrating the variability of hydrological regimes globally (Section 2.2.2) and gives an

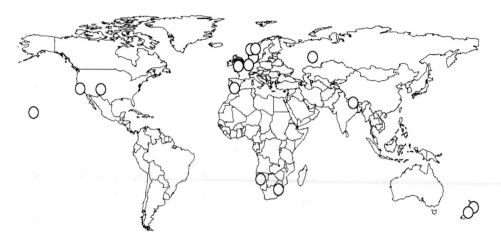

Figure 4.9 Location of the gauging stations of the Global Data Set.

opportunity for the reader to practice low flow and drought analyses on information that is, quite literally, foreign to any he or she may have encountered before. As such, the reader will be exposed to the difficulties of hydrological analysis, and the applicability (or limitations) of certain methods, with data from different regions.

The flow records were selected to typify some of the key regime types that are found in different parts of the world. The list of selected catchments, shown in Table 4.3, is by no means exhaustive because the range of possibilities is vast. Nevertheless, as can be seen in Figure 4.9, the Global Data Set comprises data from catchments in both the Northern and the Southern Hemisphere. There are catchments from cold regions (e.g. Lågen and Inva) and tropical regions (e.g. Honokohau), others from moist, temperate regions (e.g. Huruni and Bagamati) and some from dry, arid (e.g. Dawib) and semi-arid (e.g. Arroyo Seco) regions. Analysis of this data will show that low flows and droughts occur at different times and with varying intensity all over the world. While climate strongly influences such occurrences, other processes affect the low flows of rivers. The soil and hydrogeological conditions of a catchment and the degree of artificial, or human, influence modulates river response, especially during periods of low flows. To illustrate this, data from four contrasting catchments (Lambourn (permeable), Ray (impermeable), Lindenborg (mixed), Upper Guadiana and Trent (influenced)) in northern Europe are included. Illustration is also provided of the behaviour of a large river (Rhine), whose catchment comprises a variety of climatic and physiographical features.

Table 4.3 Summary of the Global Data Set

Country	River	Site	Period of record Start	End	Catchment Area (km²)	Station Altitude (m.a.s.l)	Mean flow (m³ s⁻¹)	Basis of Selection
Denmark	Lindenborg	Lindenborg Bro	1925	1997	214	5	2.34	Northern European catchment having mixed response to rainfall
Spain	Sabar	Alfartanejo	1963	1993	39	n/a¹	0.19	Southern European river, having zero flows that persist for months at a time
Spain	Upper Guadiana	Site 4008	1960	1995	16000	560	9.71	Heavily exploited catchment in southern Europe
Namibia	Dawib	Dawib	1978	1993	560	n/a¹	0.02	Ephemeral stream in southern Africa where zero flows predominate
The Netherlands	Rhine	Lobith	1901	1993	160800	9	2212.09	Large catchment in western Europe featuring many climate types
Norway	Ostri	Liavatn	1965	2000	235	733	10.52	Snow and glacier influenced catchment in northern Europe
Norway	Lågen	Rosten	1917	2003	1755	737	32.43	Snow-affected, northern European river with distinct winter low flows
Nepal	Bagamati	Sundurijal	1970	1995	17	1600	1.04	South Asian, rain-fed catchment having a monsoon dominated regime
New Zealand	Ngaruroro	Kuripapango	1963	2000	370	n/a	17.24	Southern Hemisphere river, temperate regime with no distinct dry season
New Zealand	Hurunui	Mandamus	1956	2000	1060	n/a	53.07	Southern Hemisphere river, temperate regime with no distinct dry season

(→ Table continued on next page)

Table 4.3 (continued)

Country	River	Site	Period of record Start	Period of record End	Catchment Area (km²)	Station Altitude (m.a.s.l)	Mean flow (m³ s⁻¹)	Basis of Selection
Russia	Inva	Kudymkar	1936	1995	2050	126	12.35	Continental, East European river, with low flows in summer and winter
South Africa	Elands	Elands River Drift	1963	1992	690	n/a[1]	3.10	Perennial river in southern Africa, having distinctly seasonal regime
United Kingdom	Trent	Colwick	1958	2001	7486	16	85.07	Heavily influenced catchment in northern Europe
United Kingdom	Ray	Grendon Underwood	1962	1999	19	66	0.10	Impermeable upland catchment in northern Europe
United Kingdom	Lambourn	Shaw	1962	2000	234	76	1.69	Permeable lowland catchment in northern Europe
United States of America	Pecos	Pecos	1930	1999	490	94	2.91	River in arid part of North America having Rocky Mountains tributaries
United States of America	Arroyo Seco	Soledad	1901	1999	632	103	4.84	Catchment in semi-arid part of North America having occasional zero flows
United States of America	Honokohau	Maui, Hawaii	1922	1996	11	265	1.08	Tropical island regime with no dry season

[1] n/a: not available

The data for the Global Data Set were collated from a number of disparate sources. The majority of the data were provided from databases of the regional FRIEND (Flow Regimes from International Experimental and Network Data) projects of northern Europe, southern Africa and the Hindu Kush-Himalaya (Gustard & Cole, 2002). Some of the other data, from the United States of America in particular, were obtained from the WWW, while others (e.g. United Kingdom, Spain and New Zealand) were obtained directly from national hydrometric agencies. Although only records of reasonably good quality were selected, the data is real and, as would be expected in reality, does feature errors, missing values and some imperfections. No attempt has been made to correct the data, as it was considered that exposure to such problems would be of benefit to the reader.

4.5.2   Regional Data Set: Baden-Württemberg

The flow data of the Regional Data Set are from the State of Baden-Württemberg, in southwest Germany. In total, 83 gauged stations were selected for the regional analysis and the Self-guided Tour referenced by Chapter 8. The flow data of the Regional Data Set are available on the CD under Data and the document 'Overview Regional Data Set' provides a list of the gauging stations included. In addition, data describing the physiographic features of the catchments (catchment descriptors) are available in digital format in the Self-guided Tour. The region, within which the 83 gauging stations are found, is shown in Figure 4.10. It encompasses several landscapes and exhibits a wide range of morphometric, hydrogeological, soil, land use and climatic features.

The River Rhine drains three quarters of the study area, with the remaining quarter draining to the Danube. It is difficult to map out the exact position of the groundwater divide in this area because a significant amount of water drains from the Danube catchment into the Rhine via a karst system. With the lack of substantial tributaries, and the loss of water to the river Rhine, the Danube remains a relatively small river until its Alpine tributaries add to its flow in Bavaria, east of the area shown.

Water management activities, such as water diversions and exports, storm water ponds, reservoirs for the augmentation of low flows and groundwater extraction, are examples of how the flow regimes are artificially influenced in this area. For the Regional Data Set, however, only catchments with minimal human impacts were selected.

The natural flow regimes of the region are dominated by the effects of rainfall and moderated, to some degree, by snow melt. The Swabian Alb (part of the Süddeutsche Schichtstufenlandschaft, Figure 4.10) and Black Forest areas have nivo-pluvial (snow-rain) and pluvial (rain) precipitation regimes, with

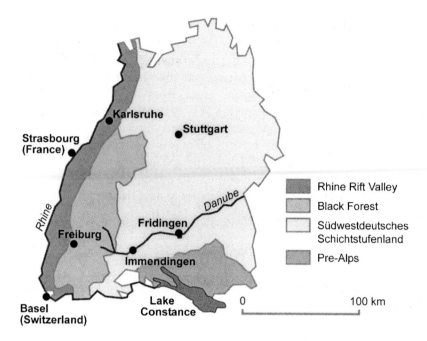

Figure 4.10 Landscapes of Baden-Württemberg.

moderate inter-annual variability and low flows occurring mainly in late summer. The Pre-Alps shift from a nival to a nivo-pluvial regime, with less seasonality and low flows in winter and in summer. The weak seasonality of flows indicates extended aquifers with high storage capacity in that region. The Rhine Rift Valley, particularly, has deep aquifers with high storage capacity and therefore only little variation in inter-annual low flows.

The climate of the study area is a result of the interaction of oceanic and continental influences. While the latter is responsible for extreme seasonality (with cold winters and hot summers), the dominant impact of the former leads to a more temperate climate. July is usually the warmest and January the coldest month. The mean annual air temperature ranges from 3.2°C at the Feldberg peak (highest elevation of the study area) to above 10°C in the Rhine Rift Valley (Table 4.4). Precipitation in the area is predominantly caused by frontal (cyclonic) storms, although orographic lifting modifies the pattern. The amount of precipitation is, therefore, a function of elevation and exposition and ranges from 2100 mm year^{-1} in the western part of the Black Forest, to 600 mm year^{-1} in the sheltered areas of the Rhine Rift Valley. During the summer, convective lifting occasionally induces the formation of short-duration high-intensity precipitation. The study area receives maximum precipitation in the summer

Table 4.4 Landscapes of Baden-Württemberg

Landscape	Elevation (m)	Geology	Drainage Density (km km^{-2})	Mean Annual Rainfall (mm)	Air Temp (°C)
Rhine Rift Valley	85–250	Fluvio-glacial deposits	-	600–900	10
Black Forest				< 2100	3
North-east	600–800	Red sandstone	Low		
South-west	1000	Granite, gneiss	1.9		
Swabian Alb	700–1000	Karstic	0.03	650–900	6–9
Pre-Alps and Lake Constance	600	Glacial sediments	-	750–1400	6–7

(June to August) and a minimum in the late winter (February and March). Snow can be present for up to 150 days a year in the Black Forest.

4.5.3   Local Data Set: Metuje catchment

The Upper Metuje catchment is mainly located in northern Bohemia (Czech Republic) with a small part in Poland. The River Metuje is a tributary of the River Elbe. The catchment area of the Upper Metuje is 73.6 km^2. Its landscape is characterized by a high diversity of deep valleys, gentle and steep slopes and uplands. Part of the catchment is protected because of its high ecological value. The catchment has a basin structure and it is filled with Mesozoic (Triassic and Cretaceous) deposits overlying impermeable Paleozoic (Permian-Carboniferous) formations, as shown in Figure 4.11. The geological layers incline towards the centre of the valley. The main aquifers are developed in the permeable Triassic and Cenomanian rocks (deep aquifer) and Middle Turonian formations (shallow aquifer). Groundwater in the deep aquifer shows semi-confined to artesian conditions, whereas in the shallow aquifer unconfined conditions occur. The Lower Turonian marlstones have a low conductivity that separates both aquifers, although they are interconnected at places where faults occur. In the southeastern part of the catchment (Teplice nad Metuje) the relatively impermeable Skalský fault drains the deep aquifer. At the site where the River Metuje crosses the fault its flow increases by some 200 l s^{-1}.

The catchment is covered mainly by sandy and sandy-loamy soils (light to medium-textured soils). The dominant soil types are acid brown soils and podsols. Gley and fluvial soil types occur along the streams. Around 40% of the

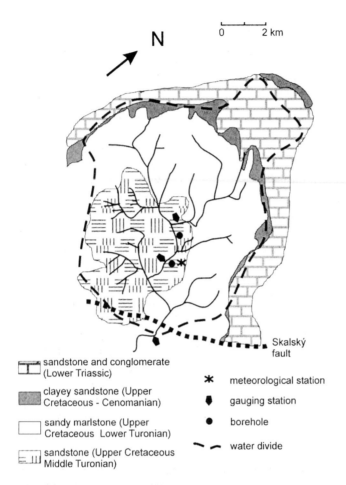

sandstone and conglomerate (Lower Triassic)

clayey sandstone (Upper Cretaceous - Cenomanian)

sandy marlstone (Upper Cretaceous Lower Turonian)

sandstone (Upper Cretaceous Middle Turonian)

* meteorological station

gauging station

● borehole

water divide

Skalský fault

Figure 4.11 Geology and monitoring network of the Upper Metuje catchment.

catchment is covered by coniferous or mixed forest. Continuous forest areas are located mainly in boundary regions of the catchment and also in the part covered by Middle-Turonian sandstone. Lakes or ponds cover no more than 1% of the catchment and it is less than 1% urbanized. About 55% of the area is used for agricultural purposes, from which two thirds is covered by arable land. The remaining 4% of the catchment is covered by meadows and bushes.

There are no significant abstractions of surface water or effluent discharges in the catchment. In the downstream part, where groundwater discharges intensively into the river due to the Skalský fault, groundwater is abstracted for drinking water supply. The long-term mean abstraction is about $50 \, l \, s^{-1}$.

Hydrological monitoring has been undertaken in the catchment for a number of reasons, including assessing long-term changes in water resources due to climate variability, for assessing the impact of groundwater abstractions on groundwater resources and ecology, and for modelling the interaction between groundwater and streamflow. The data file 'Metuje Local Data Set' (Data, CD) provides the hydrological and meteorological time series of the Local Data Set.

The Bucnice meteorological station, which is located in the centre of the catchment (Figure 4.11), has been in operation since 1963. Data on air temperature, air humidity, precipitation and the depth of snow cover have been collected. Since 1998 an automated weather station has been installed which additionally monitors wind speed, wind direction, sunshine duration and radiation. Meteorological data from 1970–2000 show that mean annual air temperature is 5.5°C, with July being the warmest month (mean of 14.5°C) and January the coldest (mean of −3.8°C). The maximum and minimum recorded mean daily temperatures are 25.6 and −28.2°C respectively. The mean annual precipitation of the catchment is 743 mm. During November to March part of precipitation is usually accumulated as snow. Maximum precipitation occurs in July, whereas the lowest values are observed during spring. Daily precipitation, air temperature, relative humidity and potential evapotranspiration for the period 1980–2000 are provided on the CD. Potential evapotranspiration is calculated with the approach included in the model BILAN (Chapter 9).

River flow data are observed at three automated streamflow gauging stations. Two of the stations are located in the Metuje mainstream: M VIII, north of the meteorological station Bucnice; and M XII, at the outlet. The third is sited in the Zdonovský Brook (Z VI). The M XII gauging station is located on a weir and water stage has been observed there continuously since 1967. The stability of the rating curve is regularly checked by discharge measurements six times a year. Changes in the rating curve have been mainly caused by changes to the river bed after high floods. For infilling gaps, and checking the discharge at station M XII, data from the M VIII station were used together with data observed at the Maršov gauging station, which is situated further downstream on the River Metuje. It was the data from these M XII and M VIII gauging stations that were used in the earlier worked example (Section 4.3.3). The mean flow at the M XII station for the period 1970–2000 was 0.86 $m^3 s^{-1}$. The maximum daily discharge of 21.3 $m^3 s^{-1}$ was observed on 8 July 1997, and the minimum daily discharge of 0.32 $m^3 s^{-1}$ on 4 January 1993. Data from gauging station M XII for the period 1980–2000 are provided on the CD.

Many boreholes have been drilled in the catchment. The data for one representative borehole (VS3) are also provided on the CD. This particular borehole, located to the north of the meteorological station at Bucnice (Figure 4.11), has been drilled into the deep aquifer to a depth of 180 m.

4.6   Summary

This chapter has provided an overview of the hydrological data that are typically used in low flows and drought analysis. Section 4.1 introduced the subject with an overview of the different uses of hydrological data, including operational, strategic and academic applications. A classification of data was given in Section 4.2, with three generic types described: time series data, thematic data and metadata. In Section 4.3 information was provided on the collection and processing of river flow data and groundwater data. Hydrometric practices were outlined, with details given on the types of gauging stations, methods of the measurement of stage and the derivation of the stage-discharge relation. Methods of obtaining groundwater data were then described, including manual and automatic measurement of groundwater levels and the monitoring of groundwater discharge. Issues of data quality control and management were also addressed at the end of the section by firstly describing errors as being random, spurious or systematic, then considering good practice, methods of data validation, approaches to the hydrological validation of data and techniques for dealing with missing data. Further detailed information was given in Section 4.4 on the thematic data that are typically available for hydrological studies at global, regional and local scales, with advice given on potential data sources and the types of data that can be obtained or derived. Section 4.5 described the example data sets of the CD, which will enable the reader to work through the exercises of subsequent chapters.

5

Hydrological Drought Characteristics

Hege Hisdal, Lena M. Tallaksen, Bente Clausen, Elisabeth Peters, Alan Gustard

5.1   Introduction

One of the first steps in a drought analysis is to decide on the hydrological drought characteristics to be studied. Drought affect many sectors in society and therefore there is a need for different ways of defining or characterizing drought. The particular problem under study, data availability and climatic and regional characteristics will influence the choice. Therefore no single hydrological drought characteristic is suitable to assess and describe droughts for any type of analyses in any region. However, it is important to be aware of how various ways of characterizing a drought might lead to different conclusions regarding the drought phenomenon. For instance, it is possible that the mean annual streamflow over a calendar year indicates no drought, whereas the streamflow in the growing season does. Furthermore, a drought in terms of, for example, a streamflow deficit, might not coincide in time with a drought in terms of a groundwater deficit. In a study of spatial patterns of drought frequency and duration in the United States, Soule (1992) showed that the type of drought characteristics analysed had a major impact on the spatial patterns found.

When calculating hydrological drought characteristics it is important to be aware of the *seasonality* in the region under study. Seasonal low flows can be caused either by low precipitation often combined with high evaporation losses, or result from precipitation being stored as snow (Section 3.3.1 and Box 3.2). It is necessary to separate droughts caused by different processes in order to characterize hydrological drought in a consistent way. A separation between 'summer drought' caused by lack of precipitation and 'winter drought' caused by frost, can be carried out by combining information about temperature and streamflow. This can be done, for example, by letting the summer season begin at the end of the spring flood and stop when the mean monthly temperature falls below 0°C (Hisdal *et al.*, 2001). In very cold climates, there might be snow melt during the whole summer season and hence only 'winter drought' exists and studies of 'summer drought' are not relevant.

General drought definitions have already been presented in Section 1.2. The main objective of this chapter is to present a range of quantitative hydrological drought characteristics, the interrelationship between them and how to derive them. Methods to describe droughts at a single site and over a region are introduced. Examples are given by using time series from the data sets described in Section 4.5. It is assumed that the time series are stationary and undisturbed by human influence. These assumptions are often not totally valid but the techniques can still be applied. However, it is then important to be aware that the less stationary and the more human influence in the catchment, the more unreliable the results. Many time series have gaps, and infilling methods are described in Section 4.3.3. If infilling is not possible, hydrological drought characterisics can still be estimated from the available data. Nevertheless, because of temporal variability of hydrologic processes, hydrologic observations for a period of years are needed for correct calculations. The hydrological regime under study and the type of hydrologic drought characteristic to be determined will influence the number of years required. Often a 30-year period is recommended (Section 2.3.3). If not available, shorter periods can be accepted, but the shorter the period the higher the uncertainty. How various hydrological drought characteristics are applied in operational hydrology is the topic of Chapter 11.

Section 5.2 describes the drought terminology applied throughout the book. Section 5.3 explains low flow characteristics for streamflow. These characterize specific features of the low flow regime. The derivation of low flow indices based on a single variable is outlined, including indices derived from the flow duration curve, mean annual minimum values and base flow and recession analyses. The CD includes spreadsheets for calculation of indices from the flow duration curve, mean annual minimum indices and a base flow index. Applications of the spreadsheets are given in the chapter as worked examples. Section 5.4 is concerned with methods to derive drought deficit characteristics. Details are given on two methods, the threshold level method and the sequent peak algorithm. How to include the areal aspect in drought characteristics is also explained. A program including the threshold level method, Nizowka, and a spreadsheet with the sequent peak algorithm are included on the CD. Both are illustrated in the chapter through worked examples. Section 5.5 describes the interrelationship between indices. In Section 5.6, the focus is on groundwater drought characteristics in terms of groundwater level as well as groundwater discharge and recharge. The derivation of groundwater drought characteristics is illustrated through a case study including spreadsheet calculations also to be found on the CD. Indices describing the low flow behaviour might be based on a combination of variables, for example precipitation and evapotranspiration; such complex indices are covered in Section 5.7. Finally, the chapter is summarized in Section 5.8.

5.2 Drought terminology

How can a hydrological drought be characterized? In a drought situation streamflows are low or even zero. Hence, indices characterizing the low flow regime of a river or time series of low flow values can be said to characterize the drought behaviour of a river. Similarly, indices or time series of low groundwater recharge, levels and discharge can be said to characterize the drought behaviour of groundwater. These time series or indices characterize droughts by considering only the flows over pre-defined durations. They do not provide a complete characterization of the drought. For instance, questions such as what was the start, the end, the total duration and the severity of the drought, are not answered. To be able to quantify the latter drought characteristics for a river or a groundwater aquifer, it is necessary to define a threshold level below which the flow or groundwater is regarded as being in a drought situation.

The derivation of *hydrological drought characteristics*, including time series and indices, can be illustrated by considering a time series of daily streamflow or groundwater recharge, levels or discharge. Figure 5.1 gives an example for streamflow. From the original time series, low flow characteristics can be derived. These describe the low flow regime of a river. They can be an index obtained using the whole time series of flow directly in its derivation, a percentile from the flow duration curve (Section 5.3.1) being one example (Figure 5.1, column I). Another low flow characteristic would be the lowest flow in a specific time period. Hence, a time series of *low flow characteristics* can be obtained, for instance the lowest daily streamflow value each year. Based on this time series, another characteristic, a low flow index, e.g. the 'mean annual minimum flow', can be derived (Figure 5.1, column II).

Drought events can be defined by introducing a threshold below which the flow is regarded as being in a deficit. Each event has a beginning and an end. The deficits can be described by different characteristics like duration, severity, time of occurrence and spatial extent. A time series of *deficit characteristics*, for example the drought deficit duration, can be derived. Finally, another characteristic, a drought deficit index, e.g. the 'mean drought deficit duration', can be estimated (Figure 5.1, column III).

To summarize, an *index* is seen as *a single number* characterizing an aspect of the drought or low flow behaviour at a site or in a region. A drought *event*

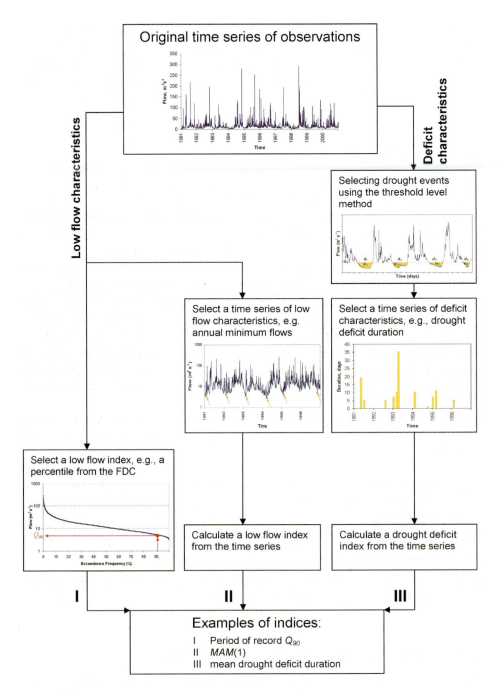

Figure 5.1 Derivation of hydrological drought characteristics.

definition implies the use of *a method* to select drought events from a time series. By introducing a threshold level, it quantitatively defines whether the site can be regarded as being in a deficit situation or not. Whether or not the drought will have adverse effects will depend on the vulnerability of the region under study. The vulnerability aspect, however, is not covered in this chapter.

5.3 Low flow characteristics

This section describes low flow characteristics, in terms of time series and indices that are derived from series of flow data. Estimation methods for ungauged sites are treated in Chapter 8. The term *low flow statistic* is equivalent to *low flow index* and both are applied in this book. In literature the terms *low flow measure*, *parameter* and *variable* are also used for *low flow index*.

The consequences of applying the same calculation procedure to data from different climatic regions are illustrated by applying the methods to the Global Data Set (Section 4.5.1). The variation in climate combined with differences in physiographical catchment properties gives a wide variety of river flow regimes (Section 2.2.2), which might require the usage of several different methods. Most methods to derive low flow characteristics have been developed and used for perennial rivers, and applying them to intermittent and ephemeral streams should be done with caution. An example would be if the mean annual minimum flow is zero.

It should be noted that many indices are derived by calculation procedures involving the whole spectrum of flows, from high flows to low flows. This includes:

a) the flow duration curve, from which the low flow percentiles are selected (Section 5.3.1);

b) the base flow separation techniques aimed at identification of different flow components, from which the base flow indices are derived (Section 5.3.3) and;

c) recession analysis aimed at characterizing the falling limb of the hydrograph, from which the recession indices are derived (Section 5.3.4).

5.3.1 Percentiles from the flow duration curve

The flow duration curve (FDC) plots the empirical cumulative frequency of streamflow as a function of the percentage of time that the streamflow is

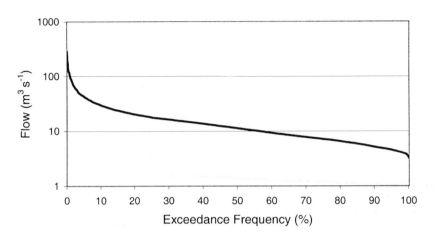

Figure 5.2 Flow duration curve for River Ngaruroro at Kuripapango, NZ.

exceeded. As such, the curve is constructed by ranking the data, and for each value the frequency of exceedance is computed. The empirical FDC for the River Ngaruroro at Kuripapango in New Zealand (NZ) is shown in Figure 5.2 (Worked Example 5.1). FDCs represent the streamflow variability of a catchment. Both high and low flows are included. To improve the readability of the curve, streamflow is often plotted on a logarithmic scale. It is also common to let the abscissa scale be based on the normal probability distribution. The transformation from exceedance frequencies to the standard normal distribution is described in Chow *et al.* (1988).

Traditionally, exceedance percentiles are obtained from FDCs based on the total period of record. Alternatively, it is possible to obtain indices for particular periods of the year considering data from, for example, the summer period only or, if the record is sufficiently long, even for a particular day of the year (Figures 5.8 and 5.9, Section 5.4.1).

The FDC can be based on daily, monthly or some other time interval of streamflow data. If daily data are analysed, it shows the relationship between the daily flow values and their corresponding flow exceedance frequencies. *As the streamflow between successive time steps generally is correlated (autocorrelation), and the streamflow characteristics are seasonally dependent, the flow duration curve cannot be viewed as a probability curve.*

Low flow indices are often derived directly from the curve as low-flow exceedances. A flow exceedance is an *index* that expresses the proportion of time that a specified daily flow is exceeded during the period of record. The flow exceedance is often given in terms of percentiles. For instance, the

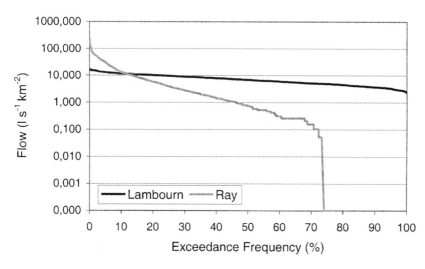

Figure 5.3 Flow duration curves for contrasting flow regimes; Lambourn: permeable catchment, Ray: impermeable catchment.

90-percentile flow, or Q_{90}, is the flow that is exceeded for 90 percent of the period of record.

Low flow percentiles from the FDC are used in various water management applications such as water supply, hydropower and irrigation planning and design, discharge permits, river and reservoir sedimentation studies and water transfer and withdrawals (Chapter 11).

Expressing flows as percentiles allows flow conditions in different rivers to be compared provided that the flow duration curves are normalized. This can be achieved by dividing by the mean or median flow of each river (e.g. Gustard *et al.*, 1992) or the catchment area. Flow duration curves for two contrasting British catchments are shown in Figure 5.3. The original time series of daily streamflow are here divided by the catchment area. Lambourn is a permeable, groundwater-fed catchment. The low variability of flows in this type of regime is reflected in a flat flow duration curve. Ray is a characteristic impermeable catchment, with a flashy flow regime, a high variability in daily flows and thus a steep flow duration curve. It is seen that many of the low flow percentiles usually applied as low flow indices for this catchment are zero (e.g. Q_{80} and Q_{90}). Since it is not possible to take the logarithm of zero, zero flows are not plotted in the figure. An even larger percentage of zero values can be found in ephemeral streams in semi-arid or polar regions, leading to Q_{50} and even smaller percentiles being zero.

Vogel & Fennessey (1994) discuss how the FDC depends on the period of record on which it is based. They suggest calculating separate FDCs for each individual year of record, which enables association of confidence intervals and recurrence intervals to the FDC.

Worked Example 5.1: Flow duration curve (CD)

Excel sheet 'Calculation'

1. Data

   Ten years of daily data from River Ngaruroro at Kuripapango (NZ) are used here to construct a flow duration curve based on a daily time step, $\Delta t = 1$ day. The total number of Δt intervals is then $N = 3653$ days. Table 5.1 lists the first seven flow values. The whole series can be found on the CD. The first two columns show the date and the corresponding flow value, Q.

2. Calculation of the FDC

   The flow duration curve is constructed following the calculation steps as given in the right part of the table:

   (a) The rank, i, of each value is calculated (using the RANK function in Excel), which means that if the list is sorted, the rank will be its position. Here the series is sorted in decending order and the ith largest value has rank i (i.e. the largest value has rank 1).

Table 5.1 Calculation of a daily flow duration curve for River Ngaruroro at Kuripapango, NZ

Data, 10-year series		Calculation of flow duration curve	
Date	Streamflow ($m^3 s^{-1}$)	Rank, i	Exceedance frequency, EF_{Q_i}
1-Jan-91	4.676	3439	0.94
2-Jan-91	4.827	3394	0.93
3-Jan-91	4.918	3364	0.92
4-Jan-91	4.634	3453	0.95
5-Jan-91	4.485	3487	0.95
6-Jan-91	4.335	3522	0.96
7-Jan-91	4.198	3547	0.97

(b) The exceedance frequency, EF_{Q_i} is calculated as:

$$EF_{Q_i} = i / N \qquad\qquad (W5.1.1)$$

which gives an estimate of the empirical exceedance frequency of the ith largest event. EF_{Q_i} designates here the observed frequency when the flow, Q, is larger than the flow value with rank i, Q_i.

Excel sheet 'Result'

3. Tabulation of the FDC

   (a) Corresponding values of streamflow (Q in $m^3 s^{-1}$) and exceedance frequency (EF_{Q_i} in %) are tabulated.

   (b) The two columns are sorted by EF_{Q_i}.

4. Plot of the FDC

   The sorted table columns are then plotted (Figure 5.2). The ordinate axis is here logarithmic.

5. Selected exceedance values

   Values for a particular frequency, for example the 90-percentile (Q_{90}), can be obtained as the value of Q corresponding to the largest value of EF_{Q_i} that is less than or equal to the value of EF_{Q_i} sought for (using the INDEX and MATCH functions in Excel). A sample of corresponding values in this range is shown in Table 5.2, and the 90-percentile flow value is taken as 5.18 $m^3 s^{-1}$. Alternatively, in case of large differences between successive values, a linear interpolation can be used.

Table 5.2 An extract of values corresponding to Q_{90}

EF_{Q_i} (%)	Q ($m^3 s^{-1}$)
89.93	5.18
89.98	5.18
90.01	5.17
90.01	5.17

In Table 5.3 Q_{95}, Q_{90} and Q_{70} for the daily streamflow series in the Global Data Set (Section 4.5.1) are displayed. Trent in the UK and Upper Guadiana in Spain are not included as these are heavily exploited catchments. The River

Table 5.3 Selected percentiles ($1\,s^{-1}\,km^{-2}$) for the Global Data Set

River, Country	Summer season	Q_{95}	Q_{90}	Q_{70}
Lindeborg, Denmark	-	7.05	7.63	9.00
Dawib, Namibia	-	0.00	0.00	0.00
Bagamati, Nepal	-	8.24	10.6	17.6
Rhine, the Netherlands	1 Mar to 30 Nov	6.10	7.00	9.61
Hurunui, NZ	1 Nov to 30 Jun	15.1	18.2	26.2
Ngaruroro, NZ	-	12.0	14.2	22.7
Lågen, Norway	15 Jun to 30 Sep	8.03	9.73	15.6
Ostri, Norway	15 Jun to 30 Sep	22.5	29.1	57.2
Inva, Russia	1 May to 31 Oct	0.93	1.13	1.84
Elandrivierie, South Africa	-	0.03	0.09	0.25
Sabar, Spain	-	0.00	0.00	0.00
Lambourn, UK	-	3.17	3.59	4.83
Ray, UK	-	0.00	0.00	0.05
Arroyo Secco, US	-	0.00	0.00	0.36
Honokohau, US	-	30.9	33.4	43.7
Pecos, US	1 Mar to 30 Nov	1.33	1.56	2.43

Rhine, the two Norwegian rivers, Hurunui in New Zealand, River Pecos in the US and the Russian river have a distinct winter low flow period. Therefore a summer season has been defined, in order to calculate indices characterizing the summer low flow caused by lack of precipitation and high evaporation losses. Interpolation has been used to correct obvious errors and fill in minor gaps. For the Norwegian river Ostri, snow melt contributes to the streamflow during the whole summer season, and high percentile values can be observed. For the ephemeral rivers Dawib in Namibia and Sabar in Spain even Q_{70} is zero. For the intermittent, impermeable River Ray in the UK, Q_{90} is zero and Q_{70} is very small. Although the zero-flow values indicate what type of river is being observed, this can cause special problems if the FDC is to be related to a probability distribution (Chapter 6). If the difference between Q_{95}, Q_{90} and Q_{70} is small, this indicates a flat FDC resulting from a river with low variability in the flows. In Croker *et al.* (2003) the theory of total probability is used to combine a model for estimating the percentage of time the river is dry with a model for estimating the FDC for the non-zero period to determine a transformed FDC.

5.3.2 Mean annual minimum flow

One of the most frequently applied low flow indices is derived from a series of the annual minima of the n-day average flow, $AM(n$-day). In its simplest form this would be the mean annual 1-day flow, hence the average of the annual minimum value. For $n > 1$, the method consists of deriving a hydrograph whose values are not simply daily flows but are average flows over the previous n-days or alternatively the previous $n/2$ days and the coming $n/2$ days. For example if $n = 7$, the entry from 1 January 2003 is in fact the average flow over the period 26 December 2002 to 1 January 2003 inclusive, or alternatively the average flow over the period 29 December 2002 to 4 January. The derived data can thus be regarded as the outcome of passing a moving average filter of n-day duration through the daily data. Based on the 'filtered' hydrographs mean annual minimum n-day indices, $MAM(n$-day), can be derived.

Worked Example 5.2: Mean annual minimum n-day flow (CD)

Excel sheet 'Calculation'

1. Data

   Ten years of daily data from River Ngaruroro at Kuripapango (NZ) are used as an example, as in Worked Example 5.1, to estimate mean annual minimum of the n-day average flow for n equal to 1, 7 and 30 days. For this station the lowest flows are observed around the turn of the calendar year. Therefore the annual minima are selected from hydrological years starting 1 September and ending 31 August. Table 5.4 lists the first flow values. The whole series can be found on the CD. The first two columns show the date and the corresponding flow value, Q.

2. Calculation

   The moving average values are calculated and displayed in the right part of the table by calculating the average of the last 7-days and 30-days (using the MOVING AVERAGE analysis tool in Excel).

Table 5.4 Calculation of n-day average flow, River Ngaruroro at Kuripapango, NZ

Data, 10-year series		Moving average calculation	
Date	**MA(1)** Q (m^3 s^{-1})	**MA(7)** Q (m^3 s^{-1})	**MA(30)** Q (m^3 s^{-1})
1-Sep-90	19.344		
2-Sep-90	17.591		
3-Sep-90	16.285		
4-Sep-90	17.231		
5-Sep-90	21.142		
6-Sep-90	30.628		
7-Sep-90	24.492	20.959	
8-Sep-90	20.646	21.145	
9-Sep-90	18.501	21.275	
⋮	⋮	⋮	
30-Sep-90	8.601	7.848	13.321
1-Oct-90	17.198	9.149	13.249

Excel sheet 'Result'

3. Selection of the annual minimum data

   Find the annual minimum (using the MIN function in Excel).

4. Calculation of *MAM*(n-day)

   Calculate the average of the annual minimum (using the AVERAGE function in Excel).

5. Tabulate the results

   The *MAM*(n-day) values for the 10-year series are given in Table 5.5.

Table 5.5 *MAM*(n-day), n=1 day, 7 days and 30 days

n	Q (m^3 s^{-1})	Q (l s^{-1} km^{-2})
1	4.14	11.2
7	4.40	11.9
30	5.44	14.7

In Table 5.6 the *MAM(n*-day) results of applying averaging periods of 1, 7 and 30 days for the time series in the Global Data Set (Section 4.5.1) are displayed. These indices are often highly correlated with the low flow statistics taken from the flow duration curve (Section 5.3.1) and have the same areas of application.

It is possible to estimate low flow frequency curves by fitting a probability distribution (Chapter 6) to the annual minimum *n*-day series. In general the frequency curves describe the annual minimum *n*-consecutive-day average discharge not lower than a given value with a specific probability. These curves allow the estimation of various return periods or so-called *T*-year events (Box 6.2), an event that *on average* will occur every *T* years. These *T*-year events can in turn be applied as low flow indices. For example, in the United States the most widely used low flow index is the 10-year 7-day minimum flow, $AM(7)_{10}$ (Riggs *et al.*, 1980). Annual minimum series often contain years with zero values. Problems related to this are further dealt with in Chapter 6.

Table 5.6 Mean annual minimum values ($1 s^{-1} km^{-2}$) for the Global Data Set

River, country	Summer season		*MAM(1)*	*MAM(7)*	*MAM(30)*
Lindeborg, Denmark	-		7.32	7.53	7.93
Dawib, Namibia	-		0.00	0.00	0.00
Bagamati, Nepal	-		9.73	10.6	11.8
Rhine, the Netherlands	1 Mar	to 30 Nov	6.75	6.92	7.78
Hurunui, NZ	1 Nov	to 30 Jun	15.3	16.1	20.0
Ngaruroro, NZ	-		11.1	11.7	14.2
Lågen, Norway	15 Jun	to 30 Sep	9.65	10.5	13.8
Ostri, Norway	15 Jun	to 30 Sep	21.0	25.1	46.0
Inva, Russia	1 May	to 31 Oct	1.19	1.34	1.72
Elandrivierie, South Africa	-		0.05	0.07	0.12
Sabar, Spain	-		0.00	0.00	0.00
Lambourn, UK	-		3.77	3.93	4.12
Ray, UK	-		0.00	0.00	0.00
Arroyo Secco, US	-		0.14	0.15	0.18
Honokohau, US	-		33.0	34.9	45.0
Pecos, US	1 Mar	to 30 Nov	1.33	1.56	2.43

5.3.3   Base flow indices

Many *hydrograph separation techniques* have been applied for identification of the different flow components of the total flow. The components are thought to represent different flow paths in the catchment, each characterized by different residence times, the outflow rate of groundwater flow being the slowest. The flow has traditionally been separated into flow that originates from overland (direct), unsaturated (throughflow) and saturated (groundwater) flow (Section 3.2). In their review Nathan & McMahon (1990a) distinguish between methods aimed at deriving the response for a given event and automated methods for a continuous separation of the different components of the flow. Isotopic and chemical hydrograph separation techniques are means of identifying the sources of river runoff on an event basis. Recession analysis has been applied for distinct events for the purpose of separating between the different flow components, as well as for the continuous separation of the total hydrograph (e.g. Bates & Davies, 1988).

Methods for continuous separation generally divide the flow into one quick and one delayed component using an automated time-based separation. The delayed flow component is thought to represent the proportion of flow that originates from groundwater and other delayed sources, by Hall (1968) defined as the *base flow*, Q_b. Time series of base flow have been seen as useful as a measure of the dynamic behaviour of groundwater in a catchment, whereas the base flow proportion of the total flow has been used as an index of the catchment's ability to store and release water during dry weather. A high index of base flow would imply that the catchment has a more stable flow regime and is thus able to sustain river flow during extensive dry periods. Base flow indices have performed satisfactorily as catchment descriptors in many low flow studies because even a rough estimate of storage properties greatly increases the performance of the estimation model. Although valuable as measures of the storage property of a catchment, response factors derived based on the different volumes identified are of limited use as indicators of the flow processes operating (Hewlett & Hibbert, 1967; Anderson & Burt, 1980; Wittenberg & Sivapalan, 1999).

Catchment descriptors derived from flow records are often referred to as indirect measures of catchment properties. In gauged catchments the record length and period of observation will influence the stability of the flow-derived indices due to natural climate variability. At the ungauged site the flow-derived indices must be estimated and this is commonly done by establishing a regional regression model between the index and catchment properties. Base flow indices are generally highly correlated to the hydrological properties of soils and geology and other storage related descriptors like lake percentage.

Consequently, information on properties like soil and geology may be used directly to derive an index of storage to be included in the regression model. However, until detailed soil and geology maps are available for larger areas and at the appropriate scale (Section 4.4.2), it is likely that flow-derived indices will be in use.

In Chapter 8 estimated mean monthly base flow is introduced as an index of base flow. The index (BASE) is calculated from ranked monthly minimum flow values, and at least ten years of observations are recommended (Demuth, 1993). A regional regression model has been developed for estimation at the ungauged site and further details are given in the Self-guided Tour of Chapter 8. Another well-known index, the Base Flow Index (BFI), is presented below. It is calculated based on a continuous separation of the base flow. Other automated separation procedures are described in Nathan & McMahon (1990a). More recent applications have attempted to apply physically-based methods (Chapman, 1999; Furey & Gupta, 2001) and observed groundwater heads (Peters & van Lanen, 2004) to perform a continuous separation of the streamflow hydrograph.

Base Flow Index

The Base Flow Index (BFI) was developed during a low flow study in the UK (Institute of Hydrology, 1980). The index gives the ratio of base flow to total flow calculated from a hydrograph smoothing and separation procedure using daily discharges. The BFI is thus considered a measure of the river's runoff that derives from stored sources and as a general catchment descriptor it has found many areas of application, including low flow estimation and groundwater recharge assessment. Values of the index range from above 0.9 for a permeable catchment with a very stable flow regime to 0.15–0.2 for an impermeable catchment with a flashy flow regime. Figure 5.4 compares the annual hydrograph and base flow separation line for the year 1996 for two UK rivers of contrasting geology; the Lambourn at Shaw located in a chalk area (BFI = 0.96) and the Ray at Grendon Underwood, an area dominated by glacial clay and mudstone (BFI = 0.20).

The BFI is closely related to other low flow indices and is frequently used as a variable for estimating low flow indices at the ungauged site. Relationships between the index and catchment descriptors have been derived for different hydrogeological regimes and used to estimate BFI, and hence flow indices, at the ungauged site (Institute of Hydrology, 1980). In a revision of the UK low flow study (Gustard *et al.*, 1992) the necessity to estimate BFI at an ungauged site was replaced by the need to obtain information on soil types, and different low flow soil groups were identified based on the HOST soil classification system (Section 8.5.4).

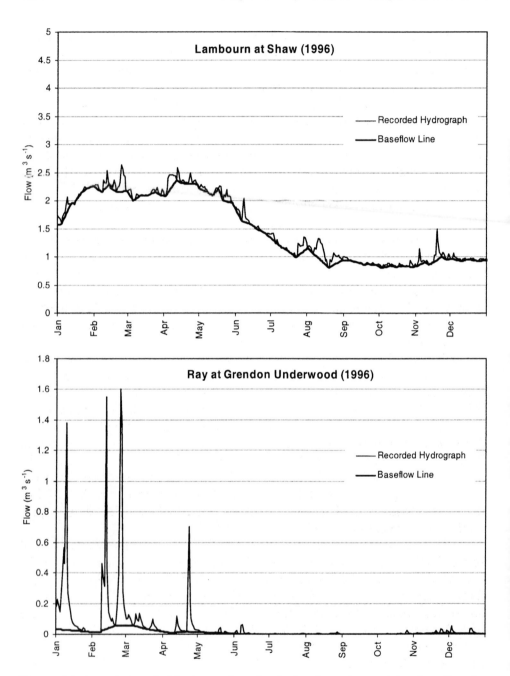

Figure 5.4 Annual recorded hydrograph and calculated continuous base flow line for two UK rivers based on the BFI separation procedure.

The base flow separation is performed on time series of daily mean flows. The minima of 5-day non-overlapping consecutive periods are calculated and turning points in this sequence of minima identified. The turning points, defined as points that belong to the base flow separation line, are connected to obtain the base flow hydrograph. The base flow is constrained to equal the observed hydrograph on any day when the base flow hydrograph exceeds the observed. A minimum becomes a turning point if its value times 0.9 is less than or equal to the neighbouring minima. In the original procedure (Institute of Hydrology, 1980) the equality test reads 0.9 'less than' the neighbouring minima. The latter does, however, not behave properly for zero flows, and it is therefore recommended to apply 'less than or equal to'. The factor 0.9 was determined by calibration on a wide range of catchments.

The base flow separation cannot start on the first day of the record, and similarly cannot finish on the last day. The start and end dates of the base flow hydrograph must therefore be used in calculating the volume of flow beneath the base flow hydrograph. For the same reason, BFI is sensitive to missing data as only one day may result in several days of data being omitted from the base flow separation. Interpolation can be used to infill missing periods providing these are of short durations (Section 4.3.3).

It is generally recommended to compute the base flow separation for the entire period of record. The BFI can then be calculated as the ratio of the volume of base flow to the volume of total flow for the whole period. Annual BFIs can also be estimated with the separation computed using the entire period of record thus avoiding the loss of some days at the start and end of each year. A period of record BFI can in this case be derived as the mean of the annual BFIs. However, the mean of the annual values will not in general equate the single value obtained from the entire record. For estimating the period of record BFI, calculating one value of BFI is generally recommended. In the UK low flow study (Institute of Hydrology, 1980) the annual BFI was found to be quite stable with a typical standard deviation of 0.04.

The BFI procedure was developed for rainfall regimes with a typical streamflow response time in days. This has guided the choice of model parameters to define the turning points in the base flow separation procedure (block size of five days and factor equal to 0.9). In lake regions or snow-dominated catchments the procedure may fail to separate larger floods as these typically last for longer periods depending on the volume of the lake or snow cover. It might therefore happen that a turning point is identified in the high flow period. In catchments with a sufficient long snow-free period, one might limit the calculations to periods without snow, i.e. to calculate annual summer BFIs. Accordingly, the calculation has to be performed for each year separately. The shorter period of observations and the loss of some days at the start and end of each year imply that longer records are necessary to obtain stable values (Tallaksen, 1987).

Worked Example 5.3: Base Flow Index

A program for calculating the Base Flow Index (BFI) can be found on the CD under Software. Data from River Ray at Grendon Underwood (UK) are used to demonstrate the procedure in the example below and in the program on the CD.

1. Data

   Three years of daily flow (1995 to 1997) from the Ray have been selected. The base flow separation is done for the whole period, whereas the BFI is calculated for the mid-year 1996. This ensures that days at the start and end of the calculation year are included. In Table 5.7 the calculation steps are illustrated using data from the beginning of the record.

2. Calculation

   (a) The daily flows, Q ($m^3 s^{-1}$), are divided into non-overlapping blocks of five days (Column 1 and 2).

   (b) Mark the minima of each of these blocks and let them be called $Qmin_1$, ... $Qmin_n$ (Column 3). Consider in turn ($Qmin_1$, $Qmin_2$, $Qmin_3$), ... ($Qmin_{n-1}$, $Qmin_n$, $Qmin_{n+1}$). In each case, if 0.9 central value $\leq$ outer values, then the central value is a turning point for the base flow line (marked in bold in Column 4). Continue this procedure until the whole time series has been analysed.

   (c) Join the turning points by straight lines to form the base flow separation line and assign to each day a base flow value Q_b, by linear interpolation between the turning points. If, on any day, the base flow estimated by this line exceeds the total flow, the base flow is set to be equal to the total flow Q, on that day.

   (d) Calculate the volume of water (V_{base}) beneath the base flow hydrograph between the first and last date of interest. The volume (m^3) is simply derived as the sum of the daily base flow values times the timespan in seconds per day.

   (e) Calculate the corresponding volume of water beneath the recorded hydrograph (V_{total}). The volume (m^3) is obtained by summing the daily flow values between the first and the last dates inclusive.

   (f) The BFI is then V_{base}/V_{total}.

Table 5.7 Calculation of the base flow separation line from time series of daily flow; non-overlapping 5-day blocks are separated with dotted lines and turning points are marked bold

1. Date	2. Daily flow, Q (m³ s⁻¹)	3. $Qmin$ (m³ s⁻¹)	4. 0.9 $Qmin$ (m³ s⁻¹)	5. Base flow, Q_b (m³ s⁻¹)
1-Jan-1995	0.109			
2-Jan-1995	0.063			
3-Jan-1995	0.043			
4-Jan-1995	0.039	0.039	0.0387	
5-Jan-1995	0.229			
6-Jan-1995	0.186			
7-Jan-1995	0.116			
8-Jan-1995	0.111			
9-Jan-1995	0.095	0.095	0.0855	
10-Jan-1995	0.123			
11-Jan-1995	0.178			
12-Jan-1995	0.091			
13-Jan-1995	0.076			
14-Jan-1995	0.073			
15-Jan-1995	0.062	0.062	0.0558	
16-Jan-1995	0.054	0.054	**0.0486**	0.054
17-Jan-1995	1.06			0.056
18-Jan-1995	0.856			0.058
19-Jan-1995	1.05			0.060
20-Jan-1995	1.34			0.062
21-Jan-1995	1.64			0.064
22-Jan-1995	1.35			0.067
23-Jan-1995	0.559			0.069
24-Jan-1995	0.255	0.255	0.2295	0.071
25-Jan-1995	0.644			0.073
26-Jan-1995	0.793			0.075
27-Jan-1995	0.896			0.077
28-Jan-1995	0.631			0.079
29-Jan-1995	1			0.081
30-Jan-1995	0.492	0.492	0.4428	0.083
31-Jan-1995	0.377			0.085
1-Feb-1995	1.67			0.087
2-Feb-1995	0.448			0.090
3-Feb-1995	0.237			0.092
4-Feb-1995	0.163	0.163	0.1467	0.094
5-Feb-1995	0.123			0.096
6-Feb-1995	0.102			0.098
7-Feb-1995	0.1	0.1	**0.09**	0.100
8-Feb-1995	0.151			0.107
9-Feb-1995	0.178			0.115
⋮	⋮			⋮

3. Results

(a) The first and second turning points are found at 16 January and 7 February 1995 (Column 4), respectively, and a linear interpolation is used to estimate the base flow between these dates (Column 5). The daily base flow separation line is subsequently calculated for the whole period by linear interpolation between all turning points.

(b) The volume beneath the base flow line, V_{base}, for 1996 is found to be 4.03 m^3, whereas the volume of the total flow, V_{total}, is 19.93 m^3. The resultant BFI is 0.20. The base flow separation line for River Ray in 1996 is shown in Figure 5.4.

5.3.4   Recession indices

The gradual depletion of water stored in a catchment during periods with little or no precipitation is reflected in the shape of the *recession curve*, i.e. the falling limb of the hydrograph. The duration of the decline is referred to as the *recession period*, and a *recession segment* is a selected part of the recession curve (Figure 5.5). The time resolution is commonly in the order of days. The recession curve describes in an integrated manner how different factors in the catchment influence the generation of flow in dry weather periods. It has therefore proved useful in many areas of water resources management; in low flow forecasting of gauged rivers (Section 2.5), as an index of drainage rate in rainfall-runoff models, as an index of catchment storage in regional regression models (Section 8.4), in hydrograph analysis for separation of different flow components (Section 5.3.3) and in frequency analysis for estimating low flow indices (Section 6.6.1).

The most important catchment properties found to affect the recession rate are hydrogeology, relief and climate (Tallaksen, 1995). Catchments with a slow recession rate are typically groundwater-dominated catchments, whereas a fast rate is characteristic of flashy, impermeable catchments with little storage. To make use of this knowledge it is necessary to express the shape of the curve in a quantitative manner. The quantification involves the selection of an analytical expression, derivation of a characteristic recession and optimization of the recession parameters. As there are many choices to be made through these calculation steps, this section does not present one particular methodology in detail. Rather an overview of different approaches and their background are given. A comprehensive review of recession analysis is provided by Hall (1968) and more recently by Tallaksen (1995).

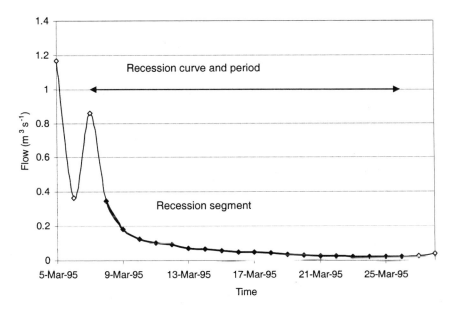

Figure 5.5 Definition of recession curve, period and segment (bold line) (flow data from River Ray at Grendon Underwood, UK).

Analytical expressions

The starting point of a recession analysis is to find an analytical expression that can adequately describe the outflow function Q_t, where Q is the rate of flow and t the time. Based on theoretical equations for groundwater flow, the decay of the aquifer outflow (groundwater discharge) rate can be modelled as a function of aquifer characteristics. This physically-based modelling approach (a) has proved successful at the catchment scale for relatively homogeneous conditions (Brutsaert & Nieber, 1977; Singh, 1969). Application in a heterogeneous catchment and at a regional scale is however, limited. Other modelling approaches include (b) modelling the recession as outflow from a reservoir, ranging from a single linear storage model to a cascade of linear or non-linear storages, (c) modelling the recession as an autoregressive process, from a simple first order autoregressive process to more complex autoregressive moving average models and finally, (d) empirical expressions have been sought.

The most commonly used function to model the recession rate is the simple exponential equation, given as Equation 5.1a or in the alternative form 5.1b or 5.1c:

$$Q_t = Q_{t=0} \exp(-t/C) \tag{5.1a}$$

$$Q_t = Q_{t=0} \exp(-\alpha t) \tag{5.1b}$$

$$Q_t = Q_{t=0} \, k^t \tag{5.1c}$$

where Q_t is the flow at time t and $Q_{t=0}$ is the flow at the start of the modelled recession. C, α and k are model parameters, often referred to as recession constants, coefficients or indices, which characterize the slope of the recession curve, i.e. the recession rate. It can be shown that this equation results as a special case of all four models outlined above. The curve plots as a straight line of slope $-t/C$, $-\alpha t$ or $\ln k$ in a semi-logarithmic plot of t against $\ln Q_t$. The constant C is the time elapsed between any discharge Q and Q/e of the recession. C is related to $t_{0.5}$ (the time required to halve the streamflow) by the equation:

$$C = - \, t_{0.5} / \ln(\tfrac{1}{2}) \tag{5.2}$$

The lack of fit of this simple equation for a wide range of flow has led to the separation of the recession curve into separate components of overland, unsaturated and saturated flow (e.g. Singh & Stahl, 1971; Klaassen & Pilgrim, 1975). The composite recession curve can be modelled as the outflow from a set of linear reservoirs and commonly two or three terms are adopted. Werner & Sundquist (1951) derived the outflow from a confined aquifer as a sum of exponential terms. Alternatively, nonlinear relationships have been sought (e.g. Brutsaert & Nieber, 1977) and shown to yield the outflow from an unconfined aquifer (e.g. Hornberger *et al.*, 1970). In either case more than one recession parameter is needed to describe the recession rate.

Characteristic recession

In a humid climate, rainfall frequently interrupts the recession period and a series of recession segments of varying duration results. It has proved difficult to define a consistent means of selecting recession segments from a continuous record. The initial discharge defining the start of the recession segment can be given as a constant value or it may vary depending on the procedure chosen. If the recession is considered a composite curve, a starting value for each flow component must be defined. It is common to disregard the first part of a recession period to exclude the influence of overland flow. A constant value restricts the recession to the range of flow below a predefined discharge. A variable starting value can be defined as the flow at a given time after rainfall or peak of discharge. Similarly, the length of the recession period can be a constant or varying number of time steps in a flow sequence. Normally a minimum length is imposed.

Several recession parameters can be calculated from the set of recession segments. It is therefore necessary to combine the information contained in the

Box 5.1

Master recession curve

Several methods have evolved to construct a master recession curve for a catchment from a set of shorter recessions. A major problem is the high variability encountered in the recession rate of individual segments, which represent different stages in the outflow process. In addition, seasonal variation in the recession rate adds to the variability. The master recession methods try to overcome the problem by constructing a mean recession curve. The most commonly used techniques are the matching strip and the correlation method.

In the *matching strip method* (Toebes & Strang 1964) individual segments are plotted and adjusted horizontally until they overlap. The master recession is then constructed as the mean line by best eye fit through the set of common lines. The method permits a visual control of irregularities in the recession curve, but as it is based on a subjective fit it might telescope or contract the true recession. In the *correlation method* (Langbein, 1938) the discharge at one time interval is plotted against discharge one time interval later and a curve fitted to the data points. If the recession rate follows an exponential decay, a straight line results and the slope of the line equals the recession parameter k in Equation 5.1c. In Figure 5.6 the method is demonstrated using data from the recession period depicted in Figure 5.5. The segment covers the period 8 to 26 March 1995. Starting one day later on the 9 March the slope changes from 0.53 to 0.70. This clearly

($\rightarrow$ Box continued on next page)

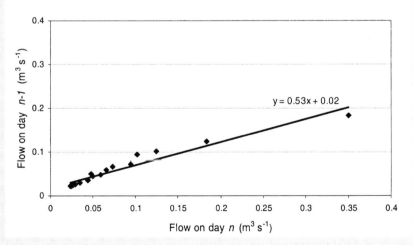

Figure 5.6 Recession rate determined by the correlation method (flow data from River Ray at Grendon Underwood, UK).

Box 5.1 (continued)

demonstrates how sensitive the model is to the selection of recession segments. Parameters of a master recession curve are derived by plotting data from several recession segments in the same plot prior to fitting a straight line.

More recent applications of the method plot the rate of change of flow, $\Delta Q/\Delta t$, against Q. The graphical analysis of the relationship is often performed by means of the upper and lower envelope of points, representing the maxima and minima observed recession rates respectively (Brutsaert & Nieber, 1977; Troch *et al.*, 1993). Application of the correlation plot requires high accuracy in the low flow measurements, and the quality of the low flow data is often a limiting factor (Section 4.3) when the time interval Δt, is chosen. If the measurement site does not exhibit sufficient precision for low flow measurements, several days may pass before a lower value is recorded. The resulting steps in the discharge series may impose difficulties in the selection and modelling of recessions.

---

recession behaviour of the individual segments to be able to identify and parameterize the characteristic recession behaviour of the catchment. This can be done by constructing a master recession curve (Box 5.1), or by obtaining average values of the recession parameters from the set of individual recession segments. If it is assumed that there are n recession segments, one from each recession period, then the j recession parameters can be obtained either from the master recession curve or for each of the n segments. In the latter case a mean value can be calculated to represent average catchment conditions and the variability can be characterized by, for example, the variance. If the simple exponential equation (Equation 5.1) is used there is only one parameter to be estimated, i.e. $j = 1$.

Optimization of recession parameters

Today, manual techniques like the subjective best eye fit of the data have been abandoned in favour of automatic curve fitting procedures to determine the functional relationship of the recession curve or plot (James & Thompson, 1970; Pereira & Keller, 1982; Vogel & Kroll, 1992). This is commonly done using a simple or weighted least-squares regression. If a separate linear regression can be fitted to each recession segment by the least squares method and the sum of squares for variation about each line obtained, both the individual lines of different slopes and the parallel lines with an average slope can be considered. In the latter case it is necessary to account for the number of observations in each recession segment to obtain an unbiased estimate of the average slope (Brownlee, 1960). The average slope equals the arithmetic mean of the individual slopes provided the segments are of equal length, otherwise a weighted average must be calculated.

Time variability in recessions

During warm and dry weather, and especially in the active growing season, water lost by evapotranspiration has a marked influence on the recession rate (e.g. Federer, 1973). Steeper recession curves are generally found during the warm season along with a reduction in base flow. The question is to what extent this can be explained by higher evapotranspiration losses during the recession period, or is the overall seasonal increase in evapotranspiration leading to a lower groundwater recharge and a general change in catchment wetness status more important. The drainage following a summer storm on initially dry soils will differ from a typically autumn storm when soils are well saturated. The variability encountered in the recession rate depends also on the particular recession model and calculation procedure adopted. The starting point and the length of the recession segment are here of importance.

The high time variability found in recession segments argues against the use of a master recession curve except as an overall approximation that might be applicable for comparative purpose at the regional scale. The choice of recession model depends on which part of the recession curve is most important. In a regional study where recession indices are thought to represent storage properties it is the time required to reach low flow that is most important. For regional analysis a simple expression is preferable as the recession behaviour generally varies considerably between sites. Low flow forecasting strives towards high precision in the lower end of the recession curve and more complicated models might be sought. Aksoy *et al.* (2001) allow seasonal effects in the recession behaviour to be considered when simulating the recession flow using an autoregressive modelling approach. Generally, variability owing to model limitations and the calculation procedure should be minimized, whereas physically-based short-term or seasonal variation should be accounted for.

5.4 Deficit characteristics

As opposed to low flow characteristics presented in the previous section, deficit characteristics are based on introducing a threshold below which the flow is regarded as being in a drought situation. Each deficit or drought event can be characterized amongst others by its total duration and its deficit volume. This section describes two methods, the threshold level method and the sequent peak algorithm (SPA), to select and characterize drought events.

5.4.1   Threshold level method

In the following a detailed description of the threshold level method for defining drought events is given. This is the most frequently applied quantitative method where it is essential to define the beginning and the end of a drought. It is based on defining a threshold, Q_0, below which the river flow is considered as a drought (also referred to as a low flow spell). The threshold level method, which generally studies runs below or above a given threshold, was originally named 'method of crossing theory' (Box 7.1), and it is also referred to as run-sum analysis. Early application of crossing theory in hydrology includes Yevjevich (1967), where the method is based on the statistical theory of runs for analysing a sequential time series. The method is relevant for storage and yield analysis, which is associated with hydrological design and operation of reservoir storage systems. Important areas of application are hydropower and water management, water supply systems and irrigation schemes.

Figure 5.7 gives an example of how drought events are identified by the threshold level method. First a threshold, Q_0, is introduced. When the flow falls below the threshold value, a drought event starts and when the flow rises above the threshold the drought event ends. Hence, both the beginning and the end of the drought can be defined. Statistical properties of the distribution of drought deficit, drought deficit duration (run-length, d_i, Figure 5.7) and volume or severity (run-sum, v_i, Figure 5.7), are recommended as characteristics for at-site drought. Simultaneously it is possible to define the minimum flow of each drought event, $Q_{min,}$ (Figure 5.7), which can also be regarded as a deficit characteristic. The time of drought occurrence has been given different definitions, for instance the starting date of the drought, the mean of the onset and the termination date, or the date of the minimum flow. Often another

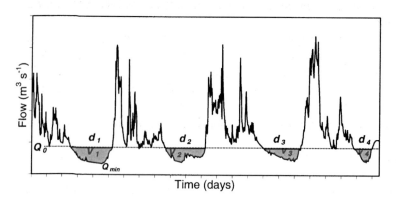

Figure 5.7 Definition of drought deficit characteristics.

drought deficit characteristic, the drought intensity (sometimes also referred to as drought magnitude) m_i, is introduced as the ratio between drought deficit volume and drought duration. Based on the time series of drought deficit characteristics it is possible to derive drought deficit indices as described in Figure 5.1, column III.

Threshold selection

The threshold might be chosen in a number of ways and the choice is, amongst other things, a function of the type of water deficit to be studied. In some applications the threshold is a well-defined flow quantity, e.g. a reservoir specific yield. It is also possible to apply low flow indices (Sections 5.3.1 and 5.3.2), e.g. a percentage of the mean flow or a percentile from the flow duration curve.

The hydrological regime will influence the selection of a percentile from the FDC as threshold level. For perennial rivers relatively low thresholds in the range from Q_{70} to Q_{95} can be considered reasonable. For intermittent and ephemeral rivers having a majority of zero flow, Q_{70} could easily be zero, and hence no drought events would be selected. For example Meigh *et al.* (2002) found Q_{70} or Q_{90} to be reasonable threshold levels for series in the southern African region; however for ephemeral rivers the mean flow was used as threshold. Woo & Tarhule (1994) tested seven thresholds, Q_5, $Q_{7.5}$, Q_{10}, $Q_{12.5}$, Q_{15}, $Q_{17.5}$, Q_{20}, for ephemeral rivers in Nigeria and argued that these are realistic levels due to the large proportion of zero flow (several months every year). Similarly, Tate & Freeman (2000) studied eight streamflow records from the southern African region and applied threshold levels ranging from $Q_{12.5}$ to Q_{90} depending on the proportion of zero flow. As an alternative, a variable threshold can be applied, as in Kjeldsen *et al.* (2000) who applied Q_{75} of the monthly FDC (variable threshold level) to select drought events from daily streamflow records in Zimbabwe. Yet another option is to calculate percentiles based on the non-zero values only.

In some areas, or in a regional study, the lack of long series over larger areas may impose restrictions on the use of very low threshold levels due to the presence of too many non-drought years at some sites (the flow never falls below the chosen threshold in a year). On the other hand, the problem of droughts lasting longer than one year (multi-year droughts) becomes more severe for increasing threshold levels and care should be taken when selecting a relatively high threshold.

The threshold might be fixed or vary over the year. A threshold is regarded as fixed if a constant value is used for the whole series. If the threshold is derived from the flow duration curve it implies that the whole streamflow record (or a predefined period) is used in its derivation. This is illustrated in

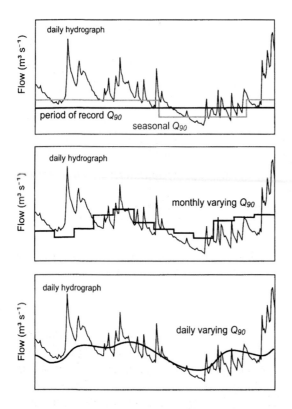

Figure 5.8 Illustration of threshold levels; fixed threshold (upper), monthly varying threshold (middle), daily varying threshold (lower) (modified from Stahl, 2001).

Figure 5.8 (upper), where the threshold is the Q_{90} for the period of record. If summer and winter droughts are studied separately the threshold can also be fixed, but is then based only on flow data from the relevant season studied. A variable threshold is a threshold that varies over the year, for example using a monthly (Figure 5.8, middle) or daily (Figure 5.8, lower) varying threshold level.

The variable threshold approach is adapted to detect streamflow deviations during both high and low flow seasons. Lower than normal flows during high flow seasons might be important for later drought development. However, periods with relatively low flow either during the high flow season or, for instance, due to a delayed onset of a snow-melt flood, are commonly not considered a drought. Therefore, the events defined with the varying threshold should be called streamflow deficiencies or *streamflow anomalies* rather than streamflow droughts.

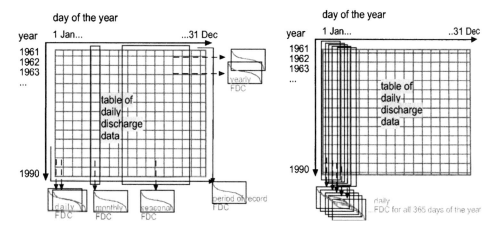

Figure 5.9 Scheme for determination of different flow duration curves for the threshold level definitions; calendar units (day, month, season) (left), moving window (daily) (right) (modified from Stahl, 2001).

A variable threshold can thus be used to define periods of streamflow deficiencies as departures or anomalies from the 'normal' seasonal or daily flow range. A daily varying threshold level can, for example, be defined as an exceedance probability of the 365 daily flow duration curves. Exceedances derived on a daily basis may be misleading where the number of observations is small because the data series are short. To increase the sampling range and smooth the threshold, daily exceedances can be calculated from all flows that occur within an n-day window. For example, applying a 31-day window, the flow exceedance on 1 June would be calculated from all discharges recorded between 17 May and 16 June in each year of the period of record.

Using a period of record of N years, and provided there are no gaps in the gauged daily flow record, the flow exceedance on any given day of the year is given by:

$$QE_d = \frac{((nN+1) - i_d) \cdot 100\%}{nN} \tag{5.3}$$

where QE_d is the flow exceedance on day d, n is the length of the window in days, N is the number of record years and i_d is the rank of the gauged daily flow on day d (flows are ranked in ascending order) in the set of nN values.

A stepwise illustration of the threshold calculation for different time resolutions is seen in Figure 5.9. Derivation of daily, monthly and seasonal period of record flow duration curves are illustrated in Figure 5.9 (left) and the

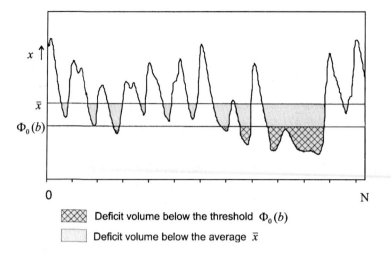

Figure 5.10 Illustration of the definition of the threshold, $\Phi_0(b)$, according to Equation 5.4 (modified from Peters *et al.*, 2003; reproduced by permission of John Wiley & Sons Limited).

exceedance percentiles for each day of the year are calculated from a n-day moving window as demonstrated in Figure 5.9 (right). One of the resulting daily exceedances (e.g. Q_{90}) can be applied as a varying threshold (Figure 5.8, lower). A streamflow anomaly is said to occur when the streamflow of a given day is lower than the threshold for that particular day.

To be able to make a comparison between drought deficits based on different groundwater variables Peters *et al.* (2003) suggest a new approach to derive threshold levels. This method is based on relating the total deficit below the threshold, Φ_0, to the total deficit below the average, $\bar{x}$. The total deficit below a threshold is the sum of the deficits of all droughts below this threshold or, in other words, the sum of the deficits in the partial duration series (PDS) (Chapter 6). This is illustrated in Figure 5.10. The threshold function, $\Phi_0(b)$ can be defined as follows:

$$\sum_{t=1}^{N}\left[\left(\Phi_0(b)-x_t\right)_+ \Delta t\right] = b\sum_{t=1}^{N}\left[\left(\bar{x}-x_t\right)_+ \Delta t\right] \tag{5.4}$$

where:

$$x_+ = \begin{cases} x & if \quad x \ge 0 \\ 0 & if \quad x < 0 \end{cases}$$

N is the length of the time series and b is called the *drought criterion*, which determines the threshold level. The drought criterion b is the ratio of the deficit below the threshold to the deficit below the average. If different variables, e.g. groundwater recharge and discharge, are compared and the drought criterion is kept constant, the ratio is the same for all variables but the threshold percentile can vary. If $b = 1$ the threshold level is equal to the average of x. If $b = 0$ the threshold level is equal to the minimum of x.

Time resolution

The question of time resolution, whether to apply series of annual, monthly or daily streamflow, depends on the hydrological regime under study and the specific problem to be solved. In the temperate zone a given year might include both severe droughts (seasonal droughts) and months with abundant streamflow, meaning that annual data would often not reveal severe droughts. Dry regions like the B- and Cs-climates (Section 2.2.2) are more likely to experience droughts lasting for several years, multi-year droughts, which supports the use of a monthly or an annual time step. Hence, different time resolutions (annual, monthly and daily) might lead to different results regarding the drought event selection.

The threshold level method introduced by Yevjevich (1967) was originally based on analysing sequential time series with a time resolution of one month or longer. The approach has also been used for analysing streamflow droughts from a daily-recorded hydrograph (e.g. Meigh *et al.*, 2002; Hisdal *et al.*, 2001; Kjeldsen *et al.*, 2000; Tallaksen *et al.*, 1997; Woo & Tarhule, 1994;

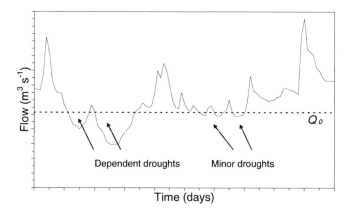

Figure 5.11 Daily time series of flow illustrating the problem of mutual dependence and minor droughts (modified from Tallaksen, 2000; reproduced with kind permission of Kluwer Academic Publishers).

Zelenhasic & Salvai, 1987). These studies demonstrate the potential of this method for a complete description of the stochastic process of seasonal (within-year) droughts. However, the use of a daily time resolution introduces two special problems, namely dependency among droughts and the presence of minor droughts. During a prolonged dry period it is often observed that the flow exceeds the threshold level for a short period of time and thereby a large drought is divided into a number of minor droughts that are mutually dependent (Figure 5.11). To avoid these problems which could distort an extreme value modelling, a consistent definition of drought events should include some kind of pooling in order to define an independent sequence of droughts (Box 5.2).

Box 5.2

Pooling procedures

Tallaksen *et al.* (1997) compared and described three different pooling procedures for dependent droughts; the moving average procedure (MA), the sequent peak algorithm (SPA) and the inter-event time and volume criterion (IC). MA simply smoothes the time series applying a moving average filter and it was recommended to apply a moving average window of 10 days. The IC is used to pool two subsequent events with characteristics (d_i, v_i) and (d_{i+1}, v_{i+1}) if (a) the inter-event time, t_i, between the two droughts is less than or equal to a critical duration, t_{min}, and (b) the ratio between the inter-event excess volume, z_i, and the preceding deficit volume is less than a critical ratio. The pooled drought is then again pooled with the next drought if (a) and (b) are fulfilled, and so on. The pooled drought deficit characteristics were calculated as follows:

$$d_{pool} = d_i + d_{i+1} + t_i \qquad\qquad\qquad\qquad\qquad\qquad \text{(B5.1a)}$$

$$v_{pool} = v_i + v_{i+1} - z_i \qquad\qquad\qquad\qquad\qquad\qquad \text{(B5.1b)}$$

The SPA can be regarded as a method for selection of drought events and is therefore described in a separate section (Section 5.4.2), where the pooling properties of the method are also explained. The results of Tallaksen *et al.* (1997) indicate that the IC method is inferior to the MA and SPA methods.

The presence of multi-year droughts restricts the use of the SPA method for analysis of within-year droughts. It can only be used for very low threshold levels. Kjeldsen *et al.* (2000) show that the SPA generally extracts larger extreme events than MA using a partial duration series approach (Chapter 6). The advantage of the SPA method is that no parameters need to be determined prior to the use of the method, as compared to the choice of averaging window in the MA method.

However, the number of minor droughts is larger in the SPA method. Kjeldsen *et al.* (2000) suggest removing the smallest droughts according to a frequency factor because these small events complicate the frequency distribution modelling of the most extreme events. The choice of frequency factor is subjective and must be regarded as a parameter of the method. The main advantage of the MA pooling method is that it reduces the problem of minor droughts at the same time as mutually dependent droughts are pooled. However it might well introduce dependency between the drought events, especially if the filter width is large. The SPA method on the other hand has a more straightforward interpretation as the observed data are used directly. The threshold can be interpreted as the desired yield from a reservoir, and the drought deficit volume defines the required storage in a given period (Section 5.4.2).

Worked Example 5.4: Threshold level method

A program, Nizowka, for selecting and analysing drought events based on the threshold level method can be found on the CD under Software. This program also allows frequency analysis to estimate extreme quantiles of drought duration and deficit volume (Worked Example 6.2). Data from River Ngaruroro at Kuripapango (NZ) are used to demonstrate the procedure in the example below. In Nizowka it is possible to analyse several of the series in the Global Data Set (Section 4.5.1) and import additional data series to be analysed.

1. Data

   36 years of daily flow (1 September 1964 to 31 August 2000) are analysed. In this river the low flow period covers the turn of the calendar year. To avoid problems with allocating droughts to a specific calendar year because of drought events starting in one year and ending in another year, the start of the year is defined to be 1 September. Periods of missing data have been interpolated if of short duration, whereas years containing long periods of missing values (> 15 days) have been removed. In total four years are omitted from the series, 1967–1968, 1978–1979, 1986–1987 and 1987–1988. As the program does not allow for missing years, the years have been connected to avoid gaps. This has to be accounted for when analysing the date of the drought events.

2. Parameter determination

   (a) Threshold selection

       A sequence of drought events is obtained from the streamflow hydrograph by considering periods with flow below a certain threshold, Q_0. In Nizowka the *threshold level* is obtained as a value from the FDC and it is determined as an exceedance percentile (%) under the menu 'Configuration'. In this example Q_{90} is used as threshold.

(b) Minor droughts

To reduce the problem of minor droughts two restrictions can be imposed:

    (i) a minimum drought duration, d_{min}, (*minimum drought length*, under the menu 'Configuration') that removes droughts with duration less than the specified number of days (here d_{min} is set equal to five days);

    (ii) a minimum drought deficit volume (*coefficient alpha*, under the menu 'Configuration'), that removes droughts with a deficit volume less than a certain fraction α of the maximum drought deficit volume in the complete series of drought events (here α is set equal to 0.005).

(c) Dependent droughts

The inter-event time criterion (IC) is used to pool dependent droughts which are separated by a short period of flow above the threshold. If the time between two droughts is less than a critical duration, t_{min}, (*minimum distance between two successive droughts*, under the menu 'Configuration') the two events are pooled. In this example t_{min} is set equal to two days.

3. Calculation

By selecting 'Tables' and then 'Droughts' from the menu, ticking the box 'All year', selecting the correct gauging station and finally pushing the button 'Droughts', the drought events are selected. A table is displayed including:

(a) the start date, defined as the first day below the threshold;

(b) the end date, defined as the last day below the threshold;

(c) the deficit volume (1000 m^3), defined as the sum of the daily deficit flows times the duration in days;

(d) the average deficit or drought intensity (1000 m^3 day^{-1}), defined as the ratio between the drought deficit volume and the number of days from the start date to the end date, the latter called *full drought duration* in Nizowka;

(e) the drought duration (days), defined as the full drought duration minus short periods (the inter-event time) above the threshold, called *real drought duration* in Nizowka;

(f) the minimum flow (m^3 s^{-1}), defined as the minimum flow within a drought event;

Table 5.8 Drought deficit characteristics

Start	End	Def. vol. (1000 m³)	Av. Def. vol. (1000 m³ day⁻¹)	Real Duration (days)	Min. flow (m³ s⁻¹)	Date min.flow	Av. Flow (m³ s⁻¹)
11.03.65	16.03.65	148.090	24.682	6	4.805	15.03.65	4.956
07.05.66	14.05.66	140.573	17.572	7	4.819	14.05.66	5.053
17.02.67	03.04.67	4953.571	107.686	45	3.210	02.04.67	4.025
26.03.69	03.04.69	312.336	34.704	9	4.404	03.04.69	4.840
06.04.69	18.04.69	1176.422	90.494	13	3.838	17.04.69	4.195

(g) the date of the minimum flow;

(h) and the average flow during the drought (m³ s⁻¹).

The drought deficit characteristics of the first five drought events are given in Table 5.8. By selecting 'Print', two tables are displayed. Either the table 'Parameters' (as in Table 5.8) or the table 'Number of droughts' which summarises the number of droughts every year sorted by drought duration. From the latter table it can be seen that there are 71 drought events in total, on average two events every year. Minor droughts are dominating with 36 events lasting less than 11 days. Only 5 events lasted more than 30 days. All tables can be written to text files. The time series of the drought duration can be seen in Figure 5.12, and the major droughts

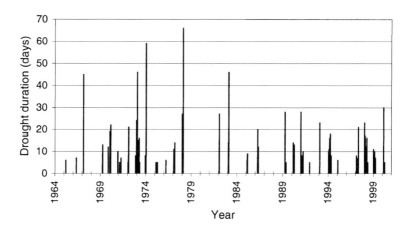

Figure 5.12 Time series of drought duration for River Ngaruroro at Kuripapango (NZ). Selection criteria: threshold level = Q_{90}, d_{min} = 5 days, α = 0.005 and t_{min} = 2 days.

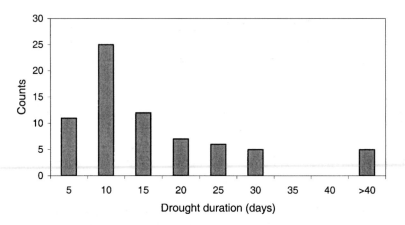

Figure 5.13 Histogram of drought duration for River Ngaruroro at Kuripapango (NZ). Selection criteria: threshold level = Q_{90}, d_{min} = 5 days, α = 0.005 and t_{min} = 2 days.

are found in 1967, 1973, 1974, 1978 and 1983. A histogram of the drought duration is seen in Figure 5.13, and a very skewed distribution is revealed.

5.4.2   The sequent peak algorithm

A procedure for preliminary design of reservoirs based on annual average streamflow data is the mass curve or its equivalent, the Sequent Peak Algorithm (SPA) (e.g. Vogel & Stedinger, 1987). Reservoir design and storage-yield analysis is treated in considerable detail in McMahon & Mein (1986). SPA can also be used for daily data to derive drought events. Let Q_t denote the *daily* inflow to a reservoir and Q_0 the desired yield or any other predefined flow, then the storage S_t required at the beginning of the period t reads (Tallaksen *et al.*, 1997):

$$S_t = \begin{cases} S_{t-1} + Q_0 - Q_t, & \text{if positive} \\ 0 & , & \text{otherwise} \end{cases} \qquad (5.5)$$

An uninterrupted sequence of positive S_t, $\{S_t, t = \tau_0, ..., \tau_{end}\}$, defines a period with storage depletion and a subsequent filling up (Figure 5.14). The required storage in that period, $\max\{S\}$, defines the drought deficit volume, v_i, and the time interval, d_i, from the beginning of the depletion period, τ_0, to the

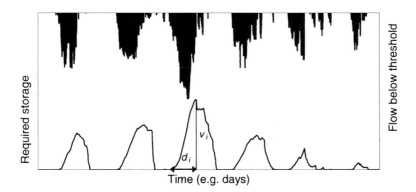

Figure 5.14 Definition of drought events using the sequent peak algorithm (SPA) method (modified from Tallaksen *et al.*, 1997; reproduced by permission of IAHS Press).

time of the maximum depletion, τ_{max}, defines the drought duration $(\tau_{max} - \tau_0 + 1 = d_i)$.

This technique differs from the threshold level method in that those periods when the flow exceeds the yield do not necessarily negate the storage requirement, and that several deficit periods may pass before sufficient inflow has occurred to refill the reservoir. Hence, based on this method, two droughts are pooled if the reservoir has not totally recovered from the first drought when the second drought begins ($S_t > 0$).

Worked Example 5.5: Sequent peak algorithm (CD)

Excel sheet 'Calculation and Result'

1. Data

    Ten years of daily data from River Ngaruroro at Kuripapango (NZ) are used as an example. Table 5.9 lists the first flow values. The whole series can be found on the CD. The first two columns show the date and the corresponding flow value, *Q*.

2. Calculation

    (a) Type the value of the desired yield (in cell number B5).

    (b) Calculate the storage according to Equation 5.5 (using the IF function in Excel).

Table 5.9 SPA calculation of drought deficit volumes and duration for River Ngaruroro at Kuripapango (NZ)

Data, 10-year series		SPA calculation
	Streamflow	**Storage (Eq. 5.5)**
Date	Q_t **(m^3 s^{-1})**	S_t **(m^3 s^{-1})**
		0.000
1-Jan-91	4.676	0.504
2-Jan-91	4.827	0.857
:	:	:
23-Jan-91	4.038	18.569
24-Jan-91	4.281	**19.468**
25-Jan-91	15.557	9.091
26-Jan-91	12.747	1.524
27-Jan-91	7.762	0.000
:	:	:
10-Feb-91	5.234	0.000
11-Feb-91	4.983	0.197
12-Feb-91	4.800	0.577
13-Feb-91	4.670	1.087
14-Feb-91	4.581	1.686
15-Feb-91	4.369	2.497
16-Feb-91	4.234	3.443
17-Feb-91	4.358	**4.265**
18-Feb-91	99.101	0.000

The calculated storage is displayed in the right column. In the calculation Q_{90} is used as the desired yield, Q_0.

3. Selection of the drought deficit volume and duration

   The deficit volume is the maximum value in an uninterrupted sequence of positive S_t, and the drought duration is the time from the beginning of the depletion period to the time of the maximum depletion. Find these values by pushing the button named 'Extract max.' (Pushing this button activates a routine programmed in Visual Basic.) The date of the maximum depletion is also displayed.

4. Results

   An extract of the drought duration and deficit volumes for the 10-year series is given in Table 5.10. The time series starts with a flow value below the threshold and the previous flows are not known. The first

Table 5.10 An extract of drought deficit volumes and durations for River Ngaruroro at Kuripapango (NZ), calculated by SPA

i	v_i $(m^3 s^{-1})$	d_i (days)
1	19.468	24
2	4.265	7
3	3.776	10
4	0.926	5
5	0.053	1

deficit volume and duration should therefore be omitted from further analyses.

As can be seen from the results displayed on the CD, even though the SPA procedure is pooling minor and dependent droughts, the obtained time series of events still contains a number of minor drought events.

5.4.3 Regional drought characteristics

There are two main categories of regional drought studies. The first category, regional methods aims at estimating low flow and drought characteristics at the ungauged site or reducing sampling uncertainty at sites with short records, is described in Chapters 6 and 8. As droughts commonly cover large areas and extend for long time periods, it is a key issue to study such events within a regional context. The consequences of a drought are certainly influenced by the extent of the area affected by it, and the second category aims at analysing regional properties of drought by:

a) analysing spatial patterns of at-site droughts either by studying historical time series (e.g. Tallaksen & Hisdal, 1997; Hisdal *et al.*, 2001) or the real time development of a drought in space (e.g. Zaidman *et al.*, 2001);

b) studying regional drought characteristics such as the area covered by drought and the total deficit over the drought area (Tase, 1976; Sen, 1980; Santos, 1983; Rossi *et al.*, 1992; Sen, 1998; Vogt & Somma, 2000; Hisdal & Tallaksen, 2003).

Both categories of regional drought analyses have been reviewed by Rossi *et al.* (1992) applying precipitation as an example. A more recent review with the focus on hydrological droughts in terms of streamflow is given in Hisdal (2002).

When analysing spatial patterns of at-site droughts, the drought analyses are often based on at-site event definitions, e.g. the threshold level method, where the areal aspect is included by studying the spatial pattern of point values and without introducing a separate, regional drought event definition. This section describes how to derive regional drought characteristics such as the area covered by drought and the total deficit over the drought area, for a given threshold level. Complex drought indices such as the Surface Water Supply Index (SWSI) (Section 5.7) are calculated for basins or regions and as such also describe regional drought aspects. They are, however, described separately as they do not include the area covered by drought as a dynamic variable (varying in time).

Precipitation or streamflow can be described as a space-time random process. Our observations of this underlying process of droughts are made at a finite number of sites. If the site coordinates are given, the phenomenon can be viewed as a multivariate stochastic process. Analysis of regional drought characteristics involves three main steps (Rossi *et al.*, 1992): development of a mathematical model for the underlying process, selection of regional drought characteristics, and analysis of the statistical properties of the drought characteristics. These properties can in principle be investigated in three ways:

a) by determining the probability density functions (pdfs) of drought characteristics from *observed time series* (Chapter 6). This technique is hampered by the fact that sufficiently long time series with a reasonable area coverage can rarely be found for the region under study;

b) by application of *analytical methods* to obtain the pdfs (or moments of the pdfs) of the indices from the statistical characteristics of the underlying process (e.g. Sen, 1980; Santos, 1983; Sen, 1998). To obtain analytical solutions it is necessary to base the calculations on simplified assumptions that often do not comply with reality. As for at-site studies, this is a major limitation of the method;

c) by application of *Monte Carlo techniques* (Chapter 7) to simulate long time series at many sites in a region and then study the statistical properties of the regional characteristics calculated from the simulated series. Examples are given by Tase (1976), Tase & Yevjevich (1978), Krasovskaia & Gottschalk (1995), Henriques & Santos (1999) and Hisdal & Tallaksen (2003).

Tase (1976) introduced the areal aspect in the definition of a drought by generating monthly precipitation at a systematic grid of points. The threshold level method was then applied to each grid cell. Three regional drought characteristics were selected. For a given threshold level, Φ_0 (equivalent to Q_0 for streamflow), and a number of grid cells, n, these characteristics at time t were defined as:

The *deficit area* being the number of grid cells with precipitation below a certain threshold:

$$A_{def} = \sum_{i=1}^{n} I_{(\Phi \leq \Phi_0)}(\Phi_i) \tag{5.6}$$

where $I_{\Phi \leq \Phi_0}(\Phi_i)$ is an indicator function described by:

$$I_{(\Phi \leq \Phi_0)}(\Phi_i) = \begin{cases} 1 & \text{if } \Phi_i < \Phi_0 \\ 0 & \text{if } \Phi_i > \Phi_0 \end{cases} \tag{5.7}$$

The *total areal deficit* being the sum of drought deficit volumes in the drought-affected grid cells:

$$V_{area} = \sum_{i=1}^{n} (\Phi_0 - \Phi_i) I_{(\Phi \leq \Phi_0)}(\Phi_i) \tag{5.8}$$

The *maximum deficit intensity* being the maximum deficit volume in one grid cell:

$$m_{max} = \Phi_0 - \min\{\Phi_1, \Phi_2, ..., \Phi_n, \Phi_0\} \tag{5.9}$$

In Santos (1983) the areal extent of a drought is included in the definition by introducing a critical area as a second threshold. The study is carried out on annual precipitation, assuming each precipitation station to represent a fraction of the total area. So-called 'auxiliary variables' and regional drought characteristics are introduced. The auxiliary variables are defined to be:

a) proportion of instantaneous deficit-area (defined equivalent to Equation 5.6);

b) proportion of instantaneous drought-affected area. Here a critical area is taken into consideration. Unless the critical area is exceeded, the proportion of drought-affected area is zero;

c) instantaneous regional areal deficit (defined equivalent to Equation 5.8);

d) instantaneous deficit-area areal deficit. This variable gives the mean of the total area deficit for a fraction of the area;

e) instantaneous drought areal deficit. The variable describes the mean amount of 'missing' water over a region if the deficit area is greater than or equal to the critical area.

The auxiliary variables are used to define the following regional drought characteristics:

a) regional drought duration. The length of the time interval where the drought covers a fraction of the area that is larger than or equal to the critical area;

b) total regional areal deficit. The total deficit volume over the time interval where the drought covers a fraction of the area that is larger than or equal to the critical area;

c) proportion of temporal drought area. The average drought area over the time interval where the drought covers a fraction of the area that is larger than or equal to the critical area;

d) intensity. This is the total regional areal deficit divided by the regional drought duration.

An application of a regional drought definition for precipitation can be found in Henriques & Santos (1999). The threshold level method is used within elementary areas. Drought occurs when the value of the variable is less than a fixed threshold, Φ_0. The stepwise selection of adjacent elementary areas (neighbourhood) in drought is represented in Figure 5.15. A drought event starts in the nucleus with maximum drought severity (according to Equation 5.9). The spread of the drought can be obtained by identifying drought-affected areas adjacent to the nucleus. This selection envisages the enlargement of the core to the detriment of dispersed areas. Areas in drought are selected, until the whole region has been selected or the threshold is exceeded in all adjacent elementary areas ($\Phi > \Phi_0$). A critical area can then be introduced as a subsequent step by only considering droughts affecting a certain percentage of the total region under study.

Tase (1976) and Santos (1983) developed their regional drought characteristics for precipitation. A similar procedure can be used for monthly streamflow thus allowing the definition of regional hydrological drought characteristics. However, two specific features of streamflow have to be considered. In many hydrological regimes streamflow is more persistent than precipitation, and this has to be accounted for when generating monthly

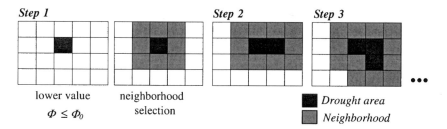

Figure 5.15 Stepwise selection of adjacent elementary areas in drought (from Santos *et al.*, 2000).

streamflow. Furthermore, whereas precipitation can be regarded as a point process, streamflow represents an integrated value over the catchment. How to estimate gridded runoff is discussed on a general basis in Gottschalk & Krasovskaia (1998).

Hisdal & Tallaksen (2003) made the first study of regional hydrological drought characteristics based on a monthly time step. Streamflow in Denmark is generated in a systematic grid of points based on time series of observed streamflow. As the catchment areas are small it is assumed that each streamflow record can be considered as representing a point in space. Persistency between months is included in the streamflow generation procedure. The droughts are allowed to develop over the time step applied and, in addition to the regional drought characteristics developed by Tase (1976), the total areal drought duration is defined as: the total number of consecutive months with streamflow below the threshold level within one or more grid cells. This study is described in more detail in Section 6.8.

5.5   Interrelationships between indices

As illustrated in the previous sections, there are many indices for characterizing low flow regimes and droughts. Using a range of averaging intervals and exceedance frequencies (Section 5.3.2) to define the various indices, the count easily exceeds 50, see for example the review of low flow hydrology by Smakhtin (2001). The preceding sections describe the most commonly used low flow and drought indices.

The choice of index, or indices, for a specific study depends primarily on the purpose of the study. However even after the purpose has been well defined there often are several indices that could be used. The questions arise: How similar are these indices? Do they express different aspects of the drought, or are they just slightly different? How many indices are needed to describe the most important aspects of the drought? These questions can be answered by studying the interrelationships between indices.

Close relationships between indices have been found, for example, between percentiles from the FDC (e.g. Q_{75}) and percentiles from the annual minimum frequency curve (e.g. $MAM(7)$, Section 5.3.2) (Smakhtin & Toulouse, 1998), between the recession constant and $MAM(7)$ and $MAM(30)$ standardized by the mean flow (Tallaksen, 1989), between the recession constant and drought deficit volume (Kachroo, 1992), between drought duration and volumes (Zelenhasic & Salvai, 1987; Chang & Stenson, 1990; Clausen & Pearson, 1995), and between drought deficit duration and volume and BFI (Clausen & Pearson, 1995). However, weak relationships were found between drought duration and percentiles from the FDC (Smakhtin & Toulouse, 1998).

Because of the close interrelationships between many of the indices, often one index has been used to derive other indices. In some cases the conversion has been from a low flow index to the same type of index with a different averaging interval or exceedance frequency (e.g. Drayton *et al.*, 1980; Gustard *et al.*, 1992). In other cases a primary index, for example BFI (Hutchinson, 1993; Nathan & McMahon, 1992; Clausen & Pearson, 1995) or Q_{20}/Q_{90} (Arihood & Glatfelter, 1991), has been used to calculate other low flow indices using regression equations (Chapter 8).

5.5.1   Ranks and correlation coefficients

For preliminary data analysis it is useful to rank those low flow indices that are expressed in $(m^3\ s^{-1})$ for each site. Which one has the highest value, and which one has the lowest value? Do the low flow indices have the same ranks for all stations?

Most of the studies mentioned above used correlation coefficients to analyse the relationships between indices. This is the most direct and useful approach when just a few indices are to be compared. The *Pearson correlation coefficient*, r_P (Equation 8.4), often referred to just as the correlation coefficient r, is a measure of the linear relationship between two variables. However, sometimes it can be advantageous to use the *Spearman rank correlation coefficient*, r_S, rather than r_P. r_S is a non-parametric statistic, calculated on the ranks rather than on the actual values of the variables (i.e. x and y in Equation 8.4 are the numbers 1, 2, ... , n, where n is the number of stations). Thus, in contrast to r_P, r_S is independent of the actual values of the variables and therefore more robust for variables with very high values (outliers) for some of the sites. Examples of variables where outliers can occur are indices in $(m^3\ s^{-1})$, when the data set includes small streams as well as a few large rivers. See descriptions of the Pearson correlation coefficient and the Spearman rank correlation in, for example, McClave *et al.* (1997). The use of correlation coefficients to study the interrelationship between low flow indices is illustrated in the following worked example.

Worked Example 5.6: Ranks and correlation coefficients

1. Data

   Seven low flow indices (described in Sections 5.3.1–5.3.4) and the mean ($\overline{Q}$) and median flow (Q_{50}) were calculated for a subset of the Regional Data Set (Section 4.5.2). The values are shown in Table 5.11. The flow indices were derived from daily flow values for at least 23 years. The recession constant ALPHA is equivalent to α as defined in Equation 5.1a.

2. Ranks

   The flow indices in Table 5.11 (those expressed in $m^3 s^{-1}$) were ranked in ascending order (Table 5.12). The results show that $\overline{Q}$ has the highest value and Q_{50} the second highest value in all cases. This is typical for daily flow data (data are positively skewed, or skewed to the right). Also common for all sites is that Q_{70} is the third highest value and $MAM(30)$ is the fourth highest value, whereas the ranks of the four remaining indices ($MAM(1)$, $MAM(10)$, Q_{95} and Q_{90}) differ. Q_{95} has the lowest value for 13 of the 21 sites; for the remaining sites $MAM(1)$ has the lowest value. In four cases $MAM(1)$ has the third lowest value, with both Q_{95} and Q_{90} being lower than $MAM(1)$.

3. Correlations

   The Pearson and Spearman correlation coefficients were calculated for pairs of low flow indices (Table 5.13). Not all indices from Table 5.11 were included because of space concerns. Also included are $MAM(1)/Q_{50}$ and Q_{95}/Q_{50}, which are standardized indices indicating flow variability rather than absolute values. Q_{50} was used here for standardization rather than $\overline{Q}$ because it is less sensitive to outliers, especially high values.

4. Results

   Table 5.13 shows that all relationships are positive (the tendency is that when one index increases, so does the other). The relationship between $\overline{Q}$ and Q_{50} is, as expected, very strong ($r_P = 0.99$, $r_S = 0.98$). There are also strong relationships (significance levels $< 0.1\%$) between these two central tendency measures and the absolute low flow indices ($MAM(1)$, $MAM(30)$, Q_{95} and Q_{70}). ALPHA is only weakly related to the low flow

Table 5.11 Flow indices (Sections 5.3.1–5.3.4) for a subset of the Regional Data Set

Site No.	Period of record	$\overline{Q}$ (m³s⁻¹)	Q_{50} (m³s⁻¹)	MAM(1) (m³s⁻¹)	MAM(10) (m³s⁻¹)	MAM(30) (m³s⁻¹)	Q_{95} (m³s⁻¹)	Q_{90} (m³s⁻¹)	Q_{70} (m³s⁻¹)	ALPHA -
139	1960–96	0.595	0.374	0.101	0.122	0.153	0.108	0.149	0.261	−0.0559
148	1966–96	4.528	4.180	2.177	2.457	2.697	2.050	2.350	3.400	−0.0280
153	1961–97	0.448	0.280	0.051	0.088	0.114	0.060	0.080	0.170	−0.0887
353	1962–96	0.404	0.300	0.141	0.162	0.179	0.130	0.160	0.230	−0.0442
363	1960 96	0.511	0.270	0.062	0.076	0.095	0.070	0.090	0.160	−0.0817
380	1960–96	1.670	1.060	0.283	0.354	0.418	0.310	0.400	0.710	−0.0686
407	1960–92	1.149	0.580	0.129	0.156	0.202	0.110	0.150	0.330	−0.0809
443	1960–97	0.734	0.571	0.293	0.351	0.381	0.260	0.313	0.441	−0.0404
455	1960–96	0.352	0.250	0.122	0.134	0.144	0.120	0.120	0.170	−0.0213
478	1960–97	2.291	1.720	0.901	0.996	1.070	0.680	0.850	1.280	−0.0191
480	1960–95	0.293	0.220	0.086	0.103	0.122	0.080	0.100	0.140	−0.0287
1325	1969–95	1.402	0.943	0.382	0.447	0.519	0.408	0.466	0.664	−0.0573
1421	1960–82	0.135	0.088	0.029	0.035	0.041	0.030	0.032	0.050	−0.0598
1448	1960–96	0.229	0.095	0.021	0.028	0.040	0.026	0.033	0.053	−0.0944
1476	1963–92	0.711	0.430	0.129	0.150	0.179	0.110	0.160	0.260	−0.0522
1487	1960–97	0.749	0.345	0.093	0.126	0.157	0.108	0.145	0.218	−0.0722
1489	1960–97	0.314	0.177	0.073	0.084	0.101	0.060	0.079	0.122	−0.0509
1491	1960–95	0.106	0.058	0.021	0.026	0.030	0.017	0.020	0.038	−0.0741
2457	1960–83	0.253	0.142	0.060	0.066	0.073	0.057	0.067	0.100	−0.0288
2459	1963–96	0.850	0.512	0.225	0.256	0.282	0.171	0.201	0.334	−0.0291
2463	1960–97	1.296	0.910	0.382	0.407	0.451	0.320	0.390	0.590	−0.0286

indices that are not standardized, which is expected because ALPHA does not reflect the size of the stream, whereas the low flow indices expressed in $m^3 s^{-1}$ do. ALPHA expresses the steepness of the recession and is strongly (significance levels < 0.1%) related to $MAM(1)/Q_{50}$ and Q_{95}/Q_{50}, which also expresses the flashiness of the stream. The standardized low flow indices are only weakly related to $MAM(1)$, $MAM(30)$, Q_{95} and Q_{70} with values of r_P and r_S around 0.50 (or less), and the coefficient of determination (for the linear regression case simply equal to r_P^2) around $0.50^2 = 0.25$ (or less).

Table 5.13 shows that r_S is smaller than r_P in many cases (but not all). This is a common tendency. The difference between the two

Table 5.12 Ranks of flow indices for the sites in Table 5.11 (1 indicates the lowest value and 8 the highest; site no. 455 has two identical values, which are given the lowest rank of 1)

Site no.	$\overline{Q}$	Q_{50}	MAM(1)	MAM(10)	MAM(30)	Q_{95}	Q_{90}	Q_{70}
139	8	7	1	3	5	2	4	6
148	8	7	2	4	5	1	3	6
153	8	7	1	4	5	2	3	6
353	8	7	2	4	5	1	3	6
363	8	7	1	3	5	2	4	6
380	8	7	1	3	5	2	4	6
407	8	7	2	4	5	1	3	6
443	8	7	2	4	5	1	3	6
455	8	7	3	4	5	1	1	6
478	8	7	3	4	5	1	2	6
480	8	7	2	4	5	1	3	6
1325	8	7	1	3	5	2	4	6
1421	8	7	1	4	5	2	3	6
1448	8	7	1	3	5	2	4	6
1476	8	7	2	3	5	1	4	6
1487	8	7	1	3	5	2	4	6
1489	8	7	2	4	5	1	3	6
1491	8	7	3	4	5	1	2	6
2457	8	7	2	3	5	1	4	6
2459	8	7	3	4	5	1	2	6
2463	8	7	2	4	5	1	3	6

coefficients is small (<0.05) for values close to 1. The biggest difference is for Q_{50} and Q_{95}/Q_{50} ($r_P = 0.43$, $r_S = 0.17$). For indices without outliers (close to being normal distributed), there is usually very little difference between the two coefficients. In our case the data set does hold a few relatively high values (for the large rivers), and r_P is affected by these high values, whereas r_S is not.

Table 5.13 Correlation coefficient matrix (Pearson/Spearman) for some of the indices in Table 5.11, including the standardized indices $MAM(1)/Q_{50}$ and Q_{95}/Q_{50}.

	$\bar{Q}$	Q_{50}	$MAM(1)$	$MAM(30)$	Q_{95}	Q_{70}	$ALPHA$	$MAM(1)/Q_{50}$	Q_{95}/Q_{50}
$\bar{Q}$	1.00/ 1.00	0.99***/ 0.98***	0.96***/ 0.89***	0.97***/ 0.94***	0.96***/ 0.89***	0.98***/ 0.96***	0.32/ 0.28	0.37/ 0.23	0.36/ 0.12
Q_{50}		1.00/ 1.00	0.99***/ 0.93***	0.99***/ 0.97***	0.99***/ 0.93***	1.00***/ 0.99***	0.36/ 0.33	0.42/ 0.29	0.43/ 0.17
$MAM(1)$			1.00/ 1.00	1.00***/ 0.98***	1.00***/ 0.99***	0.99***/ 0.95***	0.41/ 0.59**	0.50*/ 0.58**	0.50*/ 0.47*
$MAM(30)$				1.00/ 1.00	1.00***/ 0.98***	1.00***/ 0.98***	0.39/ 0.49*	0.48*/ 0.46*	0.48*/ 0.36
Q_{95}					1.00/ 1.00	1.00***/ 0.95***	0.38/ 0.57**	0.47*/ 0.55**	0.50*/ 0.47*
Q_{70}						1.00/ 1.00	0.36/ 0.40	0.44*/ 0.37	0.45*/ 0.25
$ALPHA$							1.00/ 1.00	0.88***/ 0.89***	0.75***/ 0.77***
$MAM(1)/Q_{50}$								1.00/ 1.00	0.90***/ 0.88***
Q_{95}/Q_{50}									1.00/ 1.00

Significance levels: * $p < 5\%$, ** $p < 1\%$, *** $p < 0.1\%$

5.5.2   Principal components analysis

If many indices are available, it can be difficult to obtain an overall picture of all the interrelationships using correlation coefficients. In this case it is helpful to use multivariable analysis, for example principal components analysis (PCA) as used by Clausen & Biggs (2000) to visualize the relationship between 35 flow indices. Multivariate methods are further elaborated in Chapter 8. Here the focus is on PCA, equivalent to empirical orthogonal function (EOF) transformation of a discrete, finite set of observations. The term EOF-analysis is used when time series are analysed (Sections 6.7 and 6.8). The method is used to represent original variables with a reduced number of variables, and in this way it can be used to identify the (most important) indices that explain a large proportion of the variation in the data set, and the similarity between indices can be identified.

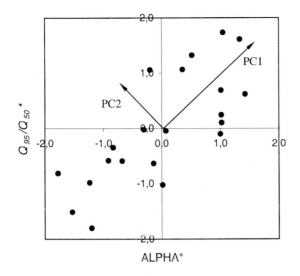

Figure 5.16 Plot of $Q_{95}/Q_{50}*$ versus ALPHA* (* indicates that the indices are standardized to have zero mean and unit standard deviation) and the two principal components, PC1 and PC2.

To explain what a PCA is, let us first imagine that there are only two indices, for example ALPHA and Q_{95}/Q_{50} from the example in Table 5.11. The values for the 21 sites are plotted in Figure 5.16, after they have been standardized to have zero mean and unit standard deviation (now referred to as ALPHA* and $Q_{95}/Q_{50}*$). A PCA rotates the axes to minimize the variation. In this way two new variables, the principal components, are defined. The principal components are linear combinations of the original variables and expressions of the new axes. The principal components are uncorrelated (the correlation coefficient is equal to zero).

Thus, from two original indices a PCA produces two principal components (PC). The two components, or axes, are rectangular or orthogonal to each other and the first (PC1) explains the largest proportion of the variance in the original data set. Generally, for a data set with n variables, the number of principal components is also n and these are orthogonal to each other. The proportion of variance in the data set explained by each component reduces with the number of components (PC1 explains most of the variance and PCn the least). Usually only the most important components are studied, and there are different ways of deciding how many to include. One method is to look at the scree plot, which graphs the eigenvalues (the variances of the principal components). Finally, the indices are plotted as a function of the most important PCs, so that the position of an index on the plot reflects the PCs coordinates (also called loadings); it is from these loadings plots that the interrelationships between indices are easily visualized (Worked Example 5.7). Indices that lie close together on the loadings plot are strongly correlated.

Worked Example 5.7: Principal components analysis (PCA)

A PCA was carried out on the low flow indices in Table 5.11, including relative (standardized by Q_{50}) flow indices. The scree plot for the PCA (Figure 5.17) shows that the first two principal components explain 94% (71% + 23%) of the total variance, and these are therefore most interesting.

Figure 5.18 shows the first two principal component coordinates of the indices (loadings), from which the underlying structure of the data can be detected. The coordinates are also the correlations between the indices and the principal components. Thus, Figure 5.18 shows that all indices are highly and negatively correlated with PC1. Some indices are positively correlated with PC2, while others are negatively correlated with PC2. The figure shows that the points generally fall in two groups, those with positive PC2 loadings and those with negative PC2 loadings. The first group consists of all the flow indices in $m^3 s^{-1}$, and because they lie close together they are also strongly correlated with each other. The second group consists of the relative indices and ALPHA. Both Q_{70}/Q_{50} and ALPHA are located a little distance from the

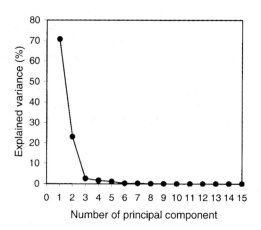

Figure 5.17 Scree plot for the PCA of the data set in Table 5.11 (including the low flow indices standardized by Q_{50}) with the variance expressed in % of the total variance.

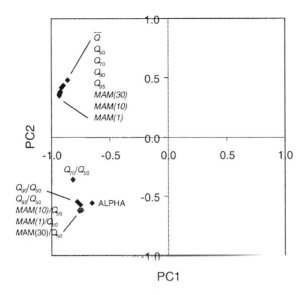

Figure 5.18 Principal component (PC1 and PC2) coordinates of the indices (loadings plot).

other indices in their group. Thus, to answer the questions raised in the introduction of Section 5.5 for this example: there are two groups of indices, which express different aspects of drought, and the various indices within each group are only slightly different from each other. This also agrees with the results from the correlation analysis (Worked Example 5.6). The loadings plot gives a good visual overview of the underlying structure of the data, and it is easier to obtain an overall picture of the structure from this plot compared to the correlation coefficient matrix (Table 5.13). However, sometimes it can be convenient to see the actual values of the correlation coefficients when interpreting the loadings plot.

Usually, when the data set has more indices than were used in this example, the analysis can be somewhat more complicated with a 'flatter' scree plot, and more principal components need to be studied.

5.6   Groundwater drought characteristics

Groundwater is seldom incorporated in drought analyses. For example, in an extensive overview of drought definitions by Wilhite & Glanz (1985) groundwater is mentioned only once as one of the variables that should be monitored in case a drought warning is issued. In the past many authors considered groundwater drought to result from over-exploitation of groundwater resources, rather than resulting from the natural variability of the climate (Day & Rodda, 1978). Thus there is only limited experience using groundwater drought characteristics that can be derived from variables such as groundwater recharge, levels and discharge (Section 3.4). Groundwater drought has been studied indirectly through base flow and recession analyses (Sections 5.3.3 and 5.3.4). The relationship between hydrogeological properties and base flow has been studied for regionalization purposes (Chapter 8).

In recent years the number of groundwater drought studies has increased and these studies cover a variety of aspects, e.g. from a water resources point of view (Robins *et al.*, 1997; White *et al.*, 1999), in connection with storage properties of chalk during drought (Price *et al.*, 2000), to analyse the non-linearity of aquifer streamflow interaction (Eltahir & Yeh, 1999), to study the regional character of groundwater drought (Chang & Teoh, 1995) and also related to climate change (Leonard, 1999). A recent study by Peters (2003) is the first that investigates groundwater drought in a systematic way. The interest in groundwater drought has mainly originated from problems with groundwater management for abstractions during drought periods. Some examples of groundwater droughts with different characteristics are presented below.

The UK suffered from a severe drought from 1988 to 1992. It was characterized by a decrease in groundwater storage (Figure 5.19, left). Four winters with abnormally low accumulated recharge resulted in unprecedented groundwater levels. Marsh *et al.* (1994) present a comprehensive overview of the development of the drought and its most important impacts: reduced streamflow and extremely low overall groundwater resources. Besides water supply, riparian ecosystems and the amenity value of the streams were affected.

In southern Africa a drought occurred in the early 1990s. It had a severe impact on the water supply of many, mainly rural, communities (Calow *et al.*, 1999). By the end of 1992 normally reliable sources of water began to fail and severe water shortage occurred in, for example, Malawi, Zimbabwe, South Africa and Lesotho. In these areas many communities rely on the water supply from a thin, but spatially extensive aquifer. During the drought many wells dried up or yielded less water (Figure 5.19, right). Not only insufficient depth of boreholes and wells, but also local depletion of the aquifer posed a serious problem.

For areas with shallow water tables the depth of the groundwater table with respect to the ground surface is very important. When the hydraulic head decreases, the aeration of the root zone will increase and this has large consequences for many chemical and biological processes (van Lanen & Peters, 2000). In The Netherlands the average groundwater level has fallen by 0.25–0.35 m and locally by more than 1.0 m over 40 years (1950 to 1989),

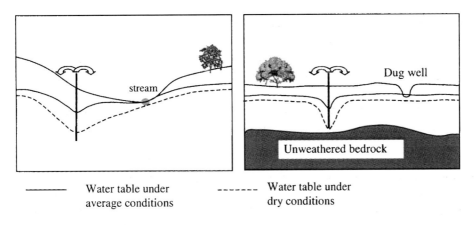

Figure 5.19 Example of water table conditions in the UK (left) and Africa (right).

mainly as a consequence of human activities (Rolf, 1989). The ensuing problems are more related to water quality than to water quantity, especially affecting natural ecosystems. During the 1988 to 1992 drought the conflict between human intervention and nature was intensified due to lack of water and deterioration of water quality.

5.6.1 Low groundwater characteristics

Currently no groundwater specific indices are commonly adapted. In principle the indices presented in Section 5.3, which are percentiles from the flow duration curve (FDC; Section 5.3.1), mean annual minimum n-day values ($MAM(n$-day); Section 5.3.2) and recession based indices (Section 5.3.4) can be calculated from time series of groundwater recharge, hydraulic heads and groundwater discharge. The application of the FDC is the same as for streamflow and gives similar information. It should, however, be noted that hydraulic head observations are not always equidistant in time, which makes calculation of the FDC more difficult. The $MAM(n$-day) can be applied to all three variables, however for groundwater recharge it will be zero in many climate types even if the averaging interval, n, is large.

Recession analysis can only be applied to hydraulic heads and discharge. For groundwater discharge the application is identical to that for streamflow, however the application is much easier because of the absence of direct runoff. The recession rate of hydraulic heads can be described by a similar relation as in Equation 5.1:

$$H_t - H_b = (H_{t=0} - H_b) \exp(-t/C) \qquad (5.10)$$

where H_t is the hydraulic head at time t, $H_{t=0}$ is the hydraulic head at time $t = 0$, H_b is the drainage base and C is a constant. The drainage base is the level below which the groundwater cannot fall and is often defined as the bottom of the stream. However it is not always possible to determine the drainage base a priori. In such a case it needs to be optimized together with the constant C. A further discussion on the use of the recession constant for hydraulic heads and the related recession of streamflow can be found in Halford & Mayer (2000).

5.6.2   Deficit characteristics

In principle the same methods of deriving deficit characteristics that were introduced for streamflow (Section 5.4) can be used for groundwater. Because of the specific properties of the groundwater variables (Section 3.4) the following additional remarks can be useful. For recharge the deviation from the expected or normal anomalies (Section 5.4.1) may be more meaningful than the absolute low level. When hydraulic head is used to describe groundwater drought it is important to remember that it is a state variable. This means, for example, that the drought deficit as defined by the threshold level approach is in units of [LT] instead of [L] or [L^3]. Some methods, such as the sequent peak algorithm (Section 5.4.2), are inappropriate and therefore the threshold level method has been used most often to derive drought deficit characteristics in groundwater levels (Chang & Teoh, 1995; Eltahir & Yeh, 1999; Peters *et al.*, 2003).

In large and slowly responding aquifers, multi-year droughts can occur frequently. Therefore the method selected to analyse groundwater drought should be able to cope with multi-year drought. Naturally, a daily time-step is usually not necessary for drought analysis in a groundwater system.

5.6.3   Case Study: Groundwater drought in the Metuje catchment (CD)

Different aspects of drought are compared for the Metuje catchment (Section 4.5.3), namely groundwater recharge, I, hydraulic head, H, groundwater discharge, Q_{gr} and streamflow, Q. All time series are 20 years long. Hydraulic head and streamflow are observed and aggregated to monthly values and recharge is simulated with the BILAN model (Section 9.2.2). The groundwater discharge is derived from the streamflow using the method developed by Kliner & Kněžek (1974) and Holko *et al.* (2002), using hydraulic heads. This method is used instead of some more well-known base flow separation procedures because, unlike most other methods, this method is physically based. As the analysis is performed on monthly values the presence of dependent droughts is much smaller, which means that droughts do not have to be pooled.

1. Derivation of drought events

The derivation of drought events from groundwater recharge, I, hydraulic heads, H, groundwater discharge, Q_{gr}, and streamflow, Q, is carried out using the threshold level approach described in Section 5.4.1. Because of the different characteristics of the time series, the threshold is derived using Equation 5.4, for b equal to 0.3, and a normalization is necessary. The normalization can be done directly on the time series of I, H, Q_{gr} and Q or on the drought deficits after the droughts have been derived. Here the last option is used and the drought deficits are normalized in two ways. Firstly, the deficit is expressed as a percentage of the mean annual flux for I, Q_{gr} and Q only. Secondly, the deficit is divided by the standard deviation. The result is the same as if the time series themselves had been normalized prior to the drought event selection. The drought deficits in the hydraulic heads are not normalized.

2. Results

In Table 5.14 summary statistics and the derived thresholds along with corresponding percentiles (derivation on CD) are presented. For I, Q_{gr} and Q the threshold derived using Equation 5.4 is approximately identical to the 55-percentile. However for H, which has a much lower skewness, the threshold equals the 79-percentile.

The distributions of the drought deficits are plotted in Figure 5.20 and 5.21 by application of Equation 6.6. A drought with a return period of T years is equalled or exceeded once every T years *on average* (Box 6.2). In Figure 5.20 the drought deficit as a percentage of the average annual flux is presented. The

Table 5.14 Average, standard deviation, skewness, threshold and corresponding percentile for I, H, Q_{gr} and Q for the Metuje catchment

	I	H	Q_{gr}	Q
Average	0.70 mm d^{-1}	485.60 m	$0.46 \text{ m}^3\text{s}^{-1}$	$0.86 \text{ m}^3\text{s}^{-1}$
Standard deviation	0.81 mm d^{-1}	0.98 m	$0.12 \text{ m}^3\text{s}^{-1}$	$0.51 \text{ m}^3\text{s}^{-1}$
Skewness	1.38 mm d^{-1}	-0.03 m	$1.56 \text{ m}^3\text{s}^{-1}$	$1.95 \text{ m}^3\text{s}^{-1}$
Auto Correlation (lag 1)	0.21	0.84	0.71	0.36
Threshold	0.24 mm d^{-1}	484.8 m	$0.40 \text{ m}^3\text{s}^{-1}$	$0.62 \text{ m}^3\text{s}^{-1}$
Percentile	0.56	0.79	0.56	0.55

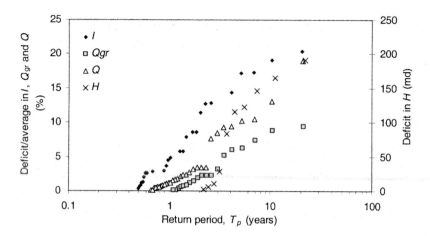

Figure 5.20 Distribution of the drought deficit as a percentage of the average annual flux for recharge, I, groundwater discharge, Q_{gr} and streamflow, Q (left axis) and drought deficit for the hydraulic head, H (right axis).

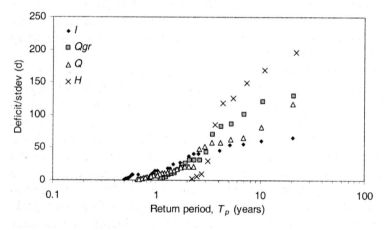

Figure 5.21 Distribution of drought deficit divided by standard deviation for recharge, I, hydraulic head, H, groundwater discharge, Q_{gr} and streamflow, Q.

droughts in I have the highest deficits and the droughts in Q_{gr} the lowest. The deficit of droughts in the hydraulic head cannot be compared, because they are not normalized. The most important cause of the difference in deficit is probably the standard deviation, which is higher for the recharge compared to the average. In Figure 5.21, where the drought deficit is normalized using the standard deviation, the drought deficits are of the same order of magnitude. For low return periods the deficits of the droughts in *H* are smallest, whereas for high return periods they are largest. In general the higher the autocorrelation the steeper the slope of the distribution.

5.7 Complex indices

Single low flow and drought indices are based on only one variable (Sections 5.3, 5.4 and 5.6), typically streamflow or groundwater. Complex indices are based on several variables, often including many elements of the hydrological cycle. For example, a complex index could be based on a combination of meteorological variables such as precipitation and evapotranspiration.

Several complex indices have been developed. They are often called drought indices, but should rather be referred to as wetness indices as they describe the whole spectre of moisture conditions from wet to dry. The most frequently applied index, especially across the US, is the Palmer Drought Severity Index, PDSI (Palmer, 1965) (Box 5.3). Although the PDSI is sometimes used as an indicator of hydrological drought, it is more correctly referred to as a meteorological drought indicator, following the disciplinary perspective described in Section 1.2.

A complex index that can be characterized as a hydrological moisture index in terms of indicating surface water availability is the Surface Water Supply Index, SWSI (Shafer & Dezman, 1982). This was developed to incorporate both hydrological and climatological features into a single index complementing the Palmer Index for each major river basin in the state of Colorado (Hayes, 1999). It was later modified and applied in other western states in the USA. Unlike the PDSI, the SWSI is designed for topographic variations across a basin and snowpack is a major component. The index values need to be standardized to allow comparisons between basins. The SWSI is computed from snowpack, precipitation, and reservoir storage data in the winter. During the summer months, streamflow replaces snowpack as a component within the SWSI equation. Each component has a monthly weight assigned to it depending on its typical contribution to the surface water within that basin. The weighted components are summed to determine a SWSI value representing the entire basin. Like the PDSI, the SWSI is centred on zero and has a range between −4.2 (dry) and +4.2 (wet).

One of the advantages of the SWSI is that it gives a representative measure of surface water availability across a region. However there are several aspects that limit its application. Changes in the water management within a basin, such as flow diversions or new reservoirs, mean that the entire SWSI algorithm for

---

Box 5.3

Palmer Drought Severity Index

The Palmer Drought Severity Index, PDSI, is based on the supply and demand concept of the water balance equation taking into consideration monthly mean precipitation and temperature as well as the local available water content of the soil. It is most effective measuring impacts on sectors sensitive to soil moisture conditions, such as agriculture (Willeke *et al.* 1994). This index generally ranges from −6 to +6, with negative values denoting dry conditions and positive values denoting wet conditions. The PDSI has also been useful as a drought-monitoring tool and has been used to trigger actions associated with drought contingency plans (Willeke *et al.* 1994). It has typically been calculated on a monthly basis, and a long-term archive of the monthly PDSI values for every Climate Division in the United States exists at the National Climatic Data Center in the US, from 1895 to date. In addition, weekly, modified PDSI values are calculated for each Climate Division during the growing season and are available in the Weekly Weather and Crop Bulletin. These weekly Palmer Index maps are also available on the World Wide Web (http://www.cpc.ncep.noaa.gov/products/analysis_monitoring/regional_monitoring/palmer.gif).

The water balance model that forms the basis of the PDSI also allows the development of other indices that have different rates of response to changes in supply and demand for moisture and therefore represent different types of drought. Examples are the monthly moisture anomaly index (ZINX), the Palmer drought hydrological index (PDHI) (Palmer, 1965; Soulé, 1992), and the Crop Moisture Index (CMI) (Palmer, 1968).

---

that basin needs to be redeveloped to account for changes in the weight of each component. Thus it is difficult to obtain a homogeneous time series of the index (Heddinghaus & Sabol, 1991).

An example of a complex index which incorporates groundwater is the index introduced by Pálfai (1990). This is based on precipitation sums, but is corrected for groundwater level amongst other corrections (Putarić, 2001). Hayes (1999) describes many other complex indices and refers to several studies where the indices are applied.

5.8   Summary

This chapter has provided an overview of the derivation of the most commonly used hydrological drought characteristics. Methods of deriving low flow and deficit characteristics both in terms of time series and single indices have been explained. For practical applications some examples are given in Chapters 10 and 11. The methods described are applicable for *observed* or *simulated* time series of streamflow or groundwater. It is also assumed that the series are stationary and undisturbed by human influence; hence there should be no trends or sudden jumps in the records. Furthermore, a certain record length is required to obtain stable and reliable results. A summary of the low flow characteristics described in Section 5.3, the deficit characteristics explained in Section 5.4 and the complex indices summarized in Section 5.7, is given in Table 5.15. Section 5.5 used parts of the Regional Data Set on the CD to illustrate the interrelationship between indices. It has been shown how the FDC based and *MAM*(*n*-day) low flow indices are closely related, whereas another closely related group is the standardized FDC based, *MAM*(*n*-day) and the recession based index. Hence, many indices express the same feature of a drought. Section 5.6 discussed the derivation of groundwater drought characteristics and it is seen that in principle the same methods as for streamflow can be applied with the part exception of hydraulic head, which is a state variable.

Table 5.15 Summary of low flow and deficit characteristics described in Sections 5.3–5.4 and 5.7

Method	Comments
Percentiles, FDC	Percentiles are used as low flow indices. They give information about the low flow regime of the river and are often used as threshold levels in the threshold level method. The FDC can be based on daily, monthly or some other time interval of streamflow data. For perennial streams the percentiles will never be zero. For intermittent and ephemeral streams even Q_{50} can be zero.
MAM(*n*-day)	*MAM*(*n*-day) are low flow indices derived from time series of low flow characteristics (annual minimum series). They give information about the low flow regime of the river. Large averaging intervals might be required in intermittent and ephemeral streams to obtain non-zero values. Time series of annual minimum values, *AM*(*n*-day), are often fitted to a probability distribution to obtain estimates of *T*-year events. Problems related to AM-series with many zero values are the topic of Chapter 6.

(→ Table continued on next page)

Table 5.15 (continued)

Method	Comments
Base flow indices	Base flow indices are derived from streamflow and groundwater data. Commonly a hydrograph separation procedure is imposed to calculate the base flow proportion of the total streamflow. An index that gives the ratio of base flow to total streamflow is considered a measure of the river flow that derives from stored sources. Accordingly, it can reflect the catchment's ability to store and release water during dry periods. The Base Flow Index (BFI) is one well-known example.
Recession indices	The recession curve expresses the rate of decay of the hydrograph during periods with little or no precipitation. It describes in an integrated manner how different factors in the catchment influence the outflow process. Parameters of the recession model, also referred to as low flow indices, represent the recession rate; fast in a flashy catchment with little storage and slow in a groundwater-dominated catchment.
Threshold level method	Drought event definition method. It derives time series of drought deficit characteristics such as the deficit duration and the deficit volume of a drought based on a predefined threshold and can be based on daily, monthly or some other time interval of streamflow data. The time interval and the threshold level is decided by the purpose of the study and the hydrological regime of the river. Drought indices can be derived from the time series of drought deficit characteristics, e.g. mean drought deficit duration.
Sequent peak algorithm	Drought event definition method. It derives time series of drought characteristics such as the duration and the deficit volume of a drought based on a predefined yield (threshold) and can be based on daily, monthly or some other time interval of streamflow data. The time interval and the yield will depend on the purpose of the study and the hydrological regime of the river. Drought indices can be derived from the time series of drought characteristics, e.g. mean drought duration.
Regional drought characteristics	These are based on at-site drought characteristics derived using e.g. the threshold level method, but allow derivation of regional drought characteristics over the area covered by drought. Require generation or observation of streamflow for the region under study.
Complex indices	Observation records of several hydrometeorological variables are required. Complex indices are often based on water balance calculations and describe the whole spectrum of wetness conditions from wet to dry.

6

Frequency Analysis

Lena M. Tallaksen, Henrik Madsen, Hege Hisdal

6.1   Introduction

"It is likely that something unlikely is going to happen" (Aristotle, 384 322 BC). For water resources management and planning it is of vital importance to assess how likely or probable it is that an extreme hydrological event may occur. Figure 6.1 shows the extreme low water levels of a mountainous reservoir during a drought in southern Norway in 1996. When a severe event like this happens, the question of assessing its frequency or probability is raised. Hydrologists often assign a return period to an extreme

Figure 6.1 The impact of the 1996 drought on Lake Hovatn in southern Norway (© Knut Gakkestad).

event, i.e. they calculate the probability that the event, for instance a low reservoir level, will occur next year or in the next ten or hundred years. If the chance is 1 in 100 that it will occur next year or in any year, the event is called a 100-year event, which *on average* will occur once in a 100-year period. In terms of water management, design events of given return periods are often the basis for operational guidelines, policy and insurance standards, and risk acceptance. Estimates of the probability of occurrence of extreme events can be derived from analyses of time series, either historical records or simulated series, by the method of frequency analysis.

Frequency analysis is one of the most common and earliest applications of statistics within hydrology. It involves (a) definition of the hydrological event and extreme characteristics to be studied, (b) selection of the extreme events and probability distribution to describe the data, (c) estimation of the parameters of the distribution, and (d) estimation of extreme events or design values for a given problem. The procedure is straightforward, but the uncertainty of the estimated extreme values depends strongly on the sample size and the basic assumptions of the models adopted. Hydrological time series typically range between 20 and 50 years, and thus hydrological design often requires extrapolation beyond the range of observations. Parametric frequency models are therefore adopted to obtain an efficient use of scarce data. One of the main purposes of frequency analysis is to find the most suitable probability distribution to describe the data in question.

The issue of climate change in particular has raised the question of whether the frequency of extreme events like flood and drought might increase as a result of an intensified hydrological cycle (Chapter 2), and thereby violate the assumption of stationarity in the time series. Independent of the origin of the change, there is a need to check whether there is any tendency for the frequency or severity of the studied events to increase or decrease during the observation period. It is also important to be aware that hydrological variables are often influenced by human activities in time of drought when the demand for water is high, for example through withdrawals for irrigation and water supply (Chapter 9). It is therefore important to evaluate the possible impacts this may have on the extreme events prior to performing a frequency analysis. In heavily modified catchments efforts should be made to obtain naturalized observation series (Chapter 4).

Many of the hydrological phenomena studied using frequency analysis are regional in nature, which means that there is a tendency for extreme values to occur simultaneously at several sites. This is reflected in a spatial correlation between the observation series. The variable of interest can often be described as a space-time random process that can only be sampled at a finite number of sites within the region under study. Regional frequency methods have been developed to utilize additional information from sites within the region to improve at-site estimates and to obtain estimates at sites without observations. Regional drought characteristics such as the spatial extent of the drought can also be included as a property of the drought in a frequency analysis covering several sites in a region. Subsequently, the probability that a certain percentage of the area considered will suffer a drought of a given severity can be estimated.

This chapter has two main objectives. First of all it gives a statistical background to frequency analysis for hydrologists, including both at-site and regional procedures. Secondly, it presents several applications of frequency analysis to time series of hydrological drought characteristics to demonstrate the procedures and to highlight methodological concerns and international experience. It should be noted that the chapter deals primarily with streamflow data, which until now have been dominant in frequency analysis of hydrological drought. Streamflow droughts are characterized in terms of minimum discharge, deficit volume and duration (Chapter 5). The methodology presented is of a general nature, although it is less suited for rivers in arid and semi-arid regions that run only occasionally. In these regions other hydrological time series such as groundwater or reservoir data may be of greater interest. Frequency analysis of groundwater data like groundwater levels and discharge is, however, limited and will only be briefly mentioned. Traditionally, graphical methods that compare the current drought situation to historical records have been applied for time series of groundwater data (Box 6.1). These methods are useful in the sense that they are simple and provide a good visual impression of the current situation in a historical context, and are thus also frequently used for streamflow. In general a graphical display of the data is recommended prior to performing a statistical analysis as it can reveal and help to explore important characteristics of the time series and the relationship between variables. Such an exploratory data analysis provides a first look at the data, and can guide the choice of which statistical procedure to apply.

Section 6.2 starts with an introduction to basic probability concepts. Section 6.3 discusses important concerns related to the data that are used for frequency analysis of extreme events, including different approaches for selecting extreme events from a time series. Emphasis is on the annual maximum or minimum series and partial duration series approach and their applicability for extreme value analysis of minimum and maximum values. In Section 6.4 the selection of probability distribution is described with emphasis on two limiting distributions: the Generalized Extreme Value distribution and the Generalized Pareto distribution. Some other distribution functions commonly applied in extreme value analysis are briefly presented along with procedures for selecting a distribution function. Section 6.5 describes the most common parameter estimation methods: the method of moments, including

Box 6.1

Analysing methods for groundwater

The method of frequency analysis as described in this chapter can also be applied to time series of groundwater data, although in practice this is seldom done. One example can be found in Peters (2003), who defined groundwater droughts using the threshold level approach, and calculated return periods for groundwater recharge, discharge and hydraulic heads. No theoretical distribution functions were, however, fitted and the return periods were derived from empirical distribution functions of the observed and simulated droughts. An example is presented in Section 5.6.3.

Analysis of the drought condition in the groundwater system usually takes a different form. Groundwater level is the most commonly used quantity for monitoring groundwater resources. It is relatively easy to observe and time series of groundwater level are usually available on a weekly or monthly time resolution. For each station the current level is compared to the historical record for the same

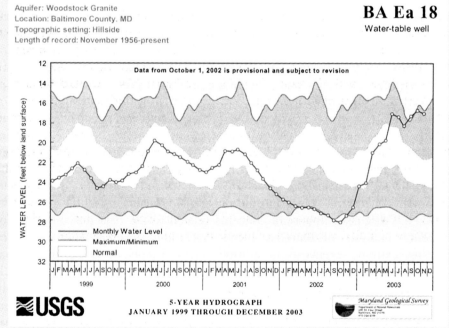

Figure 6.2 Hydrograph of monthly groundwater levels plotted against the long term minima and maxima values (http://md.water.usgs.gov/groundwater/web_wells/current/water_table/counties/baltimore/#baea18). Reproduced with permission from the US Geological Survey.

period. The result can be displayed in several ways, for instance as a plot of the current water level against the minima and maxima observed levels in the period of record (Figure 6.2). Alternatively, the observed value for each month (or any other time period used) is given a rank or expressed as percentile of all the observed values for that month.

In the USA, drought watch or water watch pages are available on the internet. Many states provide real time information on groundwater levels, see for example http://waterdata.usgs.gov/nwis/gw. A five-year time series of groundwater level for a borehole in Baltimore is presented in Figure 6.2. It is an unused water table well drilled into bedrock. The series is compared to the historical record and as can be seen from the graph a drought occurred in 2002. New record monthly low water levels were recorded from April through October 2002. These were the lowest water levels in 40 years, exceeding the previous records set during the historic drought of the 1960s. The all-time record (since data collection began in 1956) was set in October 2002 at 28.13 feet (below land surface). A similar site is available for UK at http://www.nwl.ac.uk/ih/www/research/idroughtwatch.html.

product moments and L-moments, and maximum likelihood estimators. Estimation of design values and their precision close the section. Section 6.6 presents a number of at-site studies dealing with frequency analysis of streamflow drought, covering both low flow (minimum values) and drought deficit volume and duration (maximum values). Two worked examples demonstrate the procedures using a daily streamflow series from the Global Data Set. Section 6.7 introduces regional frequency analysis, including different models for regional estimation and procedures for testing regional homogeneity and determination of a regional distribution. The regional estimation procedure is demonstrated in a stepwise manner as a self-guided tour on the CD using the Regional Data Set from Baden-Württemberg in Germany. Frequency analysis of regional drought characteristics is introduced in Section 6.8 and an approach to derive severity-area-frequency curves demonstrated for time series of streamflow. Finally, a summary of the chapter is given in Section 6.9.

6.2 Basic probability concepts

A fundamental concept in statistics is the *population*, which refers to a set of elements with measurable properties. Statistics are concerned with methods (estimators) for drawing inference about these properties based on the properties of a *sample* drawn from the population. It is therefore important to understand how the properties of the sample relate to the properties of the population. A given characteristic, e.g. the mean value, computed by an estimator is called a sample estimate or statistic. Non-representative sampling methods might result in *biased* estimates, which on average deviate from the true population value. *Probability* is related to methods for calculating the likelihood of a given sample value providing the true population characteristics are known or estimated from the sample.

6.2.1   Populations

Let X denote a random variable, and x a real number. The cumulative distribution function (cdf):

$$F_X(x) = Pr\{X \le x\} \tag{6.1}$$

designates the probability (Pr) that the random value X is less than or equal to x, i.e. the non-exceedance probability for x. The probability takes on values in the interval [0,1]. The exceedence probability, $EP_X(x)$, is defined as:

$$EP_X(x) = Pr\{X > x\} = 1 - F_X(x) \tag{6.2}$$

The probability density function (pdf) is the derivative of the cdf, and describes the relative likelihood that the continuous random variable X takes on different values:

$$f_X(x) = \frac{dF_X(x)}{dx} \tag{6.3}$$

The percentiles or *quantiles* of a distribution are often used as a design quantity. The quantile, x_p, is the value with cumulative probability p, i.e. $F_X(x_p) = p$, which will have an exceedance probability of $1 - p$. The exceedance probability is frequently expressed in terms of the *return period*, T (Box 6.2). A *design event* is the estimated extreme quantile (return level) associated with a given return period. If $(1 - p)$ denotes the exceedance probability in any time interval Δt, the return period T is defined as:

$$T = \frac{1}{1 - p} \tag{6.4}$$

where T is the mean time interval between occurrences of an event $X > x_p$.

For minimum values, the value of interest is that of the smallest value. The return period for the non-exceedance probability p is thus of interest:

Box 6.2

Return period

The concept of return period is convenient when dealing with extreme events such as flood and drought. It is commonly related to the design and operation of hydrological structures. For instance, if the size of a water reservoir is designed to withstand a T-year drought event it should be able to sustain a minimum water yield with a probability of $1 - 1/T$. Return levels are also used to set instream flow requirements, e.g. the river flow is not allowed to drop below the T-year low flow (Chapter 10).

Let X design the lowest flow observed within a time interval. If X is assumed independent and $EP_x(x)$ constant for all time intervals Δt, Equation 6.2 gives:

$$Pr\{X \leq x \text{ in the next time interval}\} = F_X(x) = 1 - EP_X(x_p) = p \tag{B6.2.1}$$

$$Pr\{X < x \text{ in all of the next } n \text{ time intervals}\} = [1 - EP_X(x_p)]^n = p^n \tag{B6.2.2}$$

$$Pr\{X > x \text{ in all of the next } n \text{ time intervals}\} = [EP_X(x_p)]^n = [1 - p]^n \tag{B6.2.3}$$

$$Pr\{X \leq x \text{ in at least one interval of the next } n \text{ intervals}\} = 1 - [1 - p]^n \tag{B6.2.4}$$

$$Pr\{X \leq x \text{ in an interval } n \text{ intervals from now (and not before)}\} =$$
$$[EP_X(x_p)]^{n-1} [1 - EP_X(x_p)] = [1 - p]^{n-1} p \tag{B6.2.5}$$

Equation B6.2.5 is known as the Geometric distribution. The mean and variance are given as:

$$E\{n\} = \frac{1}{p} \text{ and } Var\{n\} = \frac{1-p}{p^2} \tag{B6.2.6}$$

If, for instance, the probability of an event occurring ($X \leq x$ in the example above) is $1/100$ in one experiment (time interval), the mean value is 100, which means that the experiment on average has to be repeated 100 times until the event occurs. The variance equals $0.99/0.01^2 = 9900$, and the standard deviation is thus 99.5, which is large compared to a mean value of 100. It is thus nearly equally likely that the event will occur in one as in 200 experiments.

The probability that the first non-exceedance of x (event) occurs after n intervals is given by Equation B6.2.5. Its expected value, i.e. the average time interval between events, is often referred to as the return period, T:

$$E\{n\} = \frac{1}{p} = T(n) \tag{B6.2.7}$$

For maximum values the return period of the exceedance probability is the quantity of interest, and p should thus be replaced by $1 - p$ in Equation B6.2.7. It
($\rightarrow$ Box continued on next page)

Box 6.2 (continued)

should be noted that $T(n)$ has a large variance and therefore gives only an approximate description of the random variable n.

For an annual time interval T-year events are defined as events with a return period of T years. Following Equation B6.2.4 the probability that a 100-year event ($p = 0.01$) occurs in any of the next 100 years can be calculated as: $1 - 0.99^{100} = 0.634$. In general:

$Pr\{$at least one occurrence of a T-year event in N years$\} = 1 - [1 - 1/T]^N$ (B6.2.8)

This is commonly referred to as *risk*. If a design is built to withstand a T-year event, the risk that it will fail during N years equals the probability that the event will occur at least once in the N year period. For N equal to T the expression in Equation B6.2.8 approaches $1 - 1/e = 0.632$ as T increases. A T-year event will, however, *on average* occur once in a T-year period.

Similarly, it is possible to calculate what the return period of the low flow design event should be in order to be, for example, 90 per cent sure that a lower value is not recorded in the next n time intervals (years). According to Equation B6.2.3 one gets for n equal to 10 years, $0.90 = (1 - p)^{10}$, which gives a non-exceedance probability p of 0.0105 and a return period T of 95 years. In other words, to be 90 per cent sure that a value lower than the design low flow event is not recorded in a 10-year period, a 95-year event must be adopted.

$$T = \frac{1}{p} \tag{6.5}$$

where T is now the average time or recurrence interval between occurrence of an event $X \leq x_p$.

For a time interval of one year, annual exceedance or non-exceedance probabilities are defined, and the corresponding *T-year event* is given by the value for x_p (also denoted x_T). The probability of a T-year event occurring in any one year is $1/T$.

6.2.2   Samples

Summary statistics such as the mean, median and standard deviation are used to describe samples of a population (Section 6.5). If the true distribution of the population is assumed to be known, the sample statistics can be compared to the theoretical values. The range of values observed within a sample can be displayed by a *histogram*, which is a plot of bars showing the fraction of the

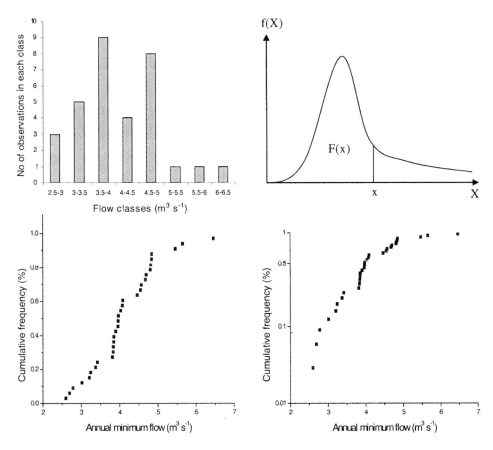

Figure 6.3 Sample annual minimum 1-day flow, $AM(1)$, for River Ngaruroro at Kuripapango (NZ); histogram (upper left), assumed pdf for the population (upper right), probability plots using a linear scale (lower left) and a normal probability scale (lower right).

total sample that falls into different class intervals. Figure 6.3 (upper left) shows the histogram of annual minimum 1-day values, $AM(1)$ (Section 5.3.2), derived from time series of daily streamflow for River Ngaruroro at Kuripapango, New Zealand (Table 4.3). The observations cover the period 1964 to 2000, and with the omission of four years with missing data 32 values result in total. The lowest observed value is 2.7 m^3 s^{-1} and, as demonstrated in the figure, the data are clearly skewed towards higher values.

The choice of class interval will influence the graphical impression; a too large interval may fail to show the real pattern, whereas a too small interval may cause a very high class-to-class variability. Iman & Conover (1983) suggest that the number of intervals, k, should be the smallest integer such that $2^k \geq n$, where n is the sample size. In our case n equals 32 and the recommended number of intervals is five. However, for small samples this may not give the desired resolution and in Figure 6.3 eight intervals are chosen. If the number of observations is assumed to be infinitely large and the class intervals are made infinitesimally small, then the limit will be the pdf for the population (Figure 6.3, upper right). The area under the probability curve is unity, whereas $F(x)$ equals the area under the curve for $X \leq x$.

Quantile plots are cumulative distribution functions (cdf) of the quantiles of sample data, also labelled *empirical distribution functions*. If a series of length N is sorted in ascending order the non-exceedance frequency, $F(x_{(i)})$, of the ith smallest event is simply equal to the fraction of values less than or equal to this value, i/N. In hydrology, empirical distribution functions are commonly labelled duration curves, in which case all the values in a time series are included. A detailed description of their construction is given in Section 5.3.1 for daily flow values; however, any time resolution and variable can be adopted.

In Figure 6.4 the quantile plots (upper part) of annual minimum 10-day, $AM(10)$, series of two UK rivers of contrasting geology are compared for the period 1963 to 1999. The y-axes show their cumulative frequencies which add up to 1 (or 100 using percentage as in the figure). Lambourn at Shaw (left) is considered a permeable catchment with a stable flow regime, whereas Ray at Grendon Underwood (right) is an impermeable catchment with a flashy flow regime. The figure clearly demonstrates the different flow regime of the two catchments, in particular the high number of years with zero flows found for the Ray catchment (27 out of a total of 37, i.e. 73%). The problem of zero values in the observation series is discussed further in Section 6.3.2.

A *probability plot* is a special form of the quantile plot as the observations are now assumed independent (Section 6.3.1). Determining which frequency or probability should be assigned to each value is now less straightforward, because it is not known which probability to assign to the largest or smallest value of the sample (0 and 1 should be avoided). This is generally done by the use of *plotting positions*, a well-known example being the Weibull formula:

$$p_i = i/(n + 1) \qquad\qquad (6.6)$$

where p_i is a plotting position which gives an estimate of the non-exceedance frequency of the ith smallest event.

The use of plotting positions requires the data to be independent and the sample representative. The Weibull formula provides an unbiased estimate of the non-exceedance probability for all distributions and is therefore often recommended. The actual exceedance probability of the largest or smallest observation has a mean value of $1/(n + 1)$ and a standard deviation of nearly the

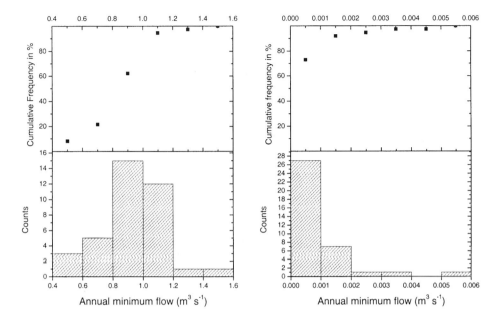

Figure 6.4 Histograms and quantile plots of annual minimum 10-day series, *AM*(10), for Lambourn at Shaw, UK (left) and Ray at Grendon Underwood, UK (right).

same size. The uncertainty associated with estimates of the exceedance probability for the largest or smallest event is therefore large, irrespectively of choice of plotting position. An overview of alternative plotting positions and their motivation is given by Cunnane (1978) and Stedinger *et al.* (1993), whereas a critical discussion of the use of plotting positions is presented by Klemeš (2000a).

Probability plots are commonly used to compare sample data to a theoretical distribution. In Figure 6.3 (lower) probability plots of *AM*(1) values for Ngaruroro are given. The spread of the data can easily be detected from the graphs. In the left figure a linear scale is adopted, whereas the scale on the right figure has been adapted to a theoretical probability distribution, here the Normal distribution is chosen. The curve will thus plot as a straight line if the data follow the Normal distribution, although some deviations might be encountered due to sampling uncertainties. In the present case, the sample deviates significantly from the Normal distribution. A straight line would also result if the probability axis is linearized. This can be achieved by introducing a reduced variate y, which is linearly related to x and substitutes the random variable x in the expression for $F(x)$. This is further elaborated in Box 6.4 and demonstrated for Ngaruroro in Worked Example 6.1.

6.3   Data for extreme value analysis

6.3.1   Basic assumptions

The basis for a drought frequency, or extreme value, analysis is a record of low flow or drought deficit characteristics that have been extracted from a time series (Sections 5.3 and 5.4). It is important that the events can be considered to be a series of extreme events, and there are several approaches to the selection of the extreme events as described in the next section. Frequency analysis requires that the data are *independent* and *identically distributed*. 'Identically distributed' implies that the data should be from the same population, i.e. homogeneous. 'Independent' means that there should be no serial correlation (short range dependence) in the time series and the hydrologic regime should remain stationary during the period of record (no long-term trend in the time series). The presence of serial correlation is often referred to as persistence. For stationary processes the classical extreme value theory still holds, but the presence of persistence influences the precision in the estimated statistics (Box 6.3). Independent, identically distributed (iid) time series have the same population characteristics independent of the time, i.e. there is no seasonality in the series.

The assumption of stationarity may be violated in catchments influenced by land use or climate change. Several tests for detecting changes in time series are available (e.g. Kundzewicz & Robson, 2000). *Non-stationarity* in frequency modelling can be accounted for by introducing time dependent extreme value parameters. Smith (1989) allowed one of the model parameters to be dependent on the year through a linear trend. Alternatively, the data can be split into two periods and a model without trend fitted to each period. In both cases, the number of parameters to be estimated will increase. Strupczewski *et al.* (2001) present a short review of existing methods for flood frequency modelling and their potential for extension to non-stationary conditions. They suggest that when investigating a long time series of unknown character it is preferable to assume the most general case with a trend, and then to proceed to a simpler stationary case based on identifying the best-fitting model. In their own work the trend was explicitly introduced in the first two moments. It is, however, questionable in terms of hydrological design whether a trend detected in the last, say, 50 years will be also valid for the next 50 years. It is therefore recommended to look for additional sources of information concerning the likely cause of a trend, not merely to extend the information of a present time series into future prediction.

A prerequisite to the assumption of identically distributed data is that the events belong to the same population, i.e. the same generating processes have caused the extreme events (Chapter 3). Some regions may experience strong seasonality with different types of weather conditions dominating in different seasons. In snow and ice-affected regions low streamflow may occur both during winter and summer although caused by different processes (Sections 3.3.1 and 5.1). The data are thus non-homogeneous and for independent, non-identically distributed data the general extreme value theory is less helpful (Smith, 1989). If the observation series can be divided into separate subsets, either by calendar season or by using meteorological information, one extreme value series can be obtained for each season or generating process. The statistical properties of the time series will subsequently depend on the season of the year (generating process), and the model parameters need to be estimated for each season (process) separately. A problem inherited in this approach is the limited length of the time series and the large number of parameters to be estimated, and Rasmussen & Rosbjerg (1991) recommend the one-seasonal

---

Box 6.3

Persistence in time series

Frequency analysis requires that the data are independent, which implies that there should be no serial correlation (persistence) in the time series. Persistence occurs as a result of a memory in the investigated system. In hydrological time series a long memory may be caused by large storages, like extensive groundwater reservoirs or lakes. The theory of extreme values in dependent stochastic processes is summarized in a review article by Leadbetter & Rootzén (1988). Persistence can be measured by the autocorrelation function (or autocorrelation coefficient) $r(k)$, which measures the degree of linear correlation between the observations at a time and the observation at k time steps later. The lag-one autocorrelation coefficient, $r(1)$, is a simple measure of the degree of time dependence of a series. Confidence limits for an estimated value of the autocorrelation function $r(k)$ can be derived provided $r(k)$ can be assumed to be approximately normally distributed and the time series assumed to be stationary. The estimated autocorrelation for a given time lag is considered not to be significantly different from zero if it is inside the confidence limits for a given significance level α. A time series can be assumed to be independent only if $1 - \alpha$ of the autocorrelation coefficients $r(k)$ for $k \geq 1$ are found to be non-significant.

While this dependence in time does not influence the process of parameter estimation (Chung & Salas, 2000), the presence of persistence influences the

(→ Box continued on next page)

Box 6.3 (continued)

precision in the estimated statistics as the series contains less information compared to the same number of independent observations. Given a hydrological time series with variance s^2 and lag-one autocorrelation $r(1)$, the equivalent number of independent observations, n', is (Hansen, 1971):

$$n' = n\left[1 + \frac{2r(1)}{n} \frac{n(1-r(1))-(1-r(1)^n)}{(1-r(1))^2}\right]^{-1}$$ (B6.3.1)

The relationship is plotted in Figure 6.5 for different autocorrelation coefficients. An observation series of, for example, 50 years with $r(1) = 0.2$ has an equivalent number of independent observations of 33. As a result the estimated variance, s_x^2, will underestimate the true variance, $s_x^{2\prime}$. A corrected estimate can be obtained from (Hansen, 1971):

$$s_x^{2\prime} = s_x^2\left[1 - \frac{2r(1)}{n(n-1)} \frac{n(1-r(1))-(1-r(1)^n)}{(1-r(1))^2}\right]^{-1}$$ (B6.3.2)

The deviation from the true variance will be largest for small samples and high autocorrelations, and decreases rapidly with increasing number of observations. For samples of size larger than 20 and autocorrelation less than 0.6 the deviation in estimated variance is less than 20%.

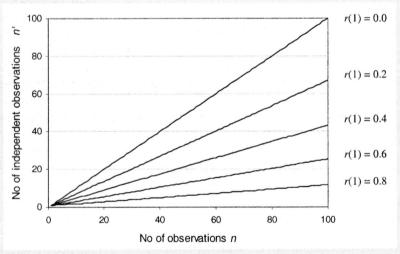

Figure 6.5 The equivalent number of independent observations in a time series, n', as a function of number of observations, n, and the lag-one autocorrelation, $r(1)$.

approach to be applied to the most dominant season. If an a priori subdivision of the data is not feasible a composite probability distribution might be applicable such as the two-component Exponential distribution (Kjeldsen *et al.*, 2000).

6.3.2   Selection of extreme events

The two most common approaches to select extreme events from a time series are the annual maximum or minimum series (AMS) and the partial duration series (PDS) method. AMS is a special case of the block maxima/minima model, which selects the largest/smallest events within each time block. In AMS the block size is chosen to be one year. The PDS approach studies exceedances over an upper limit as compared to maxima over fixed time periods in the AMS.

Annual maxima or minima series

The definition of the AMS is straightforward providing the hydrological year can be clearly defined (Section 3.2.3). In this case the maximum or minimum value is extracted for each hydrological year in the record. The hydrological year should preferably start in a period of the year where extreme events seldom or never occur to ensure that an extreme event is not split between years. A specific season might also be defined as the period of interest, e.g. the summer season, and one event selected for each season and year.

It may happen that a river runs dry. This is not only the case in arid and semi-arid regions but may also occur during the dry season in temperate regions, particularly in quickly responding headwater catchments or during winter in catchments affected by frost. Consequently, annual minimum flow series may contain zero values (zero-flow years). Such data sets are called *censored data* because not all values from a complete sample could be assigned a value as they fall below a level of observation. Streamflow is recorded as zero either because the river has run dry or because the discharge was below a recording limit (Section 4.3.3). It may also happen that a missing observation is recorded as zero, which should be avoided. The problem of zero flow is discussed amongst others by Nathan & McMahon (1990c) and Gordon *et al.* (1992). Annual maximum series of deficit volume and duration may also contain zero values, i.e. years without events, if the flow never becomes less than the defined threshold in a year (non-drought years).

A *conditional probability model* is recommended for frequency analysis at sites whose record of AMS contains zero values (Stedinger *et al.*, 1993). An additional parameter describes the probability, p_0, that an observation is zero. A continuous cdf, $G(x)$, is fitted to the non-zero values of X. The unconditional cdf, $F(x)$, for any value $x > 0$ is then:

$$F(x) = p_0 + (1 - p_0)G(x) \tag{6.7}$$

Similarly, probability plots that use plotting positions for censored data can be applied. These are useful for graphical fitting and a visual display of the data. An increase in the block size, for instance by selecting the largest event in two years, may reduce the proportion of non-drought years. The information content of the series will on the other hand be reduced.

For estimation of the T-year event the distribution F is fitted to the AMS series and the estimate given by:

$$\hat{x}_T = F^{-1}\left(1 - \frac{1}{\lambda T}; \hat{\theta}\right) \tag{6.8}$$

where $\hat{\theta}$ are the estimated parameters of the distribution F and λ the fraction of events per time block. The value of λ is thus ≤ 1 in an annual maximum or minimum series (AMS) and equal to $1 - p_0$, where p_0 is the probability that an observation is zero. It should be noted that the design quantity x_T is here used instead of the equivalent quantile x_p. The relationship between the return period T and the non-exceedance probability p is given in Section 6.2.1.

Partial duration series

Modern applications of the PDS approach have developed from hydrological studies of exceedances over high threshold, commonly denoted POT (Peak Over Threshold) methods (e.g. Todorovic & Zelenhasic, 1970). The classical PDS model assumes a Poisson distributed number of events above an upper limit and independent, exponentially distributed magnitudes, and is therefore particularly suited for modelling high flows. The PDS model ensures that only the most severe events are selected for the extreme value analysis, independent of the time of occurrence. Hence, it intuitively provides a more consistent definition of the extreme value region than the AMS approach. The assumption of a Poisson process implies that successive events are independent and that the arrivals are instantaneous, i.e. of zero duration. A similar theoretical approach for PDS of minimum values has not been developed to the same extent and is therefore not elaborated in the following. Instead the intermittent process of streamflow below a given threshold is discussed, and times series of drought deficit volume and duration investigated (Section 6.3.4). An example of PDS modelling of minimum flow can be found in Önöz & Bayazit (2002).

The PDS model includes two stochastic model components, the occurrence of extreme events and the magnitude of the events. The probability distribution of the number of events in a time period t for a Poisson process is given by:

$$P(N_u) = \frac{(\lambda t)^{N_u}}{N_u!} \exp(-\lambda t) \qquad (6.9)$$

where $P(N_u)$ is the probability of N_u events in time t.

Equation 6.9 is a Poisson distribution with parameter λt. The one-parameter Exponential distribution (Equation A6.1.63 with location parameter equal to zero) describes the time between occurrences of events of a Poisson process. Its scale parameter is equal to $1/\lambda$, where λ is the expected number of events in each time period (the average rate of occurrence):

$$\hat{\lambda} = \frac{N_u}{t} \qquad (6.10)$$

For estimation of the T-year event a probability distribution F can be fitted to the PDS of the exceedance series $\{x - u\}$, where u is the chosen upper limit and an estimate given by:

$$\hat{x}_T = u + F^{-1}\left(1 - \frac{1}{\hat{\lambda}T}; \hat{\theta}\right) \qquad (6.11)$$

where $\hat{\theta}$ are the estimated parameters of the distribution F.

The Poisson process refers to the occurrence of events along a continuous timescale. A departure from the Poisson assumption in PDS models is discussed by Cunnane (1979) and it is suggested that a rejection of the Poisson distribution reflects dependency in the occurrence of events, i.e. they are mutually dependent (Section 5.4.1). In Section 6.4.2 the theoretical background of the Generalized Pareto (GP) distribution for modelling the magnitudes of PDS is introduced. The Exponential distribution, a special case of the GP distribution, is frequently chosen to model the magnitudes of excesses over an upper limit.

Annual exceedance probabilities can be estimated from the PDS provided the average number of events per year, λ, larger than the upper limit u, is known. Let $H(x)$ be the probability that events, when they occur, are less than x and thus fall in the range (u, x). The probability of no exceedances of x in a year, $F_a(x)$, is for independent events and any level x, with $x \geq u$, given by the Poisson distribution so that (Stedinger *et al.*, 1993):

$$F_a(x) = \exp[-\lambda(1 - H(x))] \qquad (6.12)$$

The annual exceedance probability is correspondingly $1 - F_a(x)$ and the annual return period, T_a, $1/(1 - F_a(x))$. Equation 6.12 can be written as:

$$\frac{1}{T_a} = 1 - \exp\left(-\frac{1}{T_p}\right) \tag{6.13}$$

where $T_p = 1/\lambda(1 - H(x))$ is the average return period for level x in the PDS, and corresponds to the definition of T_p in Equation 6.11. It should be noted that T_p and T_a are only slightly different for $T > 10$. Equation 6.13 can also be solved for T_p:

$$T_p = -\frac{1}{\ln\left(1 - \frac{1}{T_a}\right)} \tag{6.14}$$

When the number of extreme events in the PDS is assumed to follow the Poisson distribution and the magnitudes the Exponential distribution, the corresponding AMS will be Gumbel distributed. If the magnitudes are GP distributed, the GEV distribution is obtained for the AMS (Section 6.4.1).

r-Largest events

A third and less common alternative, which partly combines the classical AMS and the PDS approach, is to select the *r*-largest events in each time interval of equal size (block), for example the three largest events each year. It is also possible to consider the whole time series as one block and extract *r* events from the total series. If *r* equals the number of observation years it is referred to as the annual exceedance series. This method is, however, not as general and flexible as the PDS method. The choice of method depends on the data available and the type of analysis to be carried out, considering both practical and theoretical aspects.

Outliers

The sample estimates may sometimes be affected by unexpectedly high or low values, called outliers. Their magnitudes deviate significantly from the rest of the sample and they do not seem to belong to the same distribution. They may result from unusual conditions or recording errors. It is therefore important to visibly detect likely outliers and identify if possible apparent errors associated with the observations. After initial detection, precise identification is possible through homogeneity tests (Kottegoda & Rosso, 1997). If the outliers are removed from the sample the series should be treated as a censored sample and a conditional probability model applied. Alternatively, statistics based on the ordered observations such as the Spearman rank correlation coefficient (Section 5.5.1), can be adopted. Estimating the probability of extreme events when outliers are present is more difficult. If some observations have been identified as outliers a mixed distribution can be adopted comprising one distribution with only one or a few observations. These observations might include some that are possibly mixed with those of the original distribution. Some approaches to this problem are presented in Kottegoda & Rosso (1997).

6.3.3   Minimum values

Traditionally, the annual minimum flow has been the variable of interest in frequency analysis of hydrological drought and the method is usually referred to as low flow frequency analysis (Section 6.6.1). It encompasses time series of annual minimum (AM) flow averaged over a range of durations, the annual minimum n-day average discharge, $AM(n$-day). Commonly 7, 10 or 30 days are considered (Section 5.3.2), but intervals as high as 180 days have been adopted. Care should be taken, however, not to introduce dependency (serial correlation) in the time series when too long averaging intervals are used. Dependency may also be present in AM series of low averaging intervals due to large storage capacity and thus large memory, which can cause even the $AM(1)$ values to be correlated. The analysis of minimum flow is generally limited to the AMS model. An example of frequency analysis of $AM(1)$ is given in Worked Example 6.1.

The extreme values of the smallest values of a random variable are found at the left hand tail of the distribution towards its lower bound. The extreme values are bounded below by zero and the number of zero values can be considerable in some catchments. If the distribution that is fitted to the non-zero values in a series of annual minimum values is not bounded below, negative estimation values may result. This problem can be dealt with by assigning any negative value to zero and thereby introducing a discrete component at zero, similar to the conditional probability model described in Section 6.3.2. The statistical model accordingly allows zero values to occur even if they are not observed. Sometimes the negative values are simply ignored, which can be acceptable provided their probabilities are very low.

The *principle of symmetry* was introduced by Gumbel (1958) to study the asymptotic distribution of the smallest value of a random variable X. It introduces a new variate, X^*, whose pdf is a mirror image of the pdf of X, so that $f_{Xmin}(x) = f_{X^*max}(-x)$. Using the principle of symmetry, the distribution of the smallest value can be determined from the distribution of its largest value by reversing the sign and taking complementary probabilities (Kottegoda & Rosso, 1997). This approach might be suitable if the extreme value distribution is bounded above.

The fact that one is forced to select one event each year in the AMS might result in inclusion of events from rather wet years. These events might not belong to the extreme low flow population and their inclusion can significantly bias the results of the extreme value modelling. Some studies have reported a break in the probability plots of the annual minimum *n*-day series, which has been interpreted as representing a point where the higher frequency flows are not considered to belong to the extreme low flow region. Institute of Hydrology (1980) suggested a break in the curve for an exceedance probability of 65 per cent, whereas Nathan & McMahon (1990c) adopted the 80 percentile. Another solution is to introduce a lower limit, below which only 'true' low flow values are selected, similar to the upper limit in PDS of maximum values. This means that the AMS may contain both zero values (zero-flow years) and years without events (non-drought years). Hence, a conditional probability procedure should be adopted (Section 6.3.2).

Both the plotting position method and the conditional probability model are considered to be simple and reasonable procedures when the majority of years have observations (Stedinger *et al.*, 1993). In arid and semi-arid regions the rivers are often of an intermittent or ephemeral character (Section 2.2.2) and zero flows may be recorded more often than non-zero flows. It might thus happen that the annual minimum values are nearly all or all zero despite a high averaging interval. A traditional frequency analysis of annual minimum flow will therefore fail or not be relevant. Instead variables like the duration of the zero flow periods can be seen as a useful indicator of the severity of a drought and subsequently be applied in a frequency analysis. Alternatively, other types of hydrological data might be used such as groundwater and reservoir levels.

6.3.4   Maximum values

Drought deficit volume and duration comprise series of maximum values that can be analysed to assess the severity of a streamflow drought. The events are selected using the threshold level approach (Section 5.4.1), which ensures that only events below a certain threshold are considered. The problem of including rather wet years in the sample as in the AMS approach for minimum flow is thus reduced provided the threshold chosen is sufficiently low. On the other hand, the flow may never become less than the threshold in a year, which means that the annual maximum series may contain years without events (non-drought years).

One characteristic of the drought event is its duration, and the definition of the hydrological year is thus important as it may happen that the event is split between years. If this happens the event may be assigned to one year following a predefined rule, e.g. belong to the year of the longest duration. In some regions drought events lasting longer than a year (multi-year droughts) are frequent, and the AMS will be difficult to establish.

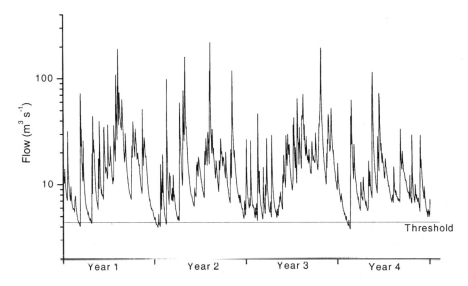

Figure 6.6 Selection of drought events using the threshold level approach on a four-year series of daily streamflow from River Ngaruroro at Kuripapango (NZ).

Both AMS and PDS can be derived for deficit volume and duration. Figure 6.6 shows daily streamflow for River Ngaruroro for the period 1990–1993. A logarithmic scale is used on the *y*-axis to better depict the lower range of flow values. The AMS consists of the largest event each year. The PDS comprises all events below the threshold, and in total five events result as two drought events are present in both the first and the second year. The flow never becomes less than the threshold in the third year (non-drought year).

A major drawback of the threshold level method for selecting drought events, which also applies to the method of selecting exceedances over a high threshold, is the probability of introducing dependency between the events. Dependence does not in this context mean that the values are correlated, but it exists as a form of persistence in the point stochastic process producing the events. A consistent definition of drought events below a threshold should therefore include some kind of restriction or pooling of successive events in order to define an independent sequence of events. This is clearly demonstrated in Figure 6.6 where the first drought in the second and fourth year is split into two *mutually dependent* droughts. These should be considered as one event, resulting in a total of five events, not seven. Without pooling the AMS may contain events that are split and consequently less severe, whereas in the PDS all events are included implying more, but mutually dependent and less severe, events. Different methods for pooling dependent droughts are presented in

Box 5.2. Analog procedures are common in POT modelling of high level exceedances. Here the events are not pooled, but cluster intervals are introduced and maxima within each cluster selected to comprise a series of independent events. There is, however, no universally-accepted method for identifying clusters (Smith, 1989).

The selection of drought events using the threshold level on a daily time series also exhibits the problem of minor droughts (Figure 5.11). These events may distort the extreme value modelling, and different procedures for removing minor events have been suggested (Tallaksen *et al.*, 1997). To smooth the time series and thus avoid a possible high day-to-day variability in the low flow records a moving average procedure can be adopted. The method has proved efficient in reducing the number of minor droughts as well as pooling dependent droughts (Box 5.2). Still, drought events with a duration of less than a given number of days might be excluded. This criterion is imposed as a second step in the selection process after the events have been identified by the threshold level method. Alternatively, an upper limit might be introduced to allow only the most extreme events to be included in the analysis (Box 6.5). A daily or monthly time step is required to study within year or seasonal droughts (Section 5.4.1).

6.4   Probability distributions

Although it is possible to perform a frequency analysis by merely plotting the data and without making any distributional assumptions, hydrological design often requires extrapolation beyond the range of observations. This can be achieved by fitting a distribution function to the sample. The record length is, however, often insufficient to precisely define the probability distribution for a data set. Given a data sample from an unknown distribution, the problem is to estimate accurately the tail of the distribution, which contains the extreme events. Three important issues in this context are (a) there are few observations in the tail of the distribution, (b) estimates are often required above the largest or smallest observed value, and (c) standard probability distributions can be severely biased in estimating tail probabilities (Coles, 2001).

An overview of the distribution functions presented in this chapter is given in Table 6.1. Further details of the distributions and their properties are given in Appendix 6.1: the probability density function (pdf), the cumulative density function (cdf), the quantile function for x_p corresponding to the non-exceedance probability p, expressions of product moments and L-moments, and finally procedures for parameter estimation by the method of product moments, L-moments and maximum likelihood. This section starts with a description of

Table 6.1 List of probability distributions

Distribution	Abbrevation	Number of parameters	Extreme value model	Details in Appendix
Generalized Extreme Value	GEV	3	AMS	A6.1.1
Gumbel (EV I)	GUM	2	AMS	A6.1.2
Frechét (EV II)	FRE	3	AMS	A6.1.3
Weibull (EV III)	WEI	3	AMS	A6.1.4
Generalized Pareto	GP	3	PDS	A6.1.5
Exponential	EXP	2	PDS	A6.1.6
Pearson Type 3/Gamma	P3/GAM	3(2)	AMS/PDS	A6.1.7
Log-Pearson Type 3	LP3	3	AMS/PDS	A6.1.8
Log-Normal	LN	3(2)	AMS/PDS	A6.1.9

probability distributions commonly applied in the analysis of extreme events in hydrology, whereas parameter estimation methods are described in Section 6.5.

6.4.1 Extreme value distributions

The probability distribution of the maximum of n iid random variables depends on the sample size n and the parent distribution of the sample. Frequently the parent distribution from which the extreme is an observation is not known, but if the sample size is large, limiting assumptions concerning the parent distribution can assist in finding the distribution of extreme values (Haan, 1977). Fisher & Tippett (1928) distinguished between three types of extreme value (EV) distributions (limiting distributions) that are based on different parent distributions:

a) The *EV I distribution*, commonly referred to as Gumbel's (GUM) extreme value distribution (Gumbel, 1958) or the double Exponential distribution. It has a parent distribution that is unbounded in the direction of the desired extreme. It is therefore frequently applied for analysis of maximum values. Examples of parent distributions for maximum values are the Normal, Log-Normal, Exponential and Gamma distribution, whereas the Normal distribution is a parent distribution for minimum values;

b) The *EV II distribution* or Fréchet type (FRE) was first developed and applied to floods by Fréchet (1927), but has found little application in hydrology. It has, like the EV I distribution, a parent distribution that is unbounded in the direction of the desired extreme, but not all moments exist;

c) The *EV III distribution* has a parent distribution that is bounded in the direction of the desired extreme. A two-parameter Weibull (WEI) distribution (Weibull, 1961) is the EV III distribution for minima bounded below by zero. It is commonly applied in hydrology since many hydrological variables are bounded by zero. The Exponential distribution is a special case of the Weibull distribution. Examples of parent distributions for minimum values are the Beta, Log-Normal, Gamma and the Exponential distribution, whereas the Beta distribution is a parent distribution for maximum values.

The Generalized Extreme Value (GEV) distribution, independently derived by von Mises (1936) and Jenkinson (1955), is a general mathematical form that encompasses the three types of limiting distributions (Coles, 2001). The three-parameter GEV distribution describes the limit distribution of normalized maxima and is therefore suited to model block maxima series (Section 6.3.2). Some important relationships exist between the Weibull, the Gumbel and the GEV distribution (Stedinger *et al.*, 1993). If X has a Weibull distribution, then $Y = -\ln(X)$ has a Gumbel distribution. This allows parameter estimation procedures for the Gumbel distribution to be applied for the Weibull distribution as demonstrated in Worked Example 6.1 (Section 6.6). Similarly, a three-parameter Weibull distribution can be fitted by the method of L-moments using the parameter expressions for the GEV distribution applied to $-X$.

Suppose that $X_1, \ldots, X_n$ is a sequence of iid random variables with distribution F. Define $M_n = \max\{X_1, \ldots, X_n\}$. Following the Fisher-Tippett theorem or limit laws for maxima (Fisher & Tippett, 1928), the GEV distribution defined by:

$$F_X(x) = \exp\left\{-\left[1 - \frac{\kappa(x-\xi)}{\alpha}\right]^{1/\kappa}\right\}$$ (6.15)

where $1 - \kappa(x-\xi)/\alpha > 0$ and $\alpha > 0$, can be applied as an approximation to the distribution of M_n for finite, but large n (Section A6.1.1).

The model has three parameters: a location parameter ξ, a scale parameter α and a shape parameter κ. The important shape parameter κ controls the tail of

---

Box 6.4

Reduced variate for the GEV and the GP distributions

For a consistent and direct comparison between different distributions in a probability plot the same reduced variate should be applied. A reduced variate y, is linearly related to x and substitutes the random variable x in the expression for $F(x)$. For comparison between the EV I or Gumbel ($\kappa = 0$), EV II ($\kappa < 0$) and the EV III distribution ($\kappa > 0$) the Gumbel reduced variate can be used. On a Gumbel probability paper the Gumbel distribution will plot as a straight line, whereas the EV II and EV III will plot as curved lines.

The probability distribution of the Gumbel distribution (Equation A6.1.15) is:

$$F(x) = \exp\left[-\exp\left(-\frac{x-\xi}{\alpha}\right)\right]$$ (B6.4.1)

A reduced variate for the Gumbel distribution can be defined as:

$$y = \frac{x-\xi}{\alpha}$$ (B6.4.2)

Substituting the reduced variate into Equation B6.4.1 yields:

$$F(x) = \exp[-\exp(-y)]$$ (B6.4.3)

and solving for y:

$$y = -\ln[-\ln(F(x))]$$ (B6.4.4)

A straight line results if the values of x and y are plotted against each other as shown in the probability plot for the GEV distribution (Figure 6.7, upper). The EV II and EV III distributions can be plotted in the same diagram provided Equation B6.4.4 is used to define y also for these distributions. As can be seen from the plot, the EV I distribution is unbounded in x, whereas EV II has a lower bound and EV III an upper bound. The figure also includes on a subsidiary axis the return period T, as an alternative to y. An estimate of the return period for a given design event $\hat{x}_T$, can subsequently be obtained from the relationship between y and $F(x)$.

For comparison of the three types of distributions encompassed in the GP distribution the exponential reduced variate can be used. The probability distribution of the Exponential distribution is (Equation A6.1.64):

$$F(x) = 1 - \exp(-y) \quad , \quad \text{where} \quad y = \frac{x-\xi}{\alpha}$$ (B6.4.5)

(→ Box continued on next page)

Box 6.4 (continued)

and solving for y:

$$y = -\ln(1 - F(x))$$ (B6.4.6)

A straight line results for the exponential distribution ($\kappa = 0$) whereas the GP distributions with positive and negative shape parameter plot as curved lines (Figure 6.7, lower).

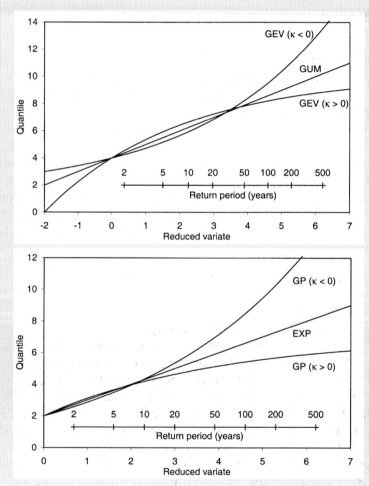

Figure 6.7 The quantile, x_T, for the three types of GEV distributions (upper) and GP distributions (lower) plotted against the Gumbel and the exponential reduced variate, y, respectively.

the distribution (Box 6.4). When $\kappa = 0$ it reduces to the Gumbel distribution (Equation B6.4.1) with arbitrary location and scale parameter and an exponential decreasing tail (Figure 6.7, upper); for $\kappa > 0$ it equals the Weibull type with a finite upper bound and therefore short-tailed; and for $\kappa < 0$ it equals the Fréchet type, which has a polynomially decreasing tail function and therefore corresponds to a long-tailed parent distribution (Smith, 1989).

The quantiles of the GEV distribution can be estimated as:

$$\hat{x}_p = \hat{\xi} + \frac{\hat{\alpha}}{\hat{\kappa}}\left\{1 - [-\ln(p)]^{\hat{\kappa}}\right\} \tag{6.16}$$

6.4.2 The Generalized Pareto distribution

The Generalized Pareto (GP) distribution, which goes back to Pickands (1975), appears as the limit distribution of scaled excesses over an upper limit u (Embrechts *et al.*, 1997). It is therefore suited to describe floods above a threshold or daily flows above zero. As there exists a limit distribution for block maxima values, PDS also have their limit distribution.

Let $X_1, \ldots, X_n$ be a sequence of iid random variables. Choose an upper limit u and denote by N_u the number of exceedances of u by $X_1, \ldots, X_n$ The corresponding excesses are denoted $Y_1, \ldots, Y_{Nu}$ The excess distribution function of X (X is a random variable $> u$) is then given by $F_u(y)$. For large u the GP distribution can be applied as an approximation to the excess distribution function. It is given by:

$$F_X(x) = 1 - \left[1 - \kappa\frac{(x - \xi)}{\alpha}\right]^{1/\kappa} \quad for \quad \kappa > 0, \ \xi \le x \le \xi + \frac{\alpha}{\kappa} \tag{6.17}$$

where $1 - \kappa(x - \xi)/\alpha > 0$ and $\alpha > 0$ (Section A6.1.5).

Analogous to the GEV distribution, there are three different cases for the three-parameter GP distribution depending on the sign of the shape parameter κ (Box 6.4). When $\kappa = 0$ it reduces to the exponential distribution with arbitrary location parameter ξ (Figure 6.7, lower); for $\kappa < 0$ the GP distribution is valid on $\xi \le x < \infty$ and the distribution has no upper limit; and for $\kappa > 0$ the GP distribution is bounded in the direction of the extreme, similarly to the Weibull type of classical EV theory.

When a predefined upper limit, u, is applied, the number of exceedances becomes a random variable and the location parameter can be set equal to u. The remaining two parameters can be estimated from the exceedance series. The choice of the upper limit is a compromise between not choosing a too low limit that might introduce model bias, whereas on the other hand, a too high limit might result in few exceedances and high variance estimators (Box 6.5). While the scale parameter α is a function of the upper limit u, κ should remain constant if u is chosen large enough. Plotting κ as a function of u or the number of exceedances can accordingly support the choice of u by illustrating the bias versus variance trade-off. The mean excess (or mean residual life) plot can also assist in the choice of upper limit u, by plotting the mean of the excesses above the upper limit against u. The mean excesses are expected to change linearly with u at levels of u for which the generalized Pareto model is appropriate (Coles, 2001). An example of application for extreme value modelling of drought deficit volume is given in Hisdal & Tallaksen (2002).

The quantiles of the GP distribution are given as (recalling that the location parameter ξ can be set equal to u):

$$\hat{x}_p = \hat{\xi} + \frac{\hat{\alpha}}{\hat{\kappa}}\left[1 - (1-p)^{\hat{\kappa}}\right]$$ (6.18)

Box 6.5

Streamflow drought in a semi-arid region

River Arroyo Seco is located in California, a semi-arid region in North America. The climate is temperate with dry summers (Köppen's Cs-climate, Appendix 2.1). Mean annual precipitation is about 850 mm year^{-1} and the vegetation is savanna and dryland crop pasture. At Soledad gauging station the catchment area is 632 km^2 and mean annual streamflow 240 mm year^{-1} or 4.8 m^3 s^{-1} (Table 4.3). In total 68 years of daily streamflow records are available, covering the period 1931 to 1998. There is no regulation or large diversion upstream from the station. The river is flashy and shows an intermittent behaviour with 12.5% of the observations being zero.

Streamflow drought events were derived from the daily streamflow series using the Q_{70} threshold level (0.26 m^3 s^{-1}). A moving average 11-day procedure was used to smooth the time series prior to selecting the drought events (Box 5.2). The events were selected from the whole year and a histogram of the resultant PDS of drought duration is plotted in Figure 6.8 (upper). At least two types of drought can be identified from the plot; minor droughts that occur during winter time and long duration droughts in summer. The sample is clearly non-homogeneous, and the two populations should be treated as separate samples in a frequency analysis. The distribution of the summer droughts further suggests a bimodal behaviour, which is even more pronounced for deficit volume (not shown). In order to fit the Generalised Pareto (GP) distribution to PDS of summer droughts a minimum duration was first introduced that excluded the minor winter droughts. The best fit was, however, obtained by introducing an upper limit that included only the most severe events, corresponding to the second peak in the distribution of summer droughts. The resultant qq-plot is shown in Figure 6.8 (lower), where the empirical quantiles are plotted against the GP distribution quantiles estimated from the PDS of deficit volume. The plot shows that the modelled GP quantiles are slightly overestimated for high return periods.

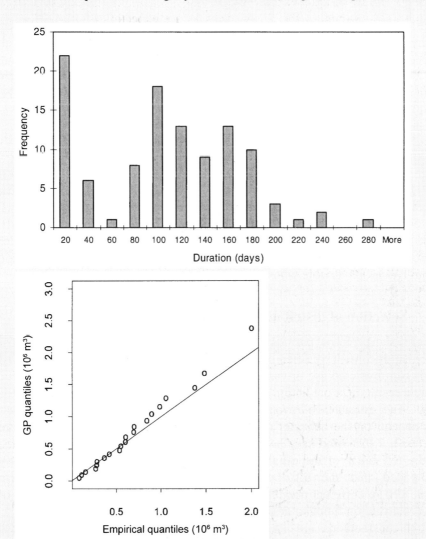

Figure 6.8 Streamflow drought in River Arroyo Seco (USA); histogram of drought duration (upper) and empirical quantiles of drought deficit volume plotted against GP distribution quantiles (lower; modified from Hisdal & Tallaksen, 2002).

6.4.3   Other distributions

In addition to the GEV and GP distributions presented above, this section briefly introduces some other commonly applied probability distributions for extreme values analysis of hydrological drought (Table 6.1). Details of the distributions are given in Appendix 6.1.

The Log-Normal distribution

Many hydrological series are positively skewed and their logarithm can be well described by a Normal distribution ($Y = \ln X$). This requires that the original series X is strictly positive, which is commonly the case for hydrological variables. As the skewness goes towards zero, the Log-Normal (LN) distribution approaches a Normal distribution. If a lower bound parameter is introduced ($X - \xi$), a three parameter Log-Normal distribution results.

The Pearson Type 3, Gamma and Log-Pearson Type 3

Another common family of distributions used in hydrology is the Pearson Type 3 (P3) distribution. The distribution is bounded below or above depending on the sign of the scale parameter α. If $\alpha > 0$ and the location parameter $\xi = 0$ (lower bound), the P3 distribution reduces to the Gamma distribution (GAM). A closed form of the cdf of the P3 distribution is not available and tables or approximations must be used. The Log-Pearson Type 3 (LP3) distribution describes a random variable whose logarithms are P3 distributed.

6.4.4   Selection of distribution function

Relatively short return periods that do not greatly exceed the length of hydrological records have often been sufficient in low flow design. This has led to a rather unrestricted use of various distributions. Such an approach may give acceptable results if a prediction of drought indices with a low return period is required. The estimates of events with higher return periods will, however, always depend on the behavior of the tail of the fitted distribution. The more parameters a distribution has, the better it will adapt to the data sample, but the lower the reliability of the estimate of the parameters will be. It is generally recommended that the distribution has no more than three parameters (Matalas, 1963). Different *goodness-of-fit tests* that can be used for selection of a proper distribution function are presented below.

   Graphical methods are an efficient way to judge whether the fitted distribution appears consistent with the data. Probability plots (Section 6.2) and qq-plots, which plot the empirical quantiles against the distribution quantiles (Figure 6.8, lower), are commonly used. The data are ranked and a plotting position formula chosen to give an estimate of $F(x)$. The points should lie close to the theoretical curve (probability plot) or the unit diagonal (qq-plot) provided the chosen model is a reasonable model for the data in question.

The plots can be judged merely by visual inspection, or statistical tests such as analytical goodness-of-fit criteria can be used to estimate whether a departure, i.e. the difference between the ordered observations and the estimated quantile from a theoretical distribution function, is statistically significant. Several measures are available (Stedinger *et al.*, 1993). The Chi-squared test statistics is based on a comparison of the number of observed and expected events in each class interval of the sample histogram (Section 6.2.2). The Kolmogorov-Smirnov test is based on the absolute deviation between the empirical and the theoretical cumulative distribution function. The conclusions that can be drawn from these tests depend on the level of confidence adopted, commonly a 95% level is chosen. In many cases several distributions will provide statistically acceptable fits to the data, and goodness-of-fit tests cannot be used to discriminate between the different distributions (e.g. Cunnane, 1985). Important in this respect is that many distributions have a similar form in their central parts, but differ significantly in the tails where the extreme values are located. The standardized least squares criterion (Takasao *et al.*, 1986) and the probability plot correlation coefficient (Vogel, 1986) are based on the correlation between the ordered observations and the corresponding fitted quantiles. The ordered observations, also defined as the *order statistics* $X_{(i)}$, is analogous to the sorted or ranked observations in a sample. These are more powerful tests, which means that they have a higher probability of correctly determining that a sample is not from a given theoretical distribution.

Still other approaches test the models according to some criterion that best describes the data and at the same time accounts for the number of parameters in the model. The likelihood ratio test (Coles, 2001) allows comparison to be made between distributions with a different number of parameters. Another flexible approach to model discrimination is the Akaike Information Criterion (AIC). It maximizes the log-likelihood over choices of parameters to form a trade-off between the fit of the model and the model complexity, measured by the number of parameters (Akaike, 1974). The model that possesses the minimum value of AIC should be selected. In hydrology it has been applied to identify flood frequency models (e.g. Mutua, 1994; Strupczewski *et al.*, 2001).

Traditionally, statistical judgment is used in combination with knowledge of the phenomenon studied to select the distribution that is best suited for the data in question. For minimum flows the distribution function should be skewed and have a finite lower limit greater or equal to zero. Likewise, a distribution that is bounded above might be preferred for the deficit volume and duration of a drought, provided no multi-year droughts are present. The families of distribution functions for extreme events also have a theoretical base as outlined in the previous sections. *The GEV distribution is adopted to model AMS, whereas the GP distribution is suited for the PDS.* This knowledge should preferably guide the choice of distribution function, despite the fact that these distributions might not necessarily give the best fit to the data. They might still provide a better prediction of a future event as there are large uncertainties related to the conclusions that can be drawn by merely comparing the distribution fit to the observations. Important factors here are the quality of the data, the representativity of the sample and the statistical uncertainty related to model fitting and testing.

It is also recommended to relate the choice of distribution function not only to the type of variable being studied, but also to take account of regional information. Often the distribution that best fit the data at a particular site is chosen. This procedure is, however, very sensitive to sample variability. The procedure adopted should yield reliable quantiles and risk estimates. Such estimators are called *robust*, which means they should perform reasonably for a wide range of catchments. This can be achieved by combining site-specific and regional information. L-moments ratio diagrams are valuable for investigating what families of distributions are consistent with a regional data set. Regional frequency methods are further discussed in Section 6.7.

6.5   Estimation methods

Having chosen a probability distribution the estimation of a design event is reduced to the problem of fitting the distribution to the sample, i.e. to estimate the parameters of the distribution. Two estimation methods are presented: the method of moments, including product moments and L-moments, and the maximum likelihood method. Other methods exist such as the restricted maximum likelihood method and Bayesian inference (Smith, 2002). This section is limited to parametric methods, although empirical distribution functions fitted with the use of plotting position formulae (Section 6.2.2) fall into the class of non-parametric methods. Parametric methods assume that a sample comes from a population with a given pdf, whereas non-parametric methods are distribution free. Recently, *non-parametric* methods for frequency analysis of hydrological variables have been introduced as an alternative to parametric methods. Non-parametric methods provide local estimates of the density function by using weighted moving averages of the data in a small neighbourhood around the point of estimation. A review of non-parametric functions and their applications in hydrology is given by Lall (1995). Examples for drought studies include Kim *et al.* (2003) for the estimation of return periods for drought characteristics derived using the threshold level method on time series of PDSI (Box 5.3) and Adamowski (1996) for frequency analysis of annual minimum series containing zero values.

6.5.1 The method of moments

In general, the use of moments covers (a) the characterization of probability distributions, (b) summarization of observed data samples, (c) fitting of probability distributions to data, (d) testing of hypotheses about distribution form, and (e) identification of homogeneous regions. The latter is further elaborated in Section 6.7.4. Here the purpose is to estimate the parameters of a distribution by applying statistical inference using sample moments. A hat (^) above a parameter indicates an estimated value (e.g. skewness in Equation 6.22). Separate symbols are commonly used for the estimated mean and variance (Equations 6.20 and 6.21). The method of moments relates the theoretical moments of the distribution (e.g. mean value, variance and skewness) to the sample estimates. Conventional product moments, probability weighted moments (PWM) (Greenwood *et al.*, 1979) and L-moments (Hosking, 1990) can be used. L-moments are weighted linear sums of the expected order statistics and are analogous to product moments used to summarize the statistical properties of a probability function or an observed data set. L-moments can be written as functions of PWMs and procedures based on PWMs and L-moments are therefore equivalent (Appendix 6.2). A distribution might be specified by its L-moments even if some of its product moments do not exist, and such a specification is always unique, which is not true for product moments (Hosking, 1990).

Product moments

The first product moment about $X = 0$ of a distribution is the *mean*, μ, or the expected value of X, $E\{X\}$. The second moment about the mean is called the *variance*. The *standard deviation*, σ, is the square root of the variance σ^2, and is a measure of the spread around the central value. A relative measure of spread is the dimensionless *coefficient of variation*, $CV = \sigma/\mu$. The third moment about the mean is a measure of the symmetry of a distribution, and is usually characterized by the dimensionless *skewness* ratio, γ_3. The fourth moment about the mean provides information about the peakedness of the central part of the distribution, or related to this, the thickness of the tail of the distribution. It is also characterized by a dimensionless ratio called *kurtosis*, γ_4. The product moments are defined as:

$$\mu = E\{X\}$$
$$\sigma^2 = Var\{X\} = E\{(X - \mu)^2\}$$
$$\gamma_3 = \frac{E\{(X - \mu)^3\}}{\sigma^3} \tag{6.19}$$
$$\gamma_4 = \frac{E\{(X - \mu)^4\}}{\sigma^4}$$

where $E\{.\}$ is the expectation operator. Based on a set of observations $x_1 \ldots x_n$, estimators of the product moments can be calculated as:

$$\bar{x} = \frac{1}{n}\sum_{i=1}^{n} x_i \tag{6.20}$$

$$s^2 = \frac{1}{n-1}\sum_{i=1}^{n}(x_i - \bar{x})^2 \tag{6.21}$$

$$\hat{\gamma}_3 = \frac{\dfrac{1}{n}\sum_{i=1}^{n}(x_i - \bar{x})^3}{\left[\dfrac{1}{n}\sum_{i=1}^{n}(x_i - \bar{x})^2\right]^{3/2}} \tag{6.22}$$

$$\hat{\gamma}_4 = \frac{\dfrac{1}{n}\sum_{i=1}^{n}(x_i - \bar{x})^4}{\left[\dfrac{1}{n}\sum_{i=1}^{n}(x_i - \bar{x})^2\right]^{2}} \tag{6.23}$$

The use of the factor $(n-1)$ instead of n in Equation 6.21 yields an unbiased estimator of the variance σ^2. The estimates of the first two moments are unbiased no matter what the distribution and the sample size, whereas the skewness and kurtosis estimator in Equations 6.22 and 6.23 are, in general, biased. Similar corrections as for the variance can be introduced for higher order moments that generally reduce, but not eliminate, the bias. The moment estimators of the distribution parameters are obtained by replacing the theoretical product moments for the specified distribution by the sample moments. An estimator is said to be the most *efficient* estimator for a parameter if it is unbiased and its variance is at least as small as that of any other unbiased estimator. Population moments and expressions of the moment estimators for the distributions presented in this chapter are found in Appendix 6.1.

Expressions for other distributions can be found in, for example, Stedinger *et al.* (1993) and Embrechts *et al.* (1997).

L-moments

Hosking's (1990) method of L-moments has found widespread application in statistical analyses of hydrological data. Let X be a real-valued random variable with cumulative distribution function $F(x)$ and let $X_{(1:n)} \leq X_{(2:n)} \leq \ldots \leq X_{(n:n)}$ be the order statistics of a random sample of size n drawn from the distribution of X. The first four L-moments are then defined as:

$$\lambda_1 = E\{X_{(1:1)}\}$$
$$\lambda_2 = \tfrac{1}{2}E\{X_{(2:2)} - X_{(1:2)}\}$$
$$\lambda_3 = \tfrac{1}{3}E\{X_{(3:3)} - 2X_{(2:3)} + X_{(1:3)}\} \quad (6.24)$$
$$\lambda_4 = \tfrac{1}{4}E\{X_{(4:4)} - 3X_{(3:4)} + 3X_{(2:4)} - X_{(1:4)}\}$$

The first moment equals the mean ($E\{X\}$), and the second moment is a measure of variation based on the expected difference between two randomly selected observations. It is common to standardize moments of higher order to make them independent of the unit of measurement of X. L-moment ratios, the L-coefficient of variation, L-CV (τ_2), L-skewness (τ_3) and L-kurtosis (τ_4), are defined as:

$$\tau_2 = \frac{\lambda_2}{\lambda_1}$$

$$\tau_3 = \frac{\lambda_3}{\lambda_2} \quad (6.25)$$

$$\tau_4 = \frac{\lambda_4}{\lambda_2}$$

L-moment ratios are bounded, and the value of $\tau_r = \lambda_r / \lambda_2$ for $r \geq 3$ is between -1 and $+1$. The L-moment estimators of the distribution parameters are obtained by replacing the theoretical L-moments for the specified distribution by the sample estimates. L-moment estimators for different distributions are given in Appendix 6.1. Appendix 6.2 provides an overview of different L-moment estimators.

Wang (1997) introduced LH-moments, a generalization of L-moments that is based on linear combinations of higher-order statistics. Generally, the moment of order r considers the rth largest value in a subsample of size $r + \eta$. When $\eta = 0$ the LH-moments become identical to ordinary L-moments. As η

increases, LH-moments (also labelled Lη-moments for η = 1, 2, ...) reflect more and more the upper part of distributions and larger events in the data. The method reduces the influence that small sample events may have on the estimation of large return period events, but is, however, sensitive to the determination of the appropriate value of η, the subsample size chosen. LL-moments have similarly been defined considering the rth smallest value in a subsample of size r + η (Bayazit & Önöz, 2002).

The main advantage of L-moments over conventional product moments is that L-moments, being linear function of the data, suffer less from the effect of sample variability as it does not involve squaring or cubing the observations, as do product moment estimators. For example, the second order product moment (variance) and L-moment both measure the difference between two randomly drawn elements of a distribution. However, in case of variance (σ^2) more weight is given to the largest differences as these are squared:

$$\lambda_2 = \frac{1}{2}E\{X_{(2:2)} - X_{(1:2)}\} \qquad \sigma^2 = \frac{1}{2}E\{X_{(2:2)} - X_{(1:2)}\}^2 \qquad (6.26)$$

The product moment-based measure of skewness is similarly very sensitive to the extreme tails of the distribution (differences are now cubed), and therefore difficult to estimate accurately when the distribution is markedly skewed.

Vogel & Fennessey (1993) demonstrated that sample product estimators of standard deviation and skewness exhibit significant bias, even for extremely large sample sizes. The study by Sankarasubramanian & Srinivasan (1999) suggests, however, that product moments are preferable at lower skewness, particularly for smaller samples, while L-moments are preferable at higher skewness, for all sample sizes. In general the performance of different estimators depends on the considered distribution. Rosbjerg *et al.* (1992) compared *T*-year event estimation for the PDS Poisson-GP distribution model. They showed that except for very large sample sizes, which are rarely available in practice, product moment estimates are more efficient than L-moment estimates for shape parameters in the range $-0.5 < \kappa < 0.5$. For *T*-year estimation based on AMS and the GEV distribution Madsen *et al.* (1997b) showed that L-moment estimates are only superior to product moment estimates for highly skewed distributions (corresponding to shape parameters smaller than -0.25), in which case product moments introduce a large bias.

Generally, L-moments are more robust to extreme values in the data and enable more secure inferences to be made from small samples about an underlying probability distribution (Hosking, 1990). Klemeš (2000b) argues that the method being less sensitive to the extremes is not an advantage as these are actually observed values. If any of the extreme values can be considered outliers, alternative procedures may be adopted (Section 6.3.2). Still, another solution to the problem of at-site sample variability is to adopt a regional procedure as less emphasis is then put on single observations. Properties of L-moment estimators are especially advantageous in regional frequency estimation (Section 6.7).

6.5.2 Maximum likelihood estimators

Another commonly used method of estimating parameters of a distribution function is the maximum likelihood approach. Assume that there are n random observations $x_1 \ldots x_n$, which have a joint probability distribution $f_X(x_1 \ldots x_n; \theta_1 \ldots \theta_m)$, where m is the number of unknown model parameters. The values of the m parameters that maximize the likelihood that the particular sample is the one that would be obtained if n random observations were selected from $f_X(x; \theta)$, are known as the maximum likelihood estimators. The probability that a particular sample would be obtained from the population is denoted the likelihood function, L. Commonly a logarithmic transformation of the likelihood function is performed, and the parameter estimators obtained by maximizing:

$$L(\theta) = \sum_{i=1}^{n} \ln[f(x_i; \theta)] \qquad (6.27)$$

The values of the parameters that maximize the likelihood function can be found by taking the partial derivative of $L(\theta)$ with respect to each of the parameters and setting the expressions equal to zero. The resultant m equations are then solved for the unknown m parameters.

The maximum likelihood (ML) method is one of the most widely-used and theoretically-sound methods for fitting probability distributions to data. In the theoretical case with a known distribution it produces asymptotically efficient and unbiased estimators of the parameter set, which is the case if a random sample is generated from a known probability density function. In the practical case, when a distribution function is assumed, the ML method produces bias in the estimators and large samples are necessary before the sample estimator becomes unbiased (Kottegoda & Rosso, 1997). The bias increases with the deviation between the distribution of the observations and the distribution being fit (Stupczewski *et al.*, 2001). For large samples the ML estimator has very good statistical properties and allows seasonality, trends or periodic dependencies to be included in the model in a flexible way. However, small samples estimators may be less efficient and sometimes the procedure becomes unstable (e.g. Madsen *et al.*, 1997b). Martins & Stedinger (2000, 2001) introduced a generalized maximum likelihood estimation method for the GEV and GP distributions, which produces more efficient and stable

ML estimates. Often explicit formulae are not available for ML estimation and numerical methods must be imposed. Procedures for calculating ML estimators for different distributions are given in Appendix 6.1.

6.5.3   Estimation of design values, uncertainties and risk

Estimates of extreme quantiles or *design values* are obtained by inverting the expression for $F_X(x_p)$ for a given value of the non-exceedance probability p. Quantile expressions for x_p (or x_T) for the distributions listed in Table 6.1 are given in Appendix 6.1, and can be determined once the parameters of the distribution are known. It is common to extrapolate this relationship beyond the range of the data to which the distribution has been fitted. Accordingly, a measure of the *uncertainty* in the quantile estimate or design event should be given, thereby allowing the value of additional data to be quantified as a decrease in uncertainty.

A simple measure of the precision in the estimate is the variance of the quantile estimator, which equals the square of the standard error (SE). Another measure of precision is confidence intervals (CI), which are often calculated using the SE of the quantile. A 90 or 95% CI is commonly used, which in repeated sampling will contain the parameter 90 or 95% of the time. Approximate expressions for the variance and CI can be derived using first order Taylor series approximations, e.g. Rosbjerg *et al.* (1992) for formulae for the *T*-year event estimator for the PDS Poisson-GP model and Madsen *et al.* (1997b) for the AMS GEV-estimator. The presence of persistence influences the precision in the estimated statistics and the estimated variance will underestimate the true variance (Box 6.3).

Other approaches for calculating the uncertainty include Monte Carlo simulations (e.g. Salas, 1993) and resampling methods (e.g. Good, 1994). In Monte Carlo simulations the bias and the standard deviation of the quantile are obtained by randomly generating a large number of samples that have the same statistical properties as the observed sample. Resampling techniques estimate the bias and the standard deviation by sampling a number of subsamples from the original data set.

The concept of *risk* has many interpretations in hydrological literature. Sometimes it is understood to be equal to the probability of exceedance (non-exceedance) of a certain maximum (minimum) level within a given design period (Box 6.2). The use of historical data, however, provides no guarantee that the design will perform satisfactorily in the future. The calculation of risk assumes that the underlying process is sufficiently stable to be considered stationary, and that the sample is representative.

A more generally accepted definition of risk as the expected loss, damage or utility (Berger, 1985) is also used in hydrologic applications (e.g. Stedinger, 1997). The international glossary of hydrology (UNESCO/WMO, 2003) similarly defines risk as the potential realization of unwanted consequences of an event, a function of the probability and the value of the consequence. Accordingly, a loss function which gives the expected damage as a function of the probability of the event has to be formulated. The latter is often referred to as the vulnerability of the system, and the reader is referred to, for example, Vose (2000) for further reading.

6.6 At-site frequency analysis

In this section a number of studies that deal with at-site frequency analysis of streamflow drought characteristics are presented. Low flow and deficit characteristics are presented separately, and each section contains a worked example that illustrates some of the many aspects of frequency analysis outlined in the previous sections.

6.6.1 Distribution functions for low flow characteristics

Low flow values generally show a range of variations of the skewness, which should be accounted for in the choice of distribution function. The WEI distribution has a parent distribution bounded at the direction of the extremes. The flexibility of the distribution, as well as its theoretical basis, makes it a favourable choice for frequency analysis of annual minimum discharge series (Worked Example 6.1). It has accordingly been applied in a large number of low flow prediction studies around the world (Tallaksen, 2000). An extensive review of low flow frequency analysis is given in Smakhtin (2001). The distribution functions found to be most frequently referred to in the literature are different forms of WEI, GUM, P3 and LN distributions.

In a drought study covering southern Africa a wide range of distributions were tested, and overall the LN distribution was found to fit the AM series best (Tate *et al.*, 2000). For sites with a vast majority of zero flows the fitting procedure failed. In Zaidman *et al.* (2003) the GEV, P3, GP and the generalized logistic (GL) distributions were evaluated on AM series from the UK. For averaging intervals less than 60 days high storage catchments best fitted the GL and GEV distributions, whereas low storage catchments were best described by P3 or GEV models.

Historical evidence suggests that extreme events occur more due to unusual combinations of factors that may cause an event, than to unusual magnitudes of the factors themselves, i.e. that they coincide in time. Klemeš (1993) therefore suggested treating the variable of interest as a compound event, based on an explicit consideration of the physical processes involved. This is also the philosophy behind the *derived distribution approach* which was first introduced by Bernier (1964) for low flow frequency analysis. The approach permits knowledge about the hydrological processes generating streamflow to be considered, and parameters of the recession curve were applied by Gottschalk *et al.* (1997) to derive theoretical expressions for low flow distribution functions. The parameters of the distributions can thus be given a physical interpretation that offers an alternative for more robust parameter estimation. The usefulness of the physical parameter interpretation for regionalization purposes remains to be evaluated.

Worked Example 6.1: Low flow frequency analysis

River Ngaruroro at Kuripapango in New Zealand (Table 4.3), has been selected for frequency analysis of annual minimum 1-day values, $AM(1)$, using the Weibull (WEI) distribution. In mid-latitudes in the Northern Hemisphere the calendar year can often be used to select the annual minimum flows. This is a suitable period for the selection of independent events as the drought or low flow period commonly occurs during the summer months. In the Southern Hemisphere the low flow season occurs at the opposite time of the year, and for Ngaruroro the lowest flows are typically found in the period November to May. As a result $AM(1)$ flows were selected for a hydrological year starting at the 1 September.

1. Data

   Ngaruroro has a flashy river regime, but has no observed zero flows. The $AM(1)$ flows are considered to come from the same population as only a minor part of the catchment is influenced by snow in winter. The observations cover the period 1964 to 2000 and with the omission of four years with missing data a total of 32 values results. A histogram of the values is shown in Figure 6.3 (upper left). To test the assumption of stationarity the $AM(1)$ values are plotted against time in Figure 6.9. No trend can be detected in the series and the data are therefore assumed to fulfil the requirement of independent and identically distributed data (iid).

2. Derivation of an empirical distribution function

   (a) The x values, $AM(1)$, are sorted in ascending order and the rank of each value calculated (using e.g. the RANK function in Excel). The smallest value equals 2.596 $m^3 s^{-1}$ and is given rank 1.

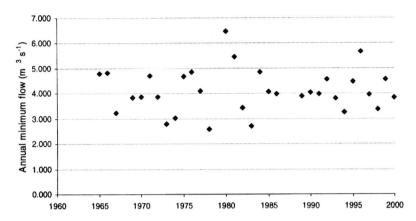

Figure 6.9 Time series of annual minimum 1-day flow, $AM(1)$, for River Ngaruroro at Kuripapango (NZ).

   (b) The non-exceedance probability, $F(x) = p$, is calculated for each x value using the Weibull plotting position formula (Equation 6.6).

   (c) A probability plot for the $AM(1)$ values is obtained by plotting the flow values against $F(x)$ as demonstrated in Figure 6.10 (upper). A staircase pattern of several nearly equal values is observed. (Note that the $AM(1)$ values now are plotted on the y-axis as compared to Figure 6.3, lower).

   (d) The return period of the smallest event can be calculated following Equation 6.5. The non-exceedance frequency, $F(x)$, of the smallest event equals 0.03 according to step 2(b), which gives a return period of 33 years.

3. Fitting the two-parameter Weibull (WEI) distribution function (Section A6.1.4) using the method of L-moments and the fact that $Y = -\ln(X)$ has a Gumbel distribution if X has a WEI distribution

   (a) The first two L-moments (λ_1 and λ_2) are estimated based on time series of $\ln(x)$:

$$\hat{\lambda}_1 = 1.391$$

$$\hat{\lambda}_2 = 0.120$$

   (b) The parameter estimates of the WEI distribution are obtained following Stedinger *et al.* (1993):

$$\hat{\kappa} = \frac{\ln(2)}{\hat{\lambda}_2} \quad \text{and} \quad \hat{\alpha} = \exp\left(\hat{\lambda}_1 + \frac{0.5772}{\hat{\kappa}}\right) \qquad \text{(W6.1.1)}$$

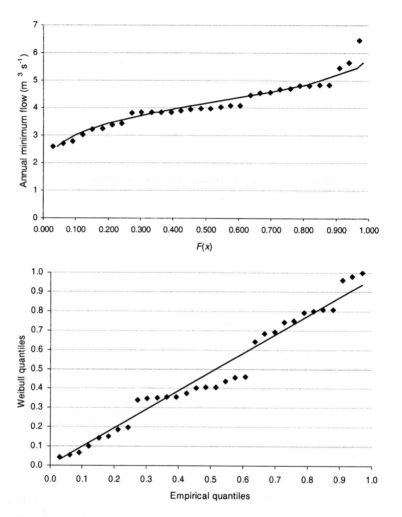

Figure 6.10 Estimated quantiles for annual minimum 1-day flow, *AM*(1), for River Ngaruroro at Kuripapango (NZ); probability plot showing the Weibull distribution (curve) as compared to the empirical quantiles (points) (upper) and qq-plot of empirical quantiles versus Weibull quantiles (lower).

which gives: $\hat{\kappa} = 5.776$ and $\hat{\alpha} = 4.441$

(c) $F(x)$ for the WEI distribution is obtained following Equation A6.1.33 (recalling that ξ is set to zero):

$$F(x) = 1 - \exp\left[-\left(\frac{x}{\alpha}\right)^{\kappa}\right]$$ (W6.1.2)

(d) The data can be plotted in a probability plot and compared to the empirical quantiles in step 2(b) (Figure 6.10, upper). The plot shows that the WEI distribution is well adjusted to the low flow extreme values, whereas the deviation in the upper range might suggest that the three highest values do not belong to the low flow population (Section 6.3.3). In Figure 6.10 (lower) the empirical quantiles are plotted against the estimated distribution quantiles (qq-plot). The points should be close to the unit diagonal if the data fit the WEI distribution well. The use of a plotting position implies that the ordered sample is plotted in regularly spaced positions. As demonstrated in the figure, the observed jumps in $AM(1)$ values are reflected in the estimated Weibull quantiles, but not in the empirical quantiles.

(e) The non-exceedance frequency of the smallest value equals 0.044 (Equation A6.1.33), which gives a return period of 23 years (Equation 6.5). This is ten years less that the empirical estimate derived in step 2(d).

(f) The 50 and 100-year events can be estimated from Equation A6.1.34:

$$x_p = \xi + \alpha\left[-\ln(1-p)\right]^{1/\kappa}$$

which gives: $\hat{x}_{50} = 2.26\, m^3 s^{-1}$ and $\hat{x}_{100} = 2.00\, m^3 s^{-1}$

4. Probability plot using the exponential reduced variate

(a) For comparison of the two-parameter EXP (Equation A6.1.64) and WEI distribution the exponential reduced variable is applied (Box 6.4). By substituting y into the expression for $F(x)$, y can be expressed as $-\ln(1 - F(x))$ (Equation B6.4.6). The parameters of the EXP distribution are estimated using L-moments (Equation A6.1.72) and used to calculate y and subsequently $F(x)$.

(b) The non-exceedance probability $F(x)$ for the observations is determined using the Weibull plotting position formula (Equation 6.6).

(c) The $AM(1)$ values are plotted against the reduced variate in Figure 6.11. A reduced variate of 0.02 corresponds to a return period of 50.5 years for minimum values (Equation 6.5). The data will plot as a straight line given they follow the EXP distribution. Again it is demonstrated that the WEI distribution fits the extreme low flow values well, and also the upper range apart from the three largest values. The two-parameter EXP distribution is less suited to model the sample. Alternatively, $-\ln(X)$ could be plotted on a Gumbel probability paper, and a straight line would result provided the data fitted the WEI distribution.

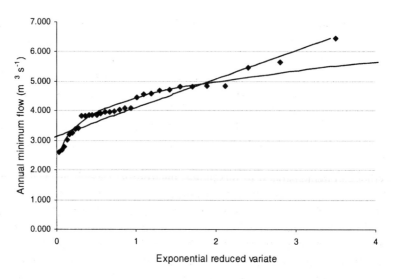

Figure 6.11 The annual minimum 1-day flow, $AM(1)$, plotted against the reduced variate, y, of the EXP distribution for River Ngaruroro at Kuripapango (NZ); the observations marked as points, the two-parameter EXP distribution (straight line) and the two-parameter WEI distribution (curved line).

6.6.2   Distribution functions for deficit characteristics

Following the theory of a random number of random variables, Zelenhasic & Salvai (1987) presented a stochastic model that considers a number of drought characteristics, including the largest drought deficit and duration in a given time interval using daily time series of streamflow. The method is based on the assumption that the droughts are iid random variables and that their occurrence is subject to the Poisson probability law (Section 6.3.2). The distribution function of the largest streamflow drought with respect to deficit or duration is derived based on the distribution of all deficits or durations in a given time interval, e.g. one year, combined with the distribution function for the number of droughts within the same time interval (Worked Example 6.2). PDS of duration and deficit volume were found to follow an EXP distribution, whereas the corresponding maximum values for a given time interval (usually chosen as the year) followed the GUM (EVI) distribution.

Clausen & Pearson (1995) found the LN distribution to be the best choice for 44 catchments in New Zealand for AMS of both deficit volume and duration. Tallaksen & Hisdal (1997) concluded that the GP distribution gave the overall best fit to the AMS of summer drought deficit and duration in a study of 52 daily discharge series from the Nordic region. Madsen & Rosbjerg (1998) found that for PDS the two-parameter GP distribution was suitable for describing drought duration for Danish catchments, but not deficits due to many small events. Meigh *et al.* (2002) fitted PDS of drought duration and found that generally the GP distribution provided the best result. The GP was also found to give the best fit to PDS of drought duration for rivers in southern Africa (Tate *et al.*, 2000). The occurrence of drought showed that they were approximately Poisson distributed, which supported the choice of the GP distribution.

Kjeldsen *et al.* (2000) tested seven distribution functions and found that the two-component exponential distribution (Rossi *et al.*, 1984) gave the best description of PDS of drought duration and deficit volume. The analysis was done on drought events extracted using a monthly varying threshold level on daily time series of streamflow in Zimbabwe. They experienced a problem in fitting a distribution due to the presence of many small events. Woo & Tarhule (1994) identified a similar problem with small droughts. They distinguished between three types of streamflow droughts extracted from daily time series of flow in Nigeria; short droughts that cluster in the rainy season, long droughts that span the dry season and multi-year droughts. They suggested separating the events into sub-populations prior to performing a frequency analysis. Tate & Freeman (2000) treated drought continuing to the end of a wet or dry season in southern Africa as censored data.

In the study of Mathier *et al.* (1992) daily values were averaged to give monthly time series, and a monthly varying threshold level was applied for selecting drought events. The threshold level was defined as a percentile of the mean discharge of a given month. A theoretical base was given for the choice of the geometric distribution for PDS of drought duration, and the EXP distribution was found to give reasonably good results for the distribution of drought deficit volumes.

Many hydrological phenomena are characterized by the joint occurrence of two or more random variables, implying that the frequency analysis of these variables must be based on their *joint probability distribution* (bivariate or multivariate). High correlations are generally found between drought duration and deficit volumes (e.g. Sen, 1977). Ashkar *et al.* (1998) demonstrated the use of two types of bivariate distributions for drought duration and deficit volume. They concluded, however, that the models possessed certain difficulties when a very strong correlation between the random variables exists, as is the case with duration and deficit. Kim *et al.* (2003) analysed the bivariate behaviour of drought deficit and duration using a non-parametric approach, and calculated bivariate return periods for duration using data from an arid region in Mexico. Frequency analysis based on a bivariate or multivariate distribution requires further empirical information regarding the drought indices (Matalas, 1991), but recent research has shown that it holds substantial promise for future application.

Worked Example 6.2: Drought deficit frequency analysis

River Ngaruroro at Kuripapango in New Zealand, applied for frequency analysis of annual minimum series in Worked Example 6.1, is here applied for frequency analysis of drought deficit characteristics. The drought events were derived using the Nizowka program (Software, CD) as described in Worked Example 5.4 (Section 5.4.1). Partial duration series (PDS) of drought events below a given threshold were selected from time series of daily discharge, and Q_{90} was selected as the threshold. Two drought characteristics are analysed in the following: *drought deficit volume* and *real drought duration* as defined in Worked Example 5.4.

1. Data

    Drought events were selected for River Ngaruroro and details of the selection criteria are given in Worked Example 5.4. The series cover the period 1964 to 2000, and the start of the year is set to 1 September. A total of four years were omitted from the series, 1967/68, 1978/79, 1986/87 and 1987/88, due to missing data. As only events below the Q_{90} percentile are selected, it might happen that the flow never becomes less than the threshold in a year (non-drought year). A total of 7 out of the 32 years with observations did not experience a drought (22%). The PDS series of drought deficit volume and real duration are plotted in Figure 6.12. Less severe values are found in the second half of the observation period for both deficit volume (upper) and duration (lower). The data are still treated as one sample as the number of observations is considered insufficient for a separate analysis of two periods. It should further be noted that a similar trend towards less severe droughts is not as pronounced for the $AM(1)$ values (Worked Example 6.1). This is likely a result of the high base flow contribution in the catchment (Figure 6.6).

2. Derivation of distribution function

    Following Zelenhasic & Salvai (1987) an estimate of the non-exceedance probability, $F(x)$, for the largest event in each time interval is in Nizowka obtained by combining the distribution for the occurrence of events and the distribution for the magnitudes of deficit volume or duration. Here a time interval of one year is chosen. Subsequently the return period in years for a given event can be calculated. The number of drought events

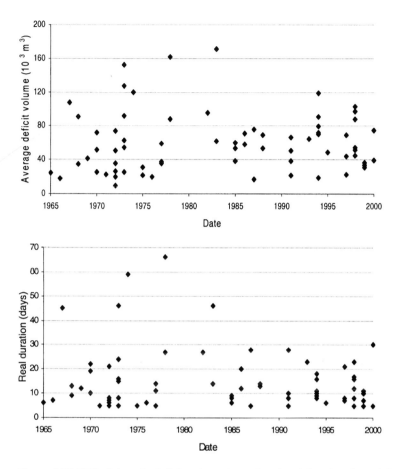

Figure 6.12 PDS of drought deficit volume (upper) and real duration (lower) for River Ngaruroro at Kuripapango (NZ) as selected by the Nizowka program.

occurring in a time period t is commonly assumed to be Poisson distributed (Equation 6.9) with parameter λt. In Nizowka the binomial Pascal distribution is offered along with the Poisson distribution as described in 'Background Information NIZOWKA' (Software, CD). The distribution that best fitted deficit volume was the Pascal distribution for the number of droughts and the GP distribution for the deficits. For duration the Pascal distribution was chosen along with the Log-Normal distribution. The $F(x)$ is for drought deficit volume plotted in Figure 6.13 together with the observed values plotted using a plotting position. The chosen distribution describes the data well, with the exception of some values in the upper range. The maximum value is, however, satisfactorily modelled.

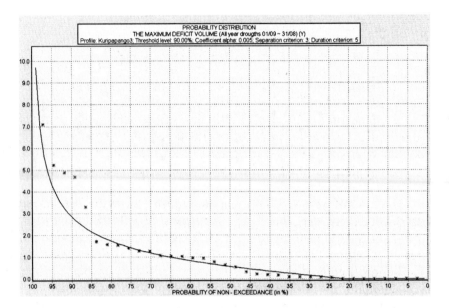

Figure 6.13 Best fit distribution for drought deficit volume for River Ngaruroro at Kuripapango (NZ) as displayed by the Nizowka program.

3. Calculation of the *T-year* event

    The return period of a given event is calculated following Equation 6.4. The relationship between the drought characteristics as defined in Worked Example 5.4 and $F(x)$ are given by the tabulated distribution functions in Nizowka. The design value for a particular return period, i.e. the *T*-year event, can be obtained from the tables for known values of $F(x)$. The estimated 10-, 100- and 200-year drought events are shown in Table 6.2.

Table 6.2 Estimates of T-year drought events for River Ngaruroro at Kuripapango (NZ)

Non-exceedance probability, $F(x)$	Return period, T (years)	Real drought duration (days)	Relative volume (days)	Absolute volume ($10^3\,\text{m}^3$)
0.9	10	42	2.84	4268
0.99	100	116	9.72	14595
0.995	200	149	13.44	20182

6.7 Regional frequency analysis

The uncertainty of a *T*-year event estimate is, in general, a combination of model and sampling uncertainty related to the choice of frequency distribution and estimation of the distribution parameters based on the available observations. These uncertainties are especially pronounced when estimating a *T*-year event with a return period *T* far beyond the observation period. To obtain more reliable estimates additional information should preferably be included. A common approach is the use of regional data where observations from different sites in a region that can be assumed to have similar drought characteristics are combined.

The use of regional information reduces the sampling uncertainty by introducing more data, i.e. space substitutes time to compensate for a short record at a site. In addition, the combination of drought statistics from different sites in a region facilitates the choice of an appropriate regional frequency distribution. Regional analysis also forms the basis for making inferences at ungauged sites by relating drought indices to catchment properties.

In this section regional estimation procedures related to frequency analysis are described. In Chapter 8 regional estimation methods for drought indices in general are discussed with focus on estimation at the ungauged site. The presentation in the following is related to the estimation of streamflow drought duration and deficit volume although with some modifications the procedures can also be applied to frequency analysis of other drought characteristics (Chapter 5).

6.7.1 Regional estimation model

When inferring drought characteristics at a site the mean value can usually be estimated adequately even if the available record is short, whereas estimates of second and higher order moments have large sampling uncertainties. This is the main motivation behind the index-flood method, or variants of that such as the station-year method, which has been widely applied in regional frequency studies. The term 'index-flood' is due to the early applications of the method in regional flood frequency studies (Dalrymple, 1960), but the method can be used with any kind of data. The method has been adopted in drought studies by Clausen & Pearson (1995) in regional analysis of AMS records of deficit volumes in New Zealand and by Madsen & Rosbjerg (1998) in regional analysis of PDS records of drought duration and deficit volume in Denmark. In the following the method will be referred to as the index method.

The basic hypothesis of the *index method* is that data at different sites in a region follow the same distribution except for scale. Data in the region are divided by an at-site scale parameter (which serves as the index parameter), and the normalized data are then jointly used to estimate the parameters of the regional distribution. The at-site quantile estimator is subsequently obtained by multiplying the normalized quantile estimator with an estimate of the site specific index parameter. Generally, the mean value of the extreme value population is used as the index parameter.

One approach for estimating the parameters of the regional normalized distribution is a simple pooling of data. In this method the normalized data in the region are treated as if they form a single random sample from the regional distribution, and this sample is then used to estimate the regional parameters. Since this method assumes independent sites, it is expected to lead to bias in quantile estimates, especially for large return periods (Cunnane, 1988). Wallis (1980) and Greis & Wood (1981) introduced an index method in which the parameters of the regional distribution are estimated from regional weighted averages of normalized probability weighted moments, or equivalently, regional weighted average L-moment ratios. By using regional averages, the problem of intersite dependence is less severe in the sense that it does not introduce any bias into quantile estimation, although it does increase the sampling variance (Hosking & Wallis, 1988; Madsen & Rosbjerg, 1997a).

The L-moment index procedure put forward by Wallis (1980) and further extended by Hosking *et al.* (1985), Wallis & Wood (1985) and Hosking & Wallis (1997) for AMS analysis and by Madsen & Rosbjerg (1997a) for PDS analysis, can be summarized as follows:

1. Consider a region of M sites with records of drought characteristics in terms of drought duration and deficit volume. The records are defined according to the PDS or AMS method, and $\hat{\lambda}_i$, $i = 1, 2, ..., M$ denotes the average annual number of drought events included in the series at the different sites.

2. At each site calculate the L-moment ratio estimates $\hat{\tau}_{2i}$, $\hat{\tau}_{3i}$, ... using unbiased probability weighted moment estimators or direct L-moment estimators (Appendix 6.2).

3. Calculate regional L-moment ratio estimates using:

$$\hat{\tau}_r^R = \frac{\sum_{i=1}^{M} w_i \hat{\tau}_{ri}}{\sum_{i=1}^{M} w_i} \quad , \quad r = 2, 3, ... \tag{6.28}$$

where w_i are weights assigned to each site. Generally, the weights are set equal to the record length n_i, since the uncertainty of the at-site estimators is inversely proportional to n_i. However, as noted by Stedinger *et al.* (1993), this weighting procedure may give undue influence to sites having much larger records than the average. Instead they suggest:

$$w_i = \frac{n_i n_R}{n_i + n_R} \tag{6.29}$$

where the parameter n_R depends on the degree of regional heterogeneity, i.e. the variability of L-moment ratios in the region (Section 6.7.2). In Section 6.7.3 weighting according to another procedure, i.e. a generalized least squares estimation method, is outlined.

4. Based on the regional L-moment ratio estimates determine the parameters of the normalized regional distribution. For a two-parameter distribution the two first L-moments $\lambda_l = 1$ and $\tau_2 = \hat{\tau}_2^R$ are used to estimate the parameters, whereas for a three-parameter distribution $\tau_3 = \hat{\tau}_3^R$ is also included in the estimation. L-moment estimates for different distributions are outlined in Appendix 6.1.

5. Calculate the normalized regional *T*-year event estimate using:

$$\hat{z}_T(s) = F^{-1}\left(1 - \frac{1}{\hat{\lambda}(s)T}\right) \tag{6.30}$$

where $\hat{\lambda}(s)$ is the average annual number of drought events at the site, and $F^{-1}(.)$ is the inverse of the cdf of the normalized regional distribution. In AMS analysis the regional average annual number of drought events is usually applied in Equation 6.30 instead of the at-site value, i.e.:

$$\hat{\lambda}^R = \frac{\sum\limits_{i=1}^{M} w_i \hat{\lambda}_i}{\sum\limits_{i=1}^{M} w_i} \tag{6.31}$$

6. Finally, calculate the *T*-year event estimate of the drought characteristic at an arbitrary site by multiplying the at-site sample mean value $\hat{\mu}(s)$ by the regional normalized quantile estimate

$$\hat{x}_T(s) = \hat{\mu}(s)\hat{z}_T(s) \tag{6.32}$$

An estimate of the associated variance of the regional estimator is given by:

$$Var\{\hat{x}_T(s)\} = \hat{z}_T^2(s)Var\{\hat{\mu}(s)\} + \hat{\mu}^2(s)Var\{\hat{z}_T(s)\} \tag{6.33}$$

where it is assumed that the at-site mean estimator is independent of the regional normalized quantile estimator.

The homogeneity assumption in the index method prescribes that dimensionless product moments, or equivalently L-moment ratios, of order two and higher (e.g. CV, skewness, kurtosis) are constant in the region and that data follow the same regional distribution. For testing regional homogeneity and determination of a regional distribution, procedures based on L-moment analysis were introduced by Hosking & Wallis (1993). These procedures are described in Section 6.7.2. For testing regional homogeneity Madsen & Rosbjerg (1997b) used a generalized least squares estimator of the regional parameters. This procedure is elaborated in Section 6.7.3.

Lettenmaier *et al.* (1987) introduced a modified index method in which the second moment is also estimated from at-site data and only the skewness and higher order moments are based on regional data. The homogeneity assumption in this case prescribes that dimensionless moments of order three and higher are constant in the region, and hence the method possesses less strict assumptions with respect to regional homogeneity than the standard index method.

A general use of the regional index method involves estimation at ungauged sites, and two aspects have to be considered in this respect. First, the ungauged site has to be assigned to a region where the regional frequency distribution is known. Procedures for grouping sites into homogeneous regions are discussed in Section 6.7.4. Secondly, since no at-site data are available the index parameter has to be inferred from regional data. Estimation of this parameter is usually achieved from regression analysis that relates the parameter to catchment properties, such as physiographic and climatic descriptors. This procedure is described in Section 6.7.3. Regression analysis procedures for regional analysis of general drought indices are presented in Section 8.3.

6.7.2   L-moment analysis

The index procedure described above requires identification and grouping of sites into homogeneous, or fairly homogeneous, regions and determination of frequency distributions for these regions. To assist with this, procedures based on L-moments were developed by Hosking & Wallis (1993). L-moment analysis has found wide applicability in regional frequency analysis and was applied in regional drought studies by Clausen & Pearson (1995) and Tallaksen & Hisdal (1997). The presentation of the following is mainly based on Hosking & Wallis (1993, 1997).

L-moment ratio diagrams

In an L-moment ratio diagram sample estimates of the L-moment ratios L-CV, L-skewness and L-kurtosis are compared to the theoretical relationships for a range of probability distributions. The L-skewness versus L-kurtosis relationship for a number of different distributions is shown in Figure 6.14, including the Generalized Pareto (GP), Generalized Extreme Value (GEV), Log-Normal (LN), Gamma (GAM), Weibull (WEI), Gumbel (GUM) and Exponential (EXP) distributions. Two-parameter distributions will show as points in the diagram, whereas three-parameter distributions will be depicted as curves. The population L-moments for these distributions are given in Appendix 6.1. Polynomial approximations to the L-moment relationships for the GP, GEV, LN, GAM and WEI distributions that provide sufficient accuracy in most applications and are easier to apply are given in Table 6.3.

Theoretical relationships between L-CV and L-skewness for different two-parameter distributional alternatives are shown in Figure 6.15. These include the two-parameter GP, LN, GAM and WEI distributions, which are obtained from their three-parameter counterparts by setting the location parameter equal to zero. Polynomial approximations of the L-CV versus L-skewness relationship for these distributions are shown in Table 6.4. In Figure 6.15 the one-parameter EXP distribution is also shown. Note that the GP, GAM and WEI distributions all include the EXP distribution as a special case.

Test of regional homogeneity

The L-moment ratio estimates from the different sites in the region can be plotted in an L-moment diagram (Figure 6.16). The dispersion of these points gives an indication of the regional homogeneity. In a homogeneous region all sites have identical population parameters except for scale, but due to sampling uncertainties some variability will always be observed. Thus the question is whether the observed variability is significant (i.e. the points form a heterogeneous sample) or can be explained by sampling uncertainties. The difference between the observed variability and the expected sampling variability in a homogeneous region is a measure of the regional heterogeneity (Figure 6.16). To quantify the significance of this difference a test statistic can be applied where the expected variability and the variance of the variability in a homogeneous region are obtained from Monte Carlo simulations.

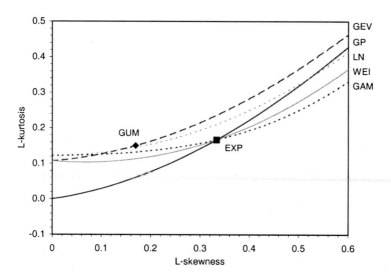

Figure 6.14 L-moment ratio diagram illustrating the relationship between L-skewness and L-kurtosis for the Generalized Pareto (GP), Generalized Extreme Value (GEV), Log-Normal (LN), Gamma (GAM), Weibull (WEI), Gumbel (GUM) and Exponential (EXP) distributions.

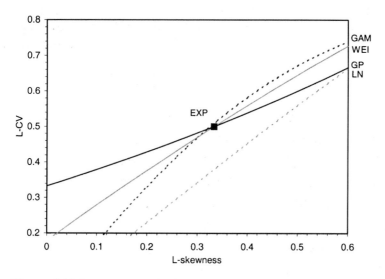

Figure 6.15 L-moment ratio diagram illustrating the relationship between L-CV and L-skewness for the two-parameter Generalized Pareto (GP), Log-Normal (LN), Gamma (GAM), and Weibull (WEI) distributions and the one-parameter Exponential (EXP) distribution.

Table 6.3 Coefficients of polynomial approximation of L-kurtosis (τ_4) as a function of L-skewness (τ_3), $\tau_4 = \Sigma A_i \tau_3^i$ for the Generalized Pareto (GP), Generalized Extreme Value (GEV), Log-Normal (LN), Gamma (GAM), and Weibull (WEI) distributions (Hosking & Wallis, 1997)

A_i	GP	GEV	LN	GAM	WEI
A_0	0	0.10701	0.12282	0.12240	0.10701
A_1	0.20196	0.11090	0	0	-0.11090
A_2	0.95924	0.84838	0.77518	0.30115	0.84838
A_3	-0.20096	-0.06669	0	0	0.06669
A_4	0.04061	0.00567	0.12279	0.95812	0.00567
A_5	0	-0.04208	0	0	0.04208
A_6	0	0.03763	-0.13638	-0.57488	0.03763
A_7	0	0	0	0	0
A_8	0	0	0.11368	0.19383	0

Table 6.4 Coefficients of polynomial approximation of L-CV (τ_2) as a function of L-skewness (τ_3), $\tau_2 = \Sigma A_i \tau_3^i$ for the two-parameter Generalized Pareto (GP), Log-Normal (LN), Gamma (GAM), and Weibull (WEI) distributions (Vogel & Wilson, 1996)

A_i	GP	LN	GAM	WEI
A_0	0.33299	0	0	0.17864
A_1	0.44559	1.16008	1.74139	1.02381
A_2	0.16641	-0.05325	0	-0.17878
A_3	0	0	-2.59736	0
A_4	0	-0.10501	2.09911	-0.00894
A_5	0.09111	0	0	0
A_6	0	-0.00103	-0.35948	-0.01443
A_7	-0.03625	0	0	0

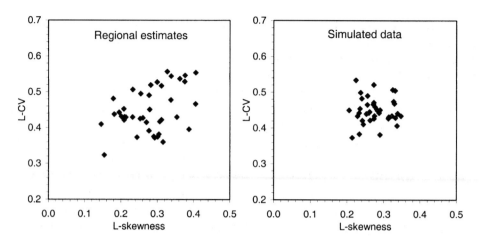

Figure 6.16 Dispersion of L-moment ratio estimates from the different sites in the region (left) compared to a single realization in a homogeneous region (right).

The test statistic (H) is based on the dispersion of the L-CV estimates in the region. The record-length-weighted standard deviation of the at-site L-CV estimates is calculated as:

$$V = \left[\frac{\sum_{i=1}^{M} n_i (\hat{\tau}_{2,i} - \hat{\tau}_2^R)^2}{\sum_{i=1}^{M} n_i} \right]^{1/2} \quad , \quad \hat{\tau}_2^R = \frac{\sum_{i=1}^{M} n_i \hat{\tau}_{2,i}}{\sum_{i=1}^{M} n_i} \tag{6.34}$$

The test statistic reads:

$$H = \frac{V - \mu_V}{\sigma_V} \tag{6.35}$$

where μ_V and σ_V are, respectively, the mean and the standard deviation of V obtained from simulations of a homogeneous region with L-moment ratios equal to the record-length-weighted averages and record lengths identical to the historical records.

In the simulations a four-parameter kappa distribution is adopted. The flexible four-parameter kappa distribution is used in the simulations to avoid choosing a particular distribution as a regional parent at this stage of the analysis. A region can be regarded as 'acceptably homogeneous' if $H < 1$; 'possibly heterogeneous' if $1 \leq H < 2$; and 'definitely heterogeneous' if $H \geq 2$.

Alternative test statistics in which V in Equation 6.35 is replaced by measures of dispersion of, respectively, L-CV and L-skewness and L-skewness and L-kurtosis can be constructed. These measures, however, were found to be less powerful in discriminating between homogeneous and heterogeneous regions. Note that in the case of the modified index method where L-CV is also site specific (Lettenmaier *et al.*, 1987) the L-skewness/L-kurtosis measure should be used to test regional homogeneity.

In general, a homogeneity test statistic is only moderately powerful in discriminating between homogeneous and heterogeneous regions (e.g. Wiltshire, 1986a; Lu & Stedinger, 1992). Hence the *H*-statistic should not be used as a strict significance test but rather as a guideline. It should also be noted that the *H*-statistic assumes independence between sites. A homogeneity measure that explicitly includes intersite correlation is presented in Section 6.7.3.

Determination of a regional distribution

Due to the small samples usually encountered in hydrology, traditional goodness-of-fit tests that are based solely on at-site data have little power for discriminating between various distributions, and hence procedures for identifying the parent distribution at a regional scale have been advocated by several researchers. In this respect the L-moment ratio diagram has been shown to be a valuable tool (e.g. Vogel & Fennessay, 1993).

For a visually-based choice of an appropriate distribution, sample estimates of L-CV, L-skewness and L-kurtosis are compared in L-moment ratio diagrams with the theoretical relationships for a number of parent distributions. Since L-moment ratio estimators are nearly unbiased, approximately half of the at-site sample points in the L-moment ratio diagram are expected to lie above the theoretical curve and half to lie below. To discriminate between different three-parameter distributions the L-skewness/L-kurtosis diagram is used, whereas the L-CV/L-skewness diagram can be used to discriminate between various two-parameter distributional alternatives.

To supplement visual judgement a goodness-of-fit test may be conducted. For a three-parameter distribution the L-kurtosis is determined by the L-skewness estimate. The difference between the regional average L-kurtosis $\hat{\tau}_4^R$ and the L-kurtosis of the fitted regional distribution τ_4^{DIST} can then be used as a measure of the goodness-of-fit (Figure 6.17). To quantify the significance of this difference it is related to the sampling uncertainty of the regional L-kurtosis estimate. The test statistic reads:

$$Z = \frac{\tau_4^{DIST} - \hat{\tau}_4^R + \beta_4}{\sigma_4} \tag{6.36}$$

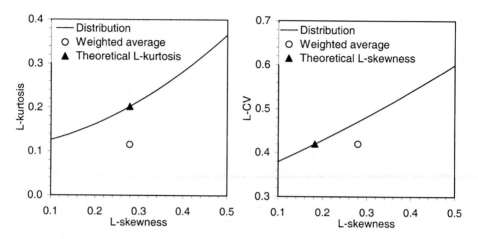

Figure 6.17 Definition of goodness-of-fit measure for, respectively, a three-parameter distribution (left) and a two-parameter distribution (right).

where β_4 and σ_4 are, respectively, the bias and the standard deviation of the regional average L-kurtosis obtained from simulations of a kappa population (similar to calculation of the homogeneity statistic).

To test the significance of this statistic it can be evaluated against the quantiles of a standard normal distribution. That is, the distribution can be accepted as giving an adequate fit to the data at significance level α if $|Z| > \Phi^{-1}(1 - \alpha/2)$ where $\Phi^{-1}(1 - \alpha/2)$ is the $(1 - \alpha/2)$-quantile in the standard normal distribution. A bias correction is applied in Equation 6.36 since the L-kurtosis estimate has a non-negligible bias for small sample sizes ($n < 20$) or large L-skewness ($\tau_3 > 0.4$).

For testing the goodness-of-fit between different two-parameter distributions an alternative test statistic was formulated by Madsen *et al.* (2002). Since the L-skewness of a two-parameter distribution is determined by the L-CV estimate, the distance between the regional average L-skewness $\hat{\tau}_3^R$ and the L-skewness of the fitted regional distribution τ_3^{DIST} can be used as a measure of the goodness-of-fit (Figure 6.17), i.e.:

$$Z = \frac{\tau_3^{DIST} - \hat{\tau}_3^R}{\sigma_3} \qquad (6.37)$$

where σ_3 is the standard deviation of the regional average L-skewness obtained from simulations of a kappa population. Since the L-skewness estimate has a negligible bias, a bias correction is not included in Equation 6.37.

6.7.3   Generalized least squares regression

An alternative procedure for estimation of regional parameters and testing regional homogeneity was introduced by Madsen & Rosbjerg (1997b). Their method is based on a generalized least squares (GLS) regression procedure that explicitly accounts for sampling uncertainty and intersite dependence. For parameters that show a significant regional variability the GLS regression procedure can subsequently be used to evaluate the potential of describing the variability from catchment properties. The GLS method provides more accurate estimates of the regression model parameters than the ordinary least squares procedure when the model residuals are heteroscedastic (have different variances) and are cross-correlated (Stedinger & Tasker, 1985). Parameterization of the drought characteristics in terms of L-moments (mean, L-CV and L-skewness) and the average annual number of drought events forms the basis for the GLS regression analysis.

The GLS regression model, based on work by Stedinger & Tasker (1985), Madsen & Rosbjerg (1997b) and Madsen *et al.* (2002), is described in Box 6.6. The special case of the so-called regional mean model is obtained when only the

---

Box 6.6

GLS regression model

Denote by $\hat{\theta}_i$ an estimate of a parameter at station i, $i = 1, 2, ..., M$. The following linear model is considered:

$$\hat{\theta}_i = \beta_0 + \sum_{k=1}^{p} \beta_k A_{ik} + \varepsilon_i + \delta_i \qquad (B6.6.1)$$

where A_{ik} are the considered catchment descriptors, β_k are the regression parameters, ε_i is a random sampling error, and δ_i is the residual model error. The sampling error and the residual model error are assumed to have zero mean and covariance structures:

$$Cov\{\varepsilon_i, \varepsilon_j\} = \begin{cases} \sigma_{\varepsilon_i}^2 & , i = j \\ \sigma_{\varepsilon_i} \sigma_{\varepsilon_j} \rho_{\varepsilon_{ij}} & , i \neq j \end{cases} \quad ; \quad Cov\{\delta_i, \delta_j\} = \begin{cases} \sigma_\delta^2 & , i = j \\ 0 & , i \neq j \end{cases} \qquad (B6.6.2)$$

where $\sigma_{\varepsilon_i}^2$ is the sampling error variance, $\rho_{\varepsilon_{ij}}$ is the correlation coefficient due to concurrent observations at stations i and j (intersite correlation coefficient), and σ_δ^2 is the residual model error variance. The assumption of homoscedasticity in
(→ Box continued on next page)

Box 6.6 (continued)

ordinary least squares regression requires that the variance of the error is the same for all i, i.e. both $\sigma_{\varepsilon_i}^2$ and σ_δ^2 have to be independent of i. In the GLS procedure only σ_δ^2 is assumed to be independent of i. It should also be noted that ordinary least squares regression does not consider any cross-correlations, i.e. it assumes that $\rho_{\varepsilon_{ij}} = 0$.

If the sampling error covariance matrix in Equation B6.6.2 is known, the GLS estimates of the regression parameters and the residual model error variance are obtained from the following set of equations:

$$\left[\mathbf{A}^T \mathbf{\Lambda}^{-1} \mathbf{A} \right] \boldsymbol{\beta} = \mathbf{A}^T \mathbf{\Lambda}^{-1} \mathbf{\Theta} \quad , \quad (\mathbf{\Theta} - \mathbf{A}\boldsymbol{\beta})^T \mathbf{\Lambda}^{-1} (\mathbf{\Theta} - \mathbf{A}\boldsymbol{\beta}) = M - p - 1 \qquad \text{(B6.6.3)}$$

where:

$$\mathbf{\Theta} = (\hat{\theta}_1, \hat{\theta}_2, ..., \hat{\theta}_M)^T$$

$$\boldsymbol{\beta} = (\beta_0, \beta_1, ..., \beta_p)^T \qquad \qquad \text{(B6.6.4)}$$

$$\mathbf{A} = \begin{pmatrix} 1 & A_{11} & \cdot & \cdot & \cdot & A_{1p} \\ \cdot & \cdot & \cdot & \cdot & \cdot & \cdot \\ 1 & A_{M1} & \cdot & \cdot & \cdot & A_{Mp} \end{pmatrix}$$

and $\mathbf{\Lambda}$ is the error covariance matrix of the total errors:

$$\Lambda_{ij} = Cov\{\varepsilon_i + \delta_i, \varepsilon_j + \delta_j\} = \begin{cases} \sigma_{\varepsilon_i}^2 + \sigma_\delta^2 & , \ i = j \\ \sigma_{\varepsilon_i} \sigma_{\varepsilon_j} \rho_{\varepsilon_{ij}} & , \ i \neq j \end{cases} \qquad \text{(B6.6.5)}$$

The solution of Equation B6.6.3 requires an iterative scheme. In some cases, however, no positive value of σ_δ^2 can satisfy Equation B6.6.3. In these instances the sampling errors more than account for the difference between $\mathbf{\Theta}$ and $\mathbf{A}\boldsymbol{\beta}$, and σ_δ^2 is set equal to zero. The GLS (regional) estimate of the parameter and the associated variance at an arbitrary site are given by:

$$\hat{\theta}^R(s) = \mathbf{A}(s)^T \hat{\boldsymbol{\beta}} \quad , \quad Var\{\hat{\theta}^R(s)\} = \mathbf{A}(s)^T \mathbf{\Sigma}_\beta \mathbf{A}(s) + \hat{\sigma}_\delta^2 \qquad \text{(B6.6.6)}$$

where $\mathbf{A}(s)^T = (1, A_1(s), A_2(s), .., A_p(s))$, and $\mathbf{\Sigma}_\beta = (\mathbf{A}^T \mathbf{\Lambda}^{-1} \mathbf{A})^{-1}$ is the covariance matrix of the estimated regression parameters. The variance of the regional estimate accounts for sampling errors, corrected for intersite dependence, and residual model errors.

The solution of the GLS regression equation, Equation B6.6.3 requires an estimate of the sampling error covariance matrix. This estimator should be independent, or nearly so, of the at-site parameter estimates $\hat{\theta}_i$. For estimation of the sampling error variances of the L-moment ratio estimators a Monte Carlo simulation procedure is applied following the lines of calculation of the heterogeneity and goodness-of-fit statistics described in Section 6.7.2. The estimate of $\sigma_{\varepsilon_i}^2$ for an L-moment ratio estimate is given by:

$$\hat{\sigma}_{\varepsilon_i}^2 = \frac{\bar{n}V^2}{n_i} \quad , \quad \bar{n} = \frac{1}{M}\sum_{i=1}^{M}n_i \qquad (B6.6.7)$$

where V^2 is the variance of the L-moment ratio estimate obtained from simulations of a kappa distribution with L-moment ratios equal to the regional averages and using the regional average number of observations $\bar{n}$. For the mean value of the drought characteristic μ a log-linear regression model is usually applied. In this case the estimate of $\sigma_{\varepsilon_i}^2$ for $\ln(\hat{\mu}_i)$ is given by:

$$\hat{\sigma}_{\varepsilon_i}^2 = \frac{\overline{CV}^2}{n_i} \quad , \quad \overline{CV} = \frac{1}{M}\sum_{i=1}^{M}CV_i \qquad (B6.6.8)$$

where CV_i is the sample coefficient of variation at site i. In the case of PDS analysis the number of drought events can be assumed Poisson distributed. In this case the sampling error variance of the average annual number of droughts is given by:

$$\hat{\sigma}_{\varepsilon_i}^2 = \frac{\bar{\lambda}}{t_i} \quad , \quad \bar{\lambda} = \frac{1}{M}\sum_{i=1}^{M}\hat{\lambda}_i \qquad (B6.6.9)$$

where t_i is the number of years of record at site i.

For estimation of the intersite correlation between parameter estimates, two types of correlation are considered: (a) correlation between concurrent drought events, and (b) correlation between the annual number of drought events. The correlation coefficient between the sample mean values is equal to the correlation coefficient between concurrent drought events ρ_{ij}. The correlation between higher order sample moments depends on the order of the moment so that the correlation between the L-CV and L-skewness estimates is equal to, respectively, ρ_{ij}^2 and ρ_{ij}^3 (Stedinger, 1983; Madsen & Rosbjerg, 1997a). Thus, the effect of intersite dependence is less severe when estimating higher order moments. The correlation coefficient between the estimated λ parameters in the PDS is equal to the correlation coefficient between the annual number of drought events (Mikkelsen *et al.*, 1996).

For estimation of the correlation between the drought events a series of concurrent events at the two stations is defined based on the time of onset and termination of the droughts. From this series a conditional correlation coefficient can be calculated (conditional upon droughts occurring at the two stations at the same time). To account for drought events that do not overlap temporally, an unconditional correlation coefficient is calculated (Mikkelsen *et al.*, 1996).

intercept β_0 is included in the regression equation. In this case the GLS regression model provides an estimate of the regional average parameter (β_0) and the associated uncertainty. The GLS estimate of the residual model error variance σ_δ^2 can then be interpreted as a measure of regional heterogeneity. This implies that if $\hat{\sigma}_\delta^2 = 0$ the region can be considered homogeneous with respect to the considered parameter (e.g. L-CV or L-skewness).

In the case of homogeneous sampling errors, an explicit solution of the GLS regional mean model exists (Madsen *et al.*, 1994):

$$\hat{\beta}_0 = \frac{1}{M}\sum_{i=1}^{M}\hat{\theta}_i$$

$$\hat{\sigma}_\delta^2 = \max\left\{0, s^2 - (1-\hat{\rho}_\varepsilon)\hat{\sigma}_\varepsilon^2\right\} \tag{6.38}$$

$$s^2 = \frac{1}{M-1}\sum_{i=1}^{M}(\hat{\theta}_i - \hat{\beta}_0)^2$$

where $\hat{\sigma}_\varepsilon^2$ is the regional average sampling error variance, and $\hat{\rho}_e$ is the regional average intersite correlation coefficient.

In this case the regional estimator $\hat{\beta}_0$ is equal to the simple average of the at-site estimates $\hat{\theta}_i$. The variance of the regional estimator is given by:

$$Var\{\hat{\beta}_0\} = \frac{M+1}{M}\hat{\sigma}_\delta^2 + \frac{1}{M}[1 + \hat{\rho}_\varepsilon(M-1)]\hat{\sigma}_\varepsilon^2 \tag{6.39}$$

In Equation 6.39 the first term accounts for the variability due to regional heterogeneity, whereas the second term represents the sampling variability of the regional mean value corrected for intersite dependence.

In general the regional estimate $\hat{\beta}_0$ is not equal to the simple average since the GLS algorithm weights the estimated parameters according to the covariance matrix of the errors. If no cross-correlation exists, the GLS estimator becomes:

$$\hat{\beta}_0 = \frac{\displaystyle\sum_{i=1}^{M}\left[\sigma_{\varepsilon_i}^2 + \sigma_\delta^2\right]^{-1}\hat{\theta}_i}{\displaystyle\sum_{i=1}^{M}\left[\sigma_{\varepsilon_i}^2 + \sigma_\delta^2\right]^{-1}} \tag{6.40}$$

which corresponds to a weighted least squares estimator with the weights $w_i = \left[\sigma_{\varepsilon_i}^2 + \sigma_\delta^2\right]^{-1}$. Assuming that the sampling variance $\sigma_{\varepsilon_i}^2$ is proportional to $1/n_i$, and that the region is homogeneous ($\sigma_\delta^2 = 0$), then $w_i = n_i$ and Equation 6.40 reduces to the record-length-weighted average estimator in

Equation 6.28. In the case of a heterogeneous region with a residual model error variance expressed as $\sigma_\delta^2 = 1/n_R$ then Equation 6.40 reduces to the weighted estimator suggested by Stedinger *et al.* (1993) (Equations 6.28 and 6.29). Thus the regional GLS estimator is a general extension of the regional weighted average estimator that explicitly accounts for intersite correlation and regional heterogeneity. In the case of a significant and highly heterogeneous intersite correlation structure the GLS and weighted average estimator may differ significantly (Madsen & Rosbjerg, 1997b).

Estimation of index parameter by combining regional and site specific data

For estimation at ungauged sites with the index method the mean value is estimated from catchment descriptors using regression analysis (Box 6.6). In the case where some at-site data are available the sample mean and the mean value obtained from the regression equation can be combined. An efficient weighted average estimator is given by (e.g. Madsen & Rosbjerg, 1997b):

$$\hat{\mu}(s) = w\hat{\mu}^R + (1-w)\hat{\mu}^{AS} \quad , \quad w = \frac{Var\{\hat{\mu}^{AS}\}}{Var\{\hat{\mu}^{AS}\} + Var\{\hat{\mu}^R\}} \tag{6.41}$$

The weighting factor w expresses the relative weight assigned to the regional ($\hat{\mu}^R$) and at-site ($\hat{\mu}^{AS}$) estimates, depending on the uncertainties (variances) of the two. If $Var\{\hat{\mu}^{AS}\} > Var\{\hat{\mu}^R\}$ more weight is given to the regression estimate, and vice versa. The variance of the regional estimator is obtained from the GLS regression model, Equation B6.6.6. The variance of the sample mean is estimated by:

$$Var\{\hat{\mu}^{AS}\} = \frac{\hat{\sigma}^2(s)}{n(s)} \tag{6.42}$$

where $\hat{\sigma}^2(s)$ and $n(s)$ are, respectively, the sample variance and the record length at the considered site. The variance of the weighted estimator in Equation 6.41 is:

$$Var\{\hat{\mu}(s)\} = wVar\{\hat{\mu}^R\} \tag{6.43}$$

6.7.4 Delineation of homogeneous regions

An important aspect in regional frequency analysis concerns the delineation of homogeneous regions. The sites should be grouped to satisfy the homogeneity assumption of the index method, i.e. sites within the group should follow the same distribution except for scale. Traditionally, geographically coherent regions have been applied in regional frequency studies. Geographical regions are convenient for administrative reasons and practical applications, however geographical proximity does not necessarily imply similarity with respect to extreme drought characteristics. Climate conditions are essential to describe drought characteristics, but also more local catchment properties such as land use, morphometry, soil type and hydrogeology are important.

Several approaches have been proposed for grouping of sites that define hydrologic similarity in a multi-dimensional space of extreme value statistics and physical catchment descriptors instead of geographical coherence. Since homogeneity is defined in terms of statistical measures, extreme value statistics are required for a correct grouping of sites into homogeneous regions, whereas inclusion of catchment properties is important for a physical understanding of hydrologic similarity. Evidently, assignment of an ungauged site to a region requires the use of physical catchment descriptors.

In the following a brief description of the methods that have been most widely applied in regional frequency studies is given. All methods are based on partitioning the available sites into a number of fixed, disjoint groups. An alternative method, known as the region of influence approach (Burn, 1990), uses a dynamic partitioning where each site has its own region. An application of this approach for a general regionalization of drought indices is described in Sections 8.5.4 and 11.4.2.

Cluster analysis

A widely-used regionalization technique is cluster analysis. In this method a data vector is defined for each site that consists of selected catchment descriptors and/or extreme value statistics. Regions are formed according to the similarity between the data vectors. Most cluster algorithms measure similarity by the Euclidian distance in the data vector space. Since the distance measure depends on the scale of the different variables included in the data vector, a normalization is usually done whereby all variables are giving equal weights in the clustering. However, if some variables are more important for measuring similarity these should be given a larger weight. The choice of appropriate weights is a complicated task, which always involves subjective judgements.

Hierarchical clustering procedures are used to identify groups of similar sites based on a distance measure between each pair of re-scaled data vectors. These algorithms start with each site in separate clusters and combine clusters until only one is left. The most appropriate grouping of sites can then subsequently be analysed using the *H*-statistic and the GLS regional variability estimator for the different clusters in the hierarchy. In this respect a proper balance should be sought between regions including only a few sites and regions with a large number of sites. If only a few sites are included the advantage of regional estimation compared to at-site estimation is reduced. On the other hand, if very large regions are considered there is a high risk of introducing significant bias in quantile estimation at some sites. This is the well-known bias-variance trade-off problem in statistical analysis.

Wiltshire (1986b) applied cluster analysis to define groups in a two-dimensional space of the coefficient of variation and the specific mean of annual flood data. Subsequently, discriminant analysis was employed to relate these groups to catchment descriptors. While this method virtually ensures a homogeneous grouping of sites, the efficiency of the discriminant analysis is limited, reflecting the, in general, great difficulties of relating extreme value statistics to catchment descriptors. Acreman & Sinclair (1986) used cluster analysis on the basis of catchment descriptors. Their method is more convenient for allocating ungauged sites to predefined clusters, but homogeneity may be more difficult to attain. Nathan & McMahon (1990b) considered various problems associated with cluster analysis with special emphasis given to the important issue of selecting and weighting variables used to assess similarity between sites. Other application examples of cluster analysis include Burn (1989) and Gustard *et al.* (1989). Hosking & Wallis (1997) generally recommend using cluster analysis with catchment descriptors in regional frequency studies.

Split-sample regionalization

Wiltshire (1985) proposed a split-sample regionalization approach for defining groups according to catchment descriptors. At its simplest, the method splits a set of catchments into two groups based on a single partitioning value of one chosen catchment descriptor. For instance, the catchments can be divided into wet and dry groups according to mean annual precipitation. Measures of variability in each group are evaluated and aggregated into one statistic, and the optimum grouping for the chosen catchment descriptor is then achieved at the partitioning point where the variability statistic is minimum.

To measure the variability Wiltshire (1985) used different statistics based on fitting the GEV distribution to each group, while Pearson (1991) and Madsen *et al.* (1997a), more generally, used a function of L-moment ratios:

$$V = \frac{\sum_i n_i \sqrt{3e_{2,i}^2 + 2e_{3,i}^2 + e_{4,i}^2}}{\sum_i n_i} \qquad (6.44)$$

where $e_{j,i}$ is the deviation of the jth L-moment ratio estimate at site i from its group record-length-weighted average. In this case the L-CV is weighted ahead of the L-skewness, which in turn is weighted ahead of L-kurtosis, so that homogeneity is primarily influenced by L-CV and less so by L-kurtosis.

The partitioning process is repeated for other descriptors, and the optimal two-way grouping is achieved for the catchment descriptor that provides the smallest variability measure at the optimum point. In the next step the two defined groups are then analysed individually and further sub-divided. At each partitioning level the *H*-statistic and the GLS regional variability estimator is used to assess regional homogeneity. If these measures indicate regional heterogeneity further sub-divisions should be analysed.

Empirical orthogonal functions

There is a long tradition of applying the method of empirical orthogonal functions (EOF) in statistical analyses. The method, equivalent to principal components analysis (PCA) when restricted to a discrete, finite set of observations, is a general method suitable for studies of variation patterns in time and space and for concentration of information in large data sets (Section 5.5). The EOF method can also be applied for delineation of homogeneous regions with respect to statistical properties of time series. The method was used for this purpose by Gottschalk (1985) for analysing daily observations of streamflow in Sweden and by Tallaksen & Hisdal (1997) for analysing annual maximum drought duration series in the Nordic countries.

The basic principle behind the EOF method applied to time series is a linear transformation of spatially correlated series from a region into two sets of series, one describing the temporal variations in the original data set (named amplitude functions) and one describing the spatial variability in the region (named weight coefficients) from which the original series are collected. The weight coefficients describe the contribution of each amplitude function to the original time series. They vary between the time series but are constant in time. The main common property of the time series is reflected in the first amplitude function. The proportion of this main feature that is needed to obtain the original series at individual observation points is reflected in the first weight coefficient. The second largest contribution to the total variability of the original series is described by the second amplitude function, and its contribution to the original series is reflected through the second weight coefficient. If a reasonable proportion of the total variability is described by the two first amplitude functions, a two-dimensional plot of the corresponding weight coefficients for all sites in the region in a scatter diagram can serve to identify groups of stations with equal properties. If more than two weight coefficients are included, other graphical presentations might allow for a proper regional grouping. Further details of the method are presented in Section 6.8.

The EOF method has certain similarities with cluster analysis. Both methods use a distance measure to measure similarity between sites. In the case of the EOF method the correlation coefficient is used as the distance function.

However, unlike cluster analysis, no objective means of grouping sites exists for the EOF method. The *H*-statistic and the GLS regional variability estimator can be used to assist in defining an appropriate grouping.

Since the EOF method uses hydrological time series for the definition of regions, the method does not directly support assignment of ungauged sites into regions unless geographically coherent regions can be identified. For assigning an ungauged site into a region it is thus necessary to relate the EOF based grouping to catchment descriptors.

6.7.5 Self-guided Tour: Regional frequency analysis (CD)

The regional frequency analysis procedure outlined in Section 6.7.1 was applied to the Regional Data Set with daily streamflow series from Baden-Württemberg, Germany (Section 4.5.2). From this data set stations with a record length larger than 20 years were included in the regional analysis, which comprises 46 stations with recording periods ranging between 20 to 37 years with an average of 30.2 years. For definition of drought events the threshold level approach was applied using Q_{70} as the threshold level (Section 5.4.1). From the drought series annual maximum series of drought duration and deficit volume were extracted.

On the accompanying CD the application of the regional procedure to this data set is included as a Self-guided Tour. The analysis includes (a) delineation of the sites into homogeneous regions based on the split-sample regionalization approach (Section 6.7.4), (b) analysis of regional homogeneity of the defined regions using the *H*-statistic (Section 6.7.2) and the GLS regional variability estimator (Section 6.7.3), (c) identification of a regional distribution for each region using the L-moment goodness-of-fit statistic (Section 6.7.2), (d) estimation of the normalized frequency distribution, and (e) determination of a regression model that relates the mean value (index parameter) to catchment descriptors (Box 6.6).

The analysis shows that the sample of all 46 sites comprises a definitely heterogeneous region with respect to both deficit volume and drought duration. Division of the sites into wet and dry catchments according to the mean annual precipitation and a further division of the wet catchments with respect to the average hydraulic conductivity provide three acceptably homogeneous regions. The L-moment distribution analysis reveals that the two-parameter Gamma distribution is adequate for describing both deficit volume and drought duration in all three regions. The dry catchment region has a more skewed (heavier tailed) distribution than the wet catchment region. In each region the deficit volume has a more skewed distribution than the duration. The GLS regression analysis is applied to data in the three regions to provide regression equations for the index parameter. Land use, morphometry and soil characteristics as well as the mean annual precipitation are found to be important descriptors. The resulting model allows estimation of the *T*-year drought duration and deficit volume for an arbitrary catchment (gauged as well as ungauged) in the Baden-Württemberg region.

6.8   Severity-area-frequency curves

There are two main approaches to studying regional drought characteristics. The first approach studies regional properties of drought by analysing the spatial patterns of at-site droughts. The information can be presented as different types of maps describing drought characteristics over a region (Section 2.3.2). A second approach is to study regional drought characteristics (Section 5.4.3) such as the area covered by a drought and the total area deficit. This also includes frequency analysis of the *regional drought characteristics*. Santos *et al.* (2002) present a method to estimate return periods of droughts based on observed annual precipitation. The following example outlines a similar procedure introduced by Hisdal & Tallaksen (2003) for estimating the probability that a drought of a specific severity will cover an area of a specific extent. Streamflow is used as an example, but the method can also be applied to precipitation.

The method consists of three main procedures: (a) simulation of long time series in a net of grid cells, (b) calculation of the regional drought characteristics, and (c) calculation of drought severity-area-frequency curves. To simulate long time series the following steps are carried out:

1. *Subtraction of the mean value ($\overline{Q}_i$)*

    Let $Q(o_i,t)$ refer to the observed streamflow record at a station *i* among *N* in a certain domain, *o* corresponds to a two-dimensional plane $o = (x,y)$ and *t* is time. The series $Q'(o_i,t)$ is expressed as:

    $$Q'(o_i,t) = Q(o_i,t) - \overline{Q}_i \quad , \qquad i = 1, ..., N \tag{6.45}$$

2. *Expansion of the streamflow series into empirical orthogonal functions, EOFs*

    The EOF-method, equivalent to PCA for time series, linearly transforms spatially-correlated time series from a region into two sets of orthogonal and thus uncorrelated functions. The result is a set of series $\beta_j(t)$ describing the temporal variations common to all streamflow series (here named amplitude functions) and a set of series $w_j(o)$ describing the spatial variability in the region (here named weight coefficients). The linear transformation of the streamflow series can be described as:

$$Q'(o,t) = \sum_{j=1}^{N} w_j(o)\beta_j(t) \qquad (6.46)$$

The flow record at station i can then be estimated by:

$$Q'(o_i,t) = \sum_{j=1}^{M} w_j(o_i)\beta_j(t) \qquad (6.47)$$

where M is the number of amplitude functions considered. Using $M = N$ functions gives a complete description of the original data. A few of the amplitude functions, $\beta_j(t)$, will contain most of the variations in the original streamflow series. The amplitude functions are arranged in descending order according to the proportion of variance explained by each function. Hence, redundant information can be removed using only a few of the amplitude functions ($M < N$). The amplitude functions describe the temporal variation in streamflow around the mean over the region considered and are common to all the initial series. The weight coefficients describe the spatial variation over the region and hence vary in space.

3. *Generation of new streamflow series that are Y years long in an evenly-spaced grid*

   The M selected amplitude functions combined with spatially-interpolated mean values and weight coefficients permit the generation of flow records of length Y in each grid cell in the region studied, according to:

$$\hat{Q}(o_i,t) = \overline{\hat{Q}}(o_i) + \sum_{j=1}^{M} \hat{w}_j(o_i)\hat{\beta}_j(t) + \varepsilon \qquad (6.48)$$

   To preserve the variance in the original data it is necessary to add a random error term, ε, assumed to be normally distributed with zero mean and variance equal to the proportion of variance not explained by the M amplitude functions.

The monthly mean flow, $\overline{\hat{Q}}(o_i)$, and the required weight coefficients, $\hat{w}_j(o_i)$, are interpolated using ordinary kriging (e.g. Cressie, 1993). The Monte-Carlo technique (Chapter 7) is applied to simulate M amplitude functions of length Y. Persistency, reflected as autocorrelation, can be

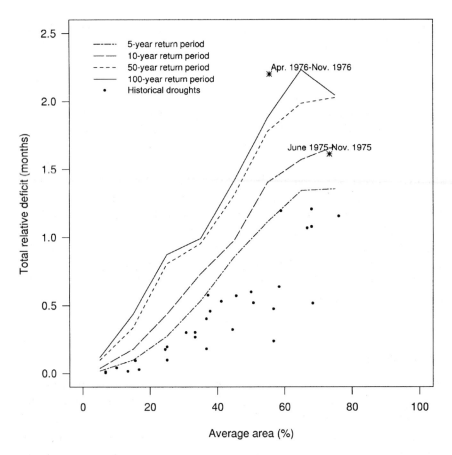

Figure 6.18 Severity-area-frequency curve for streamflow droughts in Denmark (modified from Hisdal & Tallaksen, 2003; reproduced with permission of Elsevier).

accounted for by applying a classical linear autoregressive moving average (ARMA) time-series model (e.g. Hipel & McLeod, 1994). Finally, the time series can be estimated for each grid cell following Equation 6.48. The regional drought characteristics are calculated as described in Section 5.4.3 with a time invariant but spatial varying truncation level $q_0(o_i)$, for each grid cell o_i.

For all simulated drought events the average area covered by the drought can be estimated for an interval scale defining the percentage of the total area covered (1–10%, 11–20%, and so on). Within each area class the selected quantiles and corresponding return periods of the relative drought deficit can be calculated (Section 6.2.1). Hence, return period curves of a drought of a specific severity and area coverage can be constructed, i.e. severity-area-frequency curves. This allows return periods to be assigned to a given event, including historical droughts. Similarly, the duration of a drought instead of its volume can be used to construct duration-area-frequency curves.

In Hisdal & Tallaksen (2003) the method is demonstrated using Denmark as a case study. The basic data set comprised 15 monthly streamflow records for the period 1961 to 1990. Monte Carlo simulations were applied to simulate long time series of streamflow in a net of grid cells (~14x17 km). The drought characteristics were derived for the simulated series in each grid cell applying Q_{80} as threshold level. The estimated severity-area-frequency curves are shown in Figure 6.18. The curves indicate the return periods of droughts with a total relative drought deficit volume (ordinate) covering a certain percentage of the area (abscissa) of Denmark. The deficit in each grid cell has been standardized by its threshold level to obtain relative deficits to allow for a regional comparison. The standardization implies that the deficit volumes, originally given in mm, when divided by the threshold given in mm month^{-1}, get the unit of month. As can be seen, the 1975 and 1976 droughts (Section 2.2.3) are the most severe in the 30-year record 1961 to 1990.

6.9 Summary

Estimates of the probability of an extreme hydrological event and the corresponding return period that is associated with the event are important for water resource management and prediction. Frequency analysis of historical events is a well-known tool for assessing the severity of an extreme event and the probability that the event will occur in the future. Estimation of a design event is based on the lower or upper tail of the probability distribution, which implies large uncertainties in its estimation. However, there are few alternatives in operational practice and probably more important is the awareness of the limitation inherited in the methodology.

The chapter has given a statistical background to parametric methods for frequency analysis, both at-site and regional procedures, and has described in detail the different calculation steps. The at-site procedure is straightforward; however, the reliability of the sample parameters and estimated extreme values depends strongly on the sample size and to what degree the basic assumptions of independent and identically distributed (iid) data are fulfilled. Hydrological data feature many of the problems common in applying extreme value methods to environmental time series, such as short-range correlations, seasonal variation, non-stationarity, missing values and data of dubious quality. However, often the main problem is the information content of the data series, which are of short length. In a regional analysis homogeneous regions are identified and data from several sites combined to yield a regional frequency curve. The use of regional information reduces the sample uncertainty by introducing more data. In addition, to provide more robust estimates, regionalization forms the basis for estimation at the ungauged site and allows the spatial aspects of the drought to be investigated.

The selection of extreme events is related to two limiting distributions: the Generalized Extreme Value (GEV) distribution for annual maximum/minimum series (AMS) and the Generalized Pareto (GP) distribution for partial duration series (PDS). Emphasis is on their applicability for extreme value analysis of time series of hydrological drought characteristics, both minimum and maximum values. Streamflow droughts are characterized in terms of annual minimum discharge and annual maximum or PDS of deficit volume and duration derived using the threshold level approach. Several procedures for selecting a distribution function were presented along with experience from international studies. It is generally recommended to let the choice of distribution be governed by knowledge of the phenomenon studied and the theoretical base of the family of distribution functions adapted to extreme value analysis, i.e. the GEV for AMS and the GP distribution for PDS. This might provide a better prediction of future events as there are large uncertainties related to the conclusions that can be drawn by merely comparing the distribution fit to the sample observations. The use of regional information will further reduce the sensitivity to sample variability and is therefore recommended.

The chapter has illustrated at-site frequency analysis for a streamflow series from the Global Data Set through two worked examples; AMS of the lowest flow each year and PDS of drought deficit volume and duration. The more advanced procedures introduced for regional frequency analysis have been applied to the Regional Data Set from Baden-Württemberg in Germany and are demonstrated as a self-guided tour on the CD. Homogeneous regions are here defined based on the split-sample regionalization approach and regional homogeneity tested using the *H*-statistic and the GLS regional variability estimator. A regional distribution for each region is identified using L-moment statistics, and a regression model that relates the mean value to catchment descriptors allows estimation of the *T*-year drought duration and deficit volume for an ungauged catchment in the region. Finally, frequency analysis of regional drought characteristics such as the area covered by a drought and the total area deficit has been presented. The procedure allows the probability that a drought of a given severity will cover an area of a specific extent to be estimated, and the result can be presented as drought severity-area-frequency curves.

The methodology presented is of a general nature, although less suited for rivers in arid and semi-arid regions that only run occasionally. In these regions other hydrological time series such as groundwater or reservoir data should be used instead as they are likely to provide more information on the drought situation in the hydrological system. Frequency analysis is a flexible method that can be performed on a wide range of data; here the focus is on daily streamflow series. An alternative to frequency analysis for estimating return periods and levels is theoretical derivation of the return periods of hydrological drought with a certain severity based on the concept of stochastic processes, often referred to as run theory. Run theory is commonly applied to time series of discrete data, i.e. monthly or annual time series of streamflow, and is further discussed in Chapter 7 (Time Series Modelling).

7

Time Series Modelling

Lars Gottschalk

7.1   Introduction

Observations on a phenomenon which is moving through time generate an ordered set known as a time series. Daily discharge observations at a gauging site in a river are a typical example of a time series. These observations can be denoted $x(t_1)$, $x(t_2)$, ..., $x(t_n)$, where n is the total number of observations. Their characteristic feature is that the order of the set t_1, t_2, ..., t_n is material, and not accidental as it would be for a random sample $x_1, x_2, ..., x_n$. The suffixes are only used for identification in this latter case.

Random models are able to reproduce times series that are statistically indistinguishable from observed ones. However, it is neither an exact reproduction of past events nor a prediction of the future. The similarity lies in the statistical parameters of the series. The series generated by a random model can only reproduce a limited set of statistics of the historical series. In the case of low flow and drought studies these can be (a) mean value, (b) variance (standard deviation), (c) skewness, (d) lag-one autocorrelation, and (e) number and durations of non-exceedance of given threshold discharges. The rationale behind the use of random models and simulated series is to obtain a derived distribution of important characteristics (e.g. drought characteristics) for a natural and/or manipulated hydrological system. This can be achieved either by analytical derivation for simple cases or more often by Monte Carlo methods. This chapter is mainly concerned with the derived distribution of the so-called crossing properties of time series (item (e) from the list above), establishing a model for whatever limited data are available. Under the assumption that this model preserves the basic properties of importance for the problem at hand the solution is found – the distribution and/or moments of crossing properties. The critical point is not how to find the solution to the problem: analytical and Monte Carlo methods are well established. The problem stems from how to select and validate a model against what are usually short samples of

Box 7.1

Crossing properties

A *downcrossing* is a sequence of data when the time series crosses a predetermined threshold level from high to low values (Figure 7.1). Accordingly an *upcrossing* is a sequence when the time series crosses this threshold level from low to high values. A *run* is the sequence of data below (above) the threshold level bounded by a downcrossing and an upcrossing (an upcrossing and a downcrossing).

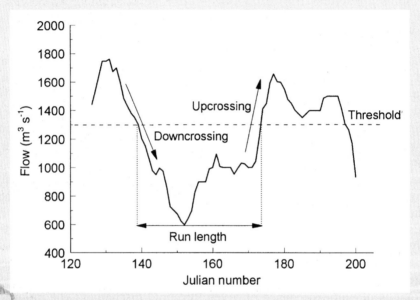

Figure 7.1 Crossing properties of a time series.

hydrological data. It is important to remember that the use of simulated series does not provide any new information. The information available is the observed series. On the other hand it is important not to lose any information when searching for the solution to the problem at hand and that is why random models and simulated series are important.

The chapter starts simply and gradually increases in the complexity of the time series models explained. This increase in complexity is motivated by taking better account of the physical processes involved in runoff formation. The complexity also increases when turning from stationary time series, which can be adequate at the annual time scale, to non-stationary time series with a higher time resolution. The latter are applicable when seasonal variations occur, for example at monthly and daily timescales. Section 7.2 describes how a random sample is generated (a sequence of independent random numbers belonging to a given theoretical distribution function) and gives an evaluation of the statistical properties of this sample. In Section 7.3 the role of storage (e.g. lakes or groundwater) is investigated leading to the identification of a first order autoregressive model. Data generated from this model show dependence reflected by the autocorrelation in the data which also influence their statistical properties. In Section 7.4 the properties of the two models (the random sample versus the first order autoregressive model) are evaluated against an observed data set, and the section examines how to test a model against observed data and eventually reject it. An overview of simple autoregressive models used in hydrology in Section 7.5 completes the treatment of stationary time series. Section 7.6 deals with the modelling of seasonal variations and develops a model for monthly data with parameters that change by season. As in Section 7.4, two models are tested with respect to how well the statistical properties of an observed series are preserved. Section 7.7 then provides an overview of the use of non-stationary models in hydrology. Random models for daily data are briefly discussed in Section 7.8 and the chapter ends by describing analytical calculations of low flow and drought characteristics such as run lengths and number of upcrossings or downcrossings (Box 7.1). Finally, Section 7.9 provides a summary of the chapter.

7.2 A random sample

Let us start by studying a random sample $x_1, x_2, \ldots, x_n$. We say that such a sample is a realization of a random variable X. This variable is fully described by its cumulative distribution function (cdf) $F_X(x)$, which expresses the probability of the event that the random variable X is less than or equal to some fixed number x (Section 6.2.1):

$$F_X(x) = Pr\{X \le x\} \tag{7.1}$$

The probability takes values in the interval between $0 \le Pr \le 1$. If this theoretical distribution is known a random sample of any length can easily be generated. Figure 7.2 shows the first part of a random sample of $n = 1000$ for a Normal distribution. This distribution has two parameters, namely the mean value μ_X and the standard deviation σ_X and it has the expression:

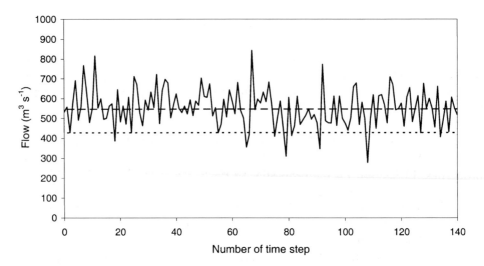

Figure 7.2 A random sample of streamflow of size $n = 140$ (of a total of 1000), normally distributed with the parameters $\mu_X = 543$ and $\sigma_X = 92$. The horizontal lines represent the median (upper line) and the 10% quantile (lower line), respectively.

$$F_X(x) = \frac{1}{\sigma_X \sqrt{2\pi}} \int_{-\infty}^{x} \exp\left[-\frac{(\varsigma - \mu_X)^2}{2\sigma_X^2} \right] d\varsigma \tag{7.2}$$

In the example the parameters were set equal to $\mu_X = 543$ and $\sigma_X = 92$ and it is assumed that the random sequence represents streamflow with units in $m^3 s^{-1}$. In this case the two parameters represent those of the theoretical model $F(x)$. Later, when estimated values of the mean value and the standard deviation are considered they will be denoted $\bar{x}$ and s_X, respectively. The generation was done in two steps. Firstly random numbers $p_1, p_2, ..., p_{1000}$ were generated from a uniform distribution over the interval $[0, 1]$. This random generator for uniform numbers can be found in any worksheet or statistical software. These values correspond to probabilities. By taking the inverse of Equation 7.1 the corresponding x values, $x_1 = F_X^{-1}(p_1)$, $x_2 = F_X^{-1}(p_2)$, ..., $x_{1000} = F_X^{-1}(p_{1000})$, can be found. These are the values presented in Figure 7.2. An alternative would be to directly use a generator of random numbers ε with a standardized Normal distribution Φ, i.e. a Normal distribution with zero mean and a standard deviation equal to one. The cdf for $\Phi(\varepsilon)$ is in this case:

$$\Phi(\varepsilon) = \frac{1}{\sqrt{2\pi}} \int_{-\infty}^{\varepsilon} \exp\left(-\frac{\varsigma}{2} \right) d\varsigma \tag{7.3}$$

Simulations of the original data are now derived from the relation $x_i = \mu_X + \sigma_X \varepsilon_i$, i = 1, ..., 1000.

$F_X(x)$ represents a random model, although for the time being a very simple one, and the data in Figure 7.2 are a realization, a simulated series, generated by this model. Both the model and the simulated series are very useful when analysing low flow and drought and have two implications:

a) Both the model and the simulated series can be used to derive properties from time series (with a random sequence as a special case) which are of interest in low flow and drought analysis. Derivation of the theoretical distribution and/or their moments of some property from our basic random model might lead to theoretical or numerical difficulties. The Monte-Carlo technique then offers an alternative, working on the simulated data generated by a basic model. The theoretical derivations are thus replaced by a direct statistical analysis of the properties of interest from long and/or many simulated series.

b) Hydrological observations are a realization of a natural phenomenon. From this manifestation it is necessary to establish a model of this natural phenomenon. Such a model is always able to reconstruct such properties that are explicitly parameters in the model (in the simple example above, the mean value and the standard deviation). However, it is never certain that other properties are exactly preserved (within the accuracy of statistical errors). It must be accepted that a model, by definition, is a simplification. It is important to create models that are able to mimic or describe reality as well as possible. Other properties than those directly parameterized in the model can with advantage be used to validate the model performance. Construction of a model should start from a simple one. If this can be rejected it is possible to increase the complexity, bearing in mind the Principle of Parsimony as formulated by Tukey (1961): "It may pay not to try to describe in the analysis the complexities that are really present in the situation". He stresses the importance of reconsidering a model structure towards a simpler representation, which might improve the performance of the estimation method.

Let us explore the random sequence and its statistical properties from the model and from the 1000-value simulation. The relevant data are presented in Table 7.1. An inspection of the table shows a good agreement between the statistics of the theoretical model and those of the simulated series. This is, of course, obvious and the possible deviations should be in the order of statistical

Table 7.1 Statistical properties of a random sequence from a theoretical model and estimated from a simulated series of this model

Statistics	Random sample model	Simulated series
Number of observations	1000	1000
Mean	543	540
Standard deviation	92	94
Median x_{50}	543	541
10% quantile x_{10}	424	418
Lag-one autocorrelation	0.000	−0.07
Number of upcrossings of x_{50}	250 ($= 1000 \cdot 0.5 \cdot 0.5$)	250
Number of upcrossings of x_{10}	90 ($= 1000 \cdot 0.1 \cdot 0.9$)	92
Number of downcrossings of x_{50}	250 ($= 1000 \cdot 0.5 \cdot 0.5$)	251
Number of downcrossings of x_{10}	90 ($= 1000 \cdot 0.1 \cdot 0.9$)	92
Number of runs of x_{50}	501 ($= 2 \cdot 1000 \cdot 0.5 \cdot 0.5 + 1$)	501
Number of runs of x_{10}	181 ($= 2 \cdot 1000 \cdot 0.1 \cdot 0.9 + 1$)	184
Mean run length of x_{50}	2.00 ($= 1/0.5$)	1.99
Mean run length of x_{10}	1.11 ($= 1/0.9$)	1.09

errors. It is known, for instance, that the standard error of the mean value is $\sigma_{\bar{x}} = \sigma_X / \sqrt{n}$. Thus in the example $\sigma_{\bar{x}} = 92 / \sqrt{1000} = 2.9$. A 95% confidence interval around the theoretical mean is then [537, 549]. The mean of the simulated series is 540, which is well within the interval. Based on the mean value it is not possible to reject the hypothesis that the realization originates from the theoretical model. In the present case such a test is simply a check that the numerical algorithm is correct and/or that no other mistake has been made. Even if the model had not been known, this test has no other value as the mean value is a parameter of the random model itself.

Let us turn to the theoretical estimation of crossing properties (Box 7.1) and start with the number of upcrossings of a quantile level x_p, where p is the probability $p = Pr\{X \le x_p\} = F_X(x_p)$. When analysing data it simplifies the task if we transform the original sequence to a series of *0* and *1*, so that the data point x is replaced by *0* if $x \le x_p$, and *1* if $x > x_p$. The probability of observing a *0* is thus p and of observing *1* is $(1 - p)$. An *upcrossing* is a sequence of *01* and as it is a random sample and data in the sequence are independent, the theoretical probability of observing this is thus $p(1 - p)$. If the sequence contains n observations the expected number of upcrossing is $n_u = np(1 - p)$. A *downcrossing* is a sequence 10 and in a similar way the expected number of downcrossings is $n_d = n(1 - p)p$. The number of 0s is $n_0 = np$, the number of 1s is $n_1 = n(1 - p)$ and $n = n_0 + n_1$. The expected number of upcrossings/downcrossings can now be expressed as $n_u = n_d = n_0 n_1/(n_0 + n_1)$.

A *run* is an unbroken sequence of 0s or 1s. To be well defined it should thus be surrounded by 1s or 0s, respectively. A run of four values, for example, would then look as $...100001...$ and $...011110....$ Following the same procedure as above in the case of independent data, the probability can be deduced of a run of length one (i.e. $...101...$ or $...010$) as $p^2(1 - p) + p(1 - p)^2$ and the probability of any such sequence of a certain length can be calculated. To find the total number of runs in a series of observations is a more complicated task. It is then necessary not only to calculate the probabilities for a certain run sequence of a given length, but how many sequences of this length can be selected from the total number of observations. Combinatorics, i.e. the classical theory for calculation of probabilities of subsets of events, will help to solve this task also, but it requires a lot of work. To find the expected number of runs in a series of n observations it is possible to proceed in a simpler way. A run must start with an upcrossing (downcrossing) and end with a downcrossing (upcrossing), i.e each time a crossing is passed a new run starts and the number is thus equal to the sum of crossings plus one to account for the start (or end) run of the series. The following formula for the expected number of runs is thus valid: $n_r = 2p(1-p) + 1 = 2n_0 n_1/(n_0 + n_1) + 1$. It can also be shown that the variance of the number of runs is equal to $\sigma_r^2 = 2np(1 - p)(2np(1 - p) - 1)/(n - 1) = 2n_0 n_1(2n_0 n_1 - n_0 - n_1)/((n_0 + n_1)^2(n_0 + n_1 - 1))$. The mean run length μ_r can be estimated from the expression $\mu_r = 1/(1 - F(x_p))$ (Equation 7.40).

The statistics of the crossing properties are not directly parameters in this basic model. They are thus a good base for testing the model. Their estimated values are inferred in Table 7.1. It is possible, for instance, to use the number of runs as a test quantity. The mean (expected value) and the variance for this statistics have already been determined. If n is large the normal approximation is applicable. The expected number of runs is estimated to be 501 for the 50% quantile (Table 7.1) and the standard deviation is 15.8. A 95% confidence interval is thus [470, 532]. The number of runs for the simulated sequence is 501, i.e. it fits well within the confidence bounds. The corresponding values for the 10% quantile are: expected value 181, standard deviation 5.7, confidence interval [170, 192] and the number of runs in the simulated series 184. A similar test can be performed on the number of upcrossings/downcrossings.

Note that the statistics for the crossing properties only contains the number of observations n and probability p. A test based on these properties is thus not based on any assumption on the underlying distribution. Such tests are named distribution free (or parameter free). The assumption left, however, is that of a random sequence (i.e. independence between observations). The test on runs is thus able to be used as a test of independence. In the example above the hypothesis of independence could not be rejected as the test quantity, number of runs, was within the confidence bounds. The series in Figure 7.2 is thus independent. This is also confirmed by the low lag-one correlation coefficient. A similar test can be performed on the number of upcrossings/downcrossings. More details on the distribution free test referred to here as well as other tests of this kind are found in, for example, Bradley (1968).

7.3   A time series simulated by an autoregressive random model

The alternative hypothesis to the run test developed in the previous section is that of dependence in the series. Such a series is not described by a one-dimensional distribution function like Equation 7.1. For series with dependent observations the order in which they are observed $x(t_1)$, $x(t_2)$, ..., $x(t_n)$ becomes important. The parameter space t_1, t_2, ..., t_n is in these applications time. In this section a time step of one year is assumed and the sequence t_1, t_2, ..., t_n is thus a sequence of n years.

Hydrological time series in general show persistence giving rise to dependence between observations (Box 6.3). The persistence originates from an inner gravity or memory of the studied system from storage and transport processes. Let us elaborate on the role of a storage $S(t)$, say a lake or a groundwater reservoir, on the outflow $X(t)$ from this reservoir when the inflow is a random sequence $Z(t)$ without memory as studied in the previous paragraph. We write down the water balance equation for this reservoir as (Section 3.2):

$$Z(t) - (X(t) - \mu_X) = \frac{dS}{dt} \tag{7.4}$$

It is assumed here that the inflow has zero mean value and therefore the outflow is adjusted accordingly to be normalized with respect to its mean value. This is to assure that the mean value in X is preserved. It is also assumed that the reservoir is a linear one, i.e. the outflow is a linear function of the storage $X(t) = S(t)/k$, where k is a constant with dimension time. The bigger the storage capacity the bigger is k. Insertion into the water balance equation yields:

$$Z(t) - (X(t) - \mu_X) = k\frac{d(X(t) - \mu_X)}{dt} \tag{7.5}$$

This equation can now be rewritten in terms of finite differences as:

$$Z_t - \frac{(X_t - \mu_X) + (X_{t+1} - \mu_X)}{2} = k\frac{(X_{t+1} - \mu_X) - (X_t - \mu_X)}{\Delta t} \tag{7.6}$$

Rearrangement of the different terms results in:

$$X_{t+1} = \mu_X + b_1(X_t - \mu_X) + Z'_{t+1} \tag{7.7a}$$

where $b_1 = \left(\dfrac{k}{\Delta t} - 0.5\right) \Big/ \left(\dfrac{k}{\Delta t} + 0.5\right)$ and $Z'_{t+1} = \left(\dfrac{k}{\Delta t} + 0.5\right)^{-1} Z_t$.

    Let us stop for a while to see what we have achieved. The equation shows that the outflow at time $t+1$ is dependent on what it was at time t. The dependence is expressed by the coefficient b_1 that in its turn depends on the storage properties of the reservoir. Note also that the equation is well suited for simulation purposes as it allows the series to be generated sequentially time step by time step. Up till now no new assumptions are added to the one that Z_t is a random sequence with no memory and zero mean value. It is confirmed that the relation preserves the mean value of X by taking the expectation of both sides of Equation 7.7a:

$$E\{X_{t+1}\} = \mu_X + b_1(E\{X_t\} - \mu_X) + c\,E\{Z_t\} = \mu_X$$

This takes advantage of the fact that the time series are stationary, i.e. the mean value is constant and does not vary with time. It is also necessary that the variance of X should be preserved. For this purpose it is necessary to multiply Equation 7.7a by $(X_t - \mu_X)$ and $(X_{t+1} - \mu_X)$, respectively, and take the expectation of the results:

$$E\{(X_{t+1} - \mu_X)(X_t - \mu_X)\} = b_1 E\{(X_{t+1} - \mu_X)^2\} + 0$$
$$\Rightarrow \quad \rho_1\sigma_X^2 = b_1\sigma_X^2 \quad \Rightarrow \quad b_1 = \rho_1$$

$$E\{(X_{t+1} - \mu_X)^2\} = b_1 E\{(X_{t+1} - \mu_X)(X_t - \mu_X)\} + E\{(X_{t+1} - \mu_X)Z'_{t+1}\}$$
$$\Rightarrow \quad \sigma_X^2 = b_1\rho_1\sigma_X^2 + \sigma_Z^2 = \rho_1^2\sigma_X^2 + \sigma_Z^2 \quad \Rightarrow \quad \sigma_Z = \sigma_X\sqrt{1 - \rho_1^2}$$

This case also utilizes the stationarity property of the mean value as well as the standard deviation. Preservation of the variance of the outflow process X thus leads to that the coefficient b_1 is equal to the lag-one autocorrelation coefficient ρ_1 and also that the standard deviation of the inflow must be directly related to

that of the outflow and its autocorrelation as $\sigma_Z = \sigma_X \sqrt{1 - \rho_1^2}$. The inflow is
now written as $Z'_{t+1} = \sigma_X \sqrt{1 - \rho_1^2} \, \varepsilon_{t+1}$, where ε is an independent random variable
with zero mean and standard deviation equal to one. A final expression for the
random model for simulating the effect of a linear reservoir on an inflow
without memory is thus:

$$X_{t+1} = \mu_X + \rho_1 (X_t - \mu_X) + \sigma_X \sqrt{1 - \rho_1^2} \, \varepsilon_{t+1} \qquad (7.7b)$$

In principle it is not necessary to make any assumptions about the
distribution of Z' or ε. Section 7.2 showed how a random sample can be
generated for any random variable when the corresponding cdf is known. In the
present case the situation is, however, more complicated. If a distribution is
assumed for Z' or ε, what distribution will then apply to X? There is no standard
answer to this. Only for certain classes of distributions will X take the same
distribution as say ε.   The Normal distribution is the most well-known
distribution that preserves normality in X. In other cases the only way to find
out is by Monte Carlo experiments. The situation is thus a little out of control if
a distribution for ε is just assumed. It is not of much help to base the choice of
distribution of an empirical fit to residuals ε derived from observations of X.
The recommendation must be to use a transformation to normality if the
original data do not follow this distribution. Then the data need to be
transformed back when the simulation is performed.

Assuming that X is normally distributed, the marginal distribution of this
variable   is   thus   equal   to   the   one-dimensional   Normal   distribution
(Equation 7.2). However, this is not sufficient to describe the random process as
it contains autocorrelation. The two-dimensional Normal distribution showing
the joint distribution of X_t and X_{t+1} is:

$$F_{X_t, X_{t+1}} (x_t, x_{t+1}) = \qquad (7.8)$$

$$\frac{1}{2\pi\sigma_x^2 (1 - \rho_1^2)} \int_{-\infty}^{x_t} \int_{-\infty}^{x_{t+1}} \exp\left[-\frac{(\varsigma_1 - \mu_X)^2 - \rho_1(\varsigma_1 - \mu_X)(\varsigma_2 - \mu_X) + (\varsigma_2 - \mu_X)^2}{2\sigma_X^2 (1 - \rho_1^2)} \right] d\varsigma_1 d\varsigma_2$$

This joint bivariate distribution for X_t and X_{t+1} is related to the conditional one
$F_{X_{t+1}} (x_{t+1} | X_t)$, the distribution of X_{t+1}, when X_t is known, through the relation:

$$F_{X_{t+1}} (x_{t+1} | X_t) = F_{X_t, X_{t+1}} (x_t, x_{t+1}) / F_{X_t} (x_t)$$

where $F_{Xt}(x_t)$ is the marginal distribution of X_t (Equation 7.2). The
corresponding conditional distribution function has the expression:

$$F_{X_{t+1}}(x_{t+1}|X_t) = \frac{1}{\sigma_X \sqrt{2\pi(1-\rho_1^2)}} \int_{-\infty}^{x_{t+1}} \exp\left\{-\frac{1}{2}\left[\frac{\varsigma - \mu_X - \rho_1(X_t - \mu_X)}{\sigma\sqrt{1-\rho_1^2}}\right]^2\right\} d\varsigma \qquad (7.9)$$

When comparing this expression with that of the univariate Normal distribution (Equation 7.2) it is seen that for a fixed value X_t, X_{t+1} is also normally distributed (which is already assumed) with the conditional expected value $\mu_{X_{t+1}|X_t} = \mu_X + \rho_1(X_t - \mu_X)$ and standard deviation $\sigma_{X_{t+1}|X_t} = \sigma_X\sqrt{1-\rho_1^2}$. The expected value of X_{t+1} will thus change depending on the value of X_t, while its standard deviation is constant. A further step is to introduce the variable transformation $\varepsilon_{t+1} = [X_{t+1} - \mu_X - \rho_1(X_t - \mu_X)]/\sigma_X\sqrt{1-\rho_1^2}$. This transforms the distribution (Equation 7.9) into the standard Normal distribution (Equation 7.1). ε is thus normally distributed with a zero mean and a standard deviation equal to one. Rearranging the expression for the transformation gives back the random model (Equation 7.7b).

In a heuristic way this section has shown that the outflow from a linear reservoir resulting from a sequence of random inflow is described by a bivariate Normal distribution under the assumption that the inflow is also normally distributed. The corresponding random model (Equation 7.7b) can be used to generate simulated series from standard normal random numbers. The model preserves the mean value, the standard deviation and the lag-one autocorrelation coefficient. In statistical literature this type of model is referred to as an autoregressive first order model, AR(1), and also a Markov model as the process at time $t + 1$ is dependent on a finite number of time steps before (here the time step equals one).

The AR(1) model can be used to generate a time series of, say, streamflow. As in the previous model application (Section 7.2) the mean value $\mu_X = 543$ and standard deviation $\sigma_X = 92$ must be preserved as well as the lag-one autocorrelation $\rho_1 = 0.476$. This latter value corresponds to a value of the reservoir constant $k = 1.41\Delta t$. Assuming that the time step is one year this is indeed a large reservoir, which is also reflected by the high autocorrelation. The statistical parameters of the model and those estimated from the simulated series of $n = 1000$ are presented in Table 7.2. The first 140 streamflow values are illustrated in Figure 7.3. For the generation of the series exactly the same sample of uniform random numbers is applied and corresponding ε values, $\varepsilon_1 = \Phi^{-1}(p_1)$, $\varepsilon_2 = \Phi^{-1}(p_2)$, ..., $\varepsilon_{1000} = \Phi^{-1}(p_{1000})$, as in Section 7.2. The fact that the same set of random numbers is used means that the two series in Figures 7.2 and 7.3 are similar. In the latter case the variations are dampened due to the interdependence. The differences between the theoretical model values and those estimated from the simulated series are also in this case small. Again, the

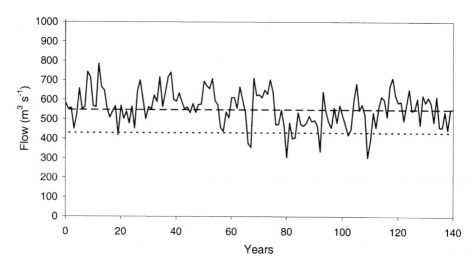

Figure 7.3 A simulated time series of streamflow of size $n = 140$ (from a total of 1000) from an AR(1) model, normally distributed with the parameters $m_X = 543$, $\sigma_X = 92.6$ and $\rho_1 = 0.476$. The horizontal lines represent the median (upper line) and the 10% quantile (lower line), respectively.

agreement for the dependent statistical parameters should be expected to be of the order of statistical errors. When the data show autocorrelation the information content in a series is reduced compared to the case of independence (Box 6.3). Standard errors of all statistical parameters are now higher and some also show systematic bias (e.g. Hansen, 1971).

The bivariate distribution $F_{X_t,X_{t+1}}(x,x)$ expresses the probability that X both at time t and $t + 1$ is less than the fixed level x:

$$Pr[\{X_t \leq x\} \cap \{X_{t+1} \leq x\}] = F_{X_t,X_{t+1}}(x,x) = p_{00} \qquad (7.10a)$$

Using earlier notations of set functions equal to 0 and 1 (Section 7.2), this situation is the probability p_{00} for a *00* event. Studying crossing properties around this fixed level x it is also of interest to determine the probabilities of all other combinations of *0*s and *1*s in a combination of two. This can be deduced from a simple geometric consideration of the two dimensional plot of $F_{X_t,X_{t+1}}(x,x)$ and the fact that it is symmetric around its principal axes. Thus:

$$Pr[\{X_t > x\} \cap \{X_{t+1} > x\}] = 1 - F_{X_t}(x) - F_{X_{t+1}}(x) + F_{X_t,X_{t+1}}(x,x) = p_{11} \quad (7.10b)$$

$$Pr[\{X_t \leq x\} \cap \{X_{t+1} > x\}] = F_{X_t}(x) - F_{X_t,X_{t+1}}(x,x) = p_{01} \qquad (7.10c)$$

Table 7.2 Statistical properties of an AR(1) model and estimated from a simulated series from this model

Statistics	AR(1) model	Simulated series
Number of observations	1000	1000
Mean	543	547
Standard deviation	92	91
Median x_{50}	543	548
10% quantile x_{10}	424	429
Lag-one autocorrelation	0.48	0.45
Number of upcrossings of x_{50}	$171 (= 1000 \cdot (0.5 - 0.329))$	177
Number of upcrossings of x_{10}	$65 (= 1000 \cdot (0.1 - 0.035))$	60
Number of downcrossings of x_{50}	$171 (= 1000 \cdot (0.5 - 0.329))$	178
Number of downcrossings of x_{10}	$65 (= 1000 \cdot (0.1 - 0.035))$	60
Number of runs of x_{50}	$343 (= 2 \cdot 1000 \cdot (0.5 - 0.329) + 1)$	355
Number of runs of x_{10}	$131 (= 2 \cdot 1000 \cdot (0.1 - 0.035) + 1)$	120
Mean run length of x_{50}	$2.92 (= 0.5/(0.5 - 0.329))$	2.80
Mean run length of x_{10}	$1.53 (= 0.1/(0.1 - 0.035))$	1.62

$$Pr[\{X_t > x\} \cap \{X_{t+1} \leq x\}] = F_{X_{t+1}}(x) - F_{X_t, X_{t+1}}(x, x) = p_{10} \qquad (7.10d)$$

Note that the last two probabilities exactly describe the ones for upcrossing and downcrossing, respectively. The expected number of upcrossings/downcrossings for an AR(1) model is thus $n_u = n_d = n(F(x) - F(x,x))$ (for convenience the indices are dropped). In the case of independence the joint distribution is derived as $F(x,x) = F(x)F(x)$. Insertion in the expression for the number of crossings results in $n_u = n_d = n(F(x) - F(x)F(x)) = nF(x)(1 - F(x))$, identical to the result in the previous section for independent data. The expected number of runs is found accordingly to $n_r = n[2(F(x) - F(x,x)) + 1]$. The mean run length μ_r (i.e. drought duration) can in this case be estimated from the expression $\mu_r = F(x)/(F(x) - F(x,x))$ (Equation 7.40). The expected number of crossings and runs for the AR(1) model is inferred in Table 7.2. The numerical evaluation of the bivariate normal probabilities is elaborated in Ambramovitz & Stegun (1972). As a special case we have that

$F_{X_t,X_{t+1}}(x_{50},x_{50}) = \frac{1}{4} + \arcsin(\rho_1)/2\pi$ when the threshold is equal to the median x_{50}.

7.4   An observed time series of annual discharge

The two random models presented in Sections 7.2 and 7.3 can now be confronted with real hydrological observations. It was necessary to first describe the models, their assumptions and their abilities and precisions in reproducing the statistical properties of data in order to understand how well a model can be expected to agree with observed data and how to test model performance. Data were chosen from the Göta älv (River Göta) in Sweden (Melin, 1954) for the period 1807 to 1937 (Figure 7.4). This is one of the longest discharge series of the world. The observations are still continuing but they have been influenced by regulations since 1938. The Göta älv drains the large Lake Vänern (5648 km^2). The time series of monthly streamflow is found on the CD under the item Data.

The statistical parameters of this discharge series (131 years of observation) are included in Table 7.3. It can be seen that the mean value, the standard deviation and the lag-one autocorrelation are exactly those used earlier for the models. The corresponding parameters for the theoretical models, a random sample (Section 7.2) and an AR(1) model (Section 7.3), now referring to a

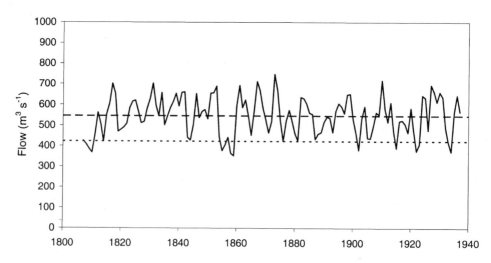

Figure 7.4 Annual discharge observations from the Göta älv (River Göta) at the outlet of the large Lake Vänern in Sweden for the period 1807 to 1939. The horizontal lines represent the median (upper line) and the 10% quantile (lower line), respectively.

period of $n = 131$ years are also included in the table. These latter parameters are represented by the mean value and the standard deviation over 1000 simulations. Both models, within the limits of statistical uncertainty (here expressed by the standard deviation over the 1000 simulations), exactly

Table 7.3 Statistical properties of the observed data series of annual discharge in Göta Älv at the outlet of Lake Vänern and corresponding ones from a random sample and an AR(1) model (1000 simulations, standard deviations shown in brackets)

Statistics	Random sample model		AR(1) model		Observed series
	Theoretical	Simulated	Theoretical	Simulated	
Number of observations	131	131	131	131	131
Mean	543	543 (8.6)	543	543 (7.6)	543
Standard deviation	92	92 (5.8)	92	92 (7.4)	92
Median x_{50}	543	543 (10.3)	543	543 (9.7)	548
10% quantile x_{10}	424	423 (14.2)	424	423 (14.9)	416
Lag-one autocorrelation	0.000	0.01 (0.08)	0.47	0.45 (0.08)	0.47
Number of upcrossings of x_{50}	33.8	33 (2.9)	22	22 (2.9)	22
Number of upcrossings of x_{10}	12	12 (1.0)	9	9 (1.5)	7
Number of downcrossings of x_{50}	33	33 (2.9)	22	22 (2.9)	21
Number of downcrossings of x_{10}	12	12 (1.0)	9	9 (1.5)	7
Number of runs of x_{50}	66	65 (5.8)	44	44 (5.7)	43
Number of runs of x_{10}	24.6	23.5 (2.0)	18	18 (3.0)	14
Mean run length of x_{50}	2.0	1.97 (0.18)	2.92	2.99 (0.41)	3.0
Mean run length of x_{10}	1.11	1.11 (0.10)	1.53	1.50 (0.28)	1.71

reproduce the mean value and standard deviation of the observed one as expected. The AR(1) model is also capable of preserving the autocorrelation of the observed series (i.e. 0.47). This is, of course, not the case with the random sample model as dependence in data is neglected.

While it is expected that the mean, the standard deviation and, in the case of an AR(1) model also, the lag-one autocorrelation are exactly preserved, this is not the case for all the other statistical parameters in Table 7.3. This fact can be exploited as a test of model performance. The median and the 10% quantile indicate whether the assumption of normality of the observed data is plausible. There is a deviation between the observed and modelled parameters, but considering the statistical uncertainty, as revealed by the standard deviation (i.e. 9.7 and 14.9, respectively), it is not large enough to be significant. Turning to crossing properties it can be noted that when the autocorrelation is neglected the number of crossings is significantly higher than in case of data with strong dependence as in the observed series and the AR(1) simulated series. It is thus possible to reject the random sample model as it is not able to reproduce the crossing properties of the observed data. The AR(1) model on the other hand, is able to preserve these properties within the statistical uncertainty, and this is thus the model of preference.

7.5   Simulation of stationary hydrologic time series

The concept of stationarity which was referred to several times in Section 7.1 means that model properties do not change with time. In the case under consideration this concerns the model parameters mean value, standard deviation and lag-one autocorrelation, which then are constants. It has been noted that this can be an acceptable approximation when analysing and modelling annual data of streamflow. The modelling of such stationary hydrological time series in hydrology already has a long tradition. The seminal work dates back to the early 1960s (Svanidze, 1961; Fiering, 1961) suggesting a first order autoregressive model for simulation of annual runoff values X_t (Equation 7.7b):

$$X_t = \overline{X} + \rho_1\left(X_{t-1} - \overline{X}\right) + \sigma_X \sqrt{1 - \rho^2}\, \varepsilon_t \qquad (7.11)$$

With reference to the derivation of Equation 7.7b it can be seen that the use of this model is straightforward in the case of normally distributed data. ε is normally distributed as well with mean zero and unit variance. When normally distributed variables are summed a Normal distribution results. This is a property of the so-called class of stable distributions to which the Normal

---

Box 7.2

Transformations to the Normal distribution

The following transformations to normality are often used in hydrology:

- $\ln(X)$ in case of lognormally distributed data.

- The cube root $(x/\bar{x})^{\frac{1}{3}}$ (Wilson & Hilferty, 1931) transformation in case of gamma distributed data.

- The more general power transformation $[(x+c)^h - 1]/h$, where h and c are parameters (Box & Cox, 1964).

- In the case of bivariate distributions, Moran (1969) shows how the two-dimensional gamma densities may be transformed to the classical two-dimensional Normal distribution.

- A final alternative is to adjust a distribution by experimental statistics (Monte Carlo methods) so that an approximately acceptable fit to the empirical distribution of X is achieved.

---

belongs. Problems arise when data do not belong to the Normal distribution. The distribution of ε is not a priori known and furthermore other frequently used distributions used in hydrology do not belong to the class of stable distributions. It is not easy to deduce the distribution of X by postulating a distribution for ε. The best solution to the problem is to replace X by a transformation of it to normal (Box 7.2).

Svanidze (1964) and Yevjevich (1964) suggest a generalized form of the model (Equation 7.11) to an arbitrary order n of the autoregressive model:

$$X_t = b_1 X_{t-1} + \ldots + b_n X_{t-n} + \sigma_\varepsilon \varepsilon_t \qquad (7.12)$$

A general formalism for this class of models is found in Box & Jenkins (1970). The basic elements in this are the autoregressive model (AR) and a moving average model (MA). Equation 7.12 is an AR(n) model. The AR and MA in combination forms an ARMA model. In case of an ARMA model, the ε in these equations should be replaced by a MA-model.

It has been questioned whether a Markov model of the kind presented above (Equations 7.11 and 7.12), i.e. with a finite memory, describes the true nature of hydrologic time series at large timescales. The discussion in the hydrological literature has been concentrated, first of all, upon the asymptotic behaviour of the range of cumulative departures from the mean for a given sequence of runoff for N years. The so-called Hurst coefficient H characterizes the asymptotic behaviour of the expected value of the rescaled range. For a Markov process the theoretical value of H is $H = 0.5$. Hurst (1951) was the first to notice that geophysical series in nature (and especially the long series of observation of water levels in the Nile) show values of H greater than 0.5. The discussions were focused on the ability of models to reproduce the so-called Noah and Joseph effects of natural series, i.e. the ability to reproduce extreme extremes and the tendency for long spells of dry and wet years. A fractional Gaussian noise model developed by Mandelbrot & Wallis (1968, 1969a, 1969b, 1969c) is able to generate synthetic data with H different from 0.5. The specific problem for the application of these models is that of parameter estimation. Discharge time series are too short to show asymptotic properties.

7.6   Observed and simulated hydrologic time series of monthly discharge

In many climates, hydrological data commonly show variations over the year in accordance with the seasons (Section 2.2.2). The statistical properties for monthly time series thus vary with the number of the month, they show non-stationarity. The principle can be illustrated by applying a non-stationary AR(1) model to the observed monthly data at Narsjø, Norway (Data, CD). This station is situated in the central mountain area of Norway with a very stable nival flow regime. In this case the model has the form:

$$X_t = \overline{X}_t + \rho_{t,t-1} \frac{\sigma_{Xt}}{\sigma_{Xt-1}} \left(X_{t-1} - \overline{X}_{t-1} \right) + \sigma_{Xt} \sqrt{1 - \rho_{t,t-1}^2} \, \varepsilon_t \tag{7.13}$$

The observed data are illustrated in Figure 7.5 (upper) and the seasonal variation in the statistical parameters in Figure 7.6. It can be seen that the statistical parameters mean value $\overline{X}$, standard deviation σ_{Xt} and lag-one correlation $\rho_{t,t-1}$ in this case strongly vary with time t (number of the month). The two first parameters show low values during winter months and very high ones during the spring flood. The autocorrelation shows a reversed pattern with very high autocorrelation during the winter low flow and very low or even negative values during the snow melt period.

Non-stationary models are able, in the same manner as the stationary models, to exactly preserve the mean, the standard deviation and in case of an AR(1) model also the lag-one autocorrelation. Table 7.4 shows the statistical properties of the observed and simulated time series, including the crossing properties. In contrast to the earlier example using annual values (Section 7.4),

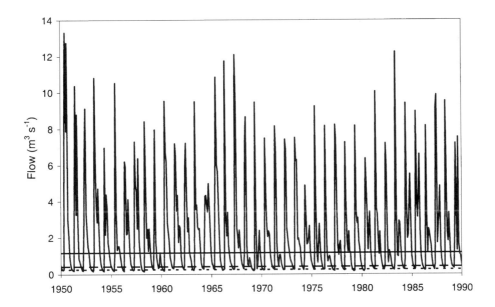

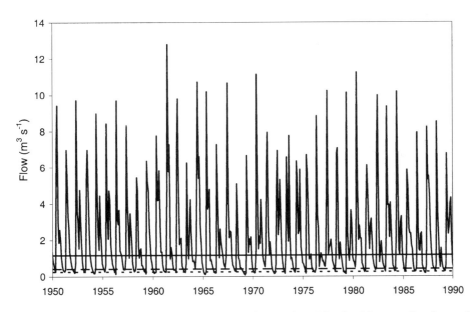

Figure 7.5 Monthly discharge observations from River Narsjø, Norway for the period 1950 to 1990 (upper) and a corresponding simulation with an AR(1) model applying a logarithmic transformation (lower). The three horizontal lines represent x_{50} (solid), x_{25} (dashed) and x_{10} (dotted), respectively.

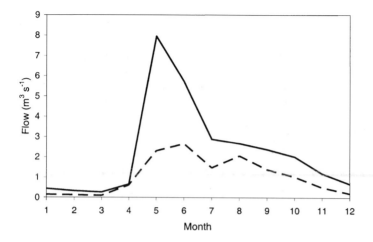

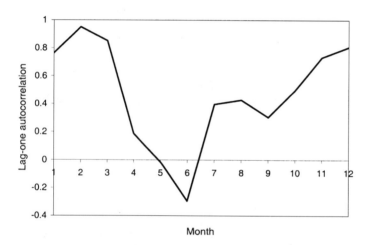

Figure 7.6 Statistical parameters for observations from River Narsjø (Norway); mean (full line) and standard deviation (dashed line) (upper), and lag-one autocorrelation of monthly discharge (lower).

an AR(1) model applied with the logarithmic transformation has now beenadded. Monthly data are positively skewed which can motivate the use of this transformation. The random sample model fails in this case giving rise to a much higher number of crossings than in the observed series. The two AR(1) models show significantly better performance, especially the one with logarithmic transformation of data. This can also be confirmed by a visual inspection of the simulated series in Figure 7.5 (lower).

Table 7.4. Statistical properties of the observed data series of monthly discharge at River Narsjø, Norway and corresponding ones from a random sample and a non-stationary AR(1) model (1000 simulations, standard deviations shown in brackets)

Statistics	Random sample model	AR(1) model	AR(1) model, ln-transf.	Observed series
Number of observations	$12 \cdot 131 = 1572$	1572	1572	1572
Median x_{50}	1.14	1.14	1.17	1.17
		(0.07)	(0.06)	(0.08)
Number of upcrossings	104	91	83	76
	(4.4)	(4.1)	(3.4)	
Number of downcrossings	104	91	83	76
	(4.1)	(4.1)	(3.4)	
Number of runs	208	182	166	152
	(8.8)	(8.2)	(6.8)	
Mean run length	3.8	4.3	4.7	5.1
	(0.19)	(0.19)	(0.19)	
25% quantile x_{25}	0.42	0.42	0.46	0.42
	(0.02)	(0.02)	(0.02)	(0.02)
Number of upcrossings	103	81	65	65
	(3.8)	(3.8)	(1.7)	
Number of downcrossings	103	81	65	65
	(3.8)	(3.8)	(1.7)	
Number of runs	206	162	130	130
	(7.5)	(7.5)	(3.4)	
Mean run length	1.93	2.44	3.0	3.1
	(0.10)	(0.11)	(0.08)	
10% quantile x_{10}	0.23	0.23	0.26	0.26
	(0.02)	(0.02)	(0.01)	(0.01)
Number of upcrossings	66	50	43	43
	(3.0)	(3.4)	(2.4)	
Number of downcrossings	66	50	43	43
	(3.0)	(3.4)	(2.4)	
Number of runs	132	99	85	86
	(6.0)	(6.9)	(6.9)	
Mean run length	1.2	1.6	1.9	1.9
	(0.05)	(0.11)	(0.10)	

7.7   Simulation of non-stationary hydrologic time series

The stationary random models treated in Sections 7.2–7.5 are relatively simple to handle from a theoretical point of view. The only controversial question refers to the long-term properties of such models. In contrast, seasonal runoff variations are obviously non-stationary, and the mathematical models used for their description are in general very complicated and cumbersome. A relatively strict mathematical theory has been developed solely for processes with deterministic seasonal variations and it cannot be directly extended to the irregularities in series of monthly runoff values. The complexity in the stochastic nature of runoff has led to numerous simplified solutions, with different degrees of detail in the properties of the real processes. Periodic AR models like Equation 7.13 are widely recommended due to the relatively simple estimation of effective parameters in this case. Early examples in hydrology are those by Roesner & Yevjevitch (1966), Reznikovskij (1969) and Beard (1965). The non-stationarity is removed by standardization of the variables for each month, resulting in a weakly stationary series (the first and second order moments are constant). This is done in the simplest way by direct use of the estimated mean values and standard deviations for each month. An alternative is to fit a Fourier series to the monthly periodic variation of these parameters. The assumption is that the hydrologic regime is stable and shows the same average pattern from year to year, which might not be the case as, for example, many European flow regimes show instability (Krasovskaia & Gottschalk, 1993; Krasovskaia, 1996). When data show non-stationarity an ARIMA model can apply in the formalism of Box & Jenkins (1970). The 'I' stands for integration. In this case the first order difference (second order, ...). of the original data is modelled by an ARMA model and then integrated back to the original data. An alternative little explored in hydrology would thus be to use an ARIMA model of order 12 to simulate monthly data.

The fact that a model preserves the moments of monthly runoff does not in any way guarantee that the corresponding moments of annual runoff, as the sum of twelve monthly runoff values, are preserved (Gottschalk, 1975; Kartvelishvili & Gottschalk, 1976). A non-stationary model for monthly values implicitly includes assumptions about the structure of the annual values and this needs to be clarified either by direct theoretical derivations or by statistical experiments. It can be noted that the statistical properties vary with the definition of the hydrological year. A direct method to overcome this scale problem is first to develop a model for annual values and then use a disaggregation approach within each year to distribute monthly values. The method of fragments by Svanidze (1977) and the disaggregation process suggested by Valencia & Schaake (1973) are examples of such approaches.

The analyses of non-stationary time series involve the following steps (Bolgov, 2002):

a) If annual discharge data can follow the Normal distribution, this is usually not the case for monthly data. The distribution can in theory differ from one month to another. The distribution can be very complex in months at the start of snow melt or start or end of a rainy/dry season. They differ significantly for months with more stable conditions. However, a mix of different distributions will lead to a situation that is out of control and it is therefore advisable to transform the data, with one and the same basic expression, to normality as the first step. Transformations mentioned in Box 7.2 for annual data are also relevant here.

b) The type and complexity of the random model to be used needs thorough consideration. A reasonable strategy is to start with the simplest possible scheme, say a non-stationary AR(1) like Equation 7.13, and if rejected in the next step, adding the complexity gradually.

c) Assessing the degree of agreement of the model properties to the observations is the final step. The way to do this is to derive confidence bands for relevant statistical properties by statistical experimentation and then perform a statistical test. Table 7.4 contains a set of independent statistical properties that can be used for such a test. A complementary check on the statistical parameters of the model is also needed as the behaviour is less controllable in the case of non-stationary simulation and especially if transformations are used as well as higher order models. If a model is rejected it is necessary to go back to step b) and select another model.

7.8 Simulation of daily hydrologic time series with sawtooth pattern

A realistic model for simulating daily runoff values, or a similar short time interval, must take into consideration two important aspects of the basic process, the non-stationarity and the very complex internal dependence in daily data. Quimpo (1967) proposed in principle an autoregressive model for transformed variables with seasonally varying statistical properties. A problem with this model is to actually reproduce the specific sawtooth pattern of daily values, illustrated by the sequence of daily streamflow in Figure 7.7, and to

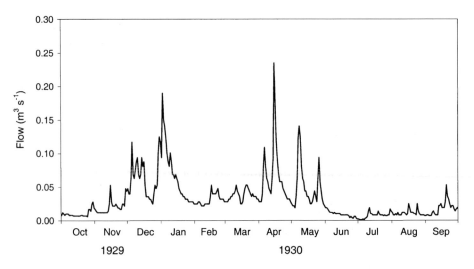

Figure 7.7 A sequence of observed daily streamflow in River Vesnaån at Hålabäck, Sweden.

decide which distribution the residual ε should have to achieve this. Bernier (1970), in search of a random model that describes this specific sawtooth pattern among other models, pointed to a so-called filtered Poisson process and more specifically a shot noise process. This idea has been further developed in hydrology by Weiss (1977), Yue *et al.* (1999) and Khristoforov & Samborski (2001). The basic theory for these types of processes can be found in, for example, Parzen (1964). The following mainly follows Weiss (1997).

In a filtered Poisson process, let $N(t)$ be a Poisson process with the intensity λ (Equation 6.9). Let Y be a random variable having an exponential distribution with mean value β and a distribution function $f(y) = 1/\beta \exp(-y/\beta)$, $y > 0$. Let the system response function have the form (in this case the hillslope response to a rainfall excess Y): $h(t) = \exp(-t/k)$, $t > 0$. The shot noise process is defined as:

$$X(t) = \sum_{i=(-\infty)0}^{N(t)} y_i \exp[-(t - \tau_i)/k] \qquad (7.14)$$

where $X(t)$ is runoff (inflow to river); t, time; N, number of events over threshold for time t; y, rainfall excess input; τ, time lag; and k, system's response time. The process is thus described by three parameters: the event intensity λ, the average jump height β of runoff events and the recession coefficient k.

With this point of departure it can be shown that $X(t)$ has a gamma distribution $X(t) \in \Gamma(\beta,\lambda k)$ and thus is non-negative and positively skewed with the probability density function for instantaneous values:

$$f_X(x) = \frac{1}{\beta^{\lambda k}\Gamma(\lambda k)}e^{-x/\beta} \tag{7.15}$$

and with the first and second order moments equal to:

$$m_X = E\{X(t)\} = \beta\lambda k$$
$$\sigma_X^2 = Var\{X(t)\} = \beta^2\lambda k$$
$$\rho(X(t)X(t+\tau)) = \exp(-\tau/k)$$

The process can also be expressed in the form:

$$X(t \mid \tau) = e^{-\tau/k}X(t) + \varepsilon(t+\tau) \tag{7.16}$$

The shot noise process is thus a first order autoregressive process in continuous time. It differs from a Normal first order autoregressive process by the fact that $\varepsilon(t+\tau)$ instead of being normally distributed is non-negative with a positively skewed distribution and with a positive probability of being exactly zero. This arises when no events occur in $[t, t+\tau]$. A realization of this shot noise process is of the form of a sawtooth curve, with vertical jumps of magnitude y_i at times τ_i and with an exponential recession between these jumps with rate k. The model for runoff formation as such is simple, but it points to important properties that control the process and that help to interpret parameters in the distribution of runoff.

For the integrated process over a discrete duration D:

$$\overline{X}_{D,j} = \frac{1}{D}\int\limits_{(j-1)D}^{jD}X(t)dt$$

the moments are derived as:

$$E\{\overline{X}_D\} = \beta\lambda k$$

$$Var\{\overline{X}_D\} = \frac{2\beta^2\lambda k^2}{D^2}\left[e^{-D/k} + D/k - 1\right] = 2\sigma_X^2\left(\frac{k}{D}\right)^2\left(e^{-D/k} + D/k - 1\right) \tag{7.17}$$

$$Cov\{\tau\} = 2\sigma_X^2\left[\frac{k}{D}\left(1 - e^{-D/k}\right)\right]^2 e^{(D-\tau)/k}$$

The mean value is constant and independent of duration.

7.9   Calculation of drought characteristics based on an analytical approach

Section 7.1 mentions that to obtain the derived distribution of important drought characteristics we can either use analytical derivations for simple cases or use Monte Carlo methods. In the latter case it is necessary to develop a model for simulation of random sequences, as previously discussed. Turning to some examples of analytical derivations, these have already been used to calculate the properties of the random sample and AR(1) models in Tables 7.1, 7.2 and 7.3, and a more complete derivation of formulae for these calculations will now be given.

The mechanisms behind a sequence of dry years are very complex involving large range atmospheric circulation (Section 2.2) as well as long time storage of water in the aquifers and in lakes (Sections 3.4 and 3.5). Applying the theory of stationary random processes will only reveal the random nature of these mechanisms. Grouping of dry years have been of interest primarily in connection with annual regulation of reservoirs and there are many works published on this topic (McMahon & Mein, 1978). Besides assuming stationarity, the applied theories are generally also based on the assumption that geophysical processes can be described as Markov processes. Based on this Markov assumption, this section will derive the probabilities for grouping of dry years and other statistical characteristics of a sequence of dry years directly from the multivariate distribution of a hydrological variable. The limitation in stationarity and Markov assumption is commented on in Section 7.5. The theoretical background for the derivations developed below can be found in Kartvelishvili (1975), Gottschalk (1976) and Rosbjerg (1977). The problem has partly been dealt with in Section 7.3 when determining crossing properties in a heuristic way, in which expressions for the number of upcrossings/downcrossings and the number of runs for the univariate and the bivariate cases were found. These results are now generalized and the following random variables added to the analysis:

K – the time interval between successive events $X > x$

M – the number of time steps in succession where $X \leq x$ (the run length)

L – the time interval until the first upcrossing of the level x occurs.

A dry year can be defined from different points of view as has already been discussed in depth in Sections 5.3 and 5.4. Choice of threshold depends on the potential uses and vary from one place to another. Here a simple approach is used and the annual runoff (calendar year) has been chosen as a basis for characterizing a dry year. To facilitate comparisons between different observation points the annual values have been standardized. A dry year is defined as the year when the hydrological variable X is below some fixed threshold x of the standardized annual values.

7.9.1 Probability for grouping of dry years

The diagram in Figure 7.8 shows a sequence of observed states of a hydrological variable where m events $X \leq x$ are preceded and followed by events $X > x$. The probability that at time 1 and $m + 2$ runoff X_i has values greater than some fixed values x_i (in the general formulation of the problem the threshold value can be allowed to vary with time step i), and for all events in between the probability that runoff is less than x_i, can be expressed in the following way:

$$\Phi = Pr\{X_1 > x_1, X_2 \leq x_2, \ldots, X_{m+1} \leq x_{m+1}, X_{m+2} > x_{m+2}\} \tag{7.18}$$

Equation 7.18 can be rewritten as:

$$\Phi = Pr\{X_2 \leq x_2, \ldots, X_{m+1} \leq x_{m+1}\} - Pr\{X_2 \leq x_2, \ldots, X_{m+2} \leq x_2\}$$
$$- Pr\{X_1 \leq x_1, \ldots, X_{m+1} \leq x_{m+1}\} + Pr\{X_1 \leq x_1, \ldots, X_{m+2} \leq x_{m+2}\} \tag{7.19}$$

The probabilities in Equation 7.19 may be replaced by the corresponding multivariate distribution functions:

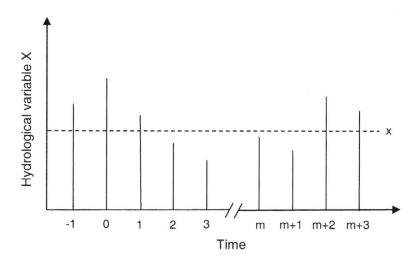

Figure 7.8 A sequence of observed states of a hydrological variable: m events $X \leq x$ preceded and followed by events $X > x$.

$$\Phi = F_m(x_2,\cdots,x_{m+1}) - F_{m+1}(x_2,\cdots,x_{m+2})$$
$$- F_{m+1}(x_1,\cdots,x_{m+1}) + F_{m+2}(x_1,\cdots,x_{m+2}) \tag{7.20}$$

It is now assumed that the annual runoff is a Markov process of order $n - 1$. It is further assumed that the distribution function F_n is known. $\Phi_s(x)$ will denote the s-variate distribution function. The problem is simplified by setting all $x_i = x$. The distribution function for $s > n$ can now be derived as:

$$\Phi_s(x) = \Phi_n(x)(\Phi_n(x)/\Phi_{n-1}(x))^{s-n} \tag{7.21}$$

Dependent on the length $m + 2$ of the considered time interval and the order n of the Markov process, three different expressions for Φ are derived by inserting Equation 7.21 in Equation 7.20:

1. $m + 2 < n$
$$\Phi = \Phi_n(x) - 2\Phi_{m+1}(x) + \Phi_{m+2}(x) \tag{7.22}$$

2. $m + 1 = n$
$$\Phi = (\Phi_{n-1}(x) - \Phi_n(x))^2/\Phi_{n-1}(x) \tag{7.23}$$

3. $m \geq n$
$$\Phi = (\Phi_{n-1}(x) - \Phi_n(x))^2/\Phi_{n-1}(x)(\Phi_n(x)/\Phi_{n-1}(x))^{m-n+1} \tag{7.24}$$

When $n = 1$ and $n = 2$, respectively, the following relations are found:

$$n = 1: \quad \begin{aligned} &\Phi_{n-1}(x) = 1; \quad \Phi_n(x) = F_1(x) \\ &\Phi = (1 - F(x))^2 F(x)^m \quad \text{for all } m \geq 1 \end{aligned} \tag{7.25}$$

$$n = 2: \quad \begin{aligned} &\Phi_{n-1}(x) = F_1; \quad \Phi_n(x) = F_2(x,x) \\ &\Phi = (F_1(x) - F_2(x,x))^2/F_1(x)(F_2(x,x)/F_1(x))^{m-1} \quad \text{for all } m \geq 1 \end{aligned} \tag{7.26}$$

The following is limited to two cases. Using the bivariate distribution, the Normal distribution or its transformations can be a reasonable model for the annual values. Using standardized values X, $F_1(x)$ and $F_2(x,x)$ are then expressed by:

$$F_1(x) = 1/\sqrt{2\pi} \int_{-\infty}^{x}(-z^2/2)dz \tag{7.27}$$

$$F_2(x,x) = \frac{1}{2\pi\sqrt{1-\rho^2}} \int\limits_{-\infty}^{x}\int\limits_{-\infty}^{x} \exp\left\{-\left[z_1^2 - 2\rho z_1 z_2 + z_2^2\right]/\left[2(1-\rho^2)\right]\right\}dz_1 dz_2 \quad (7.28)$$

where ρ is the autocorrelation between adjacent years. Ambramovitz & Stegun (1972) demonstrate how to numerically evaluate the bivariate Normal distribution, as previously mentioned in Section 7.3, with $F_{X_t,X_{t+1}}(x_{50},x_{50}) = \frac{1}{4} + \arcsin(\rho_1)/2\pi$ as a special case when the threshold is equal to the median x_{50}.

7.9.2   A discrete formulation – A Markov chain

The situation of a Markov chain, i.e. a discrete random process in discrete time, can be simplified by introducing two states: '0', a dry year ($x \le X$), and '1', a not-dry year ($x > X$). The probability that the process at time $t = n$ is in a certain state can be written as:

$$p(n) = Pr\{X(n) = i\}, \qquad i = 1,2 \tag{7.29}$$

which is called the absolute probability. Also of importance are probabilities of the type:

$$p_{ij}(n) = Pr\{X(n) = i | X(n-1) = j\}, \qquad i, j = 1,2 \tag{7.30}$$

These are named transition probabilities and thus give the conditional probability that a process is at state i at time $t = n$ when it was in state j at time $t = n - 1$. Assuming stationarity the following can be written:

$$p_{ij}(n) = p_{ij} \tag{7.31}$$

The matrix

$$\mathbf{P} = \begin{pmatrix} p_{00} & p_{01} \\ p_{10} & p_{11} \end{pmatrix} \tag{7.32}$$

is called the transition matrix. The sum of rows in this matrix is always equal to 1. $\mathbf{P}$ can thus be written in the following form:

$$\mathbf{P} = \begin{pmatrix} 1-\beta & \beta \\ \alpha & 1-\alpha \end{pmatrix} \tag{7.33}$$

It can be shown that the absolute probabilities after the process has been observed during a long time period are:

$$p_0 = \alpha/(\alpha + \beta), \quad p_1 = \beta/(\alpha + \beta) \tag{7.34}$$

Using the cumulative distribution functions of X in one and two dimensions the theoretical absolute and transition probabilities are given by:

$$
\begin{aligned}
p_0 &= F(x) \\
p_1 &= 1 - F(x) \\
p_{00} &= F(x,x)/F(x) \\
p_{01} &= (F(x) - F(x,x))/F(x) \\
p_{01} &= (F(x) - F(x,x))/(1 - F(x)) \\
p_{11} &= (F(x) - 2F(x) + F(x,x))/(1 - F(x))
\end{aligned}
\tag{7.35}
$$

The probability Φ that a period of m years will be observed in state '0' can now be calculated, and is given by:

$$\Phi = p_1 p_{10} (p_{00})^m p_{01} \tag{7.36}$$

Inserting the expressions for the absolute and transition probabilities of Equation 7.34 into this formula gives Equation 7.26.

7.9.3   Distribution of run length M and dry period K

The probability density function (pdf) $f_M(m)$ of the run length M, the number of successive time steps where $X \leq x$ follows directly from Equation 7.26. This latter is written down as the joint probability:

$$Pr\{M = m, X_1 > x, X_2 \leq x\} = \frac{(F(x) - F(x,x))^2}{F(x)} \left(\frac{F(x,x)}{F(x)} \right)^{m-1} ; \quad m \geq 1 \tag{7.37}$$

The pdf now follows immediately as the corresponding conditional probability:

$$f_M(m) = Pr\{M = m | X_1 > x, X_2 \leq x\} = \frac{(F(x) - F(x,x))}{F(x)} \left(\frac{F(x,x)}{F(x)} \right)^{m-1} ; \quad m \geq 1 \tag{7.38}$$

The corresponding cumulative distribution function (cdf) then becomes:

$$F_M(m) = \sum_{j=1}^{k} f_M(j) = 1 - \left(\frac{F(x,x)}{F(x)} \right)^m \tag{7.39}$$

Note that this is a geometric distribution with parameter $p = (F(x) - F(x,x))/F(x)$. Accordingly the mean and variance of this distribution become:

$$E\{M\} = \frac{F(x)}{F(x) - F(x,x)}$$

$$Var\{M\} = \frac{F(x)F(x,x)}{(F(x) - F(x,x))^2}$$

(7.40)

These relations have already been used when deriving crossing properties in Sections 7.2 and 7.3 for the Göta älv data. This set of data can now be used to calculate the pdf of run lengths. The result is shown in Figure 7.9 where the theoretical pdfs of run lengths related to the 50% and 10% quantiles, respectively, are compared with those derived from the observed data. From a statistical point of view the comparison might be unfair as the total number of

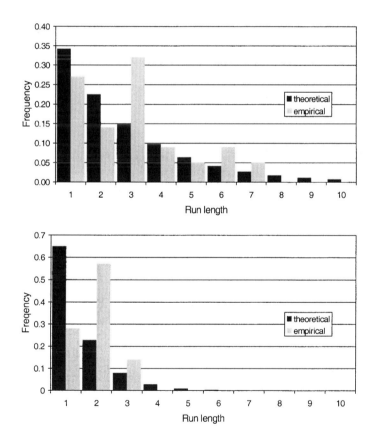

Figure 7.9 Theoretical and empirical distribution functions of run length for a threshold equal to the median (x_{50}, upper) and the 10% quantile (x_{10}, lower), respectively, based on the 131 years of data from Göta älv (River Göta).

observed runs is very small, 22 and 7, respectively. The deviations can be interpreted as a sign of the Joseph effect quoted earlier (Section 7.5), i.e. the tendency towards long sequences of dry (and wet) years.

The probability density function (pdf) $f_K(k)$ for the time interval $K (= M + 1)$ between successive events $X < x$ is easily derived from the general expressions above (Equations 7.26 and 7.34) as follows:

$$f_K(k) = Pr\{K = k | X_1 > x\} = \begin{cases} \dfrac{(1 - 2F(x) + F(x,x))}{1 - F(x)} & ; \quad k = 1 \\[3mm] \dfrac{(F(x) - F(x,x))^2}{F(x)(1 - F(x))} \left(\dfrac{F(x,x)}{F(x)} \right)^{k-2} & ; \quad k > 1 \end{cases} \qquad (7.41)$$

The corresponding cumulative distribution function (cdf) then becomes:

$$F_K(k) = \sum_{j=1}^{k} f_K(j) = 1 - \frac{(F(x) - F(x,x))}{(1 - F(x))} \left(\frac{F(x,x)}{F(x)} \right)^{k-1} \qquad (7.42)$$

and the mean and variance are:

$$E\{K\} = \sum_{k=1}^{\infty} k f_K(k) = \frac{1}{1 - F(x)}$$

$$Var\{K\} = \sum_{k=1}^{\infty} (k - E\{k\})^2 f_K(k) = \frac{2F(x)^2}{(F(x) - F(x,x))(1 - F(x))} - \frac{F(x)}{(1 - F(x))^2} \qquad (7.43)$$

7.9.4   Time to first upcrossing L

To calculate the probability distribution of the time period L to the first upcrossing of the level x one can proceed in the following way. Starting from the trivial situation when $X_1 > x$ at time step 1, the corresponding conditional probability is then expressed by:

$$Pr\{L = l | X_1 > x\} = \begin{cases} 1; & l = 1 \\ 0; & l > 1 \end{cases} \qquad (7.44)$$

For the case when $X_1 < x$ the fact that the remaining part of the run length, denoted M^*, has the same distribution as the entire length can be used. The conditional probability for this case becomes:

$$Pr\{L = l | X_1 \le x\} = \begin{cases} 0; & l = 1 \\ Pr\{M^* = l - 1\} = f_M(l-1); & l > 1 \end{cases} \qquad (7.45)$$

The pdf is now obtained as:

$$
\begin{aligned}
f_L(l) &= Pr\{L = l\} \\
&= Pr\{L = l | X_1 > x\}Pr\{X_1 > x\} + Pr\{L = l | X_1 \le x\}Pr\{X_1 \le x\} \\
&= \begin{cases} 1 - F(x); & l = 1 \\ (F(x) - F(x,x))(F(x,x)/F(x))^{l-2}; & l > 1 \end{cases}
\end{aligned}
\tag{7.46}
$$

The corresponding cdf, mean and variance are:

$$
F_L(l) = Pr\{L \le l\} = 1 - F(x)\left(\frac{F(x,x)}{F(x)}\right)^{l-1}
\tag{7.47}
$$

$$
E\{L\} = 1 + \frac{F(x)^2}{F(x) \quad F(x,x)}
\tag{7.48}
$$

$$
Var\{L\} = \frac{F(x)^2 \left(F(x) + F(x,x) - F(x)^2\right)}{(F(x) - F(x,x))^2}
$$

In case of independence between successive events the bivariate distribution is found as $F(x,x) = F(x)^2$. Insertion into Equation 7.45 gives the following pdf for this case:

$$
f_L(l) = \begin{cases} 1 - F(x); & l = 1 \\ (1 - F(x))F(x)^{l-1}; & l > 1 \end{cases}
\tag{7.49}
$$

This is identified as the geometric distribution, which has the mean and variance:

$$
E\{L\} = \frac{1}{1 - F(x)}
\tag{7.50}
$$

$$
Var\{L\} = \frac{F(x)}{(1 - F(x))^2}
$$

Note that the mean value $E\{L\}$ by definition is the return period for the event $X > x$. The equivalent expression in case of dependent data is $E\{K\}$ (Equation 7.42), which has the identical expression. This is in spite of the introduced persistence of the random process. The return period of an event $X > x$ can only be a function of the marginal distribution $F(x)$.

7.10   Summary

A first introduction to the art of random simulation was presented, containing several illustrations of observed and simulated time series of steamflow. It is very important to examine data before turning to a more formal analysis and test of performance. The questions posed in the chapter were: (a) in what manner are the properties of time series similar or even identical, and (b) which properties are we able to reproduce by a random model? It was initially stated that random models are able to reproduce times series that are statistically indistinguishable from observed ones, however this is neither an exact reproduction of past events nor the prediction of the future. The similarity lies in the statistical parameters of the series. The series generated by a random model can only reproduce a limited set of statistics of the historical series. In this textbook the focus is directed towards the ability of a model to preserve the moments (mean value, variance, lag-one autocorrelation) and number and durations of non-exceedance of given threshold levels (crossing properties).

Hydrological time series show different patterns of variations in different timescales – from long-term patterns in annual data, seasonal variations in monthly data and down to the sawtooth pattern of daily data. The text has been structured in accordance with this division into timescales. This also implies a gradual increase in the complexity of the time series models developed in order to take account of the physical processes involved in runoff formation. The complexity also increases when turning from stationary time series, which can be adequate at the annual time scale, to non-stationary time series. The latter are applicable when seasonal variations occur at monthly and daily timescales. Basically only two simple random models are treated – a random sample model generated by a sequence of independent random numbers belonging to a given theoretical distribution function, and a first order autoregressive model. The latter was identified from an investigation of the role of storage. Data generated from this model show dependence reflected by the autocorrelation in data which also influence their statistical properties. Random models for daily data were only briefly dealt with.

The random sample model and the first order autoregressive model correspond to a representation of a random process by univariate and bivariate distribution functions respectively. Finally, analytical expressions of the distributions of low flow and drought characteristics were derived such as run lengths and number of upcrossings (downcrossings) by the use of this latter representation.

8

Regionalization Procedures

Siegfried Demuth, Andrew R. Young

8.1 Introduction

Knowledge of the spatial distribution of low flow and drought is crucial to understanding the regional context. This know-how is also essential for predicting low flow and drought at the ungauged site, or at sites where data are incomplete or unavailable for a failure of instrumentation. Methods for extrapolation in space are commonly termed *regionalization procedures*. Considerable effort has been made to elaborate regionalization procedures to estimate low flow indices at the ungauged site. Within the framework of the FRIEND (Flow Regimes from International and Experimental Network Data) project, regionalization procedures have been developed and tested at regional and global scale. In this project, models and analysis techniques were exchanged to interpret the results using a common approach to analysing data derived from the different hydrological regions in Europe. The first phase of the project was completed in 1989 when the European Water Archive (EWA) was established containing flow data from about 1500 gauging stations. Today there are over 5600 stations available to study the variability of hydrological regimes across Europe (Gustard, 1993; Rees & Demuth, 2000; Gustard & Cole, 2002).

Regionalization is introduced in Section 5.4.3; it focuses on analysing regional properties of drought. Section 6.7 continues by describing regional estimation procedures related to frequency analysis. This chapter focuses on regional estimation procedures in general, and more specifically on their application to estimation of low flow indices at the ungauged site. A prerequisite for the application of a regionalization procedure is the homogeneity of the region under consideration; that means that the catchment response (flow) is similar with respect to climate, geology, land use and soil. The various homogeneity approaches are described in Section 6.7.4. The identification of homogeneous regions is only appropriate when large regions are considered. In practice the application is dependent on the number of gauging stations available.

The most popular regional estimation tools are based either on empirical methods or statistical approaches (Stall, 1962; Goddard, 1963; Speer *et al.*, 1964; Demuth, 1993; Schreiber & Demuth, 1997; Smakhtin, 2001). In addition, detailed knowledge of the catchment properties such as morphometry, surface cover, geology, soil, and climate is necessary. The empirical methods require additional information on the hydrological regimes. The statistical approaches can be separated into correlation analysis, factor analysis, multiple regression analysis and hydrological interpolation procedures.

In many practical cases simple estimation methods are used to estimate low flows at the ungauged site. This chapter elaborates on two simple estimation procedures (Section 8.2), specific discharge estimation methods and the flow correlation approach. Both methods are, in general, used because of a lack of alternatives. Multivariate analysis is the topic of Section 8.3, which describes factor analysis as a tool to depict catchment descriptors from a large data set, multiple regression analysis, including model development and the application of the model. A self-guided tour is provided to show the reader how to estimate low flow indices at the ungauged site by means of multiple statistical regression procedures. A description is given in the book and the tour is available on the CD.

The choice of both catchment descriptors and low flow indices predicted in regional estimation methods is very variable and is dictated by several factors, such as *tradition in the country* (e.g. DVWK, 1983; Gustard *et al.*, 1992), *user requirements, objective of the study and data availability*. A comparative study was thus carried out at a global scale to present an overview of catchment descriptors commonly used (Section 8.4). Various mapping procedures are described in Section 8.5. First the topic is introduced and the scientific literature is briefly reviewed. Then a grid-based river network approach which is based on nominally-scaled catchment descriptors is described. The recession routing river network approach uses the multiple regression method to estimate recession parameters at the ungauged site. Finally, the region of influence (ROI) approach, used in the UK, is described. The chapter concludes with a summary (Section 8.6).

8.2   Simple estimation methods

Simple estimation methods comprise various empirical methods for estimating drought characteristics. These consist of basic mathematical equations, which are based on experience, i.e. empirical data. They also require detailed knowledge of the region, e.g. morphometry, surface cover, geology, soil, climate (Section 4.4.2), as well as knowledge of the system response of the region (Hayes, 1990).

Parameters used most frequently in low flow regionalization studies are *catchment area* and *stream network characteristics*. The significance of the individual descriptors depends on the objective of the study and the time and areal scale requested for the analysis. Among *geomorphologic attributes* of a watershed the *delineation* or the *average basin slope*, the *relative gradient*, the *catchment width*, the *mean basin elevation* and the *catchment length* are essential to understand the river system behaviour under drought condition. Another geomorphologic descriptor generally used as an independent parameter for low flow estimation at the ungauged site is the *stream slope*. Besides the *mainstream length* and the *length of the river network*, the *drainage density* plays an important role. The *drainage density* reflects the geological and petrographical conditions of a catchment and acts as an index representing the infiltration capacity and the transmissivity of the subsoil. In addition to physical catchment descriptors, climatic descriptors, especially rainfall are also of use in prediction equations. Here *annual average rainfall* (AAR) is mainly appropriate. Short-term rainfall measures or index of *rainfall intensity* is usually used in flood studies and not relevant if dealing with low flows.

The equations developed from the empirical procedures describe neither an exact physical relationship (as is the case with physically-based models) nor do they take into account the error term common in statistical procedures. Nevertheless, the empirical procedures attempt to grasp the deterministic influences on streamflow in order to graphically or analytically deduce the desired low flow indices from them.

The selection and the application of empirical procedures depend not only on the research objectives, the water-resources-planning tradition of a particular country, and the data available, but also on the experience of the user. An empirical procedure is often chosen for a lack of alternatives. If there are no statistical transfer functions available for the study site and if it is impossible to use other procedures (such as correlation analysis) due to financial or time constraints, an empirical procedure is often applied.

8.2.1 Specific discharge estimation methods

The specific discharge (discharge standardized by catchment area) is an important hydrological variable. Estimation of specific discharge at the ungauged site based on a nearby gauged site is a frequently used empirical procedure for estimating long-term statistics, e.g. mean, Q_{95}, $MAM(7)$. It is assumed that adjacent catchments exhibit similar behavioural patterns with

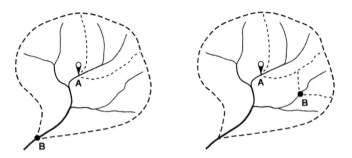

Figure 8.1 Estimation of specific discharge for a location B along the river reach (left) and for a location B in the adjacent catchment (right).

regard to runoff per unit area. Using the principle of direct proportionality, the estimated runoff at a site is a function of its corresponding catchment area. The empirical equation is:

$$Q_B = AREA_B \, Q_A \, AREA_A^{-1} \qquad\qquad (8.1)$$

where Q_B is the estimated discharge at site B, Q_A is the observed discharge in a similar (analogue) catchment, $AREA_B$ is the catchment area for which the discharge is to be estimated and $AREA_A$ is the respective gauged catchment area (Section 4.3.3). This procedure produces reliable results for intermediate flows. When it is applied to extreme events, the estimated values may differ significantly from the observed ones. The accuracy of the estimations can be improved if the site considered is located along the same stream or in a catchment with similar catchment descriptors and climate input (Dyck, 1976; Figure 8.1).

Using the least-square method, Hayes (1990) gives an example for the State of Virginia (USA) for which an empirical equation was developed, relating low flow indices to catchment area. These indices are deduced from theoretical distributions at gauged and ungauged channel sites. The equation takes the form of:

$$AM(7)_{Tj} = AM(7)_{Ti} \, [AREA_j/AREA_i]^{1.2} \qquad\qquad (8.2)$$

where $AM(7)_{Tj}$ is the T-year 7-day minima (Section 5.3.2) for site j. Site i is a site where direct measurements are taken and j is a site without direct measurements. The interval of validity for this empirical equation is given as $0.65 < [AREA_j/AREA_i] < 10.36$.

The less similar the respective catchments are with respect to their hydrological behaviour the more important the consideration of other factors in addition to catchment area becomes. The integration of average annual rainfall

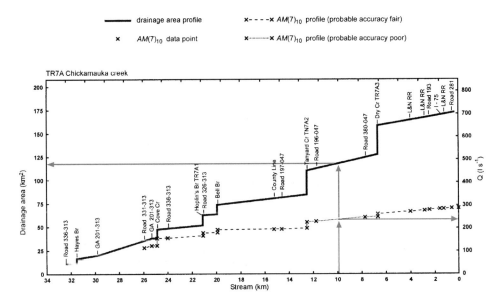

Figure 8.2 Flow accretion diagram (modified from Carter *et al.*, 1988).

into the formula generally improves the estimation. The empirical relationship can now be written as:

$$Q_B = AAR_B \ AREA_B \ Q_A \ AAR_A^{-1} \ AREA_A^{-1} \qquad (8.3)$$

where AAR_B is the average annual rainfall in the ungauged catchment and AAR_A is the average annual rainfall in the gauged catchment.

Equations 8.1–8.3 allow for estimation of low flow indices at ungauged sites both upstream and downstream. It is based on the assumption of similar physiographical properties in the catchments under study. If the areas to be investigated differ significantly from each other with regard to these properties, installing a temporary gauging station is recommended or the use of spot gaugings. Nevertheless, the comparability of the low flow regimes would no longer be ensured.

To obtain information about the low flow regimes of the tributaries of the Tennessee River in Georgia (USA), Carter *et al.* (1988) conducted a study in which the index $AM(7)_{10}$ was calculated for different locations along the river and plotted versus distance along the channel. Interpolating the individual $AM(7)_{10}$ values as a function of catchment area created a continuous low flow profile (flow accretion diagram). From this profile it is possible to deduce streamflow from the corresponding area of the catchment, and a statement regarding the reliability of the available data for every point along the channel

(Figure 8.2). Several authors report similar ways of deriving low flow indices by plotting low flows versus distance along the channel (Riggs, 1972; Browne, 1980; Carter *et al.*, 1988; Domokos & Sass, 1990; Gottschalk & Perzyna, 1993). The methodology is referred to as 'flow-line technique' or 'low-flow profiles'.

Fuchs & Rubach (1983) and Browne (1980) present examples of runoff-per-unit-area estimations for low flows in a catchment of the River Leine in Lower Saxony (Germany) and of a region in southwest England, respectively. By plotting runoff versus channel length they produced a specific low flow profile, from which the specific low flow values for all cross sections of the channel network can be extracted. Both authors applied multiple regression models and concluded that the specific low flow $AM(7)_2$ is directly proportional to the catchment area. Laaha (2003) and Laaha & Blöschl (2003b) present a regional regression model of specific low flow discharge q_{95} (Q_{95} standardized by catchment area) in Austria, based on 325 catchments which had been grouped into eight hydrologically similar regions. In all regions, no significant influence of catchment area on specific low flows was found. The occurrence of regional flow systems (Section 3.2) is likely to cause this.

8.2.2   Flow correlation

The flow-correlation method is applied when low flow indices cannot be estimated by means of other parameters, such as catchment descriptors. The flow correlation links long-term to short-term observations. As a result, one gets an estimate of low flow indices for a river for different locations. The correlation coefficient is used as a measure of the relationship between the observations. The most commonly used measure is Pearson's correlation coefficient, r_P, which is defined as:

$$r_P = \frac{1}{n-1}\sum \left[\frac{(x_i - \bar{x})}{s_x} \cdot \frac{(y_i - \bar{y})}{s_y} \right] \tag{8.4}$$

where s_x and s_y are the sample standard deviations of x and y, respectively.

Since r_P measures the linear association between two variables it is also called the linear correlation coefficient. The coefficient gives a measure of the strength of the linear relationship between a random variable y and a second variable x and is invariant to scale changes. If the data lie exactly on a straight line with positive slope, r_P equals 1. The assumption of linearity makes inspection of a plot important because a non-significant value of r_P may not only result from independence of the variables but may also be due to curvature or a great number of outliers. Pearson's r_P is not resistant to outliers because it is computed using non-resistant measures, such as means and standard deviations (Section 5.5.1).

A more robust method is Kendall's τ, which measures the strength of the monotonic relationship between x and y. The method is used for ordinal scaled variables. Kendall's τ uses a comparison of different pairs of parameters and looks into the ranking and tests whether the rank of the pairs are concordant or disconcordant, whereas Spearman's rank correlation is based on the differences between the ranks for the variables x and y. Since the procedure is based on a ranking of the values it is not sensitive to the effect of a small number of outliers. It is especially suited for variables which are skewed. Since τ depends on the ranks of data only, it can be implemented even in cases where some of the data are censored. Strong linear correlations of 0.9 or above correspond to τ values of about 0.7 or above. The lower values of τ do not mean it is less sensitive than r_P, but that a different scale of correlation is being used. Kendall's τ is advantageous since it is easy to compute, resistant to outliers and suitable for linear and non-linear correlations (Kendall, 1975).

According to Dyck (1976), the following aspects have to be considered when using correlation analysis: (a) missing low flow data for sub-catchments, (b) reliable low flow measurements with adequate accuracy, (c) knowledge of possible hydrogeological variability within the area (since correlations should only be performed for hydrogeologically homogeneous regions), and (d) location of the permanent gauges and evaluation of their suitability as reference gauges. Prior to these steps a close inspection of the data with respect to inconsistencies, errors, and stationarity has to be carried out. If data errors are obvious, external information, e.g. from the data-holding institution, should be requested and the errors should be corrected. In certain cases the rating curve can expose inconsistencies in the data set, while in other cases the gauged station may have to be excluded from the data set (Section 4.3.1).

After temporarily supplementing existing flow measurements by installing additional gauges the established correlations can be used to estimate the desired statistics. The number of temporary gauges may exceed the number of existing structures by several multiples. The number and spatial distribution of additional gauges should be based on local conditions and spatial distribution of the existing gauges. For the estimation of low flow indices Dyck (1976) recommends temporary gauging stations to be run between three and five years for humid regions. In other climates of the world it might be different (Section 2.2.3). If two or three measurements during low flow periods are taken each year, eight to ten values will be available for correlation analysis. When the measurements are complete a correlation analysis is performed on the basis of the measured data. This analysis allows low flow indices for the respective areas to be estimated. Further details regarding the selection of temporary

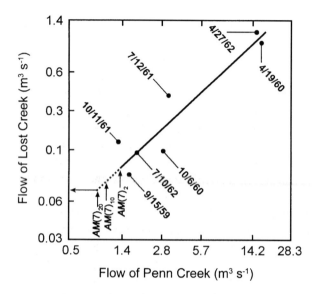

Figure 8.3 Estimation of low flow indices (from Riggs, 1985).

gauging stations and their operation are found in Riggs (1985). The correlation can also be analysed graphically by plotting the two datasets logarithmically. Dyck (1976) and Helsel & Hirsch (1992) provide further details on correlation analysis and the required preparatory work.

If partial data sets are available, e.g. samples from a larger data set or data sets of an observation period of less than ten years, low flow indices can be determined by correlating these data sets with longer observation periods. An example is given by Riggs (1985), who studied a river system in the United States. He plotted streamflow from partial observation series for Lost Creek versus the continuous data set for Penn Creek (Figure 8.3). A straight line was graphically fitted to the data to transfer the low flow with a return period of ten years from the gauging station with continuous observation data to the station where only samples were taken.

The position of the transfer line can also be determined by mathematical fitting procedures (Friel *et al.*, 1989; Hayes, 1990; Helsel & Hirsch, 1992). The accuracy obtained with these methods increases with increasing similarity of the flow regimes of the rivers. The greater the variation of the low flow indices, the less reliable the transfer function is (Figure 8.3). The results remain questionable, particularly when an extrapolation beyond the observed interval is required to determine e.g. $AM(7)_2$ and $AM(7)_{10}$ (annual minimum 7-day flow with a return period of 2 and 10 years, respectively).

In their investigation regarding the estimation of low flow indices by means of correlation methods for three test sites in former East Germany, Glos & Lauterbach (1972) indicate the magnitudes of the estimation errors. The average error for the estimation of the annual minimum 7- and 15-day flow with a return period of 2 and 20 years was found to range from 15 to 35%. Generally, the average error related to the estimation of low flows increases as the return period increases. Riggs (1972) found that the error related to the estimation for shorter intervals, such as the *AM*(7) (annual minimum 7-day flow), was greater than the error related to the estimation for longer intervals, such as the *AM*(30).

Dyck (1976) emphasized that it is difficult to interpret the correlation coefficients between the variables. Due to imprecise knowledge about process-response relationships, careful analysis of the established correlations and regressive connections is indispensable. Among other factors, the success of correlation analysis depends on whether or not the observation period contains low flow events. If this is not the case, the low flow indices calculated from the observed data tend to be too high. Among the disadvantages of flow correlation are the extensive financial and manpower requirements as well as the fact that these estimations only become available after a certain observation period.

8.3   Multivariate analysis

Multivariate analysis, and especially regression analysis, has frequently been applied in the field of hydrology since the mid 1960s. The estimation of flow indices by means of regression techniques is performed using catchment descriptors, which can be deduced from appropriate maps (Section 4.4.2). In contrast to the flow correlation, results can be obtained after a relatively short time period. The regression method seeks to deduce a model by the linear combination of catchment descriptors which maximizes the amount of explained variance. The choice of catchment descriptors depends on the resources available (e.g. availability of digital maps) and the experience of the person in charge. It is in most cases subjective and therefore open to criticism.

An objective method to select catchment descriptors from a large data set is factor analysis. The resulting data sets can then be used in multiple regression models. The method is described in Section 8.3.1. Section 8.3.2 describes the multiple regression approach together with the assumptions and the different algorithms needed to build the best model. Examples from Germany and Austria are given to demonstrate the power of the methods. Section 8.3.3 introduces a self-guided tour, in which multiple regression model building, calibration, validation and finally application to a case are described (Self-guided Tour, CD).

8.3.1   Selection of catchment descriptors

Factor analysis is an objective method of selecting catchment descriptors from a larger sample (Section 5.5.2). The main objectives of the technique are (a) to *reduce* the number of variables, and (b) to *detect the structure* in the relationships between variables, that is to *classify variables*. Therefore factor analysis is applied as a data reduction or structure detection method. The idea is to identify inter-correlated groups of variables and assign these groups to independent factors. Every variable will be more or less strongly correlated with the assigned factor. The strength of the relationship can be quantified by a correlation coefficient and is expressed as factor weight. This allows for an evaluation of the variables with respect to their grouping with linearly independent factors. Based on the different factor weights, it is possible to determine a representative catchment descriptor, which can be used as a substitute for this factor and for the variables assigned to this factor (Überla, 1968; Bahrenberg & Giese, 1975; Haan, 1982). This association, however, may not necessarily lead to model improvement. The method is also subject to debate because its results are difficult to interpret (Matalas & Reiher, 1967; Dyck, 1976).

Several studies used factor analyses to estimate low flows and groundwater heads at the ungauged site (e.g. Dreher *et al.*, 1985; Nathan & McMahon, 1990a; Demuth, 1993). Demuth (1993) applied factor analysis to a case study in Germany to identify catchment descriptors to be used in a regression model for the estimation of base flow, BASE (Section 5.3.3), at the ungauged site. Here, log-transformed catchment descriptors were used, and an overview of the descriptors is given in Table 8.1. The study was carried out for ten catchments in the south of the Black Forest in Germany.

To extract the factors a principal component analysis (PCA, Section 5.5.2) was applied, followed by a rotation according to the 'varimax' criteria (variance maximizing) of the original variable space. PCA is one method of factor analysis, which looks at the total variance among the variables. The criterion for the rotation is to maximize the variance (variability) of the 'new' variable (factor), while minimizing the variance around the new variable. In general the most significant loadings are on one factor and therefore the interpretation is very difficult. The rotation helps to distribute the loadings between the factors and eases the interpretation. Basically there are several steps for factor analysis: (a) data collection and generation of the correlation matrix, (b) extraction of the initial factor solutions, and (c) rotation and interpretation. Table 8.1 shows the calculated and rotated factor pattern for the four extracted factors (matrix of the factor weight). The factor weights are used for interpretation of the factors.

Table 8.1 Rotated factors and eigenvalues (from Demuth, 1993)

Catchment Descriptors	Acronym	Factor 1 'Scale'	Factor 2 'Climate'	Factor 3 'Relief'	Factor 4 'Cover'
Catchment length	LE	0.92992	−0.30718	0.08455	−0.08629
Main stream length	MSL	0.90046	−0.34330	−0.03129	−0.15026
Catchment area	AREA	0.79323	−0.53937	0.01352	−0.26733
Length of the river network	LG	0.69094	−0.62141	0.10969	−0.30688
Drainage density	DD	−0.83665	0.04251	0.32575	0.02491
Annual average rainfall	AAR	−0.12135	0.95914	0.14337	0.17250
Mean basin elevation	HMEAN	−0.35005	0.90777	0.05606	−0.04796
Catchment width	BE	0.54221	−0.68670	−0.05794	−0.40485
Mean slope of the catchment	NEIG	0.03837	0.02361	0.92989	0.03560
Relief ratio	RELIEF	0.44570	−0.13153	0.86337	0.04029
Stream slope	SL	−0.38499	0.20994	0.85978	0.22859
Relative gradient	RR	−0.51466	0.18611	0.79729	0.13152
Percentage of forest	FOREST	−0.17603	−0.17845	0.22708	0.92897
Eigenvalues		4.558	3.235	3.190	1.328
Proportion of total variance (%)		35.1	24.9	24.5	10.2
Cumulative proportion of total variance (%)		35.1	60.0	84.5	94.7

Factor 1 has a high weight with respect to the catchment descriptors LE, MSL, AREA, LG, and DD, which describe scale-related attributes and are therefore characterized as 'scale factor'. The catchment descriptors AAR and HMEAN have a high weight with regards to Factor 2, which is allocated the title 'climate factor'. Factor 3 is dominated by NEIG, HE, SL, and RR and is therefore considered the 'relief factor'. Since FOREST has the highest load in Factor 4, it is called 'cover factor'. The four factors explain about 95% of the total variance of the variables analysed (catchment descriptors), of which about 35% of the variance is explained by the scale factor, 25% by the climate factor, 25% by the relief factor, and finally 10% is explained by the cover factor.

For each factor an appropriate descriptor was identified, which represents the factor in the regression model. As Factor 1 (scale) the drainage density DD was selected, because the high loads of catchment length, LE, main stream length, MSL, and catchment area, AREA, are influenced by the standardization of the base flow. The average annual rainfall, AAR, represents the Factor 2, the mean slope of the catchment, NEIG, Factor 3, and the percentage of FOREST Factor 4. The factors are each represented by the one descriptor, which has the highest weight within the respective factor groups. Finally, a regression analysis was carried out based on the descriptors found by factor analysis. The result is summarized in the equation:

$$BASE = 1.08 \cdot 10^{-6} \, AAR^{1.7} \, FOREST^{0.98} \, DD^{-0.29} \, NEIG^{-0.07} \qquad (8.5)$$

The coefficient of determination, R^2, is 84%. R^2 measures the proportion of total variation about the mean explained by the regression. BASE is derived from monthly minimum flows through a separation procedure developed by Demuth (1993). AAR explains 55%, FOREST 24%, DD 4%, and NEIG 1% of the total variance. In this example, factor analysis proved a statistically-objective tool to select appropriate catchment descriptors for multiple regression analysis.

8.3.2   Multiple regression analysis

Multiple regression analysis is a simple method, which allows for the estimation of streamflow at ungauged sites. The method is among the more frequently-applied statistical procedures in hydrology and water-resources management. The objective of this method is to estimate a flow index at the ungauged site through establishing a relationship between the respective index and catchment descriptors. The value to be estimated could be mean flow, a flood index, or, as in this context, a low flow index. This is called the dependent variable, or the response variable. The catchment descriptors comprise the independent variables (or predictor variables) of the relationship to be established. They account for various attributes of the catchment, such as morphometry, surface cover, geology, soil and climate (Section 4.4.2). The prerequisites for the application of multiple regression analysis are listed in Box 8.1. If large regions are under investigation, methodologies to delineate homogenous regions are used (Section 6.7.4) and regression analysis is applied to the regions found. However, in most practical applications the number of stations is limited and therefore a subdivision in homogenous regions is not appropriate.

The relationship between low flow indices and catchment descriptors is described in a mathematical equation, which is also called the transfer function. The general form of a regression equation reads:

$$Y_i = b_0 + b_1 X_{1i} + b_2 X_{2i} + \ldots + b_p X_{pi} + e_i \qquad (8.6)$$

Box 8.1

Basic requirements for multiple regression models

The constants or coefficients of a multiple regression model are usually estimated by the least-square method. This requires:

a) *Adequate description*
   An adequate description of the relation between the flow index and the catchment descriptors *through Equation* 8.6.

b) *No specification errors*
   The assumption that the parameters are free of specification errors presupposes that the selected model, as represented by the regression equation, is correct.

c) *No measurement errors*
   This requirement is obvious. If the measurements are already inaccurate, it is very likely that the estimated values will be inaccurate as well. The reliability in the estimation of stream-flow indices depends, among other factors, on the quality of the measurements at the gauging station (e.g. hydraulic conditions of the cross-section) and on the length of the observation period (Chapter 4).

d) *Homoscedasticity*
   Homoscedasticity is an important requirement, which means that the variance of the error attached to the estimated values is constant. In other words, the variance of Y_i must remain constant for all i. If the requirement of homoscedasticity is violated, minimal variance has not been attained for the estimated parameters, when a suitable model is being built. Therefore the general procedure with regard to the t-test and the F-test as well to the confidence intervals are no longer valid. The t-test is used to test the correlation coefficient against 0. The F-test is a useful criterion for adding or removing terms from the model. A closer look at the estimation error and residuals helps to evaluate the regression model (Draper & Smith, 1998).

where Y_i is the value to be estimated at catchment i, usually a low flow index (Section 5.3), given as $m^3 s^{-1}$ or $1 s^{-1} km^{-2}$, and $b_0, b_1, ..., b_p$ are constants or coefficients. The constants are determined based on the properties of the sample. X_{pi} indicates the catchment descriptor number p for catchment number i; n is the total number of catchments in the sample and p is the total number of catchment descriptors. Finally, e_i is the error term. In practice, the equation often takes either a logarithmic or an exponential form, which will be discussed later. Furthermore, the interpretation of the multiple regression equation is simplified if the predictor variables are independent, which means

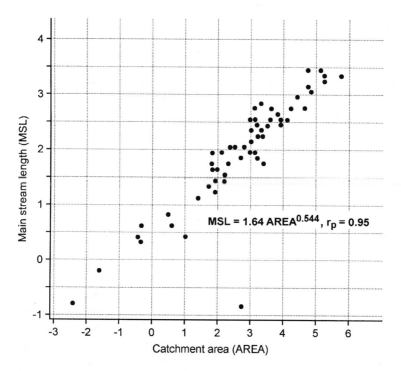

Figure 8.4 Relationship between main stream length, MSL, and catchment area, AREA, (logarithmic scale, from Demuth, 1993).

that the correlation coefficient between any pair of predictor variables equals 0. The problem of intercorrelation has been the subject of many studies (e.g. Draper & Smith, 1998; Tallaksen, 1989; Demuth, 1993). The mutual dependency between different features of a catchment, as expressed in the intercorrelation, as well as in the underlying factors which affect these features, form a multi-level, complex structure of interactions. Therefore, it is very difficult to ensure the independence of predictor variables when estimating regression coefficients (Weisberg, 1985). An example of the intercorrelation between catchment descriptors is shown in Figure 8.4. The data are taken from 58 small catchments in Western Europe, selected from the European Water Archive.

The correlation coefficient, r_P, describing the relationship between the logarithm of the main stream length, MSL, and the logarithm of the catchment area, AREA, is 0.95. Since the two variables are not independent, they should not both be used in the same regression equation. Regression equations, which contain intercorrelated variables, produce unreliable parameter estimates.

Lewis-Beck (1986) suggests not only checking for inter correlation between two individual variables but also performing a multi-colinearity test among all the predictor variables. As the upper limit for multi-colinearity, Demuth (1993) used a coefficient of determination of 0.80. If a combination exceeds this limit, at least one of the predictor variables must be excluded from the predictor set.

Streamflow data are rarely normally distributed, which may lead to unsatisfactory residual diagnostics. In such cases, a logarithmic transformation may help to satisfy the regression assumption (Box 8.1). The log-transformed regression equation has the form:

$$\log Y_i = b_0 + b_1 \log X_{1i} + b_2 \log X_{2i} + \ldots + b_p \log X_{pi} + \varepsilon_i \tag{8.7}$$

The algebraically equivalent form when logarithms of base 10 are used in the transformation and Equation 8.7 is re-transformed to original units, is:

$$Y_i - 10^{b_0} (X_{1i}^{b_1})(X_{2i}^{b_2}) \ldots (X_{pi}^{b_p}) 10^{\varepsilon_i} \tag{8.8}$$

Based on the multiple regression approach, Demuth (1993) developed regression models to estimate base flow, BASE, at ungauged sites for several regions in Europe. In one experiment the regression approach was based on a total number of 54 small research basins. Prior to the experiment the quality of both flow and thematic data had been checked (Section 4.4.2). The resulting model was derived by a stepwise procedure (Box 8.2) and the data were log-transformed to account for the deviation from normal distribution. The model has the form:

$$BASE = 4.2 \cdot 10^{-5} \, HMEAN^{0.2593} \, DD^{-0.308} \, AAR^{1.477} \tag{8.9}$$

where HMEAN is the mean altitude of the catchment; DD is the drainage density and AAR is the average annual rainfall. The coefficient of determination, R^2, is 81%, which means that 81% of the total variation of the base flow is explained by the three catchment descriptors. AAR is the catchment descriptor with the highest weight, whereas drainage density and mean altitude are only of minor importance.

A different implementation of the regression approach to predict Q_{95} in Austria was carried out by Laaha (2003) and Laaha & Blöschl (2003b). Available data consist of daily streamflow and catchment descriptors from 325 catchments. Low flow indices were standardized by catchment area to eliminate the predominant effect of catchment size on low flow discharges. Initially, a global regression model was fitted to the data using a stepwise variable selection. The model includes eight variables representing topography, precipitation and catchment geology. The coefficient of determination, R^2, is 60%. Model quality was assessed by cross-validation, leading to a cross-validated coefficient of determination, $Cv\text{-}R^2$, of 57%. Cross validation generally denominates procedures by which data are divided into two subsets, one data set is used to calibrate the model, and the other one to validate the data set and assess the model performance. One approach, especially efficient for small data sets, is leave-one-out cross-validation. For n data points, the procedure consists of n steps. For each step, one data point is factored out (this observation will be reserved for validation), the model is refitted to the remaining $n - 1$ data points and used to estimate the value of the one remaining observation. The assessment of model performance from the residuals between so obtained estimates and the original measurements fully emulates the case of ungauged catchments.

Since the study area shows a strong hydrological heterogeneity, regional regression (i.e. a set of regional restricted regression models, one for each group of hydrologically similar catchments) appeared more suitable than a global regression model. Different classification techniques included (a) the residual pattern approach based on residuals of a global regression model (e.g. Hayes, 1990), (b) weighted cluster analysis based on the most relevant catchment descriptors (Nathan & McMahon, 1990b), (c) the regression tree (Laaha, 2002), and (d) seasonality-based grouping into eight homogeneous regions (Laaha & Blöschl, 2003a). These were compared to obtain the optimal grouping which yielded the best regional regression model. Among all the techniques, seasonality-based grouping performed best, leading to a rise in model quality to $Cv\text{-}R^2 = 70\%$.

The residual pattern approach ($Cv\text{-}R^2 = 64\%$) and the regression tree ($Cv\text{-}R^2 = 63\%$) had a somewhat smaller effect on model quality; the effect of the weighted cluster analysis was only marginal ($Cv\text{-}R^2 = 59\%$). One reason for the strength of seasonality-based grouping is the use of different information for catchment classification and regression, whereas the remaining techniques are based on flow and catchment descriptors and hence use the same information for catchment grouping and regression. Furthermore, grouping techniques are better suited for information provided by single, significant indices (e.g. seasonality indices) than for information which is extremely scattered on a large number of catchment descriptors.

In a more detailed study, the variance of the seasonality approach was examined (Laaha, 2003). Results indicated that a coarse classification of summer and winter seasonality only led to a marginal improvement in model quality ($Cv\text{-}R^2$ increased up to 60%). The reason for this is that different underlying processes of summer and winter low flows were implicitly represented in the global model as well, due to high correlation between seasonality and independent variables, such as catchment altitude. Models which explicitly account for seasonality, however, appear more plausible, since rather homogeneous landscapes are better represented than highly heterogeneous landscapes.

A finer classification into regions indicating similar seasonal distribution of low flows only had a marginal effect on model quality when it was included as one factor variable in the global regression model (Cv-R^2 = 58%). However, it had an important influence on model quality when it was used as the basis for regional regression (Cv-R^2 = 70%) (Laaha & Blöschl, 2003a). As a result, seasonality regions do not necessarily give different estimates of specific low flows, but exhibit different processes. Therefore, adequate regionalization should take seasonal differences of low flows into account.

So far, many authors have selected the set of predictor variables intuitively in order to avoid the problem of multi-colinearity (linear combination of all variables) when building the best regression model. Subjective procedures have, however, been harshly criticized. Because an objective method is recommended, numerous procedures for the extraction of a set of predictor variables have been suggested. More objective ways of selecting predictor variables through the process of model building include the step-by-step regression with various algorithms: (a) forward selection, (b) backward elimination procedure, and (c) stepwise technique (Box 8.2). Principal component analysis and factor analysis are methods which are used prior to regression analysis (Section 8.3.1).

The absence of multi-colinearity is not the only criterion for selecting a set of catchment descriptors from the pool of available predictor variables. By including as many variables as possible into the regression model the variance of the target value can be explained to a large degree. This is an acceptable method if the aim of the regression analysis is to produce a model that is as 'good' as possible (Holder, 1985). However, if the effort required for the quantification of all variables is unacceptable due to costs or time required, this approach becomes impractical and therefore insignificant.

Another negative effect of including many predictor variables is the loss of explanatory value. As soon as the number of variables equals the number of sample values all variance can be explained without gaining insight into the process. If it is possible, however, to find a reduced set of catchment descriptors which contains almost the same information as the full set, results can not only be simplified, but the transfer function also contributes towards the understanding of the effect of certain factors on the flow value. The modeller must choose the 'best' from a possible 2^k (k = number of potential predictor variables; i.e. catchment descriptors) regression equations. For example, for the Self-guided Tour based on data from Germany with its 21 explanatory variables (Section 8.3.3), a combination of 2 097 152 models would have had to be checked. This approach is not very efficient due to the high amount of

Box 8.2

Objective methods to select catchment descriptors through the process of model building

Step-by-step regression

The step-by-step procedures are based on different sets of the predictor variables. Initially, a certain set of variables is related to the target value. This model is then compared to others, which would be created by either adding or eliminating a certain variable. There are three different algorithms:

a) Forward selection

The procedure begins by creating a simple model from the one-predictor variable, which exhibits the highest correlation with Y. As a second step, an additional variable is accepted into the model based on the following criteria: (a) it correlates best with Y, (b) its addition to the model will increase the coefficient of determination more than any other variables, and (c) based on a significance test, it is superior to the other variables with regard to model improvement.

Beginning with one variable, more variables are added until a criterion for termination is reached. This is the case if either the maximum number of variables to be allowed into the model is reached, or if there is no further variable to equal or exceed a previously determined minimum significance level. Furthermore, a negative tolerance check in which the colinearising effect of a newly adopted variable is examined can lead to termination of the procedure (Weisberg, 1985).

b) Backward elimination

The method differs from the forward selection procedure in that the starting model contains all variables, which are then eliminated through a process of several steps. In every step the one variable which least fulfils the acceptance criteria as described for the forward selection procedure is excluded from further consideration. The procedure terminates once the set is reduced to a given number of predictor variables or if the variable falls short of the significance level.

c) Stepwise technique

In contrast to the forward selection procedure, variables that have been accepted into the model do not necessarily have to be kept. After adding a new variable, the partial correlation coefficients of all variables of the model are computed and those variables whose correlation coefficients prove to be non-significant are eliminated from the model. The advantage of the stepwise technique is the fact that, unlike the forward selection procedure, it does not necessarily adhere to an established model, but continuously checks all variables. According to Draper & Smith (1998) and Holder (1985) the stepwise procedure is also superior to the backward elimination procedure because it is less susceptible to rounding errors.

processing required. The algorithms described in Box 8.2 systematically and objectively determine the 'best' model (Davis, 1973; Draper & Smith, 1998; Haan, 1982; Weisberg, 1985; Holder, 1985).

Even though multiple regression analysis is an important tool for the estimation of low flow indices at ungauged sites, certain restrictions apply. The procedure may be used to quantify the relationship between variables, but it does not replace observations and experiments. The equations are based on correlation and not on the understanding of processes, which means that the models are not generally valid and cannot be extrapolated beyond their interval of validity. For example, if a regression model is based on a sample of catchments with a size varying between 20 and 150 km^2, the model cannot be applied to a catchment of a size of 1000 km^2. There is a clear scale restriction and no up- or down-scaling is allowed.

8.3.3 Self-guided Tour: Estimation of low flow indices at the ungauged site (CD)

For the purpose of better understanding multiple regression analysis, a self-guided tour is provided to teach the user how to estimate low flow indices through multiple regression using the learning by screening method. The Self-guided Tour is on the CD. To facilitate the learning, the example is menu driven, which helps to navigate through the whole procedure. The viewer works through the procedure at an individual pace and may view or skip a section.

To facilitate the learning a virtual practical objective is given: a hydropower station is planned to be built at the outlet of the small River Wiese in the Black Forest. According to the regulations of the State of Baden-Württemberg, a minimum flow requirement has to be guaranteed (Section 11.4). It is set at Q_{90}. At the site of interest flow data are unavailable so Q_{90} has to be estimated. In the Self-guided Tour the multiple regression approach is employed using catchment descriptors (morphometric, soil, geologic, climatic and land use) and flow data from 83 stations (Regional Data Set). The flow data and the catchment descriptors were not log-transformed since the frequency distribution showed that all the data nearly followed a normal distribution.

Based on the practical example the user learns (a) to develop a conceptual model, (b) to translate a conceptual model into a mathematical model, (c) to calibrate the necessary mathematical transfer function by means of linear multiple regression between catchment descriptors and low flow parameters, (d) to evaluate the model based on prior assumptions and goodness of fit, (e) to validate the model with the help of a separate dataset (data splitting), and (f) to apply the model to the River Wiese (South Black Forest) to estimate Q_{90}.

At the bottom of the main menu there are six key buttons (a) aim, (b) study area, (c) procedure, (d) application, (e) data, and (f) appendices. These buttons allow jumping between sections. At the bottom right two arrow buttons help to navigate through the procedure in a sequence. Throughout the Self-guided Tour, results are discussed and conclusions are drawn. To guarantee a broad distribution and dissemination of the product, a commercial software package was used as platform (Microsoft PowerPoint presentation). It is important to note that the Self-guided Tour is not designed to develop the statistical models by the user. The models were already computed with a commercial statistical software package (SPSS) prior to storage on the CD.

8.4   Regional regression models – A review and global comparison

The first European statistical estimation procedures for low flows were developed for catchments in the United Kingdom. It is noteworthy that most of the European Studies were carried out during the 1980s, a period in which extreme low flow conditions occurred in parts of Europe. The resulting models are based on a hydrometric area approach and are summarized in the Low Flow Study Report (Institute of Hydrology, 1980). Nathan & McMahon (1990a) estimated similar low flow variables for ungauged sites in two regions in Australia (New South Wales and Victoria). Within the FRIEND project low flows were estimated at ungauged sites for different regions in Europe (Gustard *et al.*, 1989). Demuth (1993) carried out a comprehensive study at the European scale developing regression models at continental and regional scales. This was the first study to explore and compare data from experimental and representative catchments from all over Europe (115 catchments).

In the United States, the first significant regional low-flow estimation studies were carried out by Thomas & Benson (1970). Wetzel & Bettandorff (1986) developed estimation models based on an extensive database of 489 catchments. Recently, the US Geological Survey developed a method for estimating various low flow percentiles from the flow duration curve for ungauged sites in the state of Massachusetts, USA (Ries & Friesz, 2000). It is a user-friendly system called "StreamStats", which can be accessed on the Internet and estimates statistics at user-selected sites. In contrast to the European estimation studies, the studies carried out in the USA do not include hydrogeology and soil properties as predictors.

Demuth (1993) provides a comprehensive overview of catchment descriptors used in regional estimation models. He presents a detailed literature review exploring various aspects of the application of multiple regression models for the estimation of low flow indices, which will be briefly summarized in the next section. For the purposes of this book the review has been updated. It focuses on (a) the geographical coverage of the multivariate regression models (Section 8.4.1), (b) the type of catchment descriptors used in the multivariate regression models (Section 8.4.2), (c) the type of low flow index estimated (Section 8.4.3), and finally (d) the fractional membership of catchment descriptors within different model groups (Section 8.4.4). The evaluation leads to recommendations to assist in the selection of catchment descriptors according to the low flow indices to be estimated.

8.4.1   Geographical coverage

The global coverage of statistical low-flow estimation models is limited to certain areas. Most studies are found for North America and Europe (United Kingdom, Scandinavia, Germany, France, Belgium, the Netherlands and Switzerland), and some for African countries (Malawi, Tanzania, Zimbabwe) and New Zealand. For the Eastern European countries only a few studies are available and for Asian, Central- and South-American countries and the Australian continent, low flow studies are unavailable. Figure 8.5 shows the global coverage of the studies (in total 123 different models).

Figure 8.5 Global coverage of low flow estimation studies.

Nearly half of the models (about 45) were developed in the United States where there is a long tradition and a great deal of experience in using statistical low flow estimation methods. The most extensive regional low flow studies worldwide so far have been carried out in the framework of the UNESCO FREND project (1985–1988) for the region of western Europe (Gustard *et al.*, 1989). In this study global statistical models were developed for the investigation of the whole study area as well as for individual countries, groups of countries or regions (Gustard & Gross, 1989; Demuth, 1993; Schreiber & Demuth, 1997). For the development of statistical models data for about 1500 river basins were available.

8.4.2   Model evaluation based on catchment descriptors

This section provides an overview of catchment descriptors commonly applied for the estimation of low flows at the ungauged site. This will help hydrologists to select and choose the most appropriate catchment descriptors for their studies. In numerous statistical models both first order catchment descriptors (i.e. derived directly from maps) and second order catchment descriptors (derived from the hydrograph) were used (Demuth, 1993; Schreiber & Demuth, 1997; Caruso, 2000; Young *et al.*, 2000c).

When considering all models regardless of the type of estimated low flow index, it can be seen that 73% of all descriptors used are physiographic. Within

Table 8.2 Frequency of catchment descriptors

Rank	Catchment descriptors	Acronym	Frequency
1	Catchment area (km²)	AREA	66%
2	Annual average rainfall (mm)	AAR	49%
3	Soil index [-]	SOIL	22%
4	Mean basin elevation (m)	HMEAN	15%
5	Lake percentage (%)	FALAKE	13%
6	Percentage of area above the tree line (%)	MOUNT	13%
7	Stream slope (m km^{-1})	SL1085	12%
8	Weighted lake percentage (%)	WPLAKE	11%
9	Base flow index [-]	BFI	10%
10	Drainage density (km km^{-2})	DD	10%

the physiographic category, morphometric descriptors make up the highest proportion (46%), followed by surface cover (17%), geology and soil (10% each). The other two categories, climatic and hydrologic descriptors, amount to 22 and 5% respectively. This suggests that the physiographic catchment descriptors are the most important parameters influencing the low flow indices. Climatic and hydrologic catchment descriptors constitute a far smaller part of the total number of significant predictor variables. They are far less often used as relevant parameters for the estimation of low flow indices.

Table 8.2 summarizes the ranking of the ten most frequently-used catchment descriptors. Catchment area, AREA, and average annual rainfall, AAR, dominate with a frequency of 66% and 49%, respectively. A soil index, SOIL, follows with 22%. The remaining catchment descriptors appear with a frequency of between 10% and 15%. They play only a minor role in low flow estimation models.

8.4.3 Model evaluation based on low flow indices

The selection of catchment descriptors for low flow estimation depends not only on the objective of the study, but also on the type of indices to be estimated. About 50% of the studies estimate extreme values, driven by the requirement from the water-management practice. The first models for estimating extreme value statistics for different regions in the United States were introduced by Thomas & Benson (1970) in the late 1960s. Both the hydrograph ($MAM(n$-day)) and the flow duration curve (Q_x) were each used in 20% of the studies.

The first studies estimating low flow indices from the recession curve were carried out in the early 1980s by Pereira & Keller (1982) based on eleven catchments located in the pre-alpine region in Switzerland. In recent years, the estimation of base flow indices has become attractive, especially in the framework of the Hydrological Atlas of Germany (HAD, 2001) and the Water and Soil Atlas of the State of Baden-Württemberg (Armbruster, 2002). Demuth (1993) introduced models which estimate base flow indices and indices of the base flow recession curve within the framework of the FRIEND project (1985–1993). Section 5.3.4 gives a brief review of recession indices.

8.4.4 Fractional membership of catchment descriptors

To be able to evaluate the predictor variables (catchment descriptors) with regard to their significance for specific estimation models, individual predictor groups were investigated (e.g. morphometric descriptors, cover descriptors, geologic and soil descriptors). Here the fractional membership of catchment descriptors within several index groups was calculated (Table 8.3).

The percentages given in Table 8.3 refer to the proportion of models within the index group in which the respective descriptor could be found. Models estimating extreme value indices, hydrograph indices and indices derived from the flow duration curve (index groups I–III) show a similar composition of catchment descriptors. Catchment area, AREA, is the most significant physiographic descriptor in these groups. Its fractional membership ranges from 59% (index group II) to 77% (index group I). None of the other physiographic descriptors appears in all three index groups (I–III) (e.g. RB appears only in index group V with 14%).

Besides AREA, the soil index, SOIL, and the average annual rainfall, AAR, were also applied quite frequently in the models. AAR is used mostly in index group I (53%) and II (48%). SOIL is mostly applied in index group I. Armbruster (1976) used SOIL only in twelve models for the estimation of different extreme values statistics in basins located at the east coast of the United States.

Among the first three index groups, the composition of the catchment descriptors representing surface cover (descriptors number 11–17) is very heterogeneous. The implementation of these catchment descriptors into a set of predictor variables depends on the natural properties of the region for which the models were developed. If many studies were carried out in a certain region, the composition of catchment descriptors in this model group reflects the predominance of descriptors in that specific region. For example, in regions with a high percentage of forest, FOREST appears as a significant catchment descriptor in the models. The observation that FALAKE and WPLAKE have a relatively strong appearance in index groups I–III supports the knowledge of their important role as descriptors accounting for the retention in a catchment and the smoothing of flow variations (Section 3.5).

Groups IV and V differ distinctly from index groups I–III in terms of their predictor composition. For example, average annual rainfall, AAR, is used in all index group IV models (AAR = 100%) and constitutes 31% of all predictors used in index group IV models. In contrast to index groups I–III, the predictor that dominates index groups IV and V is drainage density, DD, (as opposed to catchment area, AREA). The soil index, SOIL, does not appear at all in index groups IV and V. Only once was a hydrological predictor used in an index group IV model (base flow index, BFI, 14%).

In index group V, 95% of the predictors can be classified as physiographic predictors. Those used in index group V are catchment area (AREA), bifurcation index (RB), percentage of forest (FOREST), percentage of lake (FALAKE), volume of trees (VGS), percentage of pasture (PAST), drainage density (DD), and the geologic index (GEO). DD and GEO dominate the models with GEO exhibiting the highest frequency (71%) within index

Table 8.3 Percentage of catchment descriptors in index groups I–V

| Predictors | Acronym | Frequency of catchment descriptors within different index groups (%) | | | | |
		Extreme values I	Hydro- graph II	Flow duration curve III	Base flow IV	Recession curve V
Morphometric descriptors						
1. Catchment area	AREA	77	59	62	43	43
2. Channel slope	SL1085	23	-	-	-	-
3. Main stream length	MSL	-	-	-	-	-
4. Mean basin elevation	HMEAN	21	-	19	29	-
5. Maximum basin elevation	HMAX	-	19	-	-	-
6. Catchment length	LE	15	-			
7. Relief ratio	RELIEF	-	15	-	-	-
8. Catchment width	BE	-	15	-	-	-
9. Drainage density	DD	-	-	-	86	57
10. Bifurcation index	RB	-	-	-	-	14
Surface cover						
11. Percentage of forest	FOREST	-	-	-	43	29
12. Weighted lake percentage	WPLAKE	-	30	-	-	-
13. Lake percentage	FALAKE	15	-	19	-	14
14. Percentage of basin above the tree line	MOUNT	19	22	-	-	-
15. Sealed basin area	URBAN	-	11	12	-	-
16. Volume of trees	VGS	-	-	-	-	14
17. Percentage of pasture	PAST	-	-	-	-	14
Geological and soil descriptors						
18. Soil index	SOIL	28	19	23	-	-
19. Geological index	GEO	-	-	-	-	71
Climatic descriptors						
20. Average annual rainfall	AAR	53	48	38	100	14
21. Shortest distance to the western coast line	WSEA	-	11	-	-	-
Hydrological descriptors						
22. Base flow index	BFI	-	15	27	14	-
23. Runoff ratio (Q20/Q90)	RATIO	17	-	-	-	-
24. Recession coefficient	ALPHA	-	15	-	-	-

group V. It is a significant predictor in five of the seven models analysed. Apart from index groups IV and V, only two models (Wright, 1970, 1974) were found in which GEO was used as a predictor. The results of the investigation (Table 8.3) facilitate the selection of catchment descriptors for future studies.

8.5   Hydrological mapping procedures

8.5.1   Background

Mapping or interpolation of hydrological variables is probably the most concise way of displaying and summarizing the spatial and temporal variation of hydrological variables (Section 2.3.2). McKay (1976) and Arnell (1995) stated that *hydrological maps* are an effective *multi-purpose-communication tool* for *decision makers*, e.g. *politicians and water managers* (to indicate the quantity of water resources available in a region for planning purposes), *scientists* (to validate regional hydrological and climatic models), and *educators* (to inform about water-related issues). It can contribute to new insights into possible solutions to environmental and social problems.

The derivation of maps has a long tradition. This is especially true for the deduction of runoff maps. In the USA the development of this method dates back to around 1890 (Herschy & Fairbridge, 1998). One can distinguish between manually- and digitally-derived maps. When deriving a map manually, the hydrologist relies both on gauged data and local precipitation patterns and topographical information, which he combines subjectively into the respective map (Krug *et al.*, 1990; Domokos & Saas, 1990).

In general, the reliability of flow estimation from maps depends on the density of gauging stations, the quality of flow data, the scale of the map and the contour interval, and the spatial and temporal variability of the flow statistics being mapped (Smakhtin, 2001). In addition to the density of gauging stations and the accuracy of the measurements, the quality of manually-constructed flow-contour maps also depends on the expertise of the cartographer. According to Gottschalk & Krasovskaia (1998), several considerations have to be made when choosing a method to construct a runoff map: (a) choice of interpolation method, (b) definition of the scale of the units of the map, and (c) acquisition of measurement data, which account for the variability at different scales. A comprehensive overview of different approaches is given in Sauquet *et al.* (2000).

Flow maps are constructed on the basis of gauged flow statistics as contour maps and the individual values are estimated from gauged stations. The mapping procedure is therefore similar to, for example, the regression procedure, which seeks to statistically link response variables to catchment descriptors (Drayton et al., 1980; Belore et al., 1990; Telis 1992; Smakhtin et al., 1995; Pearson, 1995; Caruso, 2000). Schreiber (1996) applied kriging to map $MAM(10)$, specific discharge and recession coefficients. Caruso (2000) stated that contour maps could be useful for exploring general spatial trends across regions and provides rough estimates of low flows at ungauged sites.

The automated contour methods for deduction of flow maps employ computer algorithms to estimate flow statistics and to create flow surfaces. The automated methods have the advantage of being efficient and giving easily-reproducible results. Furthermore, they do not depend on expert knowledge for their application, which is not beneficial in all cases. The automated interpolation procedures can be divided into deterministic and stochastic approaches. Both account for weighted averages by different methods, ranging from simple averaging point information to stochastic interpolation.

A deterministic approach is usually applied to account for empty grid cells, such as the triangulated-irregular-network method (TIN) (Bishop & Church, 1992; Arnell, 1995). This method reflects the differences among geographical locations. The catchment size should reflect the zonal type of flow regime and therefore, for example, very small rivers, where flow regime is usually a result of small-scale local properties, should not be selected for the purpose of mapping flow indices (Arihood & Glatfelder, 1991; Ludwig & Tasker, 1993; Risle, 1994; Smakhtin, 2001). In addition, water-balance constraints have to be respected when mapping flow statistics.

A different aspect to be considered is the fact that gauged flow data are based on point measurement, integrating the flow from the entire catchment above this point. A flow statistic estimated at any ungauged site in a region is usually assumed to be representative of the entire catchment above and calculated flows are assigned to the centroids of gauged catchments (Smakthin, 2001).

An alternative method of producing flow maps is to delineate spatial units (grids at different sizes). In this method, the constant flow values, ranges of flow values, or statistical values from a distribution function are deduced (e.g. by means of aggregation) and assigned to an individual grid cell (Arihood & Glatfelter, 1991; Gustard et al., 1992; Smakhtin et al., 1998; DVWK, 1996). Smakhtin (2001) lists examples of mapped low flow indices. These include Q_{75} (Drayton et al., 1980), Q_{20}/Q_{90} (Arihood & Glatfelter, 1991), $AM(7)_2$ and $AM(7)_{20}$ (Belore et al., 1990), $AM(7)_{10}$ (Telis, 1992), Q_{95}(7-day), and $AM(7)$ (Gustard et al., 1992). Gustard (1993) mapped Q_{95}(1-day) in Great Britain based on the proportion of HOST classes on a 1x1 km^2 grid.

8.5.2 River network approach: grid-based

Schreiber & Demuth (1997) investigated the spatial behavior of *MAM*(10) in the State of Baden-Württemberg (southwest Germany) by introducing a new procedure to estimate *MAM*(10) not only for point values but also for grid cells. The procedure is a river-network approach (or line-grid technique) as opposed to the classical catchment approach. In this technique, the regression model was based on nominally-scaled catchment descriptors.

The river network was divided into five sections upstream of the gauging stations. The main properties of the 1x1 km^2 cells along these sections were related to the calculated *MAM*(10) by means of multiple regression. The reasons for choosing five sections were (a) to exclude those descriptors from the model which would only be included due to imprecise data, and (b) to eliminate catchments less than 5 km in length. Runoff measurements tend to be inaccurate due to low absolute flow, especially for low flows of very small catchments. There may also be local discontinuities where, despite their significance for deducing the low flow indices, certain catchment descriptors cannot be represented at the scale of the thematic database.

The qualitative thematic data were dummy coded (e.g. geology with its 11 levels was divided into 10 dummy-coded indicator variables with either the value 0 or 1) and included into a multiple regression model:

$$MAM(10) = f(G_1, ... , G_{10}, H_1, ... ,H_5, M_1, ... ,M_{10}, U) \qquad (8.10)$$

where G_i is an indicator variable of geology, H_i is an indicator variable of hydrogeology, M_i is an indicator variable of geological deposits (e.g. alluvium, moraines, clay) and U is an indicator variable of land use. A multiple regression model was adopted to estimate the *MAM*(10) based on different combinations of geology, hydrogeology, petrography and land use. The hydrogeological and land-use descriptors were found to be significant in explaining the variance of *MAM*(10) in southwest Germany. The coefficient of determination was 56%. A comparison with earlier studies in sub-regions, which was carried out with the classic catchment approach (coefficient of determination, $R^2 = 86\%$), showed that the new approach led to similar estimation results (Wesselink *et al.*, 1994; Demuth & Hagemann, 1993).

This analysis illustrated that the modified multiple regression approach based on the river network is appropriate to develop a regional procedure to estimate *MAM*(10) at the ungauged site. The advantage of the method is the use of only a few catchment descriptors along a river stretch. In addition, it does not require knowledge about the catchment area and the calculation of the proportions of catchment descriptors. This is especially relevant in those regions where catchment areas are difficult to specify, or where other catchment descriptors for some reason cannot be calculated or properly included into the regression model. Another advantage of the model is the straightforward estimation process, which is based directly on hydrogeologic properties (as opposed to surrogates, such as drainage density, DD, or the base flow index, BFI). Finally, the approach provides a tool which allows estimation of low flow indices not only for entire catchments but also for any individual grid cell in the region under consideration.

8.5.3   River network approach: recession-curve routing

Demuth *et al.* (2000) present the flow-line-recession-routing technique to estimate recession parameters at the ungauged site and to calculate runoff-drought vulnerability along a river system. This method accounts for drought-predisposition of a river for the time period, T_D, until the threshold for drought flow is reached under dry-weather conditions. Recession parameters, initial runoff and drought threshold runoff were estimated for ungauged sites. Supported by GIS functions and network topology, aggregated recession curves were derived for every 2.5 km stretch along the river network so that the time period T_D could be estimated. The application of the method in the State of Baden-Württemberg (southwest Germany) and a comparison with results from earlier studies was satisfactory. The results were mapped and illustrate the degree of vulnerability of river stretches to drought.

Streamflow and recession indices were estimated for ungauged sites in areas of equal size to account for scaling problems. Technically supported by GIS functions, recession curves for evenly-spaced points along the river network were then aggregated from the upstream areas following the idea of multi-storage recession. An automated river-coding system was applied to assign the topological information of the river network to the individual river stretches (Lehner, 1998a; 1998b).

The vulnerability of a river to the occurrence of streamflow drought is defined by the time period T_D it takes for the streamflow to decrease from an initial discharge Q_{MS} to below the threshold level for drought, Q_D (Figure 8.6). The shorter the time T_D, the higher is the risk that a region may be affected by drought if a meteorological dry signal prevails for a long period of time.

This description of vulnerability is based on the presumption of the initial discharge representing a hydrological state at which the catchment's natural groundwater reservoir is no longer filled up to its highest level (Q_{MS} applies to conditions with mean storage) when an extended atmospheric dry signal occurs over the whole catchment. Besides Q_{MS} and Q_D, the typical mean recession behaviour (master recession curve, Section 5.3.4) had to be indexed and

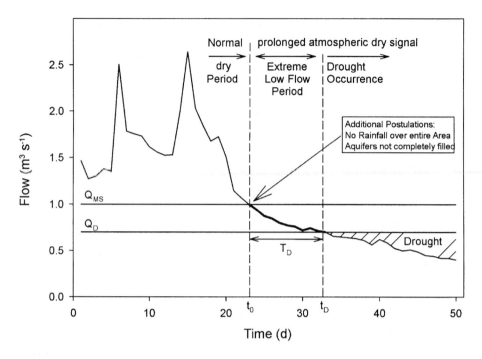

Figure 8.6 Definition of drought vulnerability (from Demuth *et al.*, 2000).

determined from gauged discharge series in order to estimate the drought vulnerability at the ungauged site.

Attention has to be paid to the analysis of the characteristic runoff behaviour during a rainless period. As the approach treats low flow as sustained groundwater flow, the outflow function $Q(t)$ describes the outflow from groundwater storage (Gottschalk *et al.*, 1997), which reflects the natural storage behaviour of a catchment in a generalized form. A simple exponential storage model has been used to depict the recession parameters. The recession model is defined by the initial discharge Q_{MS} and the recession coefficient α and according to Equation 5.1b can be written as:

$$Q(t) = Q_{MS} \exp(-\alpha t) \tag{8.11}$$

Hence, T_D can be calculated as:

$$T_D = (\ln(Q_{MS}) - \ln(Q_{(t)}))/\alpha \tag{8.12}$$

For the study presented, Q_{60} was chosen as Q_{MS}. All runoff recessions at a station starting at this value ($\pm 10\%$) and continuously decreasing for seven or more consecutive days were extracted from 35-year streamflow series. The

Table 8.4 Results of multiple regression analysis (from Demuth *et al.*, 2000)

Parameter	Variables in the model	R^2	Adjusted R^2	Standard error of the estimate	Significance (F-test)
$\overline{Q}$	Lithology, slope, river length	0.96	0.94	0.64	>99%
$Q_{MS}(Q_{60})$	Lithology, slope, river length	0.96	0.95	0.37	>99%
$Q_D(Q_{90})$	Lithology, slope, river length	0.95	0.93	0.25	>99%
α	Lithology, slope, MQ/Q_{90}	0.83	0.77	0.007	>99%

exponential recession model (Equation 8.11) was fitted to determine α. In this study a threshold level of Q_{90} was chosen as it has proven to be appropriate in earlier studies (Demuth & Heinrich, 1997).

The multiple regression equations were derived estimating $\overline{Q}$, Q_{MS}, Q_D and α as function of catchment descriptors. The results of the multiple regression analysis are summarized in Table 8.4. The results are maps which show the spatial variation of the estimated index, determined for the scale of the individual polygons. However, for the identification of streamflow drought, the integrated runoff of the whole catchment draining to the point of interest at the river has to be considered.

This method is based on the assumptions that (a) the time delay due to runoff concentration can be neglected during stable low flow conditions, and (b) the initial runoff, expressed as a percentile from the flow duration curve, occurs simultaneously within the area. As a result, the recession curve for any particular node (e.g. node 9 in Figure 8.7) within the river network could be calculated by adding the individually-estimated curves along the upstream river stretches. Finally, T_D was determined from the curves at all nodes and was visualized along the river network.

The maps clearly illustrate the typical hydrological regions within the study area and are comparable with the patterns obtained in previous studies. A map showing the drought vulnerability in terms of the time T_D is shown in Figure 8.8. Most of the study area exhibits medium T_D values of around ten days with a tendency towards increasing T_D values for larger catchment areas

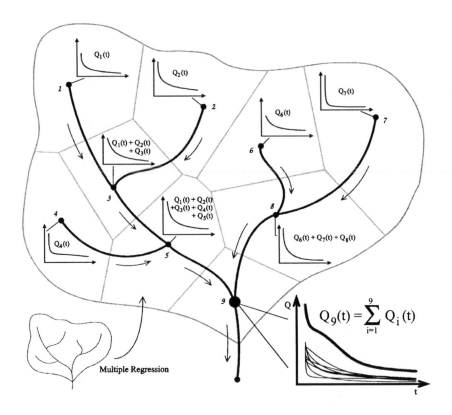

Figure 8.7 Aggregation of estimated recession curves to determine the characteristic runoff recession for streamflow drought (from Demuth *et al.*, 2000).

and distances to the spring. The shortest times were found in an area characterized by sandy soil characteristics.

Neither very low Q_{90} flows nor very fast recession behaviour were estimated for the polygons in this area. The reason for the small T_D is a stable runoff regime and thus a narrow Q_{MS}/Q_{90} relationship. The opposite is found in the karstic limestone area of the Swabian Alb. Here, runoff recession is generally fast, but the large ratio of Q_{MS}/Q_{90} results in a long T_D. If this is the case, it can be assumed that both the ecosystem and surface-water management systems are adapted to high discharge variability, whereas in the case of a more stable regime, a dry period may soon cause an unexpected water shortage (Chapter 10).

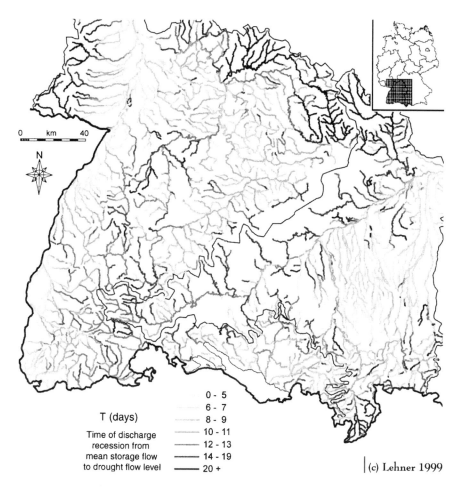

Figure 8.8 Map of T_D, derived from the recession curve and visualized along the river network (from Demuth *et al.*, 2000).

8.5.4 Estimation of low flow statistics in the United Kingdom

Regionalization of low flow statistics in the United Kingdom (UK) began with the Low Flow Studies report (Institute of Hydrology, 1980), which established relationships between low flow indices and physiographic and climatic catchment descriptors. This study showed that when low flows are expressed as a percentage of the long-term mean flow (standardized), the dependencies on the climatic variability across the country and on the scale effect of catchment area are minimized. The shape of the standardized flow duration curve indicates the characteristic response of a catchment to rainfall. Impermeable catchments have high gradient curves reflecting a very variable flow regime; low storage of water in the catchment results in a quick response to rainfall and low flows in the absence of rainfall (Section 3.4, Figure 5.3). Low gradient flow duration curves indicate that the variance of daily flows is low, due to the damping effects of groundwater storages provided naturally by aquifers or large lakes.

Gustard *et al.* (1992) built on the previous work and developed models that estimated flow duration statistics at ungauged catchments. The multivariate linear regression models used the fractional extents of soil groupings to explain the variability of standardized flow duration statistics. The Hydrology Of Soil Types (HOST) soil classification system, derived by Boorman *et al.* (1995), was chosen as the explanatory variable. This 29 class system was grouped into 12 Low Flow Host Groups (LFHG). The final model took the form shown in Equation 8.13.

$$Q_{95\,PRED} / \overline{Q} = \sum_{i=1}^{12} a_i \mathrm{LFHG}_i \qquad\qquad (8.13)$$

where Q_{95PRED} is the estimate of the Q_{95} standardized by the mean flow $\overline{Q}$, LFHG_i is the fraction of Low Flow HOST Group i occurring in the catchment and a_i are parameter estimates.

The model parameters were derived by linear regression analysis for a data set of good quality gauging station records of predominantly natural catchments drawn from across the UK. Hence, this represents a regionalization technique where the region is defined geographically and encompasses the UK.

The regionalization techniques for UK low flow statistics were revised by Holmes *et al.* (2002b) in a study that made use of an extended data set. An alternative approach to regionalization was also adopted. The regression model developed by Gustard *et al.* (1992) sought to optimize a fit across a wide range of observed values of Q_{95}, adopting the entire UK as a 'region', by a relatively complex process of minimizing the sum-of-squares differences between observed and predicted Q_{95} using multivariate soil characteristics as predictor variables. The weakness of this approach was that the extremes in Q_{95} values were poorly modelled. A 'Region of Influence' (ROI) regionalization strategy was adopted to address this issue.

The ROI-based model reduces the variability of the dependent variable within the data set by reducing the data set to a much smaller 'region' of catchments that are 'similar' to the 'target' catchment. Similarity is assessed with respect to hydrogeological setting of the target and data set catchments, building on the observed correlation between this catchment characteristic and observed low flow statistics. An arithmetic, weighted averaging process is then used to predict a value of Q_{95} for the 'target' catchment, from the observed

Q_{95} values. This approach reduces the within-region variability at the expense of the complexity of the predictive model. The final ROI methodology developed for estimating the Q_{95} flow statistic at an ungauged 'target' catchment in the UK involves three steps:

a) Catchment similarity is assessed between the target catchment and every other catchment within a large data pool of natural gauged catchments, by calculating a weighted Euclidean distance (Equation 8.14). The weightings for each HOST class are determined by a combination of linear regression modelling, observed flow behaviour in single/dominant HOST class catchments and a subjective assessment of the relative responses of different hydrogeological classes. This type of assessment is required to address the issue of poorly represented HOST classes, while not grouping classes together and consequently losing the unique responses of individual classes.

b) A 'region' is formed around a target catchment by ranking all of the catchments in the data pool by their weighted Euclidean distance in HOST space and selecting a set of catchments from the pool that are 'closest' to that target catchment.

c) An estimate of Q_{95} for the target catchment is calculated from a weighted average of the observed Q_{95} values for the selected catchments in the region. The weight is based on the reciprocal of the Euclidean distance measure.

For each catchment within the data pool, the weighted Euclidean distance from the target catchment in HOST space is calculated using:

$$de_{it} = \sum_{m=1}^{M} w_m \left(X_{mi} - X_{mt} \right)^2 \qquad (8.14)$$

where de_{it} is the weighted Euclidean distance from the target catchment t to catchment i from the data pool, w_m is the weight applied to catchment descriptor m, and X_{mi} is the standardized value of catchment descriptor m for catchment i. The catchment descriptors, X_m, used are the fractional extents of the 29 HOST classes within a catchment, which will vary between zero and unity.

A region size of ten catchments was ultimately identified as optimal. This represented a partially subjective decision based on minimizing the unexplained variance across the catchment data set and minimizing the tendency for the model to over-predict for catchments with very low Q_{95} and under-predict for catchments with very high Q_{95}. Hence, the predicted value of Q_{95} for an ungauged catchment is derived using Equation 8.15.

$$Q95^t_{PRED} = \sum_{i=1}^{n} \left[\frac{\left(\dfrac{1}{de_{it}}^{0.5} \right)}{\sum_{i=1}^{n} \left(\dfrac{1}{de_{it}}^{0.5} \right)} \right] \times Q95^i_{OBS} \tag{8.15}$$

where $Q95^t_{PRED}$ is the estimate of the Q_{95} flow for the target catchment, t, $Q95^i_{OBS}$ is the observed value of Q_{95} for the *ith* source catchment in the region of n (here n equals 10) catchments closest to the target catchment.

The improvements in the new regionalization methodology show that the ROI model performs better than the linear regression based approach across the entire range of Q_{95} values encountered in the data set. In particular, the analysis shows that the overestimation of low Q_{95} values has been reduced as has the underestimation of high Q_{95} values. The technique is now implemented nationally in the UK for low flow estimation at the ungauged site using Low Flows 2000 software (Chapter 11).

Compared with regional techniques such as cluster analysis (Section 6.7.4), the ROI model has the benefit of forming a region about the target catchment in question. Hence no regions are defined *a priori*. New data catchments may be added directly to the data pool without the difficulty of assigning regional memberships.

8.6   Summary

This chapter has explored the fundamental principles of low flow estimation procedures in ungauged catchments. Different approaches were introduced focusing on simple estimation methods (empirical methods and flow correlation) as well as multivariate techniques such as statistical regression methods. The chapter also introduced advanced estimation procedures, such as hydrological mapping methods and various river network approaches and a region of influence approach. To various degrees, these both help to identify the types of catchment descriptors relevant for the estimation at the ungauged site, and statistically relate catchment descriptors with various low flow indices.

In general, the choice of an appropriate method is based on the objective of the study, the economic resources, and the availability of both flow and catchment data. Most of the catchment descriptors are measurable and can be derived from observed data or information provided by digital or hard copy maps. Hydrogeological descriptors, however, often require additional field investigations or special studies at a regional scale.

The empirical and multivariate statistical models are only valid for the range for which they are calibrated. Application of the models is not appropriate beyond this range and would lead to misinterpretation. Empirical methods are usually used within a catchment where the general assumptions of similarities with respect to catchment descriptors are valid. Flow correlation has broader application opportunities. It is not only used to estimate low flow indices at the ungauged site but also to fill in gaps in time series (Section 4.3.3). Factor analysis is introduced here as a method of selecting catchment descriptors.

The multivariate regression analysis is a time-consuming tool, but in spite of general restrictions its advantages – namely objectivity, simplicity and directness – prevail. All of the procedures presented are scale-restricted, which means that neither down-scaling nor up-scaling are possible and that the estimates are point estimates (as opposed to grid estimates). The review of statistical regression models at a global scale provides an overview of the type of catchment descriptors that are most commonly used, depending on the type of low flow index to be estimated.

Advanced estimation methods, such as hydrological mapping (river-network approaches, region of influence approach), share the advantage of indicating both the spatial and temporal variation of low flows. With these methods the constant flow values, ranges of flow values, or the statistical values from a distribution function, are usually assigned to an individual grid cell.

A recent global initiative by the International Association of Hydrological Sciences (IAHS) has announced a decade of 'prediction in ungauged basins' (PUB). Key aspects of the programme are the assessment and comparison of models in terms of the uncertainty with respect to accuracy and precision in their predictions of hydrological variables at both gauged and ungauged sites (Littlewood *et al.*, 2003). The aim is to help soften the problem of network decline and data scarcity, especially in developing countries. For this reason regionalization tools will be even more necessary in the future to guarantee a sustainable water resources management.

Part III

Human Influences, Ecological and Operational Aspects

9

Human Influences

Henny A.J. van Lanen, Ladislav Kašpárek, Oldřich Novický, Erik P. Querner, Miriam Fendeková, Elżbieta Kupczyk

9.1   Introduction

Large basins without human influence hardly exist. This fact imposes constraints on drought analysis. Human activities make the multifaceted relationship between meteorological and hydrological droughts (Chapter 3) even more complicated. They may enhance *natural hydrological drought*. For instance, a soil moisture drought in a semi-arid region requires additional irrigation. This water may come from surface water, and the abstraction of water implies low reservoir levels or streamflow enhancing surface water drought. Irrigation water can also be abstracted from groundwater leading to lower water tables and reduced groundwater discharge. This may enhance an already existing groundwater drought and also contribute to surface water drought.

Human activities can cause drought, which was not previously reported (*man-induced hydrological drought*). Groundwater abstractions for domestic and industrial use are a well-known example of such an environmental change. These permanent abstractions lead to lowering of water tables and reduced groundwater discharge. A hydrological drought can even develop, which would not have shown up without abstractions. Construction of a reservoir could also induce development of streamflow drought downstream of the dam. Some human activities are meant to reduce water excess, but unintentionally they contribute to drought development. An example is land drainage. Like groundwater abstractions, land drainage causes permanently lower water tables making the region more susceptible to drought.

This chapter starts with an introduction to model concepts (e.g. physically-based models) that are used to assess the impacts of human influences (Section 9.2). They are described first because of their importance in understanding the impact assessment methodology. Sections 9.3–9.7 present examples of human influence on the development of hydrological drought. This chapter does not seek to cover all human activities that may lead to hydrological droughts. A few examples are presented in detail to make the reader aware of the different types of impact that *environmental change* can have on drought development. As a first example, the impact of land use is described (Section 9.3) followed by climate change (Section 9.4) and groundwater abstractions (Section 9.5). The next section elaborates the impact of urbanization on droughts (Section 9.6). The chapter concludes with some examples of the impact of human activities on rivers (Section 9.7). Examples in Sections 9.3–9.6 primarily deal with groundwater and its effect on streamflow. Section 9.7 focuses on surface water. Each of the example sections starts with generally observed effects, followed by modelling results. The model results mainly focus on a temperate humid climate as this occurs in west and central Europe. It is important to note that the emphasis is on modelling in this chapter, because the effect of a particular human activity is difficult to detect from observed time series only, mainly due to a relatively large within-year and inter-annual variability, as well as other human activities. Readers need to be aware that it is not easy to distinguish between natural variability and human influences. The models applied in this chapter appear to be comprehensive, but they also study the impact of one isolated human activity. Feedback of the system is not taken into account. Simpler methods have already been described in Section 4.3.3, including methods for naturalization, which also distinguish between artificial and natural effects.

This chapter explores the human impact on hydrological drought and does not try to explain how operational water management might try to prevent (pro-active approach) or to mitigate (reactive approach) drought. These operational aspects are described in Chapter 11. Human activities certainly affect water quality as well as water quantity. Concentrations of many critical water-quality constituents, such as nutrients and dissolved oxygen, are related to discharge. Soluble pollutants often show a negative correlation with discharge, whereas for dissolved oxygen it is the opposite implying that drought leads to poor water quality. The impact on water quality is not dealt with in this book because of its comprehensive nature. Chapman (1992) and Reeve & Watts (1994) provide more details. Human actions also influence ecosystems in periods of drought and their effects on in-stream ecology are explained in Chapter 10. For a description of water shortage effects on terrestrial ecosystems readers are referred to Ravera (1991).

9.2   Model concepts

The large number of models and different human activities encountered which might affect drought raise the important question of which model to apply to analyse the impact of human influences in a particular catchment. Various modelling concepts are available. The models *predict* drought characteristics and indices in a catchment. In the context of this book we distinguish between two main groups of approaches:

a)   multiple regression models;

b)   physically-based models.

A *multiple regression model* is a mathematical representation of the relationship between hydrological indices and catchment descriptors. It excludes description of hydrological processes. The nature of the regression model determines the kind of environmental changes that can be evaluated. For example, the method presented in Section 8.3 can be applied to elaborate the impact of some aspects of climatic change (average annual precipitation), or land use change. However, most regression models are unable to evaluate a wide range of environmental changes as introduced in the previous section, and therefore *physically-based models* are applied. In these models one or more hydrological processes, e.g. interception, soil moisture flow, saturated groundwater flow, surface water flow (Chapter 3) are included. The simplest models describe one hydrological process and use an analytical expression that can easily be solved. A spreadsheet usually facilitates the computation. However, most physically-based models use a more complex description of hydrological processes involved and apply numerical solution techniques. With these techniques the solution of water flow equations is approached (i.e. finite difference, finite element techniques, e.g. Wang & Anderson, 1982). Numerical techniques are also needed to account for spatial variability of catchment properties (e.g. precipitation, land use, hydraulic conductivities) or to simulate transient processes.

Physically-based models simulate time series of hydrological variables, for example groundwater hydraulic head, groundwater discharge, surface water level and streamflow. They generate time series like random models (Chapter 7) and as such they comprise another group of methods to estimate drought characteristics and indices (Part II). They differ from random models because uncertainty in model parameters is not generally considered.

Not surprisingly, a wide range of physically-based models is available to simulate groundwater recharge, groundwater hydraulic heads, groundwater discharge and streamflow. Thus they can also be applied to assess human impact on drought. One way of classifying them is:

a) conceptual physically-based models;

b) lumped physically-based models;

c) distributed physically-based models.

Physically-based models are typically applied at catchment scale. *Conceptual physically-based models* use physically sound hypotheses describing the main interactions in the catchment. It is mostly assumed that the catchment is homogeneous and processes governing transient groundwater flow, storage, and groundwater-stream interaction are taken into account. Section 9.2.1 first describes some conceptual physically-based models. Next, an example of a *lumped physically-based model*, BILAN, is described, followed by a representative *distributed physically-based model*, SIMGRO (Sections 9.2.2 and 9.2.3). These two models are used in the following sections to illustrate the effect of some selected environmental changes.

9.2.1   Conceptual physically-based models

Conceptual physically-based models for solving the groundwater-stream interaction have been proposed by, for example, Brutsaert & Nieber (1977), Miles (1985), Szilagyi & Parlange (1998) and Vogel & Kroll (1992). The models use several simplifying assumptions both for the catchment setting and for the equations governing the hydrological processes. An example of a conceptual model is the linear reservoir model used in Section 3.4 to simulate groundwater discharge.

Another conceptual model describes the groundwater flow in a cross section perpendicular to a stream. The most commonly-used equation for describing transient flow in the vertical plane (x-z) with a horizontal impervious bottom is a linear partial differential equation, derived from the Boussinesq equation (Boussinesq, 1877):

$$k_h \bar{H} \frac{\partial^2 H}{\partial x^2} = \eta_e \frac{\partial H}{\partial t} \tag{9.1}$$

where H is groundwater hydraulic head [L], $k_h \bar{H} = kD$ represents transmissivity of the unconfined aquifer [LT^{-2}], k_h is hydraulic conductivity [LT^{-1}], $\bar{H}$ is the average hydraulic head [L], introduced with the assumption that spatial and temporal variation of H can be neglected, and η_e is the effective porosity [-], which is approximately equal to the specific yield of the aquifer.

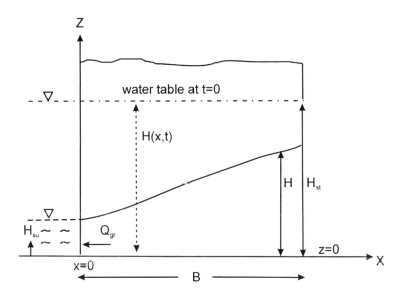

Figure 9.1 Schematic illustration of an aquifer-stream system.

The schematized flow situation is presented in Figure 9.1. The solution of Equation 9.1 for all $t > 0$ with respect to $H(x,t)$ is:

$$H(x,t) = t^{\frac{1}{2}} \exp(-\eta_e x^2 / 4k_h D_a t) \qquad (9.2)$$

where x is horizontal distance [L], and t is time since the catchment has started to be desaturated [T], i.e. at $H = H_{st}$ (Figure 9.1).

Following Equation 9.2 it can be found that the amount of water released at $x = 0$ and at time $t = 0$ per unit length of stream is:

$$q_{dr} = 2 \eta_e (\pi k_h / \eta_e)^{\frac{1}{2}} \qquad (9.3)$$

Application of the superposition principle facilitates the determination of the volume of water released from the aquifer by integrating Equation 9.2 with respect to x and t. The volume of water drained per unit length of stream at time t, when hydraulic head falls down to H and the water level in the stream is H_{su}, is given by the expression:

$$q_{dr} = \frac{2k_h H(H - H_{su})}{B} \exp\left[\frac{-\pi^2 k_h H t}{4\eta_e B^2} \right] \qquad (9.4)$$

where B is the average width of the aquifer [L].

Equation 9.4 has similarities with the recession analysis (Section 5.3.4). Total aquifer contribution to the channel of the length L is:

$$Q_{gr} = 2 L q_{dr} \tag{9.5}$$

Vogel & Kroll (1992) applied such a conceptual stream-aquifer model for prediction of low flows for rivers in Massachusetts. Many groundwater-streamflow simulation models, developed on the basis of the Boussinesq-Dupuit equation, are successfully applied for prediction of groundwater outflow to streams draining small agricultural catchments (e.g. Czamara, 1998).

The term Q_{gr} is equal to the groundwater discharge described in Section 3.4.2. Equation 9.5 has only limited potential to evaluate the impacts of human measures on droughts. For example, the effect of certain drainage measures, e.g. the lowering of the surface water level, H_{su}, can be analysed. However, more powerful models are the numerical physically-based models that are described in the next sections.

9.2.2   Lumped physically-based models

Lumped physically-based models consist of a number of mathematical equations that describe transient hydrological processes in a specific catchment, e.g. evapotranspiration, unsaturated flow, saturated groundwater flow and streamflow. Lumped models assume a homogeneous catchment, meaning, for instance, that precipitation does not vary over the catchment and only one land use type and soil type occurs. Furthermore, groundwater flow within the catchment, groundwater discharge to particular stream reaches and surface water flow inside the catchment are not considered in lumped models. An example of a lumped physically-based model is the BILAN model used in this chapter.

The BILAN model was developed to simulate components of the water balance in a catchment (Kašpárek, 1998; Kašpárek & Novický, 1997). Besides the spatially-lumped concept, the time is also lumped, i.e. the time resolution is one month. The BILAN model is based upon a set of relationships, which describe the basic principles of the water balance both in the unsaturated and saturated zone. The main structure of the model is given in Figure 9.2.

Time series of monthly precipitation, air temperature and relative air humidity are input data. The model simulates time series of monthly potential evapotranspiration, actual evapotranspiration, infiltration into the soil, q_s (Figure 3.3), and recharge, I, from the soil to the aquifer. The computed potential evapotranspiration is based on a relationship between the potential evapotranspiration, temperature and saturation deficit, which is derived for individual months from observed data. The relationship is different for various

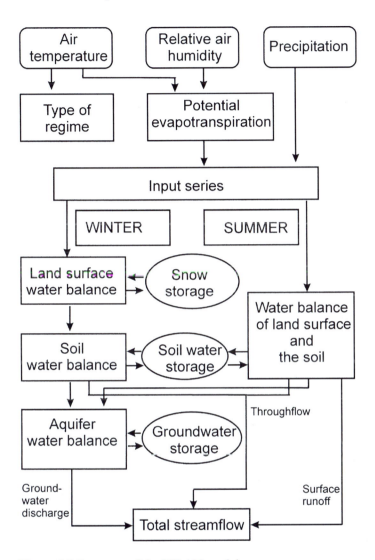

Figure 9.2 Structure of the BILAN model.

bioclimatic zones (tundra, coniferous forest, mixed forest, deciduous forest and steppe). The mean annual air temperature is used for interpolating between the bioclimatic zones at a continuous scale meaning that the potential evapotranspiration is derived from two bioclimatic zones. Potential evapotranspiration is calculated based upon Russian experience in the Northern Hemisphere. For other regions, time series of potential evapotranspiration are used as input data. Besides the potential evapotranspiration, the amount of water

Box 9.1

Model reliability

To what extent are hydrological models able to simulate the real-world hydrological system of a catchment? A hydrological model, either stochastic or physically-based, is by definition a simplified representation as it is impossible to reproduce the exact behaviour of hydrological systems that are very complex by nature. This implies that 100% reliability cannot be reached. In this context, model reliability is the ability to simulate observed time series, including its statistical moments (Section 7.1). The goals in developing a model are several, for example (a) to better understand a hydrological system, (b) to investigate how the modelled hydrological system behaves if different stresses are applied (e.g. abstraction), (c) to study effects of changes in the hydrological system (e.g. reservoir building), and (d) forecasting (short- and medium-term).

Model reliability depends on, amongst other factors:

a) hydrological processes included;

b) numerical approximation of equations describing the hydrological processes;

c) available data, representativity, quality, and

d) spatial and temporal schematization.

The first two are features relevant for all catchments to be modelled, whereas the last two depend on the catchment to be modelled and the choices of the model developer. Models are mostly calibrated, meaning that model parameters are adapted to get a better agreement between simulated and observed hydrological variables. The reliability of a model is tested through model validation, i.e. confrontation of simulated and observed time series of hydrological variables, preferably from many locations in the catchment and from a period not used for model calibration.

Clearly, it is impossible to give a general statement about the reliability of a particular model or group of models. In general, models that include only one domain of the hydrological cycle (e.g. soil, surface water) with proper boundary conditions simulate that domain more reliably than comprehensive, integrated 3D hydrological models. Simulation of hydrological extremes is more difficult than reproduction of average hydrological conditions. For example, van Lanen *et al.* (1997) present the reliability for a range of models that simulate the flow duration curve, the drought deficit duration and severity (Chapter 5). Usually the reliability for estimating drought characteristics is lower for Q_{90} as a threshold than for Q_{70}.

Note that a good fit does not automatically mean that model reliability is high for prediction of changes in the system. This requires validation of the change, which is often impossible.

that is stored in the snow pack, in the soil and aquifer is also simulated for each month. All these hydrological variables apply to the whole catchment. Furthermore three runoff components, i.e. surface runoff, throughflow and groundwater discharge, are calculated at the outlet of the catchment.

Temperature is also used to distinguish between winter and summer conditions (regime type). If a snow pack is present, a snow melting algorithm is applied for temperature above 0°C. Melting snow water and rainfall infiltrate the soil. In the soil the infiltrated water is stored and can be extracted by the vegetation. The vegetation extracts soil moisture at the potential rate (potential evapotranspiration) as long as there is sufficient water in the soil. If the amount of water in the soil is insufficient the actual evapotranspiration will drop below the potential rate (Section 3.3). During wet months, when the precipitation exceeds the potential evapotranspiration, the surplus is used to replenish the soil water storage. Subsequently, percolation from the soil occurs if the soil moisture storage reaches its maximum. The percolation from the soil can follow a quick flow path towards the stream as throughflow, q_{if}, or a slow path through the aquifer as recharge, I. A third streamflow component, i.e. overland flow, may also occur if both the rainfall intensity and the soil moisture are high (Section 3.5).

Eight model parameters have to be identified to simulate streamflow with the BILAN model. These parameters are obtained by applying an optimization algorithm that compares monthly time series of simulated and observed streamflow. The BILAN model assumes that the hydrological year starts on 1st November. A description of the model is given in 'User's Guide BILAN' and 'Background Information BILAN' (Software, CD). The BILAN program is also stored on the CD under the item Software.

In following sections the BILAN model is applied to investigate the impact of climate change (Section 9.4), groundwater abstraction (Section 9.5), and surface water augmentation (Section 9.7). The assessment of the effects of climate change is elaborated using the River Metuje catchment (Czech Republic) as a case study.

9.2.3   Distributed numerical physically-based models

Distributed hydrological models account for the spatial variability of, for example, land use, soil properties, hydrogeological conditions and stream network. Moreover, complicated boundary conditions can be handled, such as measured precipitation time series, abstraction rates, river stages. These capabilities make the models very powerful, but also data-demanding. Examples of distributed physically-based model are SHE (Singh, 1995), MODFLOW (McDonald & Harbaugh, 1988) and SIMGRO (SIMulation of

GROundwater and surface water levels). The latter model is used in this chapter and explained below.

SIMGRO is a distributed physically-based model that simulates regional transient saturated groundwater flow, unsaturated flow, actual evapotranspiration, sprinkler irrigation, streamflow, hydraulic heads and surface water levels as a response to, for example, rainfall, reference evapotranspiration, and groundwater abstraction (Querner, 1988; 1997). SIMGRO simulates water flow for a particular period, usually a series of years. This period is subdivided in shorter periods (typical length: day, week, 10-day or month) that are used for output and specification of boundary conditions (e.g. precipitation). The computation time step for some model modules is less than one day. The main feature of this model is the integration of the saturated zone, unsaturated zone and the surface water systems within a sub-region.

A system of nodal points is superimposed over the catchment delineating so-called elements. This allows input of spatially-distributed catchment properties (e.g. land use, soil type, transmissivity, stream network). SIMGRO simulates multiple aquifer systems (Figure 9.3 and Section 3.4) including interaction between aquifers (leakage)(Figure 3.4).

SIMGRO assumes steady-state in the unsaturated zone for each time step. Important controlling factors are the storage coefficient, S_y, and the capillary rise (negative percolation, q_r) (Figure 3.3). These depend on soil properties (e.g. thickness of rootable zone, moisture retention data and unsaturated hydraulic conductivity curve, Section 3.3) and crop data (rooting depth). SIMGRO simulates for each node and time step the recharge, I, the groundwater hydraulic head, water-table depth and soil moisture storage in the root zone, S_{sor}, and in the unsaturated subsoil, S_{sos} (Appendix 3.1).

The model uses time series of daily evapotranspiration for a reference crop (grass) and for woodland as input. This might be output from any evapotranspiration model, e.g. the Makkink approach or the Penman-Monteith equation (de Bruin, 1987). The potential evapotranspiration for other crops or vegetation types are derived in the model from the values for the reference crop by converting with known crop factors (Feddes, 1987). Subsequently, SIMGRO simulates actual evapotranspiration per node using the relationship between soil moisture storage and the relative evapotranspiration (actual/potential evapotranspiration, Figure 3.6).

SIMGRO distinguishes three drainage subsystems to simulate the groundwater discharge from the aquifer. These subsystems represent ditches, and tertiary and secondary water courses. The aquifer-stream interaction is calculated for each drainage subsystem separately using the drainage resistance and the difference between water level and surface water level (Ernst, 1978). A fourth subsystem, i.e. the primary system, can also be included in specific nodes

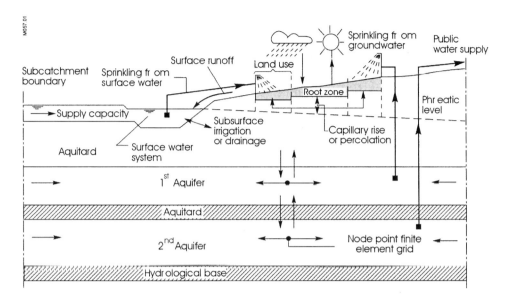

Figure 9.3 Schematization of water flows in the SIMGRO model.

(river nodes) to represent the larger streams in a catchment. Their main function is to convey water, although they also drain groundwater or infiltrate surface water into the aquifer. SIMGRO simulates the surface water system as a network of reservoirs. The inflow of a reservoir may be the discharge of the four drainage subsystems, overland flow and/or water from a sewage treatment plant. The outflow from the reservoir is the inflow to the next reservoir.

The CD gives a comprehensive description of the SIMGRO model (Supporting Document 9.1). The model is applied to the Poelsbeek and Bolscherbeek catchments (The Netherlands) to investigate the impact of human activities on droughts. The Poelsbeek and Bolscherbeek are adjacent catchments. Shallow water tables and wet soils are common. About 75% of the catchment is used as agricultural land; the remaining land is covered by woodland (16%) and residential area (8%). An unconfined aquifer with a thickness of 10 60 m occurs that overlies impermeable deposits. A dense network of streams can be observed that drain into the Poelsbeek and Bolscherbeek, which flow into the Twente canal. A full description of both basins is included on the CD (Supporting Document 9.2). In Sections 9.3–9.6 the results of the model application are presented. The results for the Bolscherbeek are given only if they substantially deviate from those of the Poelsbeek. Also on the CD, results of the calibration and validation of the SIMGRO model for the Poelsbeek and Bolscherbeek, and results for the reference situation (period 1951–1998), including drought characteristics are presented (Supporting Documents 9.3 and 9.4 respectively). Validation results for the models BILAN and SIMGRO used in this chapter are presented in Sections 9.4.4, 9.5.3 and 9.7.2 and Supporting Document 9.3. The results of the models agree reasonably well with observations.

In this chapter the two physically-based models (BILAN and SIMGRO) are used to illustrate the impact of human influence on drought. These are just examples and many different models exist that deviate, for instance, in (a) the hydrological processes included, (b) data requirements, (c) steady-state or transient conditions, (d) lumped or distributed, (e) 1-D, 2-D or 3-D, (f) numerical solution method, (g) pre- and postprocessors, and (h) computation time. The objective of the study, biophysical conditions, data availability, time, budget, skills and tradition determine which model is most appropriate. In any project sufficient time needs to be allocated to model selection, together with investigation of model reliability (Box 9.1), before it is applied to assess drought characteristics under different conditions.

9.3   Impact of land-use change

Population growth has led to extensive land-use change. This section describes the impact of some changes in land use on hydrological drought development. Section 3.3 has already introduced the possible effect of land-use changes on hydrological processes and the development of drought. The effects may be manifold. Firstly, land cover contributes to the amount of moisture in the atmosphere affecting precipitable water (e.g. Blyth $et\ al.$, 1994). Secondly, land-use change influences interception, potential evapotranspiration (PE), rooting depth and partitioning between overland flow (q_{of}) and soil infiltration (q_s). All these processes have an influence on actual evapotranspiration and recharge and finally on groundwater response and streamflow (Calder, 1992). Several feedback mechanisms exist between climate and land cover, although observational evidence is still poor (Moors & Dolman, 2003). It is likely that land-use change affects water tables and streamflow, thus affecting hydrological drought. This section first describes some observed impacts of land-use change followed by an illustration of the simulated effects of afforestation scenarios in a temperate, humid climate using SIMGRO. The lumped physically-based model BILAN is not used in this section because land use is not explicitly considered in the model.

9.3.1   Observations

There is a wide range of experience of the effect of land use on hydrological conditions, although this is often hard to distinguish from other human influences in the catchment, especially from climate forcing. Large-scale persistent droughts were reported in the Sahel region for the period 1960–1990 (Box 2.2). The lower rainfall led to a considerable decrease in river discharge in the region (e.g. Savenije, 1995; Oki & Xue, 1998). For example, the annual average discharge of the River Niger at Niamey over the period 1929–1968 decreased from 1060 to 700 $m^3 s^{-1}$ over the following 26 years. Land-atmosphere coupling (Section 9.4), including degraded land covers, is the likely reason for this major reduction by 34% (Hutjes *et al.*, 2003).

Paired catchment experiments showed that clear cutting of a natural oak-hickory basin resulted in a streamflow increase. The increase was most pronounced in the first year and then decreased logarithmically until full cover was reached again (Hibbert, 1967). Catchment properties determine which flow components, i.e. overland flow or base flow (Section 3.2), are most affected. Furthermore, forest harvesting may change soil properties, such as soil crusting and unsaturated hydraulic conductivity through disappearance of macropores. For example, Mahe *et al.* (2002) estimated that the soil moisture capacity decreased by 33% in the basin of the River Nakambe (Burkina Faso) due to clearance of 87% of the natural vegetation. This meant that, despite the reported reduction of rainfall and construction of dams, the discharge increased probably leading to smaller hydrological droughts. In Niger, intensive clearing of native savanna over the last few decades has resulted in a recharge of 1–5 mm month^{-1} in the 1950s to over 20 mm month^{-1} in the 1990s, implying a 10% increase in groundwater resources (Leduc *et al.*, 2001; Favreau *et al.*, 2002).

Afforestation seems to give a similar, but opposite response to deforestation (Andréassian, 2004). Forest treatments in New Zealand confirm this (Rowe *et al.*, 1997). In the Glendhu catchment (southeast on the Southern Island) the *MAM*(7) decreased by 16% as a result of the reforestation of tussock grasslands. In the Moutere catchment (north on the Southern Island) planting of pine trees on grasslands increased the number of zero-flow days per year from 64 to 157. Bosch & Hewlett (1982) found for higher-altitude headwater catchments that streamflow decreases as vegetation cover changes from scrub, deciduous hardwood, pine to eucalypt. The planting of tall natural grass in a wetland area in Canada unintentionally led to drought enhancement (Box 9.2).

Forest fires are a hazard that is related to drought (Section 1.3) and which causes unintended temporary change of land use. It reduces the protection of land cover resulting from the fires and it tends to increase the water repellency (hydrophobicity) of the soil surface, which might lead to more surface runoff.

The impacts can be manifold. Aronica *et al.* (2002) showed that the annual *runoff coefficient* (ratio of runoff depth to precipitation depth) of the Oreta

---

Box 9.2

Natural vegetation induces droughts in North America

In the cold semi-arid plains in North America (P = 300–500 mm year^{-1}, PE = 600–1000 mm year^{-1}) there are millions of small wetlands. These provide a critical breeding habitat for migratory fowl. The wetlands are found in the middle of extensive areas of dry-land farming. Thus, a catchment consists of a number of wetlands without an outlet and agricultural land. The wetlands are vulnerable to land-use change and global warming. An important input for a wetland is the transport of snow-derived water from the uplands of the catchment as blowing snow and overland flow over the frozen soil (25–50%). Another input is direct precipitation on the wetland. Groundwater discharge to the wetland is negligible through the thick clay-rich glacial till with low hydraulic conductivity. In the St Denis National Wildlife Area (SDNWA) in central Saskatchewan (Canada) several wetlands occur. In the SDNWA about one third of a particular catchment was converted from dryland cultivation to tall broom grass (natural grass) in the 1980s. The purpose was to improve wildlife habitat, i.e. to provide dense nesting cover and better living opportunities for mammals. The land-use change meant that some wetlands in the SDNWA were entirely surrounded by natural grass. After a few years, the wetland in the natural-grass areas completely dried up, whereas in the cultivated regions of the SDNWA the wetlands filled up every snowmelt period. Van der Kamp *et al.* (2003) found the following reasons for this unintended result:

   a)   a considerable decrease of wind-transported snow in the catchment with natural grass, and

   b)   an enhanced infiltration in the frozen grassed soils.

In winter the tall natural grass traps most of the snow, preventing the snow from being transported to the wetlands in these areas. Experiments in a particular year showed that the spatially-averaged snow water equivalent was about 100 mm in the grass areas, whereas it was not more than 25 mm on the cultivated land, which is crop-free in the winter. Ring infiltrometer tests proved that the infiltration capacity of soils in both the frozen and unfrozen grass areas increased after a few years through development of macropores. The infiltration capacity of the frozen soil in the grassland is sufficient to absorb most of the snowmelt, whereas in the cultivated areas the infiltration is small and significant overland flow occurs. So, conversion of arable land to natural grassland in snow-affected climates may lead to dried-up wetlands contributing to severe hydrological droughts.

catchment (Sicily, Italy) was not significantly affected, whereas the coefficient for another catchment in the region, i.e. the Asinaro catchment, was about three times higher than for the reference period. The reasons for the different runoff coefficient are not yet known.

9.3.2   Afforestation of the Poelsbeek catchment

The SIMGRO model is applied to the Poelsbeek catchment to illustrate in a systematic way the simulated impact of land-use change on drought development. Meteorological data from the period 1951–1998 are used to assess what would happen if the predicted land-use change took place. Current land use in the Poelsbeek catchment (Supporting Document 9.2) is changed in the model. The urban area remains unchanged, whereas most agricultural land (grassland and maize) is converted into deciduous forest (scenario 1), and coniferous forest (scenario 2). In both modelled scenarios about 80% of the catchment area is assigned forest cover.

Recharge

Effects of land-use change on groundwater drought can be analysed by investigating time series of groundwater recharge, water tables and discharge (Sections 3.3, 3.4 and 5.6). First the effect on recharge is described for the reference situation and two scenarios in which current land use (reference situation) is converted into forest. Supporting Document 9.4 describes the reference situation. *Recharge indices* are derived from the empirical frequency distribution (Table 9.1). I_{70} is the recharge that is exceeded in 70% of the time. The results presented describe a representative subcatchment in the centre of the Poelsbeek catchment (Supporting Document 9.4).

Conversion into deciduous forest results in slightly lower groundwater recharge compared to reference conditions, except for extremely wet and dry years. For instance, the I_{70} and I_{90} decrease by about 10%. The average recharge

Table 9.1 Recharge indices for the reference situation and for scenarios with 80% deciduous forest and 80% coniferous forest

Scenario	I_{70} (mm year^{-1})	I_{90} (mm year^{-1})
Current situation	256	177
80% deciduous forest	235	160
80% coniferous forest	153	90

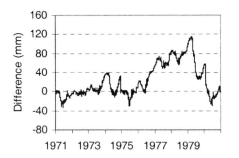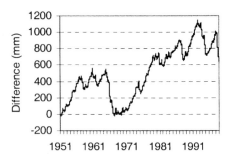

Figure 9.4 Difference in recharge deficit volume due to conversion of current land use into 80% deciduous forest (period 1971–1980, left), and 80% coniferous forest (period 1951–1998, right) in the Poelsbeek catchment (please note difference in scale of y-axis).

decreases by 4%. Coniferous forest, however, has a considerable effect on recharge. In all years recharge decreases. The average recharge is 34% lower, and the I_{70} and I_{90} decrease by 40–50%. The main reason for the low recharge is the high potential evapotranspiration. As discussed in Section 3.3 this leads to high actual evapotranspiration losses in temperate, humid climates.

Drought in recharge (recharge drought) is identified using the sequent peak algorithm (SPA, Section 5.4.2) for the reference situation and for the forest scenarios. The outflow of the reservoir is set at the I_{70}. Figure 9.4 presents the time series of the difference in the recharge deficit volume due to afforestation indicating a change in drought conditions. For deciduous forest only part of the times series is given.

In general, conversion from agricultural land into deciduous forest does not result in large differences in the recharge deficit volume (Figure 9.4, left). For about 50% of the time the recharge deficit volume for the deciduous forest is even smaller (negative value) implying less severe drought. However, during the dry periods, i.e. the late 1950s, the second part of the 1970s, and the early 1990s, the recharge deficit volume for the catchment is higher than for the agricultural catchment. In 10% of the time the recharge deficit volume is larger than 50 mm.

Change of agricultural land into mainly coniferous forest leads to a completely different recharge regime (Figure 9.4, right). Most of the time (i.e. 80%) the difference in recharge deficit volume is more than 150 mm. The wet, late 1960s are the only exception. In this period the recharge deficit volume for the agricultural catchment is larger (about 2.5% of the time).

To conclude, in dry years afforestation leads to more severe recharge drought both for deciduous and coniferous forest in the selected temperate, humid catchment. Recharge drought is more severe (higher deficit volume) and prolonged (longer deficit duration) in the scenario with coniferous forest.

Water tables

The effect of afforestation can also be analysed by comparing simulated time series of water tables for the reference situation and the scenarios with forest. The analysis uses the deviation of the simulated groundwater level from a threshold level (H_{70}, Section 5.6). The deviation is a drought characteristic that indicates the state of the groundwater system. If multiplied by the storage coefficient (S_y, Section 3.4) the state can be converted into a volume-oriented characteristic. The time series of the difference in deviation due to conversion of current land use into coniferous forest is given in Figure 9.5. The graph applies to a representative location in the Poelsbeek catchment.

In this case a dominant cover with deciduous forest in the catchment instead of agricultural land leads to hardly any differences in deviation of the groundwater levels for most of the years. This is not surprising because the water tables are strongly controlled by the recharge, which did not alter much. Dry years (i.e. 1959 and 1976), however, show up clearly as isolated peaks. The maximum deviation is 0.2 m implying a six times higher deviation compared to reference conditions (Supporting Document 9.4), which points to more severe drought.

The conversion of an agricultural catchment into a basin with mainly coniferous forest leads to clear differences in the deviation of the groundwater level (Figure 9.5). The differences are larger than for deciduous forest, and in dry periods the drought is more pronounced. Differences up to 0.4 m occur in dry years.

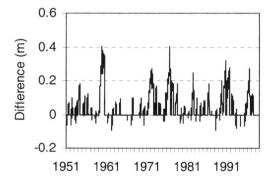

Figure 9.5 Difference in deviation of the water table due to conversion of current land use into 80% coniferous forest at a representative location in the Poelsbeek catchment.

To summarize, analysis of drought in groundwater levels supports the conclusion derived from analysis of recharge deficit volume that afforestation leads to more severe drought in the example catchment. The planting of coniferous forest in particular causes significantly more, and more severe, groundwater droughts.

River flow

Land-use change may affect streamflow (sum of groundwater discharge and overland flow simulated with SIMGRO) in the example catchments due to different recharge and associated water tables, as illustrated in the previous paragraphs. Figure 9.6 gives the empirical frequency distribution (Section 6.2.2) of the daily streamflow (FDC, Section 5.3.1) for the current situation and for scenarios with 80% forest.

The frequency distribution of the daily discharge hardly deviates for catchments with agricultural crops or covered with deciduous trees. Therefore, streamflow drought does not substantially change because of planting of deciduous trees. This implies that the groundwater system is able to attenuate minor changes in recharge drought (Sections 3.4 and 5.6).

Conversion of agricultural land into coniferous land leads to a substantially different frequency distribution of the daily discharge (Figure 9.6). The low recharge for coniferous forest results in considerably lower discharge (75 mm year^{-1}) compared to the reference situation with mainly agricultural land. The graph shows that low flows also decrease for a catchment with mainly coniferous forest. It is clear that replacement of agricultural crops by coniferous trees has longer and more severe streamflow drought as a consequence.

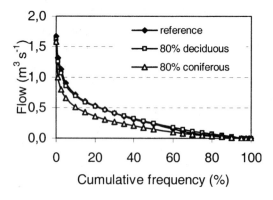

Figure 9.6 Distribution of the daily discharge of the Poelsbeek catchment for the reference situation (period 1951–1998), and for situations with 80% deciduous forest and 80% coniferous forest.

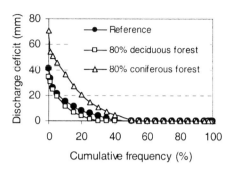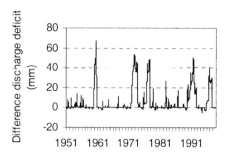

Figure 9.7 Discharge deficit volume for the Poelsbeek catchment: cumulative frequency of the daily discharge deficit volume for current land use, 80% deciduous forest and 80% coniferous forest (left), and difference in discharge deficit volume due to the conversion of current land use into 80% coniferous forest (right).

Therefore the discharge deficit volume and duration is elaborated in the following.

Discharge deficit volume is computed using the SPA and simulated daily discharge. The Q_{70} (0.086 m^3 s^1) is used as outflow from the reservoir (Section 5.4.2). Figure 9.7 presents the empirical distribution for the discharge deficit volume due to afforestation of the Poelsbeek catchment. The time series of the difference in discharge deficit volume is also given for conversion of agricultural land into predominantly coniferous forest.

As discussed earlier, replacement of agricultural crops by deciduous forest has hardly any influence on deficit volumes. However, replacement of agricultural land by mainly coniferous forest leads to a substantial increase of the severity of the streamflow drought (Figure 9.7, left). In 5% of the time discharge deficit volume exceeds 46 mm. Figure 9.7 (right) shows that the difference in discharge deficit volume is mainly bounded to the 1950s, the 1970s and the 1990s. These deficit volumes are triggered by increased recharge drought and hence lower water tables. The differences in discharge deficit volume in the dry periods reach values up to 40–70 mm.

Besides the severity expressed as discharge deficit volume, the duration also provides information about the impact of land use on streamflow drought. In catchments with mainly deciduous forest the streamflow drought is slightly shorter in dry years than in agricultural catchments. As explained above, drought is more severe in the catchment with coniferous forest. In 5% of the time even multi-year droughts develop.

9.3.3   Discussion

The simulation results for the example catchment in the temperate, humid climate confirm that afforestation mostly leads to lower average and dry-season streamflow (e.g. Dingman, 2002; Rowe *et al.*, 1997) implying a higher risk of drought, whereas deforestation has the opposite effect. This especially applies to coniferous forest. The reason for this is that forests consume more water by evapotranspiration than other land covers. High interception losses occur, especially for evergreen forest, and deeper rooting systems increase evapotranspiration. Note, that reduction of low flow due to afforestation is somewhat alleviated because the forest litter layer and topsoil tend to store and infiltrate more water than agricultural land, resulting in less overland flow (Section 3.2).

Modelling of afforestation in other regions, e.g. Scotland (UK), Poland (PL) and the Netherlands (NL), confirms the previous findings. In the River Jegrzina basin (PL) complete bush expansion was simulated resulting in low water tables in late summer (Mioduszewski *et al.*, 1997). In the Hupsel catchment (NL) Q_{70} decreased by 14 and 50% dependent on the investigation period and type of model. Drought lasted 8 to 31% longer. In the Monachyle catchment (UK) the change was less substantial, i.e. the Q_{70} reduced by 3% and the drought duration was 11% longer (Querner *et al.*, 1997).

Sellers & Lockwood (1981) modelled the influence of pine, oak, wheat and grassland cover for a certain year. They found an increasing interception loss going from grass to wheat, oak and pine cover. Transpiration losses in decreasing order were grass, pine, oak and wheat. The resulting streamflow for wheat was 417 mm year^{-1}, grass 318 mm year^{-1}, oak 279 mm year^{-1}, and pine 152 mm year^{-1}. Peters *et al.* (2001) found for the Pang catchment (UK) that oak had lower annual recharge (1960–1996) than agricultural crops. Permanent grassland, spring wheat and forage maize had similar recharge amounts, whereas winter wheat had lower recharge. Long-term average recharge was substantially lower (about 150 mm year^{-1}) than the results given by Sellers & Lockwood (1981) for these crops. In semi-arid climates recharge is much smaller. Long-term average recharge in the Upper Guadiana (ES) was 23 and 46 mm year^{-1} for deep-rooting crops (e.g. vines) and medium-rooting crops (e.g. cereals) respectively, implying that large-scale conversion from conventional dry-land arable crops to vineyards would lead to an increase in drought (Peters & van Lanen, 2001).

The simulation results in the previous section showed that drought in recharge due to land-use change is more pronounced than in water tables and discharge. Storage characteristics of the groundwater system cause this attenuation (e.g. Peters *et al.*, 2003).

Model studies in this and the previous section assume that precipitation does not change due to land-use change. However, for large-scale changes, e.g. the Amazon basin, it is expected that precipitation typically decreases by 10–20% due to deforestation. A reduction in evapotranspiration and an increase in upward long wave radiation and sensible heat flux cause the decline of precipitation. On the other hand, small-scale clearing of forest can lead to enhancement of precipitation (Moors & Dolman, 2003). Usually, the effects of land-use change and climate change are treated separately (Section 9.4), despite the known atmosphere-land coupling.

9.4   Impact of climate change

In many regions climate change affects precipitation, temperature and potential evapotranspiration (Section 2.6) having an effect on meteorological drought. An important question is how changes in meteorological drought affect groundwater and streamflow droughts. Will hydrological drought be enhanced by climate change or is it reduced due to characteristics of the groundwater system? Will the response be identical for all catchments? This section on the influence of climate change is a further elaboration of Section 2.6, which describes climate change and climate variability in a hydroclimatological context. As stated in Section 2.6 it is hard to distinguish between climate change and multi-decadal climate variability, a view supported by Berdowski *et al.* (2001). In general, two approaches can be followed to assess the possible impact of climate change. First, time series of observed hydrological data are analysed to detect trends (e.g. Hisdal *et al.*, 2001), and second, simulation with physically-based models is carried out. The latter comprise GCMs that simulate the atmosphere including a more or less simple land surface module. Section 2.6.1 summarizes the main trend in observed changes. In Section 9.4.1 some more detailed observations are added. Next, GCM scenarios as introduced in Section 2.6.2 are further elaborated for the example catchments (Section 9.4.2). The example catchments include a temperate, humid climate type (Poelsbeek catchment) and another catchment with snow in the cold season (Metuje catchment). The impact of climate change for the example catchments is described in Sections 9.4.3 and 9.4.4. The impact assessment in the Metuje catchment (Section 9.4.4) is elaborated as a case study. The section on climate change concludes with a discussion (Section 9.4.5).

9.4.1   Observations

Climate change is expected to intensify the whole hydrologic cycle leading to more frequent floods and droughts. Section 2.6.1 lists some studies of trends in observed low flow and drought. Reliable detection of trends is cumbersome and Ziegler *et al.* (2002) estimated that between several decades and more than a century of measurements are required to detect climate change with high confidence.

Rivers in Africa show the following patterns. The River Congo-Zaire in Brazzaville had a stable annual average discharge around $40\,000\ \text{m}^3\,\text{s}^{-1}$ over a long period from 1960–1990 (Laraque *et al.*, 1997). The 1960s, 1970s and the period 1980–2000 had a high, average and low discharge, respectively. In the latter drought period the annual average discharge deviated 8% from the long-term average. The River Oubangui in Bangui, which is a northern tributary of the River Congo-Zaire, had a similar streamflow pattern. The reduction in streamflow over the period 1980–2000, however, was more distinct, i.e. it deviated 49% from the average. The River Nile in Dongola had a more or less stable annual average discharge of about $3000\ \text{m}^3\,\text{s}^{-1}$ until 1970 followed by a clear decrease to about $2000\ \text{m}^3\,\text{s}^{-1}$ over the period 1970–1990 (Grove, 1998). The annual average discharge of the River Niger in Koulikoro varied from 1000 to $2000\ \text{m}^3\,\text{s}^{-1}$ with three long low flow periods, from which the one in 1970–1995 was the most extreme. Levels of Lake Chad show a similar pattern. Amani & Nguetora (2002) analysed the monthly average discharge of the lake for two periods, i.e. 1941 to 1969 and 1970 to 2000. In the second period the three months low flow period indicates a change in hydrological regime probably related to climate change. The River Zambesi also had the lowest discharge in the period 1980–1990. Climate change seems to apply to all the above examples. According to Grove (1998) there is connection between fluctuations in the Southern Oscillation Index (SOI), El Niño (Section 2.3) and rainfall in Africa. Low SOI is accompanied by reduced rainfall and river discharge in east Africa. It is obvious that more research needs to be undertaken to find the causes of variations in time series of lake and river flow that have been observed over the last decades in this region.

In the United States streamflow from catchments without evident anthropogenic influence shows an increasing trend in 16 out of 20 water regions over the period 1929–1988 (Hubbard *et al.*, 1997). Base flow in most of the selected stations also increased. It is likely that meteorological or climatic forces cause the increase.

Large areas of Asia, North America and Europe experience snow cover during a significant part of the year. The presence of snow has a considerable influence on the atmosphere-land interaction through the strong reduction of the terrain roughness and the extremely high albedo affecting the radiation balance. Shifting of snowmelt to earlier in the year may have a marked influence on the water balance of a catchment (e.g. IPCC, 2001a, 2001b; Cutforth *et al.*, 1999), as will be illustrated in Section 9.4.3. Another effect of global warming is the more intensive melt and retreat of glaciers (e.g. Berdowski *et al.*, 2001). For example, Hasnain (2002) reports the increase in discharge downstream of the Dokriani glacier snout in one of the headwaters of River Bhagirathi, a tributary of the River Ganges. Average discharge increased from 5.8 m^3 s^{-1} in 1994–1995 to 35–46 m^3 s^{-1} in 1999–2000. The increase can be related to a temperature rise.

The observed daily temperature and precipitation in 34 European countries showed a positive trend in annual precipitation between 1946 and 1999 mainly for northern stations, whereas negative trends concentrate in the south (Klein Tank *et al.*, 2002). At 45% of the stations a significant decrease in frost days occurred. The Europe-averaged increase in number of warm days is similar to the decrease in cold days. The dominant warming trend was generally associated with a slight increase in wet extremes. However, there are large trend differences between nearby stations.

Majerčáková *et al.* (1997) show a remarkable decrease in the annual streamflow since 1980 for 64 Slovak rivers covering a range of hydrological regimes. The decrease is most prominent in autumn and winter months and smallest in May and the following summer months. The highest decrease in streamflow is in the southeast (30–40%) and the lowest in the west, northwest and northeast (5–10%) of Slovakia. Analysis of maximum, mean and minimum monthly spring yields gives similar results, although the decreasing trend starts a few years later probably due to groundwater storage effects (Section 3.4). Decrease of spring yields strongly varies over the country. Karst springs in the south and flysch springs in the north and northeast are most affected.

Not all studies present clear change in hydrological drought conditions over a large region (e.g. Hisdal *et al.*, 2001). They show that frequency of drought, in terms of annual numbers of droughts, has a different pattern in Europe. Central Europe (Czech Republic and Slovakia) has a trend towards more droughts, which are more severe, whereas most of the UK tends to fewer droughts which are more severe. Based on an analysis of the discharge of the River Meuse over the period 1911–1998, it is impossible to conclude that drought has become more severe or frequent (Uijlenhoet *et al.*, 2001; De Wit *et al.*, 2001). No trend can be identified, although the River Meuse regularly suffers from drought (10 out of 100 years the river cannot meet the minimum supply criteria for some time of the year). Scenario studies show that climate change in the Meuse basin is expected to lead to lower average discharge by the end of the summer. This decrease, however, is small compared to the natural variation between the years.

9.4.2   Climate-change scenarios for the example catchments

The most common method of developing climate scenarios for hydrological impact assessments is to use results from GCMs (Section 2.6). The majority of assessment studies use time series of observed climate data to characterize the present-day situation. Next, the results of the GCMs are applied to create adapted climate data series. These perturbed data are then used as input to hydrological models, and the resulting changes in groundwater, river flow and hydrological drought are evaluated. This section also follows this approach, implying a separate simulation of the climate and hydrology. Most GCMs have a land surface model, including a simple hydrological module that has been used to study hydrological response to climate change at global and continental scales (e.g. Arnell, 1999; Ziegler *et al.*, 2002). The land surface module in these models is, however, insufficient to model hydrological effects at the catchment scale. Therefore, more comprehensive hydrological models, e.g. SIMGRO and BILAN, with perturbed time series of meteorological data are applied. Results obtained using these models are discussed in the following.

Central Europe: River Metuje catchment

The climate change scenarios used for the Metuje catchment (Section 9.4.4) are based on the HadCM2 (Johns, 1997) and ECHAM4 (Roeckner *et al.*, 1996) GCM simulations. Nakicnovic *et al.* (2000) defined a so-called pessimistic CO_2 emission development (A2) and an optimistic development (B1). These developments imply high and low climate sensitivity, respectively. Table 9.2 lists the climate change scenarios that are defined for the Metuje catchment.

The climate scenarios in the Czech Republic have 2050 as the reference year. The climate scenarios are implemented in the BILAN model (Section 9.2.2) by adapting the original time series of air temperature, precipitation and air relative humidity for the individual months. The largest changes can be found in the air temperature. The increase in the mean annual air

Table 9.2 Climate change scenarios for the Metuje catchment (Czech Republic)

Scenario	GCM	CO_2 emission scenario	Climate sensitivity
EC1L	ECHAM4	B1	low
EC2H	ECHAM4	A2	high
Ha1L	HadCM2	B1	low
Ha2H	HadCM2	A2	high

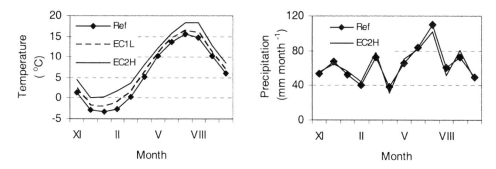

Figure 9.8 Meteorological data for the reference situation and for some climate scenarios in the Metuje catchment: mean monthly air temperature (left), and mean precipitation (right).

temperature is about 1°C for scenarios EC1L and Ha1L, whereas it is almost 2.5 and 3°C for scenarios Ha2H and EC2H, respectively. For most scenarios, the largest increase occurs during winter months (XI–II). In Figure 9.8 (left) this is shown for scenarios EC1L and EC2H.

A minor decrease in annual precipitation is reported for all scenarios. The decrease does not exceed 1% for any of the scenarios. The Ha2H scenario gives the largest decrease in precipitation. The water vapour pressure is calculated from air temperature and relative air humidity. For all scenarios it decreases, particularly during late summer (data not shown). Scenario Ha2H has the maximum decrease.

Table 9.3 Climate scenarios for the Netherlands for 2100 (Beersma *et al.*, 2001)

| | Temperature rise | |
| | Intermediate | Severe |
Variable (average	2°C	4°C
Summer rainfall	2%	4%
Winter rainfall	12%	25%
Extreme rainfall events[1]	20%	40%
Potential evapotranspiration	8%	16%

[1] for events > 20 mm d^{-1} in summer, and > 10 mm d^{-1} in winter

West Europe: Poelsbeek and Bolscherbeek catchments

The Royal Netherlands Meteorological Institute has translated the outcome of the IPCC-ECA project in climate change scenarios for the Netherlands (Beersma *et al.*, 2001). Table 9.3 lists the main features.

For the Poelsbeek and Bolscherbeek catchments both the intermediate and severe scenarios were chosen. The temperature rise is 2 and 4°C for the intermediate and severe scenarios respectively. Note that the precipitation in the Poelsbeek and Bolscherbeek catchments falls predominantly as rain, contrary to the Metuje catchment.

9.4.3   Climate change in the Poelsbeek catchment

The distributed physically-based model SIMGRO (Section 9.2.3) is used to assess the impact of climate change in the Poelsbeek catchment on hydrological drought. Meteorological data from the period 1951–1998 are used. First the effect of climate change on groundwater drought (recharge and water tables) is discussed followed by the effect on streamflow drought.

Recharge

Section 3.3 illustrated that net precipitation, potential evapotranspiration, land use and soil type are the main factors controlling groundwater recharge. Climate change modifies the first two processes and possibly land use. This affects actual evapotranspiration. Clearly the increase in summer and winter precipitation (Table 9.3) leads to a substantial increase of the average annual precipitation. The increase in the Poelsbeek is about 7 and 13.5% for the intermediate and severe climate change scenarios. The higher temperatures throughout the year cause higher evapotranspiration rates. Both the high precipitation and potential evapotranspiration lead to an increase in the actual evapotranspiration in temperate, humid climates (Section 3.3.1). Thus the average actual evaporation in a representative sub-basin in the Poelsbeek catchment increases by 33 and 44 mm year^{-1} for the two climate-change scenarios, respectively.

Groundwater recharge is positively correlated with precipitation and negatively correlated with actual evapotranspiration. Hence, high precipitation could be counteracted by high actual evapotranspiration. In the Poelsbeek catchment, however, the precipitation increase is clearly larger than the enhancement of the actual evapotranspiration. This implies that both climate change scenarios lead to higher groundwater recharge. The average annual recharge increases by 20 and 65 mm year^{-1} in the representative sub-basin. For the intermediate scenario the annual recharge does not deviate much from the reference situation in dry years. This indicates that drought does not become

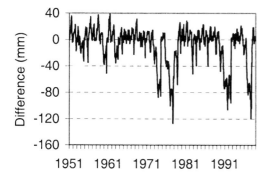

Figure 9.9 Difference in recharge deficit volume for the severe climate change scenario in the Poelsbeek catchment.

less severe in response to intermediate climate change. For the severe scenario the groundwater recharge is 20 to 30 mm year^{-1} higher in years with recharge lower than 300 mm year^{-1}, which might result in a smaller number of drought events.

The SPA is used to identify recharge droughts (Section 5.4.2). Figure 9.9 presents the time series of the difference in recharge deficit volume for the severe climate change scenario for a representative subcatchment in the Poelsbeek. A negative deficit volume implies a less severe drought compared to the reference situation.

For the intermediate scenario, about 60% of the time the recharge deficit volume is positive, implying more severe droughts (not shown). However, the deficit volume difference does not exceed more than about 30 mm meaning that these are minor droughts. Larger and negative differences occur during dry periods, i.e. 1970s and 1990s, which means that these major droughts are less severe. The severe climate change scenario shows more pronounced deviations (Figure 9.9). Particularly during dry periods the recharge deficit volume is up to 120 mm lower. This is a significant decrease, because the recharge deficit volume for the reference situation in 90% of the time is not more than about 180 mm (Supporting Document 9.4).

Water tables

Groundwater droughts due to climate change are also identified by analysing water tables. The deviation from the threshold water level is used as a measure (Section 5.6) for a representative location in the centre of the Poelsbeek catchment. The dry-season water tables are not substantially affected in either climate scenario. For the intermediate scenario there are a few dry periods, i.e. 1959, 1970s and 1990s, when the deviation is smaller pointing towards less severe drought. The pattern of the deviation for the severe scenario is more or less similar.

River flow

The slightly higher groundwater recharge causes an increased discharge for the Poelsbeek, i.e. 7 and 45 mm year^{-1} for the intermediate and the severe climate scenarios, respectively. Figure 9.10 (left) gives the empirical frequency distribution of the daily discharge.

In years with a low discharge, the annual discharge for the intermediate scenario does not differ from reference conditions. For the severe scenario the annual discharge is higher in these years (not shown). However, the distribution of the daily discharge hardly shows any difference in the low flow range, indicating that streamflow droughts are not very different (Figure 9.10, left).

Streamflow drought as discharge deficit volume is computed with the SPA using daily discharge data. Figure 9.10 (right) presents the time series of the difference in discharge deficit volume due to severe climate change in the Poelsbeek catchment. Negative differences indicate smaller discharge deficit volumes due to climate change, i.e. less severe droughts. For the intermediate climate change scenario (not shown) the discharge deficit volume most of the time is only slightly larger (maximum about 15 mm) than the current conditions. This indicates that climate change does not reduce drought severity. However, differences are relatively small. For the severe scenario some major droughts are clearly less severe (e.g. 1990s).

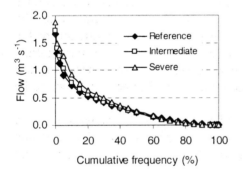 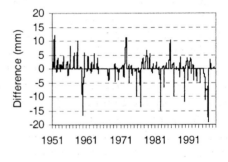

Figure 9.10 Daily streamflow for the Poelsbeek catchment for the reference situation and for climate change situations: cumulative frequency distribution (left), and difference in discharge deficit volume for the severe scenario (right).

9.4.4 Case Study: Climate change in the Metuje catchment

The lumped physically-based model BILAN (Section 9.2.2) is used to assess the impact of climate change on streamflow drought in the Metuje catchment for the period 1991–2000. The results are more extensively explained than those of SIMGRO, as the program is available on the CD. The manual and the comprehensive model description are also on the CD under item Software. A description of the Metuje catchment can be found in Section 4.5.3.

For the case study the BILAN model is used with monthly streamflow data from the Upper Metuje River Basin (the Metuje at the outlet, Figure 4.11) for the period 1991–2000. In this period the average observed flow was relatively small because of low flows in several years. The time series of monthly precipitation, air temperature and air relative humidity are from the Bucnice meteorological station in the centre of the catchment (Local Data Set).

Input file

As a first step an input file needs to be prepared. The three records in the beginning of the file contain information on the number of months to be simulated, the number of meteorological and hydrological variables per month and the hydrological year. The main part of the file consists of records to be filled in with the required meteorological data and observed streamflow. Each row contains the information for a particular month. The input file for the Metuje catchment is stored on the CD under item Software.

Initial values of model parameters

The model uses eight model parameters. For example, one parameter defines the soil moisture capacity, and another controls the outflow from the groundwater reservoir. Three other parameters specify the distribution of the percolation (Figure 3.3) to the groundwater recharge and to throughflow (Section 3.2) for different conditions (winter, snow melt and summer). For each of the eight parameters an initial value and a lower and upper limit need to be given to guide the calibration process.

The model also requires an estimate of the initial groundwater storage. A good estimate is needed for a proper simulation of groundwater discharge and consequently streamflow in the first months. The estimate of the storage is not altered during the calibration.

Calibration

After compilation of the input file and setting initial values of model parameters, the model uses an optimization algorithm to calibrate the eight model parameters within the prescribed range. The model tries to find a set of parameter values that gives the best agreement between simulated and observed streamflow. Different criteria can be used to focus on a good agreement either of low flows or total flow.

Reference situation

The model generates the following time series of monthly values of hydrological variables:

a)   streamflow (runoff)

b)   groundwater discharge (base flow)

c)   overland flow (direct runoff)

d)   throughflow or interflow

e)   potential evapotranspiration

f)   actual evapotranspiration

g)   recharge

h)   snow water storage

i)   soil water storage

j)   groundwater storage

The simulated discharge agrees reasonably well with the observed one (Figure 9.11). Some peaks are not well simulated in the period 1995 to 1997. However the focus was on low flows during the calibration. The model is unable to adequately simulate some low flow periods in the period 1998–1999. Comparison between the observed and simulated mean monthly discharge (not

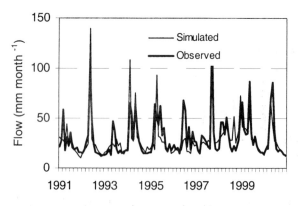

Figure 9.11 Observed and simulated monthly streamflow for the River Metuje at the outlet.

shown) shows that simulated flow is on average underestimated in the spring period when the snow melts (March and April). This can be explained by a systematic error in the observed precipitation series, which is higher for snow than for rainfall data. In the Metuje catchment the model underestimates annual flow by 19 mm (i.e. 5% of the mean annual observed streamflow of 363 mm).

Assessment of climate change

The BILAN model is run for the four climate change scenarios, and the four files with the new input data are given on the CD (Software, CD). The values of the reference situation are used for the eight model parameters and the initial groundwater storage. Note that no calibration is allowed for scenario studies! It is assumed that climate change does not change the model parameters. Simulated streamflow series for each of the four climate scenarios are compared with the flow simulated for the reference conditions. Figure 9.12 gives the flow for one of the high-sensitivity scenarios.

The graph shows that the flow simulated for the EC2H scenario is below the level for the reference situation most of the time. The higher temperature and slightly lower precipitation result in lower mean annual flow for all scenarios. The decrease for the low-sensitivity scenarios (EC1L and Ha1L) is 8%, and for the high-sensitivity scenarios (EC2H and Ha2H) it is 21–23%. Mean monthly flows in Figure 9.13 (left) show that the decrease is smaller in the winter period (November–March, month XI–III). In a number of months the flow is even higher for some scenarios (e.g. Ha2H in December; not shown).

The largest decrease in the streamflow occurs in the period April–September (month IV–IX). The decrease in mean monthly flow is up to 30–40% for the high-sensitivity scenarios. The decrease in flow is lower for the low-sensitivity scenarios (10–20%).

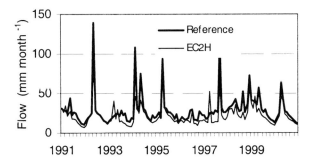

Figure 9.12 Simulated monthly streamflow of the River Metuje at the outlet for the reference situation and the EC2H climate scenario.

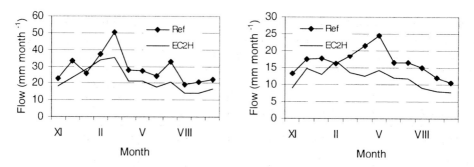

Figure 9.13 Simulated average monthly streamflow of the River Metuje at the outlet for the reference situation and the EC2H climate scenario: total discharge (left), and minimum discharge (right).

Figure 9.13 (right), which shows minimum monthly flows, illustrates that the minimum flow substantially decreases for the EC2H scenario both during the winter and summer seasons. The decrease is greater than 40% in some months. All scenarios show this pattern, although the decrease for the low-sensitivity scenario is significant smaller. Clearly the decreased flow due to climate change increases the risk of hydrological drought.

The threshold level concept (Section 5.4.1) is applied to derive streamflow drought from monthly time series of discharge. The Q_{70} for the reference situation is taken as a threshold value, i.e. 19.0 mm month^{-1}. Figure 9.14 gives the drought severity expressed as deficit volume and duration for the reference situation and for the climate change scenarios.

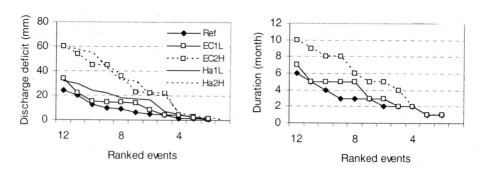

Figure 9.14 Streamflow droughts derived from monthly streamflow data from the River Metuje at the outlet for the reference situation and for four climate scenarios: deficit volume (left) and duration (right).

Table 9.4 Mean annual precipitation (*P*), potential evapotranspiration (*PE*), actual evapotranspiration (*ET*), discharge (*Q*) and percentage contribution of the three flow components to total flow for the reference situation and the four climate scenarios.

					Percentage of Q		
	P	PE	ET	Q			
Conditions	(mm)	(mm)	(mm)	(mm)	Q_{of}	Q_{if}	Q_{gr}
Reference	765	480	424	344	10	23	66
EC1L	765	535	455	318	12	21	67
EC2H	763	641	508	265	12	20	68
Ha1L	763	540	456	315	11	23	66
Ha2H	760	628	495	273	14	20	66

Eleven drought events are observed in the catchment for the reference situation in the 1990s. In Figure 9.14 the deficit volume and duration of these drought events are presented. The events are ranked in decreasing order. Climate change causes an increase of the drought severity (Figure 9.14, left). The number of droughts increases to 12 events. The discharge deficit volume for the low-sensitivity scenarios increases from about 20 mm (reference situation) to about 30 mm for the severe droughts. For the high-sensitivity scenarios the deficit volume increases to 60 mm. The drought duration shows a similar pattern. The duration for the severe droughts increases from 6 months (current situation) to 7–10 months (Figure 9.14, right). In the following the reason for the increase of drought in the Metuje catchment is elaborated.

In the summer period the higher temperature and lower relative humidity result in an increase in potential evapotranspiration (Table 9.4) and subsequently this leads to a general decrease in the minimum flows in the Metuje catchment (Figure 9.13, right).

The actual evapotranspiration increases by 7–8% for the low-sensitivity and 17–20% for the high sensitivity scenarios due to the high potential evapotranspiration. Precipitation in the Metuje catchment is relatively high, but cannot always keep up with the potential evapotranspiration, and thus part of the time the actual evapotranspiration is lower than the potential rate. The overall large actual evapotranspiration leads to additional depletion of the soil moisture storage (Section 3.3) and lower groundwater recharge (Figure 9.15).

Lower recharge negatively affects water stored in the aquifer and hence the groundwater discharge. The effect is most pronounced in the summer and

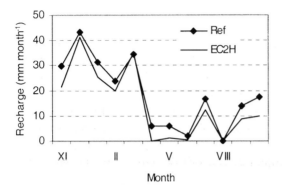

Figure 9.15 Mean monthly groundwater recharge in the Metuje catchment for the reference situation and for the EC2H climate scenario.

autumn season. In the winter period the increased temperature results in a shallower snow cover and reduced period with snow (Figure 9.16, left).

For the EC2H scenario snow only occurs in January, whereas under current conditions precipitation falls as snow in the period December to March. The effect on the snow cover is less pronounced for the low-sensitivity scenarios. More precipitation as rain and earlier snowmelt cause high surface runoff (Figure 9.16, right) and a reduction in groundwater recharge, which in turn leads to lower groundwater storage and groundwater discharge in the winter season. The effects are, however, not restricted to the winter season. As River Metuje is mainly fed by groundwater discharge, drought increases (Figure 9.14). The minor decrease in precipitation (Figure 9.8, right) enhances the effect of the above-mentioned processes.

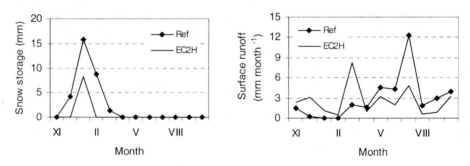

Figure 9.16 Simulated mean monthly water storage and flow in the Metuje catchment for the reference situation and for the EC2H climate scenario: water storage in the snow pack (left), and surface runoff (right).

9.4.5   Discussion

In a global-scale assessment Arnell (1999) uses a macro-scale hydrological model to simulate streamflow across the world. The results show that the patterns in change of annual river flows are generally similar to the change in annual precipitation. According to his study, the flow is expected to increase in high latitudes and many equatorial regions, but it is anticipated to decrease in mid-latitudes and some subtropical regions. The detailed model study in the Poelsbeek catchment (Section 9.4.3) confirms this, meaning that droughts become less severe. In some regions the streamflow will reduce because of a general increase in evaporation that counteracts the increase in precipitation. For example, catchments in Norway and on the Belgium-Dutch border show this trend in simulated series (Querner et al., 1997; Peters & van Lanen, 2001). In snow-affected climates the temperature increase can lead to a high portion of quick flow which does not feed the aquifer (Section 9.4.4) or to a shift in the timing of the snowmelt from spring to winter implying a reduction in the summer low flow and higher drought risk. This is confirmed by modelling results from Eckhardt & Ulbrich (2003), who found a reduction of mean monthly groundwater recharge and streamflow of up to 50% for a catchment located in the central European low mountain range.

In the Upper-Guadiana catchment (ES) where semi-arid conditions prevails, climate change ('worst' case) is expected to lead to a small water-table decline compared to seasonal and inter-annual variability (Cruces et al., 2000). The 'best' scenario predicts an increase of the water level of up to 7.5 m. The influence of climate change in the Guadiana is not very significant compared to the effects of land use and it is in a beneficial direction in terms of water resources.

At the global scale Arnell (1999) explores the change in the minimum annual flow with a return period of 10 years as an indicator for hydrological drought. He shows that this indicator changes in a similar way to the average annual streamflow. However, the proportional change tends to be larger. This conclusion is also supported by assessments at the catchment scale, e.g. Dvorak et al. (1997) show that changes in low flow characteristics tend to be proportionately larger than changes in annual, seasonal or monthly flows.

Several studies illustrate that the effect of climate change on low flow intensity and frequency is significantly affected by the geological conditions in the catchment, particularly by storage capacity of the aquifers (Sections 2.6 and 3.4). Results, for example, of low flow changes simulated for several Belgian catchments by using GCM scenarios show how the same scenario could produce rather different changes in different catchments, depending largely on the catchment geological conditions (Gellens & Roulin, 1998). Catchments with greater groundwater storage capacity tend to have higher summer flows under the climate change scenarios considered because additional winter rainfall tends to lead to larger groundwater replenishment. Low flows in catchments with low storage capacity tend to be reduced because these catchments cannot take advantage of increased winter recharge.

Box 9.1 addresses model reliability, including the need to include all relevant processes into the model. Impact of global warming response is extremely hard to assess because of several feedback mechanisms in the atmosphere-land system, the large spatial coverage and low data availability (Berdowski *et al.*, 2001; Dolman *et al.*, 2003). Most hydrological models do not include resilience of the system, i.e. the capability of living systems to adapt to changes. The BILAN model attempts to include some adaptation features. The simulated potential evapotranspiration depends on characteristics of the two bioclimatic zones, which are derived from the mean annual temperature (Section 9.2.2). If the temperature changes, potential evapotranspiration might be derived from two other bioclimatic zones. For instance, for the reference situation and the low-sensitivity scenarios (Table 9.2) potential evapotranspiration is interpolated from the coniferous/mixed forest bioclimatic zones, whereas for the high sensitivity scenarios it is derived from the mixed/deciduous forest zones. The SWAT model proposed by Eckhardt & Ulbrich (2003) includes the effect of enriched CO_2 concentrations on stomatal conductance. They showed for a German catchment that increased atmospheric CO_2 levels reduced stomatal conductance, thus counteracting increasing potential evapotranspiration induced by the temperature rise and decreasing precipitation.

9.5   Impact of groundwater abstraction

Groundwater abstractions may initiate or enhance hydrological drought (Section 9.1). Abstraction leads to lower water tables and consequently to lower spring yields and groundwater flow to streams. It may cause both groundwater and streamflow droughts. In the following section the impact of groundwater abstraction in the Poelsbeek and Bolscherbeek catchments is evaluated using the distributed physically-based model SIMGRO (Section 9.5.2) and meteorological data from 1951–1998. This study is also included on the CD as a self-guided tour (Section 9.5.3). The impact of abstraction is also assessed for the Svitava catchment using the lumped physically-based model BILAN (Section 9.5.4). Methods presented in this section can also be applied to flow naturalization (Section 4.3.3).

9.5.1 Observed groundwater drawdown and groundwater discharge

Fowler (1992) reported on the 1988 drought in Indiana (USA). Water tables declined by up to 6 m in many areas of the state. In 60% of wells the lowest water levels on record were observed. Many domestic wells ran dry and needed to be deepened. Carter (1983) described similar results for the 1980–81 drought in Georgia (USA). Mean annual water levels were up to 5 m lower than in the previous year. Levels declined for as many as 20 consecutive months. In the north of the Northern Island (New Zealand) groundwater abstraction during the 1982–83 drought led to a water-table decrease of up to 4 m below sea level for a period of 4 months (White, 1997). This allowed the intrusion of sea water into the aquifer, which water managers seek to prevent.

Lowering of water tables can be even more pronounced in semi-arid and arid regions. For example, in the La Mancha Occidental aquifer unit (Upper Guadiana, Spain) the water level has dropped about 50 m in some places since the early 1970s (Cruces *et al.*, 2000) due to over-pumping of the aquifer for irrigation. Meddi & Hubert (2003) describe the continuous decline of water levels of the upper aquifer from 441 to 431 m in a well in Bekkad (Algeria) as a response to decreasing rainfall. Clearly, it is hard to separate the effects of abstraction from the lack of recharge under these conditions (e.g. van Lanen & Peters, 2002).

Since 1950 water tables have dropped in the Netherlands as a response to land reclamation and groundwater abstraction (abstraction volume has increased by about 270%). The water level drawdown reached 0.35 m on average, implying that the levels were lowered by more than 0.50 m at many places. Drawdown is relatively large especially in the surroundings of well fields. Note that the impact of these drawdowns needs to be evaluated in the context of the originally shallow water tables and the small natural fluctuation.

Since 1973 the Podzamcok well field (central Slovakia, south of the city of Zvolen) has been developed. The abstractions had a severe impact on the groundwater discharge to the Neresnica Brook (Figure 9.17, left). Initially groundwater was extracted by horizontal pumps. The drawdown in the wells reached 8 m with an abstraction rate of about 0.15 $m^3 s^{-1}$ on average. The saturated hydraulic connection between groundwater and the brook stopped, and some of the natural springs dried up. Discharge of the Neresnica Brook started to decline. Since 1975 submersible pumps have been installed in wells. The highest abstraction rates were reached in 1981–1989 (maximum 0.19 $m^3 s^{-1}$). The water table dropped up to 20–22 m, and monthly minimum discharges reached the lowest values (Figure 9.17, left). Average minimum monthly streamflow decreased from 0.33 $m^3 s^{-1}$ (1962–1972) to 0.12 $m^3 s^{-1}$ for the period

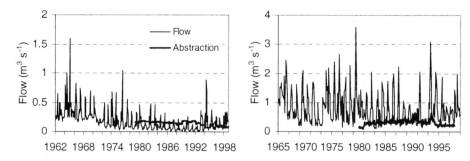

Figure 9.17 Monthly minimum streamflow of the Neresnica Brook and monthly mean groundwater abstraction from the Podzamcok well field (left), and monthly mean streamflow of the Starohorsky Brook and monthly mean withdrawal of spring yields (right).

with maximum exploitation (1973–1993). Since 1993 pumping rates have been decreased to $0.077 \ \mathrm{m^3 \, s^{-1}}$ leading to a recovery of the water levels. The brook started to drain groundwater and some of the springs began to flow again. In the late 1990s the monthly minimum discharge was about equal to the abstraction rate.

Tapping of springs in karstic areas has similar effects on groundwater flow and consequently on groundwater drought. In central Slovakia (northeast of the city of Banska Bystrica) four large limestone springs were gradually tapped for drinking water supply. The springs used to feed Starohorsky Brook. In 1978 the first spring (Podzemny tok) was tapped, followed by two other springs (Jergaly and Stary mlyn) in 1982 and by the last one (General Cunderlik) in 1983. From 1982 onwards water withdrawals were higher or equal to the monthly minimum discharges of Starohorsky Brook (Figure 9.17, right). The maximum withdrawal of $0.775 \ \mathrm{m^3 \, s^{-1}}$ was reached in November 1991. Average minimum monthly streamflow decreased from $0.84 \ \mathrm{m^3 \, s^{-1}}$ (1965–1977) to $0.56 \ \mathrm{m^3 \, s^{-1}}$ for the period with maximum exploitation (1982–1994). After 1994 water withdrawals decreased resulting in a recovery of the monthly minimum discharges of the Starohorsky Brook.

9.5.2   Groundwater abstraction in the Poelsbeek and Bolscherbeek catchments

In the Poelsbeek and Bolscherbeek catchments hardly any groundwater is abstracted. The well field for the public water supply of Haaksbergen city (24 000 inhabitants) is situated outside the catchments. Groundwater abstractions for industrial use are minor in these basins, the maximum extraction being about $138 \ \mathrm{m^3 \, day^{-1}}$.

Two groundwater abstraction scenarios are defined for the Poelsbeek catchment:

a) constant extraction rate of $2050 \, m^3 day^{-1}$ during the winter season (October–March) (scenario 05Y), and

b) constant extraction rate of $2050 \, m^3 d^{-1}$ throughout the whole year (scenario 1Y).

Total annual abstraction is 0.37 and $0.75 \cdot 10^6 \, m^3$, respectively. The abstraction volumes are typical for hydrogeological settings similar to those in the modelled area. Groundwater is abstracted from one location in the centre of the Poelsbeek catchment. This means that the expected impact on water tables in the Poelsbeek catchment will be larger than on the Bolscherbeek catchment.

Water tables

At a representative location in the Poelsbeek catchment water tables vary between 18.27 and 19.84 m.a.s.l. under natural conditions (reference situation), implying a groundwater depth between 0–1.58 m below surface. The location is at 500 m distance from the well field. Clearly groundwater extraction leads to larger groundwater depths, especially close to the pumping site (Figure 9.18).

Figure 9.18 (right) shows that shallow water levels are hardly affected by groundwater abstraction, whereas the deep levels during dry periods differ substantially. The maximum groundwater depth increases from 1.58 m to 1.95 and 2.46 m for the 05Y and 1Y scenarios respectively. The difference between groundwater depth for the reference and for the abstraction scenarios (drawdown) is not regular in time (Figure 9.18, left). Permanent groundwater abstraction (1Y) leads to larger drawdown of the water level. The drawdown is usually larger in dry periods, e.g. the 1970s. In dry periods (i.e. summer season) the abstracted groundwater can only be partly compensated for by a reduction of groundwater discharge to surface water, meaning that the extracted groundwater predominantly comes out of groundwater storage resulting in lower water tables. In wet periods the whole surface water system (streams, ditches) carries water in lowland catchments with shallow water tables. In these periods extracted groundwater reduces groundwater discharge to surface water, which may lead to the drying up of trenches. Under these conditions drawdowns are relatively small.

The combined effect of lower groundwater tables and reduction in groundwater discharge is also the main reason for the completely different drawdowns experienced for the two abstraction scenarios. The seasonal abstraction (05Y) takes place during the wet period in lowlands with a temperate humid climate, meaning that abstraction can be balanced by reduction

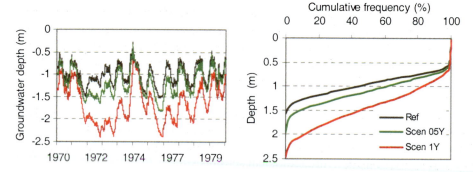

Figure 9.18 Water levels for the reference situation and for the groundwater abstraction scenarios 05Y and 1Y at a given location in the Poelsbeek catchment: water tables (period 1970–1980, left), and cumulative frequency distribution (period 1951–1998, right).

in groundwater discharge. Thus, as long as wet seasons prevail, drawdowns are small (e.g. early 1970s). Constant abstraction (1Y) takes place both during wet and dry seasons, which results in a strong fluctuation of the drawdowns (Figure 9.18, left). The drawdowns considerably increase for this scenario if winter periods are relatively dry, e.g. 1972–1974.

The groundwater system of the Poelsbeek catchment also balances extractions by reduction of capillary rise. This is typical for lowland catchments with shallow water tables leading to a lower actual evapotranspiration

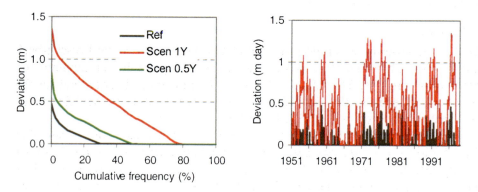

Figure 9.19 Groundwater drought for the reference situation and for the groundwater abstraction scenarios 05Y and 1Y for a representative location in the Poelsbeek catchment: empirical frequency distribution of the deviation (left), and time series of the deviation (right).

(van Lanen & Peters, 2000). For example, average *ET* in a subcatchment near the well field is 8 and 25 mm year^{-1} lower for abstraction scenarios 05Y and 1Y respectively, as compared to reference conditions. Consequently, the groundwater recharge is higher than for the reference situation (Section 3.3). At the scale of the whole Poelsbeek catchment, however, the effect on *ET* is negligible.

Groundwater drought is identified using the deviation of the daily groundwater levels from the threshold (Section 5.6). Figure 9.19 gives the deviation for the reference situation and for the two abstraction scenarios. The H_{70} is used as threshold.

Clearly, the drawdown at a location near the well field in the Poelsbeek (Figure 9.19) gives relatively large deviations pointing at more prolonged and more severe groundwater drought (Figure 19, left). For instance, a drought occurs for the reference situation in 25% of the time, whereas this increases to 40 and 75% of the time for the abstraction scenarios 05Y and 1Y respectively. Figure 9.19 (right) shows that the most severe drought (deviation over 1 m) can be observed in the 1950s, 1970s and 1990s (scenario 1Y). Furthermore, single-year droughts for the reference situation cluster to multi-year droughts in these periods for the 1Y scenario. Groundwater drought for the 05Y scenario is less severe and does not significantly deviate from the reference situation.

Groundwater drought in the Bolscherbeek basin is less severe because abstraction takes place in the Poelsbeek catchment. Water tables for a representative location in the centre of the catchment, which is at about 3 km from the pumping site, illustrate this. Differences in the deviation of the groundwater level from the threshold value due to pumping are nearly zero, indicating no increase in groundwater drought.

River flow

Daily time series of streamflow at the outlet of the Poelsbeek and Bolscherbeek catchments are analysed to assess the impact of groundwater pumping on streamflow drought. This is carried out for the reference situation and for the abstraction scenarios 05Y and 1Y. Figure 9.20 presents the empirical frequency distribution of the daily discharge for different situations.

The annual average streamflow of the Poelsbeek decreases due to pumping by 2.5 and 5.9% for the two scenarios. Figure 9.20 illustrates that low flows are hardly affected. The effect of pumping on streamflow of the Bolscherbeek is negligible. The small impact of the abstractions on the cumulative frequency curve (Figure 9.20) implies that drought severity and duration hardly change. For the 1Y scenario the differences in deficit volume for the well-known drought periods (i.e. 1959, early 1970s and mid 1990s) are between 5 and

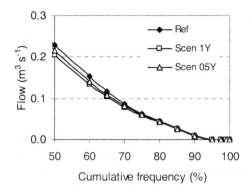

Figure 9.20 Cumulative frequency distribution of the daily discharge of the Poelsbeek catchment for the reference situation, and for the abstraction scenarios 05Y and 1Y.

10 mm. Note that the maximum discharge deficit volume for the reference situation is about 40 mm (Figure 9.7, left).

9.5.3 Self-guided Tour: Groundwater abstraction in the Poelsbeek and Bolscherbeek catchments (CD)

A self-guided tour was compiled to demonstrate for the reader how the impact of human influence can be assessed using the SIMGRO model. The example is menu-driven, and this supports the navigation through the whole assessment procedure. The tour gives guidance on how to develop a model and how it can be applied for impact studies. The reader works through the procedure at an individual pace and may review or skip a section.

The impact of groundwater abstraction in the Poelsbeek and Bolscherbeek catchments (Section 9.5.2) is used as example to facilitate the learning process. Based upon the practical example the user learns more about (a) principles of physically-based models, such as SIMGRO, (b) properties of the Poelsbeek and Bolscherbeek catchments, (c) translation of catchment properties to model input data, (d) model calibration and verification, (e) drought analysis used in combination with physically-based models, and (f) impact of groundwater abstraction as compared to the reference situation.

At the bottom of the main menu there are five key buttons: (1) aim, (2) study area, (3) method, (4) model, and (5) scenario. These buttons allow jumping between sections. At the bottom right two arrow buttons help to navigate through a particular part of the procedure. Throughout the exercise, results are discussed and conclusions are drawn. To guarantee a broad use of the Self-guided Tour a commercial software package was used as platform

(Microsoft PowerPoint Presentation). It is important to note that the Self-guided Tour is not designed to develop a SIMGRO model for the Poelsbeek and Bolscherbeek catchments. The model was already developed and applied, and only results are included in the Self-guided Tour.

9.5.4 Groundwater abstraction in the Svitava catchment

The Ústecká syncline (eastern Bohemia, Czech Republic) forms the southern part of the Upper Svitava geological basin. Hydrogeological conditions are favourable for groundwater exploitation. Large quantities of groundwater are stored in two aquifers, i.e. in the Middle Turonian sandstone (shallow aquifer), and in the Lower Turonian and Cenomanian aquifer built by sandstones with high hydraulic conductivity (deep aquifer). In this area groundwater has been abstracted for the drinking water supply of the town of Brno since 1913. Groundwater exploitation was gradually extended, and eventually groundwater was abstracted from the shallow aquifer by use of 32 boreholes and from the deep aquifer by 6 deep boreholes. After 1976 the quantity of abstracted water substantially increased and annual abstraction exceeded 40 millions m^3.

For assessment of the impact of intensive groundwater abstraction on the flows of the River Svitava, long-term series of discharge data are analysed. The discharge of the Upper Svitava is observed at Rozhraní and Letovice-Svitava gauging stations and at Letovice-Křetínka gauging station for the Křetínka tributary (Figure 9.21). In Table 9.5 the catchment area is given; the area of the

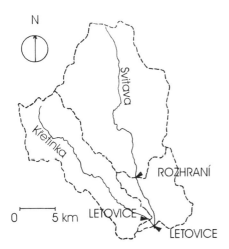

Figure 9.21 Map of the River Svitava catchment showing the subcatchments and gauging stations.

Table 9.5 Gauging stations in the River Svitava catchment

Name	Subcatchment	Area (km²)
Rozhraní	Upper-Svitava	223
Letovice-Svitava	Upper-Svitava	419
Letovice-Křetínka	Křetínka	126

Letovice-Svitava station also includes the area upstream of Rozhraní gauging station.

For the two gauging stations in the River Svitava, discharge data are available from 1926, whereas the observed discharge series of the River Křetínka are available from 1924. Thus the observations started when the water regime of the Svitava River basin was already slightly affected by groundwater abstractions for Brno. Mean annual flows observed at the three stations in the period 1926 to 1995 are shown in Figure 9.22, which demonstrates that the discharge of the River Svitava has been decreasing since the late 1970s, whereas no decreasing trend is detected in the Křetínka series.

The monthly discharge of the River Svitava, which is not affected by groundwater abstraction, is estimated using the BILAN model (Section 9.2.2). The eight model parameters are calibrated using observed discharge data observed from the period 1945–1970 when the flows were not significantly affected by groundwater abstraction. The calibrated model is used for

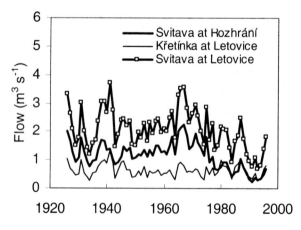

Figure 9.22 Mean annual discharge of the River Svitava (two locations) and River Křetínka for the period 1926–1995.

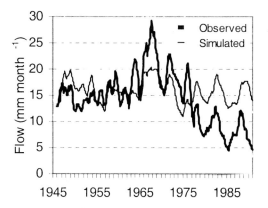

Figure 9.23 Observed and simulated discharge (12-month moving average) with BILAN for the River Svitava at the Rozhraní gauging station.

simulation of monthly flows in the period 1971–1990 (Figure 9.23). Data on precipitation, air temperature and relative air humidity are used as input data for the model.

Figure 9.23 shows that observed flows since the mid-1970s are substantially below those that were simulated with the model calibrated for conditions with minor groundwater abstraction.

In Figure 9.24 the annual groundwater abstraction volume including the trend is presented. The difference between the observed (with abstractions) and the simulated annual discharge (discharge decrease) is also calculated and the trend is given. The graph demonstrates that there is a positive correlation

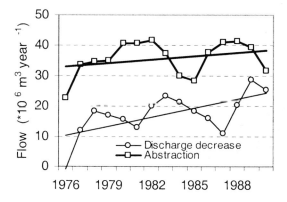

Figure 9.24 Annual groundwater abstraction and annual discharge decrease as estimated by the BILAN model for the River Svitava.

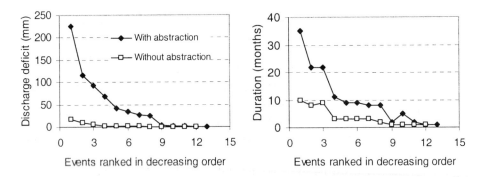

Figure 9.25 Streamflow drought derived from monthly discharge data from River Svitava for the observed situation (with abstraction) and for the simulated naturalized regime (without abstraction) for the period 1971–1990: discharge deficit volume (left), and duration (right).

between the increase in the groundwater abstractions and the reduction in annual discharge for the River Svitava. The trend in discharge reduction is even steeper than the increase in the abstractions implying that more factors than abstraction affect the discharge. The discharge decrease is equal to about one third of the annual abstractions in the mid 1970s, whereas 20 years later it exceeds half of the abstracted amount.

The threshold level method (Section 5.4.1) is applied to the time series of monthly flow data to assess the impact of groundwater abstractions on streamflow drought in the River Svitava. The threshold Q_{70} (12.7 mm month^{-1}) derived from nearly natural conditions is used. Drought deficit volume as a measure for drought severity and drought duration are analysed both for the observed and simulated series 1971–1990. The simulated series represents the natural regime without groundwater extractions (Figure 9.25).

Groundwater pumping in the River Svitava catchment has without doubt led to more severe drought. Drought deficit volume in streamflow increased up to a factor of 10. Drought duration doubles or triples.

9.5.5   Discussion

The response to groundwater abstraction on hydrological drought depends on abstraction volume (e.g. Querner *et al.*, 1997), abstraction pattern, distance to well field and hydrogeological conditions. Clausen *et al.* (1994) investigated the effect of distance to the stream and seasonality in a UK chalk catchment. They introduced the stream depletion factor (SDP) as the difference between the simulated monthly natural flow and flow affected by abstractions divided by monthly abstraction. Constant abstraction gave a strongly varying SDP for abstractions far from the stream and a nearly-constant SDP for near-stream locations due to increasing storage capacity. For two-month abstractions, maximum SDP is smaller for abstractions far from the stream.

Regional groundwater systems (Section 3.2) with large groundwater storage respond slowly to abstractions, implying that it takes several decades before reaching steady-state conditions. Drought may thus increase over time. In arid regions where recharge hardly takes place, steady-state conditions are never reached. For instance, Ebraheem *et al.* (2003) explored abstraction scenarios in arid southwest Egypt for the period 1980–2100. Simulated water level can drop from a few metres to depths up to 145 m below soil surface over the 120-year period exceeding economic lift depth (depth below which pumping costs are higher than benefits).

A particular type of shallow groundwater abstraction is *land drainage*. Implementation of drainage systems for agricultural development usually results in higher wet-season flows and lower dry-season flows (e.g. Dingman, 2002). Streamflow droughts will thus last longer and be more severe. A model experiment in the Beerze catchment (southern Netherlands) supports this (Moors & Dolman, 2003). Low flows decreased by about 25% through implementing of field drains and ditches. This is, however, not supported in a model experiment in the Poelsbeek (Querner & van Lanen, 2001). Effects of raising surface water levels of the water courses and the bed of ditches to counteract effects of drainage (i.e. restoration of natural conditions), lead to more severe streamflow droughts, which seems to contradict the general findings. The model experiment is not general because ditches still exist in the regenerated situation, meaning that excess water can quickly be conveyed to the main water courses. If all ditches were removed and no excessive overland flow took place, more water would be conserved in the catchment leading to higher low flows supporting the general experience.

Implementing drains and ditches leads to low water tables, especially at the end of the wet season. In the dry season the difference between water levels for the drained and the undrained situation decreases because the actual evaporation is higher in the latter case. However, drainage generally results in more severe groundwater drought. This is confirmed by model experiments in the River Jegrznia basin (Mioduszewski *et al.*, 1997; Querner & van Lanen, 2001).

9.6   Impact of urbanization

Emigration from rural to urban areas has occurred everywhere, and still continues in most developing countries. The provision of *water supply*, *sanitations* and *drainage* are key elements of the urbanization process. Urbanization raises the important question of what is the influence of this extreme land-use change. What processes are affected and what is the impact on drought development? Land surface impermeabilization reduces infiltration (q_s) and consequently surface runoff increases (q_{of}). It also tends to cause a decrease in evapotranspiration. Of course this depends on pre-existing land use with considerable differences between displacement of natural vegetation, irrigated agriculture or dryland farming (e.g. Foster *et al.*, 1999). Stormwater arrangements determine the fate of surface runoff. If these are not installed then water may infiltrate in soakaways or at the edge of paved surfaces, or accumulate in micro depressions (S_{su}). This depends on precipitation type and intensity, and antecedent soil moisture conditions. During and after a prolonged wet period, surface runoff may reach streams even without drainage means and thus contribute to peak flow. Where means for stormwater drainage are installed, the majority of the surface water will be routed very quickly to a neighbouring stream meaning that the streamflow regime is entirely changed.

Urbanization also involves a huge demand for water. This water may be supplied from aquifers beneath the city and its surroundings or it may be imported from other catchments at a distance. Where groundwater is abstracted in or near the city it influences hydraulic heads of aquifers beneath the city and water tables in superficial layers. Twelve megacities, whose population exceeds 10 million, rely on groundwater. These are Bangkok, Beijing, Buenos Aires, Cairo, Calcutta, Dhaka, Jakarta, London, Manilla, Mexico City, Shanghai, and Tehran. The water is distributed in pressurized mains, which are commonly prone to leakage resulting in a significant contribution to groundwater recharge. Another more important contribution to groundwater recharge takes place in unsewered cities. In these cases 90% of the water-supply provided will end up as recharge to groundwater. Urbanization can completely change the recharge of an underlying aquifer (Foster *et al*, 1999).

In Section 9.6.2 the SIMGRO model is used to explore the impact of the extension of a small Dutch city, i.e. Haaksbergen on the development of hydrological drought. Firstly, however, some observational evidence of the impact of urbanization on drought development is discussed.

9.6.1   Observations

Foster *et al.* (1999) mention a study carried out in the city of Hamburg (Germany) that showed that infiltration into the soil was reduced by up to 190 mm year^{-1} due to the increase of paved surfaces on previously permeable land. This certainly affects the underlying urban aquifer leading to more severe groundwater drought. Ahmed *et al.* (1999) illustrate the impact of urban development of the city of Dhaka on the groundwater system. Since the 1960s, the capital of Bangladesh has predominantly used groundwater from the confined Dupi Tila aquifer beneath the city. The hydraulic head of the aquifer decreased at a rate of 0.75 m per year, which has led to an extensive cone of depression. At the same time the shallow water table in the urban area showed an upward trend meaning that the increased recharge to the shallow aquifer exceeded the small leakage to the heavily-exploited deep aquifer. Beyond the city limits, the shallow water table also dropped due to low hydraulic heads in the underlying aquifer and unchanged recharge.

The conditions for the three aquifers underlying Bangkok (Thailand) are even worse than the circumstances in Dhaka (Ramnarong, 1999). Since the mid 1950s abstraction has grown from an insignificant amount to about $1.75 \cdot 10^6$ m^3 day^{-1} in 1997. The average annual decline in the hydraulic head reached a value of up to 3.6 m. This has resulted in a dramatic subsidence in the urban area. Results for the Indonesian cities Jakarta and Bandung are similar (Soetrisno, 1999). For example, in Jakarta the population has increased from less than 1 million in 1945 to 8.2 million in 1990. The annual decline of the hydraulic head was 2–4.6 m in Jakarta and 2–4 m in Bandung. In both cities downward leakage takes place considerably reducing the discharge. The subsidence rate in Jakarta equals 0.34 cm year^{-1}, whereas in Bandung no evidence is found for lowering of the soil surface. Subrahmanyan *et al.* (2003) show that annual groundwater recharge decreased from $80 \cdot 10^6$ m^3 in 1987 to $20 \cdot 10^6$ m^3 in 1994 due to the unplanned expansion of the city of Hyderabad (India).

Appleyard *et al.* (1999) describe the consequences of the urban development of the city of Perth in Western Australia. Recharge, which under natural vegetation was about 135 mm year^{-1}, has been doubled by urban development. Removal of native vegetation and import of water together with local recharge of stormwater has resulted in rising water tables over a long period. Lakes have developed from swamps due to the rising water tables. High water tables also necessitated implementation of drainage systems. Since the 1977 drought, the water table rise changed to a fall due to the increase in the number of private abstraction wells in Perth, which led to concern about the water level in some recent wetlands and lakes. Since then use of shallow groundwater has been restricted. From the mid 1970s onwards the maximum rate of decline in the water table level has been about 0.1 m year^{-1}, and the hydraulic head in the confined aquifers below the city decreased to maximum rates of 0.5–0.7 m year^{-1}. This induces downward leaking from the shallow aquifer leading to higher probability of drought.

In two Russian cities, i.e. Tomsk and Irkutsk, urban development has resulted in low hydraulic heads in the exploited deep aquifer, which is semi-confined, and a rise of the water tables in the uppermost layers due to leaking pipes and impedance of groundwater flow by deep foundations (Pokrovsky *et al.*, 1999; Shenkman *et al.*, 1999). The additional recharge in Tomsk is estimated at up to 4 mm day^{-1} leading to a maximum rise of the water level of 5 m. Foster *et al.* (1999) report from a study on Long Island (New York, USA) showing that the overall effect of urbanization with 20–30% impermeabilization (low-density residential area) was a 12% increase in groundwater recharge and 1.5 m rise in water table in areas where stormwater was directed to infiltration structures, and a 10% decrease and 0.9 m fall in areas where the stormwater was routed directly to the sea. Rowe *et al.* (1997) and McConchie (1992) also summarize the effects of urbanization on hydrology.

In financially well-off regions with dry climates, stormwater is often used for the irrigation of parks, gardens and other amenity areas (e.g. golf courses). This can lead to locally very high rates of groundwater recharge (Foster *et al.*, 1999). Some cities in semi-arid and arid regions like the Arabian Gulf are suffering waterlogging because of a heavy dependence on desalinated water from the coast which is leaking and becoming trapped in the subsurface. Over-irrigation of amenity areas is reported to be an equally important source of groundwater recharge such as the leakage of the water mains and wastewater disposal in, for instance, the city of Lima and various cities in the Middle East (Section 9.6.3).

9.6.2   Extension of the city of Haaksbergen

The city of Haaksbergen is located in the east of the Poelsbeek and Bolscherbeek catchments (Supporting Document 9.2). The effect on the Bolscherbeek is also considered because the precipitation on urban areas in both catchments is channelled to a sewage treatment plant in the Bolscherbeek. The urban area of Haaksbergen is expected to grow from 8 to 14% implying that more water will flow through the Bolscherbeek stream (which receives effluent water from the sewage treatment plant) and less through the Poelsbeek stream. However, to combat droughts mitigation measures are planned allowing the rainwater from roofs and streets to infiltrate into the green areas in the city instead of being discharged into the sewers. It is assumed that the urban area in Haaksbergen consists of 50% impermeable area and 50% grass. Furthermore, it is presumed that leakage from the mains is negligible. A scenario is defined that includes increasing the urban area by 6% and at the same time allowing all rainwater to infiltrate within the city area instead of being channelled towards the sewage treatment plant. Querner & van Lanen (2001) include some more scenarios. Meteorological data from the period 1951–1998 are used.

The simulation shows that enhancement of rainfall infiltration in urban areas leads to a locally higher recharge of the aquifer. However, at some distance the effects are relatively small. At about 6 km from the urban development the water tables are hardly influenced. The higher water levels in the lowland area are counteracted by evapotranspiration increase due to larger capillary rise. Average annual actual evaporation increases by 2%. The combined effect of a locally higher recharge and higher evapotranspiration means that groundwater drought is not very different from reference conditions. The drought is expressed as deviation from the threshold water table (Section 5.6).

River flow

The effect of an increase in urban area of Haaksbergen combined with the mitigation measures to increase rainwater infiltration within the town is given in Figure 9.26 for both the Poelsbeek and the Bolscherbeek. The graph gives the cumulative frequency distribution of the annual discharge.

The opposite effect can be found in Bolscherbeek compared to Poelsbeek due to the simulated urbanization and mitigation measures. In the Bolscherbeek a decrease in streamflow can be observed in the whole range of flows, whereas in the Poelsbeek flows are generally higher. The mean annual discharge changes by −11.2% and 9.7% for the Bolscherbeek and Poelsbeek, respectively. Clearly, infiltration of rainwater in the city, which is partly located in the Poelsbeek catchment, leads to high streamflow in this catchment. In the Bolscherbeek this effect is overridden by the decrease of water outflow from the sewage plant. In the Poelsbeek the discharge is 10–20 mm year^{-1} higher in dry

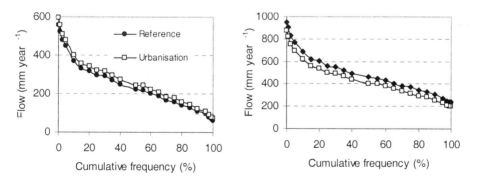

Figure 9.26 Cumulative frequency distribution of the annual streamflow for the reference situation, and for an urbanization scenario: Poelsbeek (left) and Bolscherbeek (right).

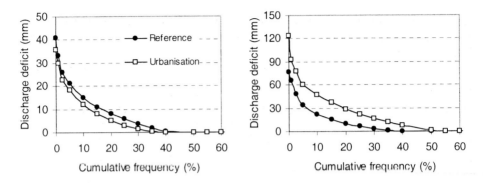

Figure 9.27 Cumulative frequency distribution of the discharge deficit volume for the reference situation and the urbanization scenario: Poelsbeek (left) and Bolscherbeek (right).

years, whereas in the Bolscherbeek it is 40–50 mm year^{-1} lower than for the reference conditions.

The discharge deficit volume is used to illustrate the impact on drought severity and drought duration. The severity and duration are derived from time series of daily flow data from the period 1951–1998. This is done for both streams and for the reference situation and the urbanization scenario. As in previous sections, the SPA method (Section 5.4.2) is applied to identify drought. Figure 9.27 presents the empirical frequency distribution of drought severity.

City expansion in combination with enhancement of infiltration within the city limits, results in less severe droughts in the Poelsbeek. The differences between the modelled and reference situation are relatively small. In the Bolscherbeek the differences are more pronounced and the severity clearly increases. The discharge deficit volume of the most extreme events increases by about 60%.

Table 9.6 gives the impact of the urban development of the city of Haaksbergen on maximum streamflow drought duration per year for the Poelsbeek and the Bolscherbeek. This table confirms the conclusions drawn for the severity (Figure 9.27). Drought duration decreases for the Poelsbeek whereas it increases for the Bolscherbeek, and the impact is larger for the latter. In extreme dry years (cumulative frequency > 90%) drought in the Bolscherbeek lasts about 50 days longer. The reduction in effluent from the sewage treatment plant results in longer streamflow drought; minor droughts merge to form longer and more severe droughts, thereby significantly decreasing the number of drought periods (Querner & van Lanen, 2001). The

Table 9.6 Cumulative frequency of the maximum drought duration per year (days) in the Poelsbeek and Bolscherbeek catchments for the reference situation and the urbanization scenario

Cumulative frequency (%)	Poelsbeek		Bolscherbeek	
	Reference	Urbanization	Reference	Urbanization
90	268	259	270	319
80	236	207	191	274
70	206	184	170	241
60	178	159	131	203
50	158	116	113	188

considerably lower number of situations of emergency overflow from the sewage plant does not subdivide major droughts into two or more minor droughts anymore. In the Poelsbeek drought is 10–30 days shorter in dry years due to urban development.

9.6.3 Discussion

Custodio (1997) concludes that it is hard to distinguish general situations for urbanization since very diverse local circumstances may play a dominant role, as has been illustrated for the Poelsbeek and Bolscherbeek catchments in the previous section. The net effect of land surface impermeabilization and stormwater drainage arrangements on groundwater recharge varies from a major reduction to a modest increase (Foster *et al.*, 1999). In many cities leakage from pressurized mains enhances this recharge Moreover in non-sewered cities a substantial amount of water needs to be added because not more than 5–10% of the water provided is actually consumed. Hence it is likely that urbanization increases groundwater recharge, especially in cities without drainage and sewage systems. Morris *et al.* (2003) described that studies carried out on data from groundwater-dependent cities show that the recharge tends to increase irrespective of climatic regime. Data are from cities in North and South America, Europe, Asia and Australia. The main reason for the increase is leakage from mains. They mention a probable minimum value of 50 mm year^{-1} for sewered cities with mains drainage. For example, they present an increase of recharge from about 50 to 200 mm year^{-1} for Los Angeles (USA).

It is possible that the water table might rise because of the enhanced recharge, and if it does it will depend on the source of the water supply. When water is imported from a catchment beyond the city limits the water table will actually be shallower, and hence groundwater drought is less severe due to urbanization. Moreover the groundwater discharge to the streams will be larger due to higher water tables, which will lead to higher low flows. However, if the water supply is from aquifers underlying the city, the rise of the water table due to enhanced recharge is counteracted by downward leakage to the exploited aquifer. Hydrogeological conditions determine exactly what will happen. When an unconfined aquifer is exploited, the water table is mostly lowered leading to more severe groundwater drought and less groundwater flow to the streams. Alternatively, when a confined aquifer underlying the city is exploited there is a high chance that shallow water tables will rise and that the hydraulic head of the exploited aquifer will show a sharp decline. In this case groundwater drought will be smaller in the city and its vicinity where the drought is related to shallow water tables. Naturally, the low hydraulic head of the confined aquifer will lead to lower groundwater discharge wherever else the confined aquifer outcrops. At many locations the net effect of groundwater abstraction in or near the city and enhanced recharge leads to falling water tables. Dhaka (Bangladesh) is an example of severe exploitation from urban aquifers. The number of abstraction wells is estimated at 1300, leading to a fall in the water table of as much as 40 m. New boreholes yield only two thirds of wells drilled before 1970 as a result (Morris *et al.*, 2003). They also report about Lima (Peru) where, despite the higher recharge, deeper and more expensive boreholes have had to be dug. The production costs of water have increased by 25%. On the other hand, Foster *et al.* (1999) also mention reports, even from dry environments (e.g. Riyadh in Saudi Arabia, Kuwait in Kuwait, and Doha in Qatar) of water levels rising. Basements and pipes have been damaged forcing the authorities to deploy extensive pumping to counteract the excessive recharge mainly coming from desalinated seawater.

When water supply systems that use groundwater are replaced by surface water resources or groundwater from other catchments to keep pace with population growth, a period of groundwater decline is often followed by a recovery period, which may eventually lead to flooding of underground spaces, such as basements and subway tunnels (Custodio, 1997).

In many cities the pre-urban natural stream network is replaced by an artificial infrastructure conveying surface runoff more quickly to the stream. The result of these processes is that a possibly slowly-responding catchment may change into a quickly-reacting catchment (Section 3.2). This implies that the number of streamflow droughts increases, but that severity and duration mostly decrease (e.g. Querner & van Lanen, 2001). In the case of major water imports the quantity of water that needs to be discharged will increase, which may very likely lead to smaller streamflow drought. Urban development using an unconfined aquifer below the city for water supply may be an exception. Here the streamflow drought probably will be more severe, especially when these cities are non-sewered and surface runoff from the paved areas infiltrates and does not reach the stream.

9.7   Impact of surface water control

Many rivers, lakes and reservoirs have been modified to serve various purposes, such as improving navigation, reducing floods, energy production, enhancing low flows, supplying drinking water, providing wildlife habitat, and increasing the possibilities for recreation (Box 9.3). Many of these modifications may affect streamflow drought.

Reservoirs have an important effect on the streamflow hydrograph. For instance, it is likely that the increase in reservoir area by about 200% from the mid 1980s to the late 1990s (Figure 9.28) has a distinct influence on the rivers regime in Burkina Faso. In the same way as lakes (Section 3.5), reservoirs mostly change a quickly-responding stream entering the reservoir into a slowly-reacting stream due to storage effects. Additionally, water abstraction from the reservoir can completely alter the outflow. This section starts with an overview of observed effects that modified surface water courses have on streamflow. Following this the impact of surface water transfers is discussed (Section 9.7.2).

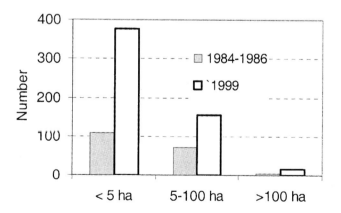

Figure 9.28 Reservoir construction in Burkina Faso (data from Andreini *et al.*, 2002).

9.7.1   Observations

The degree of change in the low-flow regime can be assessed by comparing FDCs (Section 5.3.1) compiled from present-day data (affected conditions) and naturalized time series (Section 4.3.3 and Chapter 11). Smakhtin (2002) shows how the FDC of the River Berg in South Africa has been considerably modified due to surface water abstractions with the largest relative effect in the low flow range. The Q_{75} at present is only 20% of the value of the natural condition.

Waugh *et al.* (1997) give another example for the River Waikato (New Zealand). In the natural state the river overbanked in wet periods and flooded across swampy areas in numerous lakes and wetlands. A series of disastrous floods led to *river control works* to protect urban and agricultural areas from flooding. The works seems to be successful in preventing floods. This had a significant influence on the duration curves for surface water-level. The authors present three curves for the periods 1965–1969, 1975–1979 and 1985–1989. The drop in river water levels and associated river discharge is significant. The largest differences can be found for flows in the mid range. Highest levels decreased from 5.5 to 5 m, whereas lowest levels declined from 2.2 to 1.9 m. Thus it is likely that flood protection works enhanced susceptibility to streamflow drought.

Finke & Dornblut (1998) have analysed time series of the River Elbe in Germany for the period 1936–1995. Data from the five gauging stations show that the drought duration and severity (deficit volume) changed after the confluence of the tributary Saale. The stations upstream of the confluence show a light linear time trend, whereas for those downstream a trend is not detected. It is likely that human influences cause the difference in the hydrological regime of the River Elbe.

The construction of reservoirs has a large impact on the flow regime. Water-supply reservoirs store water in the wet season and release it during the dry season to maintain downstream surface water abstractions. Water can also be directly abstracted from the reservoir (Chapter 11). Reservoirs in humid climates tend to be filled each year and thus smooth out within-year streamflow variation. Reservoirs convert a steep natural FDC to a flatter curve. Consequently, the probability of streamflow drought is lower downstream of the reservoir as the water in the reservoir is not conveyed to another catchment. In dry regions, additional carry-over storage is needed to overcome the large variations between years; reservoirs in arid regions fill only rarely (Dingman, 2002). In this case the reservoir outflow (residual flow), if any, is entirely determined by the management of the reservoir. For example, has an ecologically acceptable flow regime to be maintained or not (Section 10.4)?

Box 9.3

Lake Naivasha suffering from drought

Lake Naivasha (about 150 km^2) is situated at 1890 m.a.s.l. at the bottom of the Rift Valley in Kenya. People assume that it is a relict lake that once belonged to a larger freshwater body that comprised several rift valley lakes, including Lake Nakuru. Lake Naivasha has experienced several dry phases alternating with high water level periods. The lake is Kenya's second Ramsar site because of its international importance as a wetland (e.g Harper *et al.*, 2003). The lake supplies drinking water, water for a power plant and irrigation water to the nationally important industries of horticulture. Since the late 19th century the lake level has dropped about 10 m with some fluctuation. In the 1940s and 1950s the water level was even 3–4 m lower than in the late 1980s. The large decline of the lake levels is a major concern for lake ecology. The lake is fed by surface water inflow from the Rivers Gilgil, Karati and Malewa (70% of input) coming from higher altitudes and by direct rainfall on the lake surface. The catchment area is estimated at 2378 km^2. Lake evaporation accounts for about 80% of the losses under natural conditions. The remaining water leaks to a deep and extended volcanic aquifer. The lake has no surface water outlet. The safe yield for the lake was estimated to be $16.5 \cdot 10^6$ m^3 year^{-1}. Present-day abstractions are unknown, but simple model studies suggest that these are 3–4 times higher than the safe yield (Brecht & Harper, 2002). Over-abstraction of water is likely to contribute to the serious decline of lake levels, although some natural fluctuation should not be disregarded. There is an urgent need for accurate measurement of all abstractions, a more detailed description of the hydrological system of the lake, an improved assessment of the safe yield, and for initiation of a discussion with stakeholders on how this precious resource can be used, proposing the best balance possible between economic and ecological demands.

In Norway the residual flow from a reservoir should not be less than the 'common low flow' (Section 11.4.4). This low flow index is applied to grant licences for hydropower production. Note that reservoir performance is strongly affected by erosion in the upland catchment and silting up of the reservoir (e.g. Hicks & Davies, 1997).

9.7.2 Water transfer in the River Bílina catchment

In the north-western part of the Czech Republic large amounts of brown coal (lignite) have been extracted in open-cast mines. The mining activities were associated with the construction and operation of several power plants and chemical factories. A substantial part of the area is located in the River Bílina

Table 9.7 Mean annual values of the main water balance components of the River Bílina for the period 1932–1960

Component	(mm year^{-1})
Precipitation	617.6
Observed discharge	151.6
Simulated discharge	154.1
Potential evapotranspiration	631.8
Actual evapotranspiration	462.0

catchment (tributary of the River Elbe). The River Bílina catchment is partly located in the Krušné Mountains, whose ridges reach an altitude of 900 m.a.s.l. and the annual precipitation is about 850 mm. In the other part of the catchment where the mining is located, the altitude ranges between 200 and 300 m.a.s.l. and the annual precipitation decreases to about 500 mm. After 1960, the natural discharge of the River Bílina was insufficient to cover the demands from the fast growing energy and industrial sectors. The River Bílina had to be *augmented* with additional water transported from the River Ohře, first by an open canal (located under the Krušné Mountains) and later by an industrial water main. Additional water for drinking water supply was also transported from the Krušné Mountains. The discharge of the River Bílina was augmented by water pumped from the local mines. After 1990, the brown coal mining and electricity generation in the power plants gradually ceased and the open-cast mines were closed. One of the intended reclamation activities is to flood the mine pits, which will require large volumes of water (10^6–10^8 m^3). The River Bílina is the most important water resource for flooding. For the assessment of the amount of water available for this purpose it is necessary to know the natural flow regime, especially the low flow regime because streamflow should not become too low in the reach downstream of the mines.

Data from the Trmice gauging station for the period 1932–1990 are available to assess the flow regime. The flows that were observed prior to 1960 can be considered to be natural. From this year onwards the flows are gradually augmented with water from the River Ohře. The water management system, which has been developed since 1960, is very complicated. It is almost impossible to calculate the natural discharge from the measured flow, the water abstractions and the water inflows (naturalization, Section 4.3.3). Therefore an alternative approach is adopted using the BILAN model to simulate the natural monthly flows for the period 1961–1990.

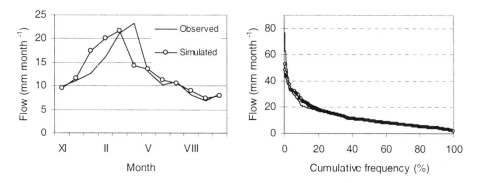

Figure 9.29 Observed and simulated streamflow of the River Bílina, period 1932–1960: mean monthly discharge (left) and cumulative frequency function of monthly flow (right).

The 1932–1960 flow series is considered to be natural and is used to calibrate the BILAN model. This assumption holds especially for the mountainous part of the catchment, which has not been directly affected by the mining. The discharge from this part of the catchment forms a significant component of the total flow of the River Bílina. The main water balance components for the calibration period 1932–1960 are given in Table 9.7. The long-term streamflow is about $152\ \text{mm year}^{-1}$ meaning that the catchment does not generate much surface water under natural conditions. The table also shows that the observed discharge is well simulated (difference < 2%). The calibrated parameters of the model were subsequently used for simulation of the natural streamflow for the period 1961–1990 using data series of observed precipitation, air temperature and relative air humidity.

Time series of observed and simulated mean monthly discharge for the calibration period (1932–1960) are presented in Figure 9.29 (left). In the late winter the simulated series shifts by one month towards the winter season, which can be explained by using the mean monthly temperature in the model for the whole basin. In reality, the air temperature is lower in the mountainous part of the catchment where snow storage occurs and the snow melts later. Figure 9.29 (right) also shows the flow duration curves for the observed and simulated monthly flow series. The duration curves illustrate that the model agrees well for a wide range of flows, including the low flow range.

The impact of all industrialization activities is reflected by the difference between the observed flow series, which represent man-affected conditions, and the flow series simulated by the BILAN model, which represents natural conditions in the same period (Figure 9.30). In most of the years, especially

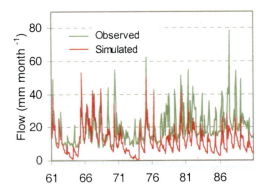

Figure 9.30 Observed and simulated discharge in the period 1961–1990 for the River Bílina catchment.

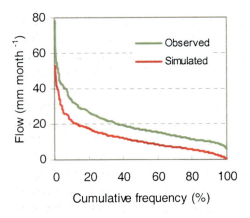

Figure 9.31 Flow duration curve of the observed and simulated streamflow for the period 1961–1990 for the River Bílina catchment.

during the low flow period, the observed (augmented) flow is higher than the simulated flow. Differences of up to 10–20 mm month^{-1} occur. The flow duration curves of the two time series also clearly show that the observed flow is higher than the simulated flow (Figure 9.31). For example, the Q_{90} of the natural streamflow equals 4 mm month^{-1}, whereas the Q_{90} of the observed flow is about 8 mm month^{-1}. This is similar to the augmentation for all industrial activities, which leads to an increase of 4 mm month^{-1}. The increase in monthly flow of the River Bílina at Trmice through water transfers from the River Ohře and other influences reaches more than 50% of the natural flow at mean flow, whereas it can be several hundred per cent during low flows.

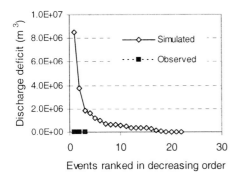
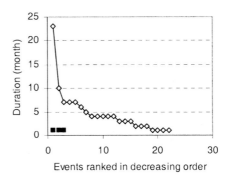

Figure 9.32 Streamflow droughts derived from monthly discharge data (period 1961–1990) for the natural situation (simulated) and the man-affected situation (observed) for the River Bílina; discharge deficit volume (left) and duration (right)

The simulated time series of monthly flow data is used to identify streamflow drought. The Q_{70} from the period 1932–1960 is applied as threshold value. Figure 9.32 gives the drought severity expressed as discharge deficit volume and duration. In total 22 drought events are derived for the natural situation from the simulated time series for the period 1961–1990. The time series of observed streamflow for the same period is used to derive droughts for man-affected conditions. In this period three drought events were observed. Clearly the increase in streamflow of the River Bílina due to water transfer from neighbouring basins is so high that it substantially changes the low flow regime. Under natural conditions seven times more drought events would have taken place than in the present situation. There are significant differences in drought severity, for instance the deficit volume of the most severe drought in the man-affected period is only 1.3% of the one under the simulated natural conditions (Figure 9.32, left). Without augmentation six droughts would have lasted for more than five months and the longest duration was almost two years. In contrast, only three drought events were observed when the flow was augmented; the duration of all these events is one month (Figure 9.32, right).

9.7.3 Discussion

The impact of surface water control is very diverse. For example, surface water abstraction without any additional measures leads to more severe streamflow drought downstream. The opposite takes place when effluent water discharged to a stream comes from water supply abstracted outside the catchment (Section 9.6.2). Reservoirs may alleviate streamflow drought downstream if water that is stored during the wet period is released during the dry period. However, when surface water is stored for irrigating crops during the dry season, then the drought downstream might be more severe. Local or regional conditions, type of surface water control and water management determine if surface water drought occurring under natural conditions increases or decreases.

9.8   Summary

This chapter has discussed the likely impact of human activities on hydrological drought. It has three main purposes: (a) to introduce concepts that can be used to predict the impact, (b) to explain in more detail physically-based models that can be applied for impact assessment, and (c) to present some examples of human influence, both observations and modelling studies. The examples are by no means complete because human activities affecting drought are numerous and very diverse.

Section 9.2 explained the different hydrological concepts that can be used for impact assessment. The effect of a particular human activity on drought is difficult to assess because it has to be distinguished from other human activities. In addition, inter- and intra-annual variability of hydrological variables complicates the assessment. Hydrological tools, either statistical approaches or physically-based models, with their own weak and strong points are important tools. Selection of these tools is more or less subjective, and as a result contradicting results can be found in the literature. Another reason for the contradiction is that catchment properties differ, leading to different impacts of a particular environmental change.

Multiple regression models that were previously explained in Section 8.3, have been only briefly mentioned and the emphasis is on physically-based models. Physically-based models simulate time series and today are the most common way to address the likely impact on the hydrological regime. Here two physically-based models, i.e. a lumped model (BILAN) and a distributed model (SIMGRO), are used to estimate drought characteristics and indices (Part II) for natural and man-affected conditions.

Land-use change affects drought development (Section 9.3). Forests, especially coniferous trees, consume more water than other land-use types through a higher interception and deeper rooting systems, and deforestation leads to less severe drought in groundwater and streamflow. Forest soils reduce generation of overland flow, and less streamflow peaks occur that can divide long low flow periods in a number of smaller ones. Cultivated catchments with spring-sown crops usually are less susceptible to hydrological drought than grassland catchments because evapotranspiration losses are lower. Exceptions do occur, as illustrated in Section 3.4 for a maize crop, which has more severe drought than grassland.

The effect of global warming on drought in a particular climate type is not unique (Section 9.4). Apparently contradicting results are found for catchments in the same climate region. Catchment properties, especially those which determine storage (e.g. soil and groundwater), cause different responses. In snow-affected catchments, the effect of climate change may be more pronounced. Temperature increase causes higher evapotranspiration and a shift of snowmelt from spring to late winter. Both processes lead to lower groundwater discharge in the dry season meaning the development of more severe hydrological droughts. The BILAN model and input files that were used for the climate impact assessment, are stored on the CD. A text file accompanies the program and explains how to get started and the results are presented as a case study.

Groundwater abstraction may enhance groundwater and streamflow droughts (Section 9.5). The impact depends on abstraction volume, climate type, hydrogeological settings, abstraction pattern (permanent or seasonal) and distance to the stream. For a lowland catchment with a modest abstraction it is illustrated that groundwater drought increases near the well field and that it has hardly any influence on streamflow drought. The consequences of abstraction only in the wet season are small compared to permanent abstraction. The whole modelling procedure with the physically-distributed SIMGRO, including results, is stored on the CD as a self-guided tour.

Land drainage is another form of groundwater abstraction. In general, drainage leads to lower flows during the dry season, which means that droughts in groundwater and streamflow increase. Removal of drainage, filling in of ditches and rising of surface water levels to improve conditions for terrestrial ecosystems have the opposite effect. Querner & van Lanen (2001) showed that incomplete regeneration (e.g. partial filling in of ditches) can have a surprising effect, i.e. groundwater drought decreases and streamflow drought becomes more severe.

Urbanization has manifold effects on hydrological drought (Section 9.6). The impact depends on surface impermeabilization, drainage of stormwater, water supply (type and resource), sewage system, and hydrogeological conditions. Groundwater recharge in cities often increases as a net effect of paved surfaces, means of drainage, and pressurized mains. Recharge is especially higher in non-sewered cities because not more than 10% of the water provided is actually utilized. The water table usually rises except for urban development in areas where water supply is from an unconfined aquifer below the city. In general, urbanization causes less severe groundwater drought due to enhanced recharge and higher water tables. As a consequence, groundwater discharge in the urban catchment is higher, which also reduces streamflow drought. Furthermore, water supply and adequate drainage and sewage systems in well-developed cities create higher flow in the streams during periods with rain (quick response) and higher base flow (increased recharge), thus reducing the likelihood of streamflow drought compared to pre-urban circumstances. A slowly-responding rural catchment may change into a quickly responding urban catchment with higher average discharge. It is worth noting that recent urban water management promotes infiltration of stormwater within city limits, meaning that response to rainfall will reduce again. The description of urban development is generally restricted to the urban catchment. However, adjacent catchments that supply the water for the city may suffer from city expansion leading to more severe drought. This is illustrated for a particular lowland catchment (i.e. the Poelsbeek).

Like urbanization, surface water control also has a large impact on drought. Commonly, river regulation without introducing mitigation measures (e.g. weirs) enhances streamflow drought. The construction of reservoirs can increase or decrease streamflow drought downstream of the dam depending on the type of reservoir and use of the resource. Water augmentation can have a considerably positive influence on streamflow drought in the catchment that receives the water. Clearly, circumstances may be worse for the catchment that provides the water. In the example for the Czech Republic drought severity and duration significantly decreases in the augmented stream.

In the next chapter human influences in terms of management of streamflow for ecological protection are introduced (Chapter 10). The textbook concludes with a description of how people try to manage hydrological drought in order to alleviate any impact that might be enhanced by human influences (Chapter 11).

10

Stream Ecology and Flow Management

Bente Clausen, Ian G. Jowett, Barry J.F. Biggs, Bjarne Moeslund

10.1 Introduction

No water – no life. One fundamental role of rivers is to provide the living space (habitat) for biological organisms (also called biota), be it plants, invertebrates (e.g. insects, snails), fish or mammals. The physical environment and the stream biota together form the ecosystem, with the physical environment having a strong influence on the biota, and in many cases vice versa also. Ecology is the study of the interrelationships between the biological organisms and their environment.

One of the most important aspects of the physical environment within a river is the flow regime. The flow regime influences the temporal variability of water depth and velocity, river morphology, sediment transport and bed sediments, and consequently the ecosystem that develops. Figure 10.1 shows

Figure 10.1 Two streams with contrasting flow regimes and ecosystems; lowland stream in Denmark with dense stands of macrophytes (flowering Batrachium peltatum; left) (photo by B. Moeslund), and salmon spawning stream at Double Hill, Rakaia catchment, New Zealand (right) (photo by N. Boustead).

two streams with contrasting flow regimes and ecosystems. One is a lowland stream with a bed of fine sediments and flows that vary little over time, which provides optimal growth conditions for aquatic macrophytes (large aquatic plants) and good life conditions for some invertebrates. The other is a gravel-bed stream, in which movement of substrate during floods prevents growth of macrophytes.

Some streams have base flow most of the year and flood only during certain seasons (for example in tropical and snowmelt climates), while others have high and low flows at any time of the year. Some streams have rare, but extreme floods interspersed with long periods of low flows, while others have more moderate, but frequent flood and no prolonged drought. Some streams are intermittent and dry up each year, which results in extreme living conditions for aquatic biota.

Flow variability and the movement of bed sediments can have profound effects on stream ecosystems. Fish and plants in stable spring-fed streams are often unable to develop or even survive in less stable environments. On the other hand, gravel-bed rivers and their aquatic biota are in a constant state of change, caused by extreme flows (flood and drought) and mobile bed sediments. Extreme flows usually have a short-term, major effect on the biota, although this might be part of a long-term 'norm'. Flood can be damaging to the biota because of the high velocities and possible associated sediment mobilization. Drought can be disastrous because of the associated physical and chemical changes; for example, the water can become so shallow that it hinders fish mobility or even survival; pollutants can become so concentrated that they are lethal for organisms; low water velocities can cause algal bloom and high temperatures. The response of river organisms to extreme flows, and the consequent impact on the ecosystem, will depend on the characteristics of the biota, the timing and frequency of extreme events, and the biotic interactions with other components of the ecosystem.

The physical characteristics and biota in rivers also show longitudinal gradients, sometimes from small steep mountain streams to large lowland rivers. This has been conceptualized in the 'river continuum concept' (Vannote *et al.*, 1980). Continuity between spatially distinct zones is often important for colonization by invertebrates and migratory fish species. Whatever the natural setting of the stream, the biota has evolved to survive in the physical environments, including the flow regimes.

Although every river or stream is unique, there is also a commonality. Rivers contain flowing water, and the distribution of water depths and velocities characterize the stream. Stream size, bed slope and bed material influence the morphology of the river, and all of these factors determine the water depths, velocities, and the type of biota that is adapted to those conditions. Many of these factors relate to flow, and changes to the flow (for example due to water abstractions) can have severe implications for the stream biota. However, stream ecosystems are also influenced by factors that may be only weakly related to flow. Water quality, temperature, and biotic interactions, such as competition and the introduction of predators, can all influence stream biota. Construction of reservoirs, with the consequent changes to flow regime and water quality, can have major effects on river morphology (shape and form) and biota. In general, human activities that change the flow regime, particularly the extremes, will have an effect on the stream ecology. The task of hydrologists, aquatic ecologists and water managers is often to predict such changes and mitigate the negative effects, if possible.

This chapter has several purposes. First of all, it is an introduction to stream ecology for hydrologists (Section 10.2). Secondly, it is concerned with the relationship between streamflow and stream biota, in particular the ecological effects of extreme flows (flood and drought) (Section 10.3). Thirdly, it describes practical aspects of flow management for ecological protection, especially during drought (Section 10.4).

Although this chapter is an introduction to ecology, it focuses on the importance of flow as a control and does not cover topics such as biotic interactions and energy and nutrient sources and cycles. For a more fully introduction to ecology, see for example Allan (1995) or Giller & Malmqvist (1998).

Until now, much freshwater ecology research has concentrated on high-flow events, while the effects of droughts have received only little attention (Jowett, 1997a). This is because floods play an enormous role in stream ecology and flow management. On a large timescale, flood frequency determines how often biota are affected, it differentiates between river types, and it is now regarded as one of the most important hydrological factors for the structure of stream communities (Resh *et al.*, 1988; Lake, 2000). Therefore, and because this chapter is also an introduction to stream ecology, descriptions of the ecological effects of floods as well as droughts have been included.

Section 10.2 introduces the stream as a living place for stream biota; it gives basic details about stream biota: periphytic algae, mosses, macrophytes, invertebrates and fish; and it describes the factors important to stream biota and the dynamic relationship between the flow regime and the biota. Section 10.3 is concerned with the effects of floods and droughts on stream biota. Section 10.4 describes the methods that water managers can use for setting instream flow requirements, especially minimum flows, for ecological protection. These methods range from simple methods based on low flow statistics (Section 5.3) to more sophisticated methods based on field data and advanced computer models. One of these models, the RHYHABSIM (River Hydraulics and Habitat

---

Box 10.1

Large, deep rivers and lakes

In ecology, a distinction is made between 'lentic' (still-water, lakes) and 'lotic' (running water, rivers) ecosystems. Basically, a river is different from a lake in that water in a river has a dominant flow direction in contrast to the fluctuating flow directions in a lake. The residence time of lakes is therefore much higher than in rivers, which is why plankton (floating organisms) can live here without being flushed away. Sometimes large, deep rivers can have lake-like characteristics, with good plankton production and pelagic (living in the open water) fish species. However, benthic (botton-dwelling) primary (plant) production in large rivers is often low because the flow is turbid and the bed substrate mobile, at least in the middle of the river. In deep sections of rivers, light at the bottom may be restricted, and this, combined with high velocities and associated sediment movement, will limit the plants and animals that live on the streambed. In large rivers, benthic life may thus be concentrated in relatively shallow areas near the banks, where the aquatic environment is more hospitable. Because large rivers are difficult working areas for biologists, less is known about the functioning of ecosystems and habitat requirements in large rivers than in smaller rivers, particularly for higher trophic (feeding) levels, such as invertebrates and fish. Important issues for large rivers are navigation and flood control. Many large rivers have reservoirs for flood control, and these can cause special ecological issues, such as blue-green algal blooms.

Simulation) model, is included on the CD (Software) and introduced in the chapter.

It should be noted that this chapter deals primarily with rivers in humid, temperate regions. Because of space limitations it was decided to comment only briefly on the ecology and management of large, deep rivers (mean flow > 100 $m^3 s^{-1}$) and lakes (Box 10.1) and on streams in semi-arid regions (Box 10.2, Section 10.3.4). Similarly, the chapter contains only limited reference to sediment and water quality issues and floodplain processes. For these topics see for example Calow & Petts (1992, 1994), Harper & Ferguson (1995), Foster *et al.* (1995) and Anderson *et al.* (1996).

10.2 Stream ecology

10.2.1 Basic biological concepts

With more than one million species of animals and over 325 000 species of
plants in the world, it is necessary to have a classification system (Gould &
Keeton, 1996). The system that is used today (Table 10.1) attempts to encode
the evolutionary history (phylogeny) and dates from the work of Linnaeus, who,
in the middle of the 18th century (a century before Darwin), grouped organisms
according to similarities and obtained results similar to those obtained today.
The system of naming species also dates from Linnaeus. Each species is named
by two Latin (or Latinised) words: the name of the genus to which the species
belongs (first letter upper case) and a species name (first letter small case), and
the standard is to print the full name in italics. For example, *Homo sapiens* and
Canis lupus (Table 10.1) both belong to the same class, but to different genera
(plural of genus). The different categories (e.g. kingdom, phylum, class) are
referred to as taxa (taxon in singular).

When describing a group of organisms, there are different levels in biology:
individual, population, community (also called 'association') and ecosystem. A
population consists of individuals of a single species, and a *community* (or
association) is an aggregation of individuals or populations occupying a specific
geographic area; thus, a community includes more than one species. An
ecosystem covers more than the biota; it consists of the biological components
as well as their physical and chemical environment.

When characterizing a stream community, three common measures are
used (a) *abundance* (or density), which is the total number of individuals, or the

Table 10.1 Seven taxa and classification of four species (examples from Gould &
Keeton, 1996)

Taxa	Red oak	House fly	Wolf	Human
Kingdom	Plantae	Animalia	Animalia	Animalia
Phylum	Tracheophyta	Arthropoda	Chordata	Chordata
Class	Angiospermae	Insecta	Mammalia	Mammalia
Order	Fagales	Diptera	Carnivore	Primata
Family	Fagaceae	Muscidae	Canidae	Hominidae
Genus	*Quercus*	*Musca*	*Canis*	*Homo*
Species	*rubra*	*domestica*	*lupus*	*sapiens*

total amount of biomass, per unit area, (b) *species richness*, which is the total number of species present in a specific area, and (c) *diversity*, which is the species richness weighted by the relative abundance of each species present. There are several diversity indices, of which the most commonly used is the Shannon (or Shannon-Weaver) Index, H' (Zar, 1999):

$$H' = -\sum_{i=1}^{s} p_i \log_{10}(p_i) \tag{10.1}$$

where p_i is the proportion of the ith species density, s is the number of species, and $\log_{10}$ is the logarithm to the base 10, although any base can be used. In general, the higher the number of species, the higher the diversity, and vice versa. For a given number of species, the diversity is relatively low if there are many individuals of one species and few of the others, and it is relatively high if there is an even distribution of individuals across the species. Thus, the diversity can be high even when the number of species is relatively low. To summarize, diversity is an expression for evenness as well as for species richness. High species richness and high diversity of a stream community is usually considered desirable.

10.2.2   The stream as a dynamic living place

A river or a stream is the living place (habitat) for many different types of living organisms. There are plants (phyton), such as periphytic algae and mosses (which are primitive, non-vascular plants) found in a wide range of rivers including steep mountain streams with a stony or rocky sediment, and macrophytes (large aquatic plants), which are more common in low-land rivers with moderate bottom slope and a soft sediment. Most streams also have a rich animal life, including insects and other invertebrates (animals without backbone and internal skeleton), and a variety of fish. Some rivers also provide habitat for larger animals such as beaver, otter and crocodile.

   With habitat meaning the place to live, micro-, meso- and macrohabitat indicate the different scales of living place. Microhabitat deals with the smallest level, such as individual stones in a river; mesohabitat usually refers to living areas that have similar characteristics (e.g. depth, velocity, substrate), such as riffles, runs and pools; macrohabitat refers to the largest level, such as river types.

   Periphyton (algae and associated micro-organisms growing attached to any submerged surface) is the main primary producer of many unshaded streams and the food base for aquatic grazing animals, typically invertebrates. In some forest streams of continental regions, inputs of leaf detritus can be high (e.g. from deciduous trees) and many stream insects have evolved to utilize this detritus as an energy source. A major portion of the periphyton in streams is composed of algae (including filamentous green algae, blue-green algae, red algae and diatoms). These algae capture the sunlight with their chlorophyll molecules, absorb carbon dioxide and other nutrients such as phosphorus and nitrogen from the surrounding water, and then synthesize organic carbon, for example in the form of new or enlarged cells. Some algae even secrete a portion of this carbon, and this food source supports other organisms such as microscopic bacteria and fungi. Protozoa (primitive animal micro-organisms) are also commonly found in the periphyton mat.

Periphyton species have a variety of growth forms, and therefore tend to have different hydraulic habitat requirements. For example, many filamentous green algae are fragile and/or weakly attached, and they are therefore confined to habitats with low water velocity ($< 0.3 \, \mathrm{m \, s^{-1}}$) (Figure 10.2); stalked diatoms and low growing filamentous algae tend to grow best in habitats with moderate water velocities (0.3–$0.7 \, \mathrm{m \, s^{-1}}$) where mass transfer is high, but drag is not excessive; while low growing and tightly adhering mucilage-producing diatoms and prostrate filamentous cyanobacteria (blue-green algae) prosper in habitats with high velocities (Biggs *et al.*, 1998a).

Bryophytes (mosses and liverworts) form an extensive cover on stable surfaces in some streams (Figure 10.3). They are generally not consumed by invertebrates and fish (at least not to the same extent as periphyton), but they are important habitats because they increase the amount of attachment surface, provide oviposition (egg laying) sites for invertebrates and fish, slow the water or create turbulence, provide shade, trap detritus, provide refuges from predators and moving sediment, and alter oxygen regimes (Suren, 1991). Mosses and liverworts have slow rates of recolonization and growth. If bed sediments are mobilized during high flows, the bryophytes are removed by abrasion and may take years to re-grow. Thus, high bryophyte cover is usually confined to streams with stable bed sediments (Duncan *et al.*, 1999).

Macrophytes (large aquatic plants, mainly flowering) are rooted in the bed sediments and have functions similar to bryophytes in streams. However they are usually larger than bryophytes (Figure 10.4) and provide a greater surface area as habitat for algae and invertebrates and better shelter for fish. They are also more susceptible to high flow events. For example, in a recent study, Riis & Biggs (2003) found a strong inverse relationship between macrophyte cover and flood frequency in streams. Significant biomass only occurred in streams with less than six floods per year (a flood was defined as a flow higher than seven times the median flow). Because of their lack of resistance to high water velocities and scour, macrophytes tend to be found in habitats with low water velocity.

Figure 10.2 Filamentous green algae (*Cladophora* sp.) (photo by B. Moeslund).

Figure 10.3 Dense growth of mosses (*Fontinalis antipyretica*) on stones in a springfed Estonian stream with stable flow (photo by B. Moeslund).

Figure 10.4 Examples of macrophytes in Danish streams; submersed macrophytes (*Batrachium* sp., *Callitriche* sp.; left), and emergent macrophytes (*Nasturtium officinale*; right) (photos by B. Moeslund).

Macro-invertebrates are insects (e.g. stoneflies, midges, caddisflies, Figure 10.5, left, and mayflies, Figure 10.5, right), snails, mussels, worms and other groups. They are benthic and live on the stream bed, macrophytes, or other stable objects in the stream such as wooden debris. Most invertebrates are macroscopic animals and thus visible to the naked eye, but early life stages of these organisms are microscopic in size. As noted above, these organisms generally feed on periphyton and other micro-organisms associated with the periphyton, and, where available, detritus derived from terrestrial vegetation, e.g. leaves from deciduous trees. They are a very important part of the food web; downwards by feeding on the primary producers, upwards by providing a large part of the diet of fish. In streams with stable flow they can, when occurring in significantly high numbers, prevent proliferations of periphyton by consuming large quantities. They are also widely used as indicators of water quality because many of the species are sensitive to common pollutants and alterations of the oxygen regime, thus reflecting the degree of pollution. For example, many species of mayflies (*Ephemeroptera*), stoneflies (*Plecoptera*) and caddisflies (*Tricoptera*) are typical clean-water species requiring high oxygen concentrations at all times. Reductions of oxygen concentrations in water due to pollution of organic matter that is degraded by bacteria and other micro-organisms under use of oxygen are therefore generally reflected in the populations of clean water organisms.

Fish are usually at the top of the food chain in streams and require the rest of the aquatic community to be in good health. In cold-water streams, salmonids (trout and salmon) tend to be the main species of interest (Figure 10.6, upper). Fish in cold-water streams tend to be carnivores, feeding on invertebrates and other fish. This makes them attractive and susceptible to anglers, whose lures often imitate stream insects or small fish. Salmonids were found originally only in the Northern Hemisphere and have been introduced to many other areas of the world to enhance recreational fishing opportunities. In warm-water streams, a variety of other species, such as cyprinids (carp-like fish) and percids (perch-like fish), tend to dominate the fish faunas. These fish are often omnivores (e.g. roach, rudd, tench, dace), feeding on both plant and animal matter, but can be algivorous, feeding on periphyton and phytoplankton (e.g. silver carp) or herbivorous, feeding mainly on macrophytes (e.g. grass carp). In fact, grass carp (often specially bred, sterile fish) is sometimes used to control the growth of aquatic vegetation.

The distribution of most fish species is closely linked with the spatial variations in depth, velocity, substrate and temperature. For example, the New Zealand native fish in Figure 10.6 (lower) favours clear and swiftly flowing water and, like most other New Zealand galaxiid fish species, it is most active at night (nocturnal). The life cycle of some species is dependent on spending some

Figure 10.5 Examples of macro-invertebrates; net spinning caddisfly (*Hydropsyche* sp., approx. 20 mm), the larva has spun a net with which it catches drifting food particles (left), and mayfly larvae (*Ephemera danica*, approx. 40 mm) (right) (photos by C.B. Hvidt).

Figure 10.6 Examples of fish species; male and female adult brown trout (*Salmo trutta*) in the Whitikau Stream, New Zealand (approx. 40 cm; upper) (photo by R. Strickland), and New Zealand native fish *Galaxias vulgaris* (approx. 9 cm; lower) (photo by A.R. McIntosh).

time at sea. Such species, which move between freshwater and the sea, are called 'diadromous' ('running', from the Greek). For example, Northern Hemisphere salmons and sturgeons spend most of their lives in the sea. When mature, they migrate back to the river to spawn (anadromy) and usually they do not feed while migrating to spawning areas. Eels do the opposite – they live (feed and grow) in freshwater, and when mature they migrate into the sea to spawn (catadromy). Some fish spawn only once (e.g. eel and most salmon species), whereas other species spawn multiple times (e.g. trout, cyprinids). Most fish species have habitat requirements for reproduction that include specific flow, substrate and temperature. For example, brown and rainbow trout usually spawn in cold-water gravel-bed streams where there is adequate water current through the gravel to carry oxygen to the eggs.

10.2.3 Factors affecting stream biota

Many factors determine whether a given species will be present in a river. One overriding, large-scale factor relates to *geography and evolution*. For example, there were no trout and salmon species in New Zealand rivers before the Europeans arrived in New Zealand around 1840, simply because they had not evolved in this region or had been unable to migrate there. Salmonids were brought from the Northern Hemisphere into New Zealand in the late 1800s, and trout fishing is now a popular undertaking with world-class fisheries. The example illustrates how suitable habitats may remain void due to natural barriers and obstacles.

On a smaller scale, there are a number of factors of importance to the distribution of species. First, *food* must be available for a species to survive in a particular area. Returning to our previous example, salmonids feed on invertebrates and smaller fish; invertebrates feed on algae or other invertebrates; and algae need nutrients (such as nitrogen and phosphorus) and light to grow. The food chain is generally more complex than described here.

Water depth and stream bed width are important factors as these define available physical space, an important part of the habitat. For example, adult brown trout prefer water depths of more than 50 cm for residence and feeding (Hayes & Jowett, 1994), and they require a minimum depth of 15 cm for passage (Bell, 1986).

Water velocity is important for the transport of resources to the organisms, be it dissolved nutrients to periphytic algae or prey items to animals. However, velocity is also a stress factor since animals must use energy to withstand the forces of the flowing water. Velocity even has a potential hazard; when water velocity exceeds specific levels, the current may sweep the organisms away, destroy them, or limit their growth. The threshold at which this happens depends on the organism's mobility, adaptation, ability and possibility to hide, or, if it is attached to the bed substrate, it will depend on the strength of its attachment. Water velocity is believed to strongly influence the adaptation of organisms in terms of distribution, shape and behaviour. For example, some species are found mainly in fast flowing water (e.g. trout, and mayfly larvae from the family *Heptageniidae*), while others are found only in slow-flowing water (for example roach, and caddis flies from the family *Limnephelidae*). Illustrative examples of body shape and other adaptations of invertebrates and fish are found in Allan (1995) and Hynes (1970).

Channel substrate is of particular importance to the biota of running water since most organisms are closely connected with the sediment (benthic biota). When small sections of stream are classified into mesohabitats, the aquatic communities within each mesohabitat are usually distinct and characteristic (Armitage *et al.*, 1995; Beisel *et al.*, 2000). Coarse substrate (such as boulders, cobbles, gravel and large woody debris) provides stable surfaces to which algae and invertebrates can attach. It also provides benthic invertebrates and fish with shelter from predators and from high velocities during floods. Sand is usually considered a poor substrate for periphyton and invertebrates, although some, especially some smaller, burrowing ones, have become specialized to exploit this habitat. However, sand may be a very suitable rooting substrate for macrophytes, which can support very high invertebrate densities (Armitage & Cannan, 2000). Some fish are less specific with regard to substrate, although salmonid spawning mostly requires a coarse substrate (mostly gravel) in which the eggs are protected and through which the water flow is high enough to supply oxygen to the eggs. Organic substrates such as submerged wood, plants, leaves, and fine particles function both as a surface for growth (especially the larger-sized particles) and as food (especially the smaller-sized particles). More on the relationships between substrate and biota can be found in Allan (1995), Hynes (1970) and Minshall (1984). Sediment movement usually takes place during floods (Section 10.3.2) and can uproot macrophytes and cause destruction or involuntary drift of organisms (for example algae, mosses and invertebrates) attached to the sediment.

The differences in adaptation with regard to depth, water velocity and substrate is used quantitatively in one of the flow management methods (the habitat method) described in Section 10.4.2.

Temperature affects the growth rate and distribution of many organisms, and extreme temperatures (high and low) can be lethal. Salmonids are generally adapted to cold-water streams and are relatively sensitive to high temperatures. For example, there are few brown trout in the northern part of New Zealand, where the average winter water temperature is higher than about 11°C, the limiting temperature for successful egg incubation (Scott & Poynter, 1991; Jowett, 1992).

In addition to the physical factors mentioned above, *water quality* is also of great importance. Dissolved oxygen, pH and ammonia are factors of importance to most organisms, and some do not tolerate concentrations below or above specific threshold values. Trout is particularly sensitive to low oxygen concentrations, low pH values and even moderate ammonia concentrations, and trout is therefore used as an indicator of water quality in some situations. Some organisms are adapted to life in well-oxygenated water and depend on high oxygen concentrations, while other organisms are able to adjust their metabolic rate to the oxygen level and even get a competitive advantage when the oxygen regime is altered, for example due to pollution with organic matter. Aquatic plants and algae are sources of oxygen during daytime, but these organisms respire and use oxygen like any other living organisms, which may cause a reduction in oxygen concentration at night. Aquatic plants and algae may therefore give rise to a distinct diurnal oxygen variation, which is not found in streams without these biological components (Section 10.3.3).

In addition, *biological interactions* such as predation and competition between species (inter-specific competition), and between individuals (intra-specific competition), may restrict the distribution or the population size of a given species.

10.2.4 Flow as a control

Flow controls many of the physical and chemical factors mentioned above, for example water velocity and depth, bed substrate composition and stability, temperature and oxygen content. However, flow is not the only determinant for these factors. Velocity and depth are also influenced by slope, substrate roughness, cross-section geometry and density of macrophytes. Bed stability is determined not only by the velocity and the associated bed shear stress, but also by the bed particle size distribution and sediment supply, which again depend on the geology, in particular the erosion, in the upstream catchment. The rate at which water temperature changes along the length of a river depends not only on water depth and velocity, but also on the amount of shade, which determines the amount of solar radiation reaching the water surface. Tributary and groundwater inflows also have an effect. Water depth influences the rate of heating, and water velocity determines the distance that the water travels as the temperature changes. In addition, the water temperature at a point will also depend on the source of the stream water, the distance from a discrete source, air temperature, wind velocity and other climatic variables. Oxygen content in water is directly (and positively) related to flow, because reaeration (the exchange of oxygen between water and atmosphere) depends on depth and velocity, and it is inversely related to temperature.

Although many of the important factors are interdependent and influenced by more than flow, there are common responses to changes in flow. An increase in flow at a particular site will usually increase water velocity, depth and width, promote sediment movement, decrease temperature and increase oxygen content. On the other hand, a reduction in flow will usually lead to a reduction in water velocity, greater sediment deposition, an increase in temperature and a reduction in oxygen content.

10.3   Ecological effects of extreme flows

10.3.1   Extreme flows as disturbances

An extreme flow is referred to as a 'disturbance' in biological terms when it causes a significant loss of individuals from a community (White & Picket, 1985; Resh *et al.*, 1988; Biggs *et al.*, 1999). However, it has been argued that the definition should be based solely on the characteristics of the flow event (not the biological response), using for example flood frequency, predictability, or a similar statistic (Section 5.3), to enable comparison of the biological responses of events (Poff, 1992; Lake, 2000).

A disturbance does not necessarily involve a change in flow. For example, a disturbance might be associated with the removal of overhanging vegetation and the transformation of a shaded stream into a non-shaded stream, with an extensive bush- or forest-fire in the catchment, a natural damming due to a fallen tree, or with pollution. However, the most common form of disturbance in natural streams is caused by changes in flow, such as floods and droughts (Lake, 2000). The effect of extreme flows on biota will depend on the duration, the magnitude and the frequency of the extreme flow, as well as on the species composition and their life history strategies.

Disturbances and biological responses can be characterized by their temporal patterns as pulses, presses or ramps (Lake, 2000) (Figure 10.7). Usually flood disturbances are pulses. The effect of a flood on periphyton attached to the streambed is a good example of a pulse response, although it may be longer than the pulse disturbance (the flood) itself. The flood pulse removes algae from the stream bed, either by shearing or by the abrasive action of moving sediment, and after the flood subsides, periphyton begins to accumulate again until the next flood (Lohman *et al.*, 1992). Other biota may react to a flood with a press response (longer-lasting effects). Droughts (both

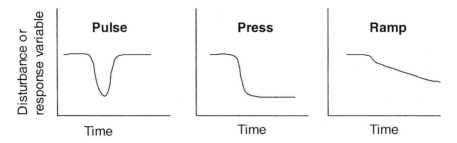

Figure 10.7 Three types of stream disturbance or response: pulse, press, and ramp.

disturbance and biological response) are typically ramps during which environmental conditions steadily decline, and there will usually be a recovery after a longer period of time (not illustrated in Figure 10.7).

Disturbances can be viewed as events that 'reset' the ecosystem's development, without necessarily changing the long-term composition or abundance of the biological community; such events may be a necessary part of the natural system. However, they might also be true 'disasters', rare events with consequences that are not part of a natural variation (Section 10.3.4).

10.3.2 Ecological effects of floods

Some floods have been shown to be disturbances, reducing the numbers of benthic invertebrates and trout (Scrimgeour & Winterbourn, 1989; Seegrist & Gard, 1972). Organisms cope with flood in different ways. A flood might devastate some biota, but not others. For example, a flood that causes high water velocities but no bed movement might tear loose green filamentous algae or rooted plants, while other biota might survive without harm (Biggs & Thomsen, 1995). Mobile invertebrates may find refuge in the bed sediments during floods, and fish seek shelter near the banks or in inundated areas outside the main channel with shallower and slower flowing water.

It is not until the bed sediments begin to move that the main ecological 'disasters' start to happen (Biggs *et al.*, 1999). When stones covered with algae and invertebrates start to roll along the bed, the attached biota is in great danger of being destroyed. When stones are moved or turned upside down, invertebrate refuges might be exposed and the biota entrained into the water column and swept away. Scouring around plant roots might lead to the whole plant being uprooted. Consequently, mountain rivers with frequent floods and unstable beds usually have 'clean' substrates and limited abundance and diversity of biota

(Biggs & Close, 1989; Clausen & Biggs, 1997). For floods involving movement of bed sediment, the term 'disturbance' seems particularly appropriate.

In many floods only a proportion of the bed particles are mobilized. Locations with unstable substrates or subject to the passage of saltating bedload are unlikely to be suitable habitats for most species of benthic invertebrates. This is either because food sources are not present where bedload movement occurs, or because the substrate does not provide a secure platform for the organisms. Substrate stability may also partly explain why invertebrate abundance generally increases with increasing substrate size, because larger stones are usually more stable than smaller ones. The adverse effects of suspended sediment on benthic invertebrates are well-documented in reviews (Newcombe & MacDonald, 1991; Waters, 1995; Wood & Armitage, 1997), which have concluded that high sediment loads may reduce the abundance and diversity of invertebrates by (a) smothering and abrading them, (b) reducing their periphyton food supply or quality, and (c) reducing available interstitial (space between substrate particles) habitat.

Recovery of biological communities following floods depends on the species and the number of refuges and adjacent water bodies from which emigration may take place, but it is usually rapid (Minshall, 1968). The first species to colonize the stream are usually small and opportunistic, called 'colonizers'. Diatoms tend to recover quickly after a flood (Fisher *et al.*, 1982). Then follow green filamentous algae, whereas mosses grow so slowly that they are usually absent in flood-prone rivers (Duncan *et al.*, 1999). Insects recolonize the stream by larvae drifting down from upstream areas or returning from the hyporheic zone (the interstitial waters below the sediment surface), and adults fly in from other areas and lay their eggs in the stream. Sometimes, when the disturbance is high enough to clean the stream from invertebrates, the algae may get a chance to bloom before the invertebrates recolonize the stream and keep the algal growth under control. Thus, a disturbance is followed by a succession of changes in the community, until some kind of near-equilibrium is obtained. The taxa in this more stable community at the end of the succession are called 'climax' taxa, or 'competitive' taxa, because they compete with each other (in contrast to the colonizers, which were the first in the succession).

The life span also determines what flood frequency a population can survive, and fluctuations in populations can often be caused by flood. For example, trout populations can be detrimentally affected by floods that occur during, or just after, egg incubation, and the effect can last for years until there is a year with good survival. A study of brown trout in a gravel-bed river in New Zealand showed that one year with successful spawning and incubation resulted in a three-fold increase in adult brown trout numbers three years later (Jowett, 1995; Hayes, 1995).

10.3.3 Ecological effects of drought

A drought is a period with low flows, during which algae have time to develop, and the invertebrate community gradually changes from one that grazes on thin periphyton films to one that can feed on or live amongst thick periphyton mats. The abundant supply of algae can increase the densities of invertebrates, and the amount of food available to fish. However, these seemingly favourable conditions can – and often will – reverse when the initial positive stages of drought are passed. If the growth conditions for periphyton are particularly favourable over an extended period, the bloom can take such dimensions that it occupies the whole stream (Figure 10.8). The abundance of plant material causes large diurnal fluctuations in oxygen content and pH, which some fish and invertebrate species do not tolerate.

Figure 10.8 Algal bloom in the Hakataramea River in Otago, New Zealand (photo by B.J.F. Biggs).

As the wetted area of the stream reduces, there will be loss of habitat (width, depth) and perhaps connectivity, and the dilution of pollutants will reduce. Eventually the physical and chemical environment may become unsuitable for some species, competition and predation may increase, and terrestrial plants can colonize the stream bed. In particular, shallow water and low velocities, combined with warm and sunny weather, can lead to such high water temperatures and wildly fluctuating oxygen concentrations that no living organism can survive.

The extreme case is when whole river reaches dry up and cause *fragmentation of the river system.* This especially affects fish species that require passage to the sea or to spawning or feeding areas. Another phenomenon disrupting the pathway for diadromous fish is restricting or closing river mouths during periods with low flows, not powerful enough to mobilize the sediments. The formation of pools following fragmentation may be beneficial for some fish species that are adapted to rivers in arid or semi-arid areas and spawn and rear young in isolated pools. However, fragmentation is often deadly to aquatic fauna. Predatory birds effectively catch fish trapped in pools and ponds and invertebrates in shallow riffles. The worse case scenario is when the whole river system dries up and the stream may, partly or totally, lose its natural content of plants and animals for shorter or longer periods.

Drought is usually easily detected in streams with macrophytes. In natural and undisturbed streams the vertical and horizontal distribution of submersed and emergent plants is strongly influenced by velocity and depth. The typical response from water plants to droughts are changes of abundance in favour of those preferring lower flow velocities and more shallow water. The response from amphibious plants (plants that can live fully submerged for shorter or longer periods, but which normally grow with the upper parts of the stems and leaves above water) is redistribution with increased density at the central parts of the stream bed, from which they are normally kept away by high velocities and depth.

Most plants and animals are adapted to cope with drought of some magnitude, especially if the droughts are seasonal and predictable. Invertebrates survive drought as dormant eggs or young larvae in the hyporheic zone. Some fish, for example trout, spend the dry summer period in lakes or the sea and return to freshwater in the autumn in some areas of the world. Some species are able to tolerate or survive events that are lethal to others.

In natural streams and rivers where droughts are frequent and predictable, biota will have adjusted to these conditions; the only species present are those that have evolved to cope with the drought conditions while other species will have been eliminated. In natural streams and rivers where droughts are rare, species with little tolerance to droughts (e.g. mayfly and stonefly larvae) will normally dominate over those species with larger tolerance (e.g. aquatic worms, midges and molluscs); if a severe drought occurs in this type of river, species may recover over a time that will depend on the longevity of the species and other circumstances, such as the possibilities for emigration from adjacent water bodies.

If the natural flow regime is disturbed by a drought, for natural as well as man-made reasons, the cascade of physical and chemical changes will cause changes in both community composition and species abundance. Knowledge about each species' response to change in flow may be used to set flow regimes to protect habitat (Section 10.4). For example, in a recent New Zealand study of a nutrient-enriched stream and a neighbouring unenriched stream, Suren *et al.* (2003) showed that mayflies and caddisflies in the unenriched stream were not adversely affected by extreme low flows, and that the invertebrate community did not become dominated by aquatic worms, snails or midges (as was expected). However, such changes did occur in the enriched stream where periphyton proliferations also occurred. This finding has major consequences for setting instream flow requirements and suggests that factors such as the degree of enrichment need to be taken into account when predicting the effects of low flow events on invertebrate communities.

Although there are some stream types with stable flow regimes and very little variation or disturbance by the flow regime (such as spring- and lake-fed streams), most streams are subject to some kind of flow variation. Moderate variations (intermediate disturbance, Section 10.3.4) are thought to have a positive influence on diversity, provided they do not exceed the tolerance of the plants and animals (Townsend *et al.*, 1997). In management of streams and rivers, the challenge is to determine the threshold values above (or below) which drought events are becoming destructive to the natural plant and animal communities (Section 10.4).

10.3.4 Flow variability and biological communities

Floods maintain river morphology by scouring fine sediment deposits from pools and by removing interstitial fine sediment from gravel-bed rivers. In this way they improve habitat quality for benthic invertebrates and some fish species and limit the encroachment of riparian or aquatic vegetation. Thus, the flushing out of biological (and non-biological) material provides a resetting and renewal of the physical habitat and the associated biological communities. However, floods are also disturbances and biologically detrimental (Section 10.3.2). Thus, although periodic flood events are desirable or even necessary, too many floods can create an unstable and inhospitable environment.

Similarly, low flow and drought reduce the sediment transport capability of the river, which may result in changes to the morphology of the river and the composition of the river bed, and it has been shown that drought reduces the abundance of fish (Closs & Lake, 1996; Bell *et al.*, 2000). However, although drought may influence the biological community, it may also be part of the natural variability that the ecosystem has evolved with. In fact, the drying up of rivers in Australia has become essential for the survival of some of the species (Meffe, 1984). Thus, the effects of flood and drought have to be evaluated in light of the evolution of the particular riverine species and the flow regime, and regular droughts and floods should be distinguished from unpredictable and more extreme events.

The intermediate disturbance theory (Connell, 1978) suggests that a moderate degree of disturbance will support the greatest species diversity. This is because streams with a high disturbance regime will be dominated by the most successful colonizers (also called weedy species), whereas streams with a low disturbance regime will be dominated by the most competitive species (or 'climax' species, Section 10.3.2). Thus, an intermediate position will create the highest diversity with a mixture of colonizers and competitive species.

As outlined in Chapter 5 there are many measures of flow variability which express different aspects of the flow regime. Poff & Ward (1989) suggested that the three aspects most important for stream communities are (a) intermittency, (b) flood frequency, and (c) predictability of extreme events, with intermittency as the most important. *Intermittent rivers* are rivers that dry up regularly and perhaps for long periods (Box 10.2). In such rivers the biological community has adjusted naturally to include only species that can survive in the harsh environment. Examples are fish with high tolerance to low dissolved oxygen and invertebrates with dormant phases (Williams & Hynes, 1977).

Flood frequency has been used in several biological models as the primary axis for classifying biological communities (Biggs *et al.*, 1998b). In streams with frequent floods, fish and invertebrates that are small and can colonize new areas rapidly are often dominant (Scarsbrook & Townsend, 1993), and periphyton community is usually sparse, with low species richness and diversity (Clausen & Biggs, 1997; Biggs & Smith, 2002). In streams with stable flow regimes, aquatic communities are thought to be influenced more by biological processes such as competition between species and grazing/predation than by external environmental factors (Poff & Ward, 1989; Biggs *et al.*, 1999). However, if a flood occurs in this type of stream, the biota is particularly vulnerable.

Less is known about the effect of the duration and frequency of drought on aquatic organisms. A reduction in flow for a short time will have less effect on aquatic life than reducing the flow for a long time, depending on species and

---

Box 10.2

Streams in semi-arid regions

In arid and semi-arid zones (Section 2.2.2), such as large areas of Australia and Africa, water in streams and rivers is a very valuable resource. Often in such places the groundwater table is deep and rivers only flow intermittently (ephemeral or intermittent rivers). This creates a highly variable flow regime with large and unpredictable floods and wide river channels or flood plains compared to the actual wetted stream channel. Because of the local scarcity of water the flood plain supports large numbers of animals and plants. Where there are braided rivers, some bird species nest on open gravel areas or use islands as refuges from predators and feed on benthic invertebrates or small fish. There may be periods when water is isolated in ponds. These will be important for the survival of animals living in and near water, although they will be highly exposed and easy targets for predators. Some streams in arid regions are saline, which has special implications for the stream biota, and many streams have limited riparian vegetation and open banks, which lead to high water temperatures and high primary production (algae and macrophytes). Some streams might dry up completely and terrestrial plants and animals can invade the stream bed. In semi-arid areas, changes to the flow and flow regime can extend well beyond aquatic fauna. This is in contrast to the more humid temperate climates, where river channels are often more confined and terrestrial communities less dependent on the river as a source of water. In arid and semi-arid areas the relationship between ecology and flow is generally complicated, and management of streamflow will invariably influence the floodplain as well as the stream. It is in these areas that management tools such as the 'holistic method' and the 'building block method' (Section 10.4.1) were developed.

---

individuals. Similarly, a low flow that occurs every ten years will have more effect than one that occurs every year. In New Zealand, the mean or median annual minimum are often taken as a measure of the worst flow condition experienced by aquatic organisms, simply because, to be present in a river, they must have survived these conditions every second year or so.

Where flood and drought are *predictable*, or seasonal, the life history of the organisms may have adjusted so that reproduction of insects and fish takes place during the flood-free periods (Gray, 1981), and young organisms are present only during these periods (Moyle & Vondracek, 1985). In such rivers, large and unpredictable floods can cause long-lasting damage (e.g. Hoopes, 1974).

Seasonal flow variation may be an essential part of the life cycle of some aquatic species, but *frequent artificial flow fluctuations* are usually detrimental. Daily and weekly flow fluctuations are often a feature of rivers downstream of hydropower stations, and it is important to understand the effects of these fluctuations on physical habitat to assess the effects on the river ecosystem, and ultimately for managing the flow. Fluctuations in flow create a zone that is wetted and dried as water levels rise and fall. With frequent flow fluctuations, this zone will not sustain immobile plant and invertebrate species. Mobile species such as fish, and probably some invertebrate species, can make some use of this zone, especially for feeding in recently inundated areas of river bed, where there may have been some terrestrial invertebrates in the substrate. However, a zone that is wetted and dried at more frequent intervals than a week is usually unproductive and can be regarded as lost habitat.

The effect of frequent flow fluctuations is greatest on *invertebrates*, because they are relatively immobile and are usually found in relatively shallow water. Regulated rivers with fluctuating flow tend to produce less diverse and abundant invertebrate communities than unregulated rivers. *Young fish* may also be affected by flow fluctuations in situations where they can be stranded by changing flow, and continual displacement may enhance downstream movements, although it is not known how stranding and displacement affect overall abundance. *Adult fish* are generally less affected by flow fluctuations. Fluctuating flows may disrupt fish spawning, and large flow fluctuations may dewater eggs, although mortality would depend on the length of time above water level. Spawning and growth of freshwater mussels may also be affected by flow fluctuations (Walker *et al.,* 2001). *Macrophyte communities* along the margins of a river may be particularly affected by variation in water level, because they cannot survive frequent wetting and drying and may be scoured or broken off if water velocities are too high. Any reduction in the macrophyte abundance will reduce the numbers of invertebrates, and ultimately fish abundance and growth rate will be affected because many fish species feed extensively on invertebrates.

10.4 Flow management

10.4.1 Introduction

During the last thirty years there has been an increasing interest in understanding stream and river environments and protecting these from harmful human impact. This has been driven, in part, by the ecological and visual effects

Figure 10.9 Skjern Å ('å' is 'river' in Danish) in Denmark during and after restoration. After nearly 40 years as a straight drainage canal, the river again flows through large meanders (photo by B. Moeslund).

of past flow regulation and canalization of streams and rivers. Rivers have been modified for hydropower, irrigation, flood control and protection, water supply schemes and reclamation of wetlands for farming. The changes have most often led to a loss of physical habitat diversity, and through that to a decrease of the biological diversity, sometimes even to a total elimination of some of the biota. The changes have been carried out to such a degree that restoration projects are now being undertaken to bring canalized streams back to their original meandering courses (Figure 10.9). At dams, flows may be deliberately released to flush fine sediments from the stream bed (flushing flows), and flood flows may be released to maintain the downstream river geomorphology (channel maintenance flows).

Instream flow management is a complex process, usually involving a combination of scientific, public and legal considerations. To be effective, the instream flow management process should consider the natural and present status of the river and its ecosystem, and then, in consultation with public and institutional organizations, set goals and objectives before establishing appropriate flow requirements (Section 10.4.5). Non-ecological issues must also be considered, for example, public water supply, irrigation, power supply, effluent dilution, amenity, recreation and provision of water for local community use (Section 11.2).

The *instream flow incremental methodology* (IFIM) is an interdisciplinary framework for the technical side of river flow management (Bovee, 1982) (Figure 10.10). It is termed a methodology because the framework encompasses more than one method. It is incremental because the procedure is to examine how stream characteristics change incrementally with flow to determine acceptable levels or to compare alternatives. The evaluation of physical habitat (usually depth, velocity and substrate) is one of the main components of IFIM, but the process of assessing the impact of flow regulation and determining an appropriate flow regime should consider whether other ecologically important characteristics change with flow. In particular, the effects of flow changes on water temperature, water quality, sediment transport and river morphology are considered parts of IFIM. Other methodologies, such as the building block method (BBM; King *et al.*, 2001), also provide a framework for the consideration of alternative flow regimes.

A large number of methods have been developed to determine flow requirements for ecological protection. These methods are called instream flow methods (because they deal with flows 'in-the-stream'). The method or methods used to develop an appropriate minimum flow or flow regime will depend on the case being considered. There is no universally accepted method for all rivers and streams. Traditionally, some of these methods have been used to define a minimum flow, below which no direct influences should take place

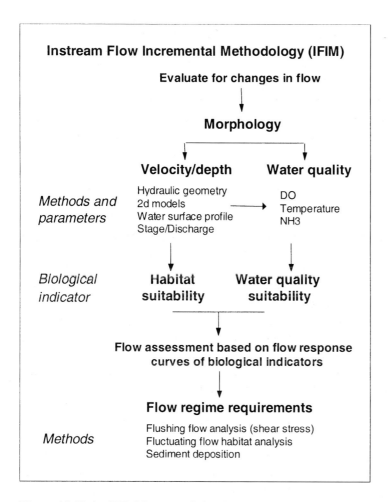

Figure 10.10 An IFIM framework for the consideration of flow requirements.

(e.g. abstractions, impacts of changing land use). However, the current trend is away from methods that set one 'minimum flow' towards methods that consider the flow regime, with some degree of flow variability, to maintain the natural morphology and ecosystem.

10.4.2 Instream flow methods

Instream flow methods can be conveniently divided into three types: historic flow, hydraulic and habitat methods. The methods are described by Jowett (1997b) and summarized in the following sections.

Historic flow methods are the simplest and easiest to apply. Stalnaker *et al.* (1995) describe these type of methods as 'standard setting' because they are generally desktop rules-of-thumb methods that are used to set minimum flows. A historic flow method is based on the historic flow record and uses a statistic to specify a minimum flow or a recommended flow. The statistic could be the average flow, a percentile from the flow duration curve (Section 5.3.1), or an annual minimum flow with a given exceedance probability (Section 6.6.1). For example, a method might prescribe that the flow should never drop to 30% of the mean annual minimum flow (*MAM*), or it could recommend that the average flow should stay above 80% of *MAM*. The percentage used is referred to as the 'level of protection'. Dunbar *et al.* (1998) gives examples on different countries 'rule of thumb' statistics.

The aim of historic flow methods is to maintain the flow within the historic flow range, or to prevent the flow regime from deviating greatly from the natural flow regime. The underlying assumption is that the ecosystem has adjusted to the flow regime and that a reduction in flow will cause reduction in the biological state (e.g. abundance, diversity) proportional to the reduction in flow; or in other words, that the biological response is proportional to flow (Figure 10.11). It is usually also believed that the natural ecosystem will only be slightly affected as long as the changes in flow are limited and the stream maintains its natural character. It is implicitly assumed that the ecological state cannot improve by changing the natural flow regime.

The most well-known historic flow method is the Tennant (1976) method, sometimes also called the Montana method, which specifies that 10% of the average flow is the lower limit for aquatic life and 30% of the average flow provides a satisfactory stream environment. The Tennant method was based on hydraulic data from eleven US streams (including streams in Montana) and considerations of what values of velocity, depth and width were needed for sustaining the aquatic life. However, many historic flow methods have been developed without any use of hydraulic data; instead, they have evolved from considerations of how frequently the flow should be allowed to go below the specified minimum flow and what would be realistically obtainable.

Historic flows can also be used to define 'an ecologically acceptable flow regime'. Arthington *et al.* (1992) describe an 'holistic method' that considers not only the magnitude of low flows, but also the timing, duration and frequency of high flows. Such a flow regime would not only sustain biota during extreme drought, but would also provide high flows and flow variability needed to maintain the diversity of the ecosystem. The building block method (King *et al.*, 2001) is a similar approach. The range of variability approach (RVA) and the associated indicators of hydrologic alteration (IHA) identify an appropriate range of variation, usually one standard deviation, in a set of

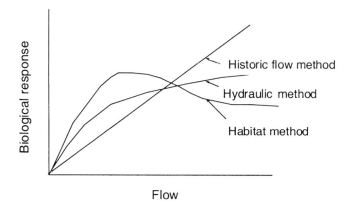

Figure 10.11 Graphs of typical biological response to flow for three instream flow methods. The biological response is assumed to be proportional to the flow (the historic flow method), the wetted perimeter or width (the hydraulic method), and the weighted usable area (the habitat method).

32 flow statistics derived from the 'natural' flow record (Richter *et al.*, 1997). The holistic, BBM and RVA methods are conservative and maintain the ecosystem by retaining the key elements of the natural flow regime. They are probably most appropriate for river systems where the linkages between ecosystem integrity and flow requirements are poorly understood or for preliminary ecological assessment.

Hydraulic methods are more time-consuming in that they are based on measurements of hydraulic data (wetted perimeter, width, depth or velocity) from one or several cross-sections in the stream. The aim of hydraulic methods is to maximize food production by keeping as much of the food-producing area below water. Because the stream bed is considered to be the most important area for food production (algae and invertebrates), it is usually the wetted perimeter or the width that is used as the hydraulic parameter.

The variation of the hydraulic parameter with flow can be found from carrying out measurements at different flows, or from calculations based on rating curves or Manning's equation. The graph of the hydraulic parameter (which is assumed to be proportional to the biological response for the hydraulic method) versus flow (Figure 10.11) is used for prescribing recommended flows, or to specify a minimum flow. The minimum flow can be defined as the flow where the hydraulic parameter has dropped to a certain percentage of its value at mean flow, or the flow at which the hydraulic parameter starts to decline sharply towards zero (the curve's inflection point). If the wetted perimeter or width is used, the inflection point is usually the point at which the water covers just the channel base. However, wetting of the channel base might not be enough to fulfil the requirements for depth and velocity of some species.

Tennant (1976) used the inflection point when he defined the minimum point as 10% of the average flow. Thus, Tennant used the hydraulic method and the suitability of water depth and velocity for aquatic organisms to develop a historic flow method. This is also referred to as a *regional method*, in that the method should be used only for the region in which the method was developed, i.e. the region that has the same type of streams as the streams used for developing the method. This has, however, not been the case with the Tennant method, which has been adopted in many different parts of the world.

Habitat methods are the most advanced, and an extension of the hydraulic methods. Their great strength is that they quantify the loss of habitat (living space) caused by changes in the natural flow regime, which helps the evaluation of alternative flow proposals. According to a review by the Environment Agency in the UK on river flow objectives, "Internationally, an IFIM-type approach is considered the most defensible method in existence" (Dunbar *et al.*, 1998). The Freshwater Research Institute of the University of Cape Town in South Africa states that "IFIM is currently considered to be the most sophisticated, and scientifically and legally defensible methodology available for quantitatively assessing the instream flow requirements of rivers" (Tharme, 1996).

Habitat models to evaluate physical habitat were developed in the 1980s along with computer technology. Current software include PHABSIM (Physical Habitat Simulation) and RHABSIM (River Habitat Simulation) used in the United States (Bovee, 1982; Milhous *et al.*, 1989), RHYHABSIM (River Hydraulics and Habitat Simulation) used in New Zealand (Jowett, 1989), EVHA (Evaluation of Habitat) in France (Pouilly *et al.*, 1995), CASIMIR (Computer Aided Simulation System for Instream Flow Requirements) in Germany (Jorde, 1997), and RSS (River Simulation System) in Norway (Killingtveit & Harby, 1994).

The latest version of RHYHABSIM is included on the CD (Software), along with the manual ('Software Manual RHYHABSIM') and a description of some of the most frequently used procedures ('User's Guide RHYHABSIM'). More details of this and the other models mentioned can be found in the relevant manuals and help functions of the programs.

The aim of habitat-based methods is to maintain, or even improve, the physical habitat for the biota, or to avoid limitations of physical habitat. They require detailed hydraulic data, as well as knowledge of the ecosystem and the physical requirements of stream biota. The basic premise of habitat methods is this: if there is no suitable physical habitat for the given species, then they

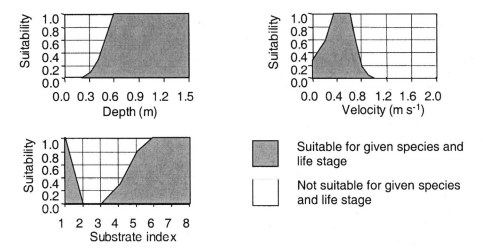

Figure 10.12 Habitat suitability curves for adult brown trout (data from Hayes & Jowett, 1994).

cannot exist. However, if there is physical habitat available for a given species, then that species may or may not be present, depending on other factors not considered in the model or not directly related to flow (e.g. biotic interactions, recruitment, water quality). In other words, physical habitat methods can be used to set the 'outer envelope' (or optimum) of suitable living conditions for the target biota.

Biological information is supplied in terms of habitat suitability curves for a particular species and life stage (e.g. adult brown trout, Figure 10.12). Other examples of suitability curves are provided in the RHYHABSIM program (Software, CD). A suitability value (ranging from 0 to 1) is a quantification of how well-suited a given depth, velocity or substrate is for the particular species and life stage. The result of an instream habitat analysis is strongly influenced by the habitat criteria that are used. If these criteria specify deep water and high velocity requirements, maximum habitat will be provided by a relatively high flow. Conversely, if the habitat requirements specify shallow water and low velocities, maximum habitat will be provided by a relatively low flow and habitat will decrease as the flow increases. The suitability curves in Figure 10.12 were developed for New Zealand adult brown trout (Hayes & Jowett, 1994). These curves specify higher depth and velocities than curves for adult brown trout developed in the US (Bovee, 1978). Whether this difference is due to different sizes of fish has not been clarified. However, it is clear that it is important to use suitability curves that were developed for the same size and life stage of fish as to which they are applied. Suitability curves or habitat criteria are usually based on observations of the given species and the available habitat.

Habitat criteria have more influence on flow assessments than any other aspect of the analysis. Failure to use appropriate criteria can result in inappropriate flow assessments. Therefore, habitat criteria need to consider all life stages and, where appropriate, include suitability criteria for the production of food for those life stages. Selection of appropriate criteria and determination of habitat requirements for an appropriate flow regime demands a good understanding of the species' life cycles and food requirements (Heggenes, 1988, 1996).

The analysis can be separated into a hydraulic component and a habitat component. The hydraulic analysis predicts velocity (most often the average velocity in the vertical) and depth for a given flow for each point, represented as a cell in a grid covering the stream area under consideration. In addition, information on bed substrate and other relevant factors, such as shade, weeds and temperature, can be recorded for each cell. A number of hydraulic methods can be used to predict water depths and velocities (Figure 10.10). The analysis is based on collected data from a number of field visits (minimum three visits) with different flows.

The habitat analysis starts by choosing a particular species and life stage (usually one that is found in the river, often the most important one for economical purposes or for anglers, or it could be a protected species), and a particular flow. For each cell in the grid, velocity, depth, substrate, and possibly other parameters (e.g. shade) at the given flow are converted into suitability values, one for each parameter (using the suitability curves). These suitability values can then be combined (usually they are multiplied) and multiplied by the cell area to give an area of usable habitat (also called weighted usable area, WUA). Finally, all the usable habitat cell areas can be summed to give a total habitat area (total WUA) for the reach at the given flow. Although WUA is often interpreted as the area of usable habitat, it only truly represents the area of suitable habitat when binary habitat suitability curves are used, i.e. habitat variables are either suitable (1) or unsuitable (0). This whole procedure is then repeated for other flows until eventually the final outcome has been produced: a graph of usable habitat area versus flow for the given species and life stage. This graph has a typical shape as shown in Figure 10.11 with a rising part, a maximum and a decline. The decline occurs when the velocity and/or depth exceed those preferred by the given species and life stage. Thus, in large rivers, the curve may predict that physical habitat will be at a maximum at flows less than commonly occurring. In contrast to the historic flow method, the habitat method does not automatically assume that the natural flow regime is optimal for all aquatic species and life stages in a river.

The relationship between habitat and flow (Figure 10.11) can be used to define a preferred flow range or a minimum flow; even a recommended maximum flow can be defined. As with hydraulic methods, the minimum flow can be defined as the inflection point or as the flow at which the habitat has dropped to a certain percentage of its value at mean or median flow. It can also be defined as the flow that has the lowest acceptable minimum amount of habitat in absolute terms.

Habitat suitability curves have been developed for threatened species (e.g. blue duck; Collier & Wakelin, 1995), for species of special interest (especially trout and salmon; Heggenes, 1996) and even for recreational activities (Mosley, 1983). When many fish species and life stages are present in a river there are usually conflicting flow requirements. For example, young trout are found in water with low velocities, and adult trout are found in deep water with higher velocities. If the river has a large natural variation with pools, runs and riffles, some of the different requirements may be provided for. However, even in these rivers, and especially in rivers with small habitat variation, one species may benefit greatly from a reduction in depth and velocity, whereas habitat for another species will be reduced. If a river is to provide both rearing and adult trout habitat there must be a compromise. One such compromise is to vary flows, if possible, with the seasonal life stage requirements of spawning, rearing, and adult habitat, with the optimum flow gradually increasing as the fish grow and their food and velocity requirements increase. Biological flow requirements may be less in winter than summer because metabolic rates and food requirements reduce with water temperature. If flow requirements of individual species are different, a solution may be found by choosing one with intermediate requirements (Jowett & Richardson, 1995) or to define flow requirements for aquatic communities.

Recently, Lamouroux & Capra (2002) demonstrated, using data from 58 streams in France, that the outputs of instream habitat models for three trout life stages and five other fish species could be predicted with dimensionless hydraulic variables (such as Reynolds number) calculated based on average characteristics of reaches, such as discharge, depth, width and bed particle size. These results should facilitate more cost-effective habitat studies in many streams, at basin or larger scales, and enhance the biological validation of habitat model predictions.

Because little is known about the effect of short-term flow fluctuations on habitat (Section 10.3.4), it is difficult to integrate this into habitat models. However, an attempt to model the potential habitat loss for macrophytes, benthic invertebrates, and thus fish food production, caused by flow fluctuations is integrated in RHYHABSIM. Because macrophytes and most benthic invertebrates are immobile, the locations within the river that remain as suitable

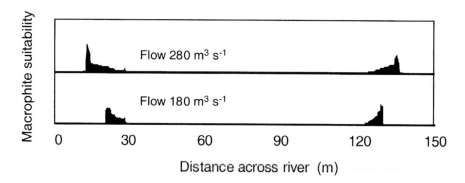

Figure 10.13 Habitat suitability for macrophytes across a river for flows of 280 and 180 $m^3 s^{-1}$. Note the loss of macrophyte habitat on the far left and far right of each cross-section as the flow reduces from 280 to 180 $m^3 s^{-1}$.

physical habitat under fluctuating flows can be identified (Figure 10.13). This analysis assumes that a location is a suitable macrophyte habitat only if it is suitable during all flows that occur frequently, say every week. As the flow decreases from the highest flow, some locations become unsuitable and some remain suitable, but none can become suitable (although this may not be the case in all situations), so that the area of suitable habitat reduces as the amount of fluctuation increases.

Habitat methods and water quality models can be integrated, although where water quality models are used, the results of hydraulic models are usually transferred into the water quality models. For example, a water temperature model (SSTemp; Bartholow, 1989) is part of RHYHABSIM; the hydraulic component of RHYHABSIM calculates water depth and velocity for each flow, and these data are then used to model how water temperature varies with distance downstream. A similar integration of a hydraulic geometry model with water temperature, dissolved oxygen and ammonia models has been implemented in the decision support system WAIORA (Jowett, 1999).

10.4.3   Case Studies: Habitat analysis

Two case studies from New Zealand (Figure 10.14) are presented. For both of these, stream habitat was analysed using the RHYHABSIM program and the results used to set a minimum flow. The RHYHABSIM program is stored on the CD (Software), and the help function of the program or the manual included ('Software manual RHYHABSIM') give detailed information on the program and its calculations and assumptions. The first case is the Whatakao Stream, for

Figure 10.14. The Whatakao Stream (Case Study 1) and the Waiau River (Case Study 2) in New Zealand.

which the collected input data are included on the CD and used as an example in 'User's Guide RHYHABSIM', also found on the CD.

Case Study 1: Whatakao Stream

Streams in the Tauranga area, New Zealand, including the Whatakao Stream, were surveyed for fish and habitat during 2001 and 2002, and the habitat area was modelled for all species present (Wilding, 2003). This work was carried out by the regional authority (Environment Bay of Plenty) to comply with the requirements of the local water and land plan. In New Zealand water is taken directly from the streams to supply irrigators, industry and municipal schemes, and these abstractions possibly conflict with the flow requirements of stream biota. It is therefore important to know what the requirements of the biota are. The results from the Whatakao Stream is presented here.

The Whatakao Stream has a median flow of $700\,\mathrm{l\,s^{-1}}$ and a mean annual minimum of $180\,\mathrm{l\,s^{-1}}$. As can be seen from the input data file on the CD, the stream was surveyed at 15 cross-sections (5 pools, 5 riffles and 5 runs), which are given percentage weights according to the proportion of area that each type represents of the total reach under consideration. The flow was gauged and water levels measured at seven different occasions. The hydraulic modelling involved adjustment of velocity distribution factors and selection of rating curves as described in 'User's Guide RHYHABSIM'.

Habitat was modelled for five native fish species: longfin eel, shortfin eel, inanga, redfin bully and common smelt. The first four species were caught during electric fishing surveys, while common smelt was thought to be possibly

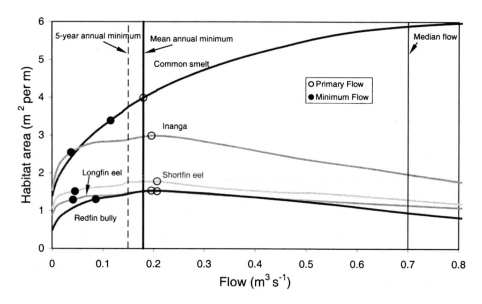

Figure 10.15 Modelled habitat area (per meter of downstream river) versus flow for five native fish species in a reach of the Whatakao Stream.

present. Suitability curves had already been established for these fish species (Jowett & Richardson, 1995). These native fish preference curves are believed to apply in all New Zealand streams and rivers. The results (Figure 10.15) show that common smelt has the largest variation in modelled habitat in the low flow range.

From these graphs a minimum flow of $85 \, 1s^{-1}$ was recommended as the minimum flow for the Whatakao Stream. This was determined using the following steps, which have been suggested for streams in this area (Wilding, 2003):

1. For each species in the stream, identify the flow that has the maximum habitat. If this flow is lower than the median flow (for example, inanga in Figure 10.15), it is called the primary flow; if it is higher than the median flow (see for example common smelt in Figure 10.15), the mean annual minimum (*MAM*) is the primary flow. The habitat of the primary flow (open circles in Figure 10.15) is here referred to as the primary habitat.

2. For each species, multiply the primary habitat by the protection level. The protection level represents the importance of the ecosystem compared to other uses and the importance of the particular species. For

example, the protection level for a species could be 85%, or, if the species is very significant and should be protected as much as possible, it would be 100%.

3. For each species, use this value (the primary habitat multiplied by the protection level) to read the corresponding flow from the habitat area versus flow curve. This is the required minimum flow for the given species (full circles in Figure 10.15).

4. Choose the highest of the minimum flows for all the species present to be the required minimum flow for the reach.

5. The allocable flow is the 5-year annual minimum flow minus the required minimum flow. If the 5-year annual minimum flow is lower than the required minimum flow, no water can be allocated for abstraction. The motivation for using the 5-year annual minimum is that the river will run dry only once every five years, if all the allocable water is abstracted.

In this case for the Whatakao Stream all species were given a protection level of 85%. The primary flows and habitats (open circles) and minimum flows (full circles) for all species are marked on Figure 10.15 and summarized in Table 10.2. Common smelt has the highest primary flow. However, because this species was not actually found in the stream, it was decided to use $85 \, ls^{-1}$ as the minimum flow for the reach, which is the next highest minimum (for redfin bully). The other species have minimum flows around $40 \, ls^{-1}$. With a 5-year annual minimum flow of $150 \, ls^{-1}$, this allows $65 \, ls^{-1}$ to be allocated for abstraction from the river upstream of the lower end of the reach.

Table 10.2 Primary flow, primary habitat and required minimum flow for all species in the Whatakao Stream

Species	Primary flow (ls^{-1})	Primary habitat $(m^2 m^{-1})$	Minimum flow (ls^{-1})
Longfin eel	207	1.52	41
Shortfin eel	207	1.78	45
Inanga	196	2.99	37
Redfin bully	196	1.53	85
Common smelt[1]	180	3.99	115

[1] Not observed, but potentially present

Case Study 2: Waiau River

The Waiau River in Southland, downstream of Lake Manapouri, is one of the larger rivers in New Zealand with a natural mean flow of around $450 \, \text{m}^3\text{s}^{-1}$. The river was diverted for power generation in 1976, and only $0.3 \, \text{m}^3\text{s}^{-1}$ was left as a minimum flow. The issue of an increased minimum flow was identified during relicensing in 1996, and a habitat analysis was carried out (Jowett, 1993). The habitat curves for adult brown trout and their food sources (Figure 10.16) show that habitat is maximum at flows higher than $15 \, \text{m}^3\text{s}^{-1}$ and drops quickly as the flow goes below $10 \, \text{m}^3\text{s}^{-1}$. The results of the habitat analysis also showed that the flow requirements for trout and their food sources are higher than for any other fish species in the river, and that a flow of $12 \, \text{m}^3\text{s}^{-1}$ or greater would provide high densities of 'clean-water' invertebrates and moderate to low periphyton biomass. Consequently a flow regime of $12 \, \text{m}^3\text{s}^{-1}$ in winter and $16 \, \text{m}^3\text{s}^{-1}$ in summer, supplemented by naturally occurring floods, was implemented in 1997. Figure 10.17 shows the minimum flow before and after implementation.

Figure 10.17 shows that the new minimum flow does not fill the channel, but it provides a good depth of water over the boulders for trout. Even with this minimum flow, flows are quite variable with flood discharges 20–30 times the minimum flow occurring several times a year (Figure 10.18). These floods have been frequent enough to keep the river bed clean and it has not been necessary to resort to flushing flows.

The improvement to the trout fishery has been outstanding. Numbers have increased from about 20 per km to 140 per km (Figure 10.19). To put this in

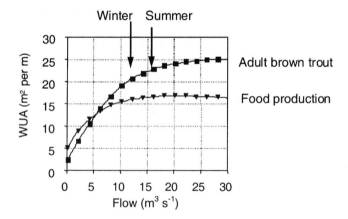

Figure 10.16 Modelled habitat area (WUA) versus flow for adult brown trout and its food sources in the Waiau River.

Figure 10.17 The Waiau River with a flow of $1 \, m^3 \, s^{-1}$ (before 1997; left) and $16 \, m^3 \, s^{-1}$ (after 1997; right) (photos by M. Rodway).

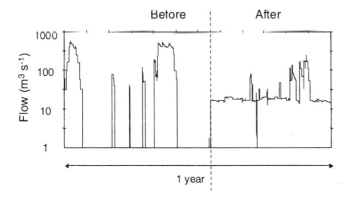

Figure 10.18 Flow (note the log scale) in the Waiau River for one year around the time when the minimum flow was increased from 0.3 to $12 \, m^3 \, s^{-1}$ (winter) and $16 \, m^3 \, s^{-1}$ (summer) in 1997.

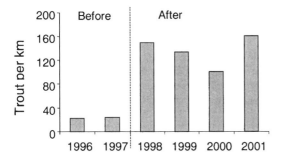

Figure 10.19 Density of trout (>20 cm) in the Waiau River for the years 1996–2001 (data from Moss, 2001).

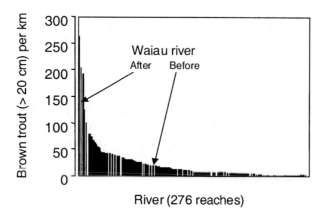

Figure 10.20 Distribution of 276 rivers with regard to trout density (data from Teirney & Jowett, 1990) and the Waiau River's position before and after the implementation of an increased minimum flow in 1997.

perspective, the river was ranked about average with respect to trout density before the implementation of the new minimum flow, whereas today it is one of the best brown trout rivers in the country (Figure 10.20). Angler data indicates that it is as popular today as it was before the diversion in 1976.

10.4.4   Monitoring and validation of methods

Many organizations spend a lot of time carrying out studies to develop instream flow guidelines or define minimum flows. However, once the decision is made, few organizations follow-up these decisions with monitoring to validate the appropriateness of the method and management objective. This leads to the frequent criticism that methods that set minimum flows or specify flow regime requirements have not been validated. For example, only 6 (1%) of 616 IFIM studies in the United States had follow-up monitoring (up to approx. 1990), and only two studies had attempted to establish relationships between habitat (WUA) and fish population (Armour & Taylor, 1991).

Some scientific studies have demonstrated a relationship between usable habitat and fish populations. Orth & Maughan (1982) found a relationship between the density of several fish species and usable habitat in Oklahoma streams. Nehring & Anderson (1993) found that the number of young rainbow and brown trout produced each year was consistently and strongly related to the amount of fry (young fish) habitat available in the same year. This study also indicated that the number of adult trout was strongly related to the number of fry 2–3 years previously. Jowett (1992) found that the density of adult brown trout in a river was related to habitat quality, as defined by the percentage of the river area that provided suitable habitat for adult trout and benthic invertebrates. In the rivers with the highest density of brown trout, low to median flows characteristically provided the greatest amount of suitable habitat. The density of adult and juvenile brown trout has also been related to the abundance of benthic invertebrates, the most common food item of trout (Jowett, 1992; Jowett *et al.*, 1996). However, trout abundance also depended on factors that were not related to flow. These included the percentage of developed land (pasture, crop or horticulture) in the catchment, the area of lake in the catchment and the percentage of sand substrate present.

Often monitoring studies are not reported in the scientific literature, but there are some studies that provide an indication of the effectiveness of flow recommendations. The first instream habitat study carried out in New Zealand was in the Tekapo River, where diversion of flow for the Waitaki Power Scheme had reduced the flow in the river from about 80 to $12 \, \text{m}^3 \text{s}^{-1}$ (Jowett, 1982). The habitat analysis showed that a flow of about $12 \, \text{m}^3 \text{s}^{-1}$ provided maximum trout and food producing habitat for that river and it was recommended that there be no additional flow diversion. Prior to diversion, the Tekapo River was hardly recognized as an angling river. Now it is one of the best in the region with high densities of brown and rainbow trout, good spawning and excellent angling. The quality of the fishery is entirely consistent with the habitat predictions for trout, and there are no known detrimental effects on any aquatic biota.

In some cases the primary objective of flow recommendations has been the maintenance of healthy benthic invertebrate communities. Below Lake Monowai in New Zealand, a lake regulated for hydropower, an increase in the minimum flow from near zero to $6 \, \text{m}^3 \text{s}^{-1}$ doubled benthic invertebrate densities (from 310 per m^2 to approx. 650 per m^2) and taxon richness (from 8 to an average of 17) (Jowett, 2000). Similarly, in the river below Moawhango Dam the minimum flow was increased from near zero to $0.6 \, \text{m}^3 \text{s}^{-1}$ in 2001 to re-establish benthic invertebrate communities. After one year with this new flow, the composition of the invertebrate community had already changed. The proportion of clean-water caddisflies had doubled from 23% in 1997 to 44% in 2002, and the proportion of the invertebrate community composed of mayflies (**E**phemeroptera), stoneflies (**P**lecoptera), and caddisflies (**T**ricoptera) (EPT), which is a measure of the relative abundance of 'healthy' invertebrates, had increased from 37% in 1997 to 57% in 2002. It is now similar to the 60% EPT composition in the river upstream of the dam (Suren *et al.*, 2002).

10.4.5   Developing instream flow guidelines

Instream flow guidelines for a particular stream or region should preferably include overall goals, specific objectives, the extent to which these are to be achieved (also referred to as protection levels), and criteria for evaluating the achievement (Beecher, 1990).

Goals for instream flow management are usually broad and general, such as "to manage sustainably", "to protect the integrity, diversity and vitality of natural systems", or "to perpetuate a diversity of species to provide sustained ecological, aesthetic, and economic benefits". For example, the EU Water Framework Directive (Section 11.2.8) aims at "maintaining and improving the aquatic environment in the Community" (Council of the European Communities, 2000). The Instream Flow Council of the United States promotes the philosophy of an overarching goal to achieve ecological integrity, "the ability to support and maintain a balanced, integrated, adaptive community of organisms having a species composition, diversity, and functional organization comparable to that of the natural habitat of the region" (Annear *et al.*, 2002).

Objectives for specific instream studies need to include details of what general policy statements like 'healthy, diverse communities' mean in terms of that stream (or streams). Objectives should be clear and measurable. The lack of clear objectives tends to lead to controversy and achieve vague results. Levels of protection can vary from enhancement at the upper end of the scale to species survival at the lower end. For example, a clear objective could be to maintain a particular population of fish in a particular section of river, and the level of protection could specify the numbers of fish to be supported.

In general, instream objectives should aim at maintaining aquatic communities rather than a target species, although non-biological (e.g. canoeing, boat passage) instream objectives can be considered. However, in practice, some species or activities within the community will be regarded as more desirable than others (e.g. mayflies are considered more desirable than aquatic worms, although both may be part of the aquatic community) and these would be used to define objectives and levels of protection. The value of out-of-stream uses, such as irrigation, water supply, effluent dilution and hydroelectric generation (Chapter 11), is also used in setting the level of protection for the environment. For example, in rivers with high-value hydroelectric systems the instream level may be relatively low. Thus, the objectives and levels of protection afforded a river will depend on the relative values of the instream resource and out-of-stream uses. However, to achieve those objectives it is necessary to understand the underlying mechanisms responsible for the biological, physical, and chemical condition of a river.

Once the objective and the level of protection have been defined they must be translated into practical operating guidelines or measures, using one of the existing flow assessment methods (Section 10.4.2). Thus, in practice, these surrogate levels of protection often take the form of the return period of the low flow (Box 6.2) or a proportion of mean annual low flow when historic flow methods are used. When habitat methods are used the level of protection may be expressed as the amount of habitat to be available after the change of flow, varying from enhancement to a proportion of the habitat at existing low flows.

The understanding of biological systems is not complete. Many factors influence stream ecosystems and, practically, flow assessments can only consider the most important and influential. Individual methods are often criticized for failing to consider some aspect of the stream environment. Hydrology, biology, fluvial morphology and water quality all influence aquatic ecosystems and should be considered in a flow assessment framework, such as IFIM (Figure 10.10). None of the methods currently available consider biotic factors (such as competition, predation, disease, recruitment) explicitly, and any change to the stream environment could potentially cause unexpected results. Flow assessments can only make use of the best available knowledge and, if necessary, be conservative. Finally, a list of key points concerning flow management is presented:

a) Flow assessment is a complicated process and involves both consultation with all stakeholders and technical assessments to provide water managers with sufficient information to make a decision.

b) Environmental response to flow is incremental.

c) Instream flow guidelines should be explicit about objectives and the levels of protection that are given to instream resources.

d) Level of protection (expressed as the return period of the low flow or proportion of mean annual low flow if historical flow methods are used to assess flow requirements, or if habitat methods are used, the amount of habitat retained) can be varied according to the value of the resource.

e) Methods of assessing minimum flow requirements are flexible and can be tailored to suit the river, management goals and objectives. No one method will provide for all needs, and minimum flow assessments can be re-evaluated with changing demands and policies.

f) One answer is clear – there would be no aquatic ecosystem or instream use without water in a river. However, because of the degree of diversity in a river and flexibility of most aquatic organisms, there is probably no sharp cut-off or single 'minimum flow'.

g) It is unlikely that the state of knowledge of biological systems will ever reach the stage where the effect of flow changes on stream populations can be predicted with absolute certainty. Experience, case studies, environmental risk, and out-of-stream benefits all play a part in the decision-making process.

10.5 Summary

This chapter has dealt with life in streams and has three purposes: (a) to introduce stream ecology, (b) to describe the ecological effects of extreme flows (flood and drought), and (c) to describe the practical aspects of flow management for ecological protection.

Section 10.2 introduced stream ecology and presented basic biological concepts such as how to name species, how to calculate diversity and commonly used terminology. It has described stream biota (periphyton, bryophytes, macrophytes, macro-invertebrates, fish) and presented example photos for each category. It has summarized how different physical and chemical factors (food, water depth, water velocity, channel substrate, temperature, water quality, biological interactions) affect stream organisms and how these factors depend on streamflow. The introduction is by no means exhaustive and references for further reading are given in Section 10.1.

Section 10.3 described the ecological effects of flood and drought, which are also called disturbances in biological terms. Extreme events can be seen as natural disasters or as a necessary part of a natural system in which they provide a resetting and renewal of the biological community. Examples were given of the coping strategies for different types of biota during floods and droughts and their recovery following the events. The intermediate disturbance theory suggests that an intermediate disturbance frequency will create the highest diversity. Other flow variability measures, such as intermittency, predictability and duration of extreme events are also important. Frequent artificial flow fluctuations, for example downstream hydropower stations, can have detrimental effects on the biota. The ecology of streams in semi-arid regions was briefly discussed.

Section 10.4 described flow management procedures for ecological protection. The instream flow incremental methodology (IFIM) is a framework for evaluating the ecological impacts of changes of flow or geomorphology, encompassing the physical and chemical changes. Instream flow methods are methods used for determining flow requirements in practice. The historic flow method is the most simple; the hydraulic method requires more field data and uses the wetted perimeter or width to assess the biological response; the habitat method is the most complex and requires biological knowledge as well as field data. The RHYHABSIM (River Hydraulics and Habitat Simulation) program uses the habitat method for assessing physical habitat and is enclosed on the accompanying CD together with a manual and an introductory description. The program was used to calculate habitat quality and recommend minimum flows (below which no direct influences should take place) in two New Zealand streams, presented as case studies. In flow management it is important to define overall goals, specific objectives and criteria for how to evaluate the achievement. It is equally important to carry out monitoring and validation of methods.

It is hoped that this chapter will contribute to an increased awareness of ecological issues among hydrologists and strengthen the interdisciplinary co-operation between hydrologists and ecologists.

11

Operational Hydrology

Alan Gustard, Andrew R. Young, Gwyn Rees, Matthew G.R. Holmes

11.1 Introduction

Information on hydrological drought and low flow is essential for designing abstraction schemes for public water supply and irrigation, for estimating the dilution of industrial and domestic effluents, for estimating the energy that can be generated from hydropower and for maintaining or improving freshwater ecosystems. This chapter focuses on the operational needs for *low flow information* and describes a number of practical examples of low flow design. In addition to providing the key information on which food production, human health and power generation depend, low flow hydrology also provides a challenging profession requiring a thorough understanding of the problem at hand, the limitations of the data available and access to and understanding of a wide range of analytical techniques.

In hydrological research, model complexity is often seen as a strength. However, legislation requirements and policy objectives often dictate that the level of sophistication should be no greater than that which is required to meet national or European water law. Indeed, the management culture of an organization is often the main factor in determining whether advanced hydrological techniques are used or simple rules of thumb.

A key issue in operational hydrology is the management unit, that is the area of river network or infrastructure facility (reservoirs, aqueducts, pumps, water treatment plants, sewerage treatment plants, sewer networks). In the operational context, the primary management unit for environmental protection is the river basin including groundwater and the importance of upstream hydrology and upstream developments on downstream issues has highlighted the need for integrated river basin management. Furthermore river basins extend across national boundaries and the co-operative management of transnational rivers present particular challenges due to the wide range of hydrological conditions and conflicting demands on the river system. For hydropower generation the management unit is often the reservoir and turbine facilities with some cognisance given to release regimes to protect downstream interests. In Europe, water supply management units are often very complex and may involve the conjunctive use of surface and groundwater systems or the re-use of sewage effluent. Such systems require complex hydrological, water resource and water use modelling for operational management.

Before embarking on any design or operational problem, it is essential to assemble all relevant data, to quality control the data, to infill gaps and to estimate data errors. These issues were discussed fully in Chapter 4, which also describes 'naturalization' techniques for deriving a natural time series from gauged observations and information on the artificial influences in a catchment. Definitions and indices of drought used in this chapter are described in detail in Chapter 5. Chapter 8 and Section 6.7 have described techniques for regionalizing drought behaviour. One of the most common applications of these techniques is in the estimation of low flows at ungauged sites and operational examples of these techniques are described here. The reader is referred to Chapter 9 where human influence on surface and groundwater is reviewed – including the impact of groundwater abstraction on river flow (Section 9.5). Chapter 9 also further elaborates on advanced naturalization methods and artificial influences (Section 9.7). An application of low flow information, which is rapidly expanding, is the issue of setting environmental flows (other terms used are instream flows, residual flows, prescribed flows, compensation flows, ecologically-acceptable flows). This is a discharge regime, which is determined by a regulatory authority, which endeavors to set a minimum discharge downstream of an abstraction or a reservoir, to provide an equitable solution to the competing demands on a river system. Chapter 10 addressed the ecological requirements of a river and operational implementation range from simple indices of low flows to more complex instream flow models.

Following this introduction the main *user requirements* for low flow information are discussed in Section 11.2. This is followed in Section 11.3 by the key principles of low flow estimation for design and management purposes. Section 11.4 describes a number of case studies which illustrate how the techniques described earlier in this book are used in practice. These include the development of national design procedures in Europe, methods for estimating hydropower potential in the Himalayas, navigation in the Meuse and mitigating drought impacts in Europe.

11.2 Operational requirements

11.2.1 Water sectors

Different low flow problems are encountered by different *sectors of the water industry*. For example, water supply, agriculture, water quality, hydropower, industrial use, ecosystem protection and amenity. Water resource problems also occur over a wide range of *space scales*. These range from detailed estimates at individual reaches of the order of 100 m with 10 km^2 catchments to the estimation of drought frequency at the pan European scale covering areas in excess of 1M km^2. At the larger catchment scale, between 10–1000 km^2, there are a wide range of operational requirements, particularly for the licensing of abstractions and a range of approaches according to the presence or absence of data. For large river basins in Europe in excess of 1000 km^2, rivers are normally gauged with long (greater than 50 years) time series of daily flow data, which enables design problems to be most commonly based on observations (Chapter 6). These larger catchments present interesting challenges of separating the myriad of artificial influences from the natural flow regime. Finally, to compare resources between countries at the continental scale, there is a need for reliable estimates of resource availability to be derived by a consistent methodology. Simple hydrological models are appropriate in these situations because the key requirement is to identify the spatial variability of the resources and a basic index is often sufficient. A wide range of different analysis methods are required for different design problems. These are summarized in Table 11.1 and described in more detail below.

11.2.2 Public water supply

The requirements for low flow information for the water supply industry using surface water resources normally fall into two categories; those with and without *storage*. For river systems without storage, the flow duration curve (Section 5.3.1) is the most frequently used low flow characteristic. However, if an estimate of the return period is required, then frequency analysis of annual minimum (Section 6.6) is required for different durations. A review of UK practice suggests that annual minimum analysis is seldom used operationally; the main application is for estimating the frequency of a drought, for durations of between 3 to 36 months.

More commonly, public water supply is derived from reservoirs with a wide range of storage types (McMahon & Mein, 1978), including direct supply reservoirs, regulating reservoirs, which maintain downstream abstractions, and

Table 11.1 Summary of drought characteristics and indices for water resources and drought assessment (Gustard *et al.*, 1992, published with the approval of CEH)

Regime measure	Property described	Data employed	Applications
Mean flow	Arithmetic mean of the flow series	Daily or monthly flows	Resource estimation
Coefficient of variation of annual mean flow	Standard deviation of annual mean flow divided by mean flow	Annual mean flow	Understanding of regime inter-annual variability; definition of carry-over storage requirements
Flow duration curve	Proportion of time a given flow is exceeded	Daily flows or flows averaged over several days, weeks or months	General regime definition; licensing abstractions (water rights) or effluents (discharge consents); hydropower design
Annual minimum series	Proportion of years in which the mean discharge (of a given duration) is below a given magnitude	Annual minimum flows – daily or averaged over several days	Drought return period; preliminary design of major schemes; first step in some storage/yield analyses
Threshold level – durations	Frequency with which the flow remains below a threshold for a given duration	Periods of low flows extracted from the hydrograph followed by a statistical analysis of durations	More complex water quality problems, such as fisheries, amenity, navigation
Threshold level – deficit volumes	Frequency of requirement for a given volume of 'make-up' water to maintain a threshold flow	As for durations, except the analysis focuses on the deficit volume below the threshold	Preliminary design of regulating reservoirs
Storage-yield	Frequency of requirement for a given volume of storage to supply a given yield	Daily flows or flows averaged over several days or monthly flows	Preliminary storage/yield design for large dams; design of small dams; review of yield from existing storage
Recession indices	Rate of decay of hydrograph	Daily flows during dry periods	Short-term forecasting; hydrogeological studies; modelling
Base Flow Index	Proportion of total flow which derives from stored catchment sources	Daily flows	Hydrogeological studies; preliminary recharge estimation

pumped storage reservoirs where water is pumped up from an adjacent river. Analysis techniques include critical period techniques, behaviour analysis, time series analysis (Chapter 7) and probability matrix approaches. Scheme applications include single reservoir, multiple reservoir and conjunctive use with river and/or groundwater abstractions.

11.2.3 Agriculture

The primary use of water for agriculture is to supply *irrigation schemes*. A key element of the planning for proposed schemes is an assessment of water availability. This assessment is used for defining the potential area that can be irrigated by the supply river or aquifer. It is carried out by identifying the volume of water in the river during critical periods of the year, as determined by the irrigation demand for different periods of the cropping calendar. Irrigation demand is calculated for each period from knowledge of rainfall, evaporation, crop type and soil type. To account for losses of water associated with inefficiencies of water supply, irrigation demand is multiplied by an efficiency factor to calculate the diversion requirement. These are calculated for each phase of the cropping calendar, normally expressed as a diversion requirement per hectare of irrigable crop. In low lying artificially drained areas it is important to maintain surface and groundwater water levels to maximize crop production through capillary rise. For example, in the Netherlands and Bangladesh a key drought issue is the need to maintain water levels during dry periods.

11.2.4 Domestic and industrial discharges

A common application in operational hydrology is to estimate the *dilution of a domestic or industrial discharge* to a river. A legal consent is frequently required to discharge a pollutant. Water quality models based on the rate and quality of the discharge and the flow and quality of the receiving stream are used to determine the frequency distribution of downstream water quality. The flow duration curve of receiving river flows is used for design with Q_{95} being the most common index. This planning or design application contrasts with the real time application of assessing the impact of specific pollution incidents, e.g. a discharge of oil following an industrial accident. Flow rates are then required to estimate the rate of dispersion and the time of travel. The latter is required to warn downstream public water supply abstractors to stop abstracting or to divert flow.

11.2.5   Hydropower and conventional power

Small-scale hydropower schemes generally have *no artificial storage*, and thus rely entirely on the flow conditions of the river to generate electricity. In small-scale hydropower design the conventional method for describing the availability of water in a river is by using the flow duration curve (FDC). For high head schemes a more traditional storage yield analysis can be carried out with discharge being routed through high head turbines. Design must accommodate fluctuating power demands and protection of downstream abstractors' interest and ecosystem health (Chapter 10). The availability of cooling water for thermal power stations is a key determinand in locating them close to large rivers. Low flow design criteria are used to ensure sufficient water availability during drought and that the increase in downstream water temperature, due to the return of water to the river, does not damage aquatic ecosystems. In extreme conditions water temperatures may exceed the acceptable limits, in which case power generation is reduced (Box 2.3).

11.2.6   Navigation

River systems provide an important transport facility for both industrial and leisure navigation. Navigation is interrupted or loads reduced during drought because of *insufficient depth* of water in natural river systems and reduction of water available to supply the up and downstream movement of vessels through locks. The critical hydrological variable is water depth and in the absence of field observations the depth profile must be estimated using time series of river flows and hydraulic models. These enable the frequency of interruptions to navigation for different size vessels to be estimated and proposals to be made for improved channel design whilst protecting the natural ecosystem.

11.2.7   Ecosystem protection and amenity

Ecosystems are most vulnerable at times of low flow due to a reduction in the availability of habitat, extremes of water temperature, reduction in dissolved oxygen, deterioration in water quality (due to reduction in effluent dilution) and habitat fragmentation (caused by natural or artificial barriers to fish movement). Chapter 10 has identified a range of modelling techniques, which can be used for *predicting* and *mitigating* the impact of low flows on *freshwater ecology*. Techniques range from simple methods based on low flow indices often known as 'standard setting' and include the mean annual minima of a given duration or a percentile from the flow duration curve. The estimated discharge is then used to set a minimum flow in a river so that when the discharge falls below this level abstractions should cease (or be reduced). In addition to supporting complex ecosystems, rivers are natural assets for sport and recreation – canoeing, fishing, ornithology and walking. Ensuring adequate water depth and or velocity even when flow rates are very low, can artificially enhance their natural attraction.

11.2.8   Policy issues

Over the last 100 years the management of water resources in Europe has progressed from a municipal level, dealing with pollution and land drainage schemes, to the development of *catchment-based* water management schemes.

Box 11.1

Timetable for implementation of Water Framework Directive

One of the main developments of the WFD is to provide a comprehensive long-term strategy for catchment management. The timetable outlined below highlights the key deadlines for publishing River Basin Management Plans (RBMPs) and the programme of follow up activities.

2003: Transpose directive into domestic law. Identify river basin districts and the competent authorities.

2004: Complete first characterization of, and assessment of impacts on, river basin districts. Complete first economic analysis of water use. Establish a register of protected areas in each district.

2006: Establish environmental monitoring programmes. Publish a work programme for producing the first river basin management plans.

2007: Publish an interim overview of the significant water management issues in each river basin district for general consultation.

2009: Finalize and publish first RBMPs. Finalize programme of measures to meet the objectives.

2014: Publish second draft RBMPs for consultation.

2015: Achieve environmental objectives specified in first RBMPs. Finalize and publish second RBMP with revised programme of measures.

2021: Achieve environmental objectives specified in second RBMPs. Publish third RBMPs.

2027: Achieve environmental objectives specified in third RBMPs. Publish fourth RBMPs.

The operational use of low flow information is now normally applied within national and international law and policy directives. In Europe the most important of these is the Water Framework Directive (WFD), which was adopted, by the European Parliament and the Council of the European Union in September 2000. This has established a strategic framework for the sustainable management of both surface and groundwater resources. Each country must set up a 'competent authority' to implement the Directive and log every significant piece of water, above ground and below it, inland and on the coast. Most of the provisions of existing water related directives, covering issues such as water abstraction, fisheries, shellfish waters, and groundwater, will be subsumed, with past legislation being repealed or modified as the new regulations incorporate them.

Implementation of the Directive will not only require the state of the water in a river or lake to be measured, but also the state of the river or lake or seashore itself. That includes the chemical purity of water and its biological condition – from bacteria to bird populations. Physical condition must be monitored, from assessing the shape of the water body to determining the physical inputs and outputs quantitatively as well as qualitatively.

The WFD requires management by river basin rather than approaching water management in terms of administrative or political boundaries. Each river basin district (RBD) will have to have a river basin management plan (RBMP). At a general level the objectives are to provide protection for the basin in terms of aquatic ecology, unique and valuable habitats, drinking water resources and bathing waters. To achieve these ambitions everyone who has an impact on a river has to be identified and engaged in planning how to meet the Directive's requirements. The implementation timetable is shown in Box 11.1.

11.2.9   Managing catchment abstractions in the UK

As an example of implementing the Directive within England and Wales, the Environment Agency has developed a number of initiatives to implement the Directive, which include the Catchment Abstraction Management Strategy (CAMS). This strategy is a sustainable, catchment-specific approach to water resource management that aims to balance the human and environmental water requirements both in the present and in the future.

The key objective of CAMS implementation in England and Wales is to provide a *consistent and structured approach to local water resources management*, recognizing both abstractors' reasonable needs for water and environmental needs. It will also make information on water resources publicly available and provide the opportunity for greater public involvement in the process of managing abstraction at a catchment level. A key element of CAMS

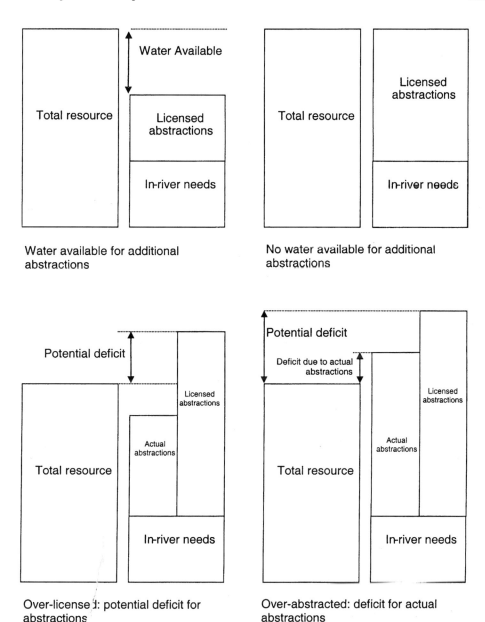

Figure 11.1 Concept of resource availability status at low flows (modified from Environment Agency, 2002b; reproduced with permission of publisher).

is to assess the low flow resource available. This will provide information on whether there is a surplus or deficit of water available to meet current licensed abstractions. It will also enable time limited (normally 12 year) abstraction licenses to be issued in contrast to the historical practice of issuing licenses for perpetuity. The CAMS approach will be to produce flow duration curves for a range of situations including the natural, current and future demand scenarios and will enable the status of a specific reach to be identified. The balance between the available resource and the environmental needs of the river, current actual abstractions and licensed abstractions determines the status of a specific location (Figure 11.1). For example in Figure 11.1 (lower left) the river is over-licensed because the sum of current licenses and in river needs exceeds the total resource. In this example although current abstraction can be met from the river, if all abstractions used their full license there would be a deficit. The figure illustrates three other situations e.g. 'water available for additional abstractions' (upper left), 'no water available for additional abstractions' (upper right) and 'over-abstracted' (lower right).

By sustainably developing water resources, the CAMS strategy will reduce conflict at the basin level between competing users of the water resource. At the international level this is of even greater importance and significant progress has been made to resolve conflict through the establishment of formal and informal international agreements. Within the context of the Water Framework Directive it will be necessary to manage European rivers as a single entity and this will require close cooperation between member states. In many cases in the past, for example in the rivers Rhine, Meuse (Section 11.4.5), Elbe and Danube, the issue of co-operation has had to be addressed to meet a common goal of minimizing the impact of pollution incidents and improving water resource planning and flood mitigation. International commissions, which currently exist to address these and other issues, will be permitted to be the competent authority in the context of the Framework Directive.

11.3   The Design Cube

There are a variety of procedures which the hydrologist can adopt for hydrological design and water management, the selection of which is determined by the nature of the output required – *the design requirement* and the risk associated with the design decision. For example, the risk associated with investment in construction of a large impounding reservoir would necessitate the establishment of gauging stations and analysis of observed river flows. These data provide the basis for hydrological design, typically the storage/yield characteristics and spillway capacity. In contrast, an application

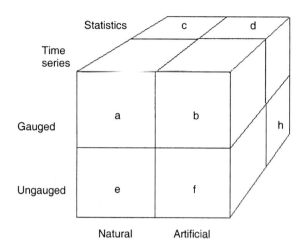

Figure 11.2 Design Scenario Cube – a conceptualization of the variety of approaches to estimate low flows (published with the approval of CEH).

for a small abstraction license would frequently be at an ungauged location, not warranting gauging station construction, and the design would often be based on flow of given probability without necessarily requiring time series analysis (Section 8.3).

The complexity of different design scenarios (UNESCO, 1997) can be simplified by conceptualization of a 'Design Scenario Cube' (Figure 11.2) defining three dimensions to the design requirement:

a) The location of the design problem may be at a site where there is no recorded hydrological data. Alternatively there may be a nearby gauging station up or downstream. This distinguishes between the gauged and ungauged cases.

b) The operational requirements of the hydrological design, for example, the water sector and magnitude of the capital investment will determine whether simple statistics are required or a long continuous hydrological time series. This is the data characteristics dimension that distinguishes between the requirement for time series or regime statistics.

c) The catchment may be relatively natural or heavily modified by water resource development in which case it may be necessary to naturalize the flow record (Sections 4.3.3 and 9.7). The catchment water use dimension distinguishes between the requirement for natural or artificially-influenced flows.

The simplistic three-dimensional cube presents eight distinct combinations of different design scenarios. Each one is summarized below and refers to one of the eight cubes a to h shown on Figure 11.2 (note that 'cube' 'g' cannot be seen in the figure).

a) Natural gauged time series – a time series of daily flows at a gauged site, which represent the river flow regime from a natural catchment. A typical application of this design requirement may be in the setting of environmental river flow regimes at a location in the vicinity of a gauging station.

b) Artificial gauged time series – a time series of daily flows at the gauged site, which represent the river flow regime from an unnatural, artificially-influenced catchment. Different definitions of 'artificial' can be adopted including historic, current or future scenario-based water use. A typical application of this design requirement may be the determination of the extent to which different upstream abstractors have diminished downstream flows over a historical period.

c) Natural gauged statistics – a flow regime statistic, such as the 95-percentile flow (Section 5.3), which represents the river flow regime from a natural catchment at a gauged site. A typical application of this design requirement may be to determine whether to approve a proposed abstraction.

d) Artificial gauged statistics – a flow regime statistic that represents the river flow regime from an artificially-influenced catchment at a gauged site. A typical application of this design requirement may be in the determination of the extent to which upstream abstractors have diminished the 95-percentile discharge at a gauged site.

e) Natural ungauged time series – a time series of river flows that represents the natural regime at an ungauged site. A typical application of this design requirement is in the use of continuous simulation models for estimating the impact of future land use change (Section 9.3).

f) Artificial ungauged time series – a time series of river flows which represents an artificially-influenced regime at an ungauged site. A typical application of this design requirement may be in the design of a conjunctive use (surface and groundwater) scheme in a catchment, which has groundwater pumping but no discharge measurements.

g) Natural ungauged statistics – a flow regime statistic that represents the natural regime at an ungauged site. A typical application of this design requirement is in regional water resource assessments.

h) Artificial ungauged statistics – a flow regime statistic that represents the artificially-influenced regime at an ungauged site. A typical application of this requirement may be in the preliminary design of a small hydropower scheme in an artificially-influenced catchment.

Added complications to this scheme include evaluating the impact on downstream river flows of artificial influences based on historic sequences, current water use or future abstraction scenarios. In most cases this will require the naturalization of a continuous flow record or low flow statistics. The *naturalization of flows* is the disaggregating of observed river flows into artificial and natural components and enables conversion of measured flow to the flow that would have occurred without man's interference. The analysis methods were described fully in Sections 4.3.3 and 9.7 and they provide a first step in resolving many operational low flow problems. In highly-developed catchments the artificial component may exceed the natural contribution particularly at low flows. For example, when there are high imports into a catchment for public water supply, which are then discharged as effluent. The impact of the cumulative uncertainties within the naturalization process must be borne in mind during the interpretation of the low flow statistics derived. In many catchments the artificial component of discharge at low flows is very significant and it is essential that this is incorporated in hydrological models.

11.4   Operational case studies

11.4.1   Establishing the operational requirement

Section 11.2 outlined the operational requirements for low flow information for different water sectors. Current techniques can be advanced through research and development but their use will be increased if applied research programmes are based on a survey of *the user requirements*. A survey should include an assessment of why drought information is required, the preferred drought or low flow analysis techniques, the number of occasions when information must be provided and the typical staff time used for design. This can then guide the development of more advanced design procedures and software. As an example a survey of eight Regions of the Environment Agency in England and Wales was carried out in 1998 when a software product Micro LOW FLOWS (Young *et al.*, 2000b) was in use. The main conclusions of the survey were:

a) Low flow statistics estimated at ungauged sites varied between the Agency Regions. The method by which statistics were estimated and the time taken in doing so was also variable.

b) On average approximately 4400 requests for low flow statistics were processed by the Agency per annum. Abstraction licensing and discharge consenting accounted for almost 90% of the total number of requests.

c) Mean flow and the flow duration curve were the most commonly requested flow statistics, accounting for 42 and 43% respectively, of the total requests.

The review identified that the Micro LOW FLOWS software was the most commonly-used software for low flow estimation. Manual methods for estimation at the ungauged site were based on national low flow studies (Institute of Hydrology, 1980; Gustard *et al.*, 1992). Continuous simulation modelling using physically-based models (Chapter 9) was carried out where a time series of flows were required for the design of major schemes. Data from nearby catchments were often transferred to a design site by applying an areal adjustment factor (Section 8.2). Taking occasional field measurements at low flows and relating the observed discharge to a flow exceedance from a nearby analogue gauging station often supported estimates at ungauged sites. Based on this review a number of recommendations were made to advance the science of low flow estimation and improve the software and associated database. These are described in the following section, and are followed by other case studies relating to operational low flow design.

11.4.2   A catchment-based water resource decision support tool – England and Wales

The current policy background to low flow estimation in England and Wales has the objective of making information on water resources availability and abstraction licensing more accessible to the public and providing a transparent, consistent and structured approach to water resource management (Environment Agency, 2002b). Within the overarching CAMS process (Section 11.2.9) the Resource Assessment and Management (RAM) framework provides the technical approach for water resource assessment on a *catchment basis* and river flow objectives are developed for catchments based on the sensitivity of the riverine environment to changes in the flow regime. The analysis is used to identify the portions of the natural flow regime, as defined by the natural flow duration curve (FDC), that are available for abstraction. The impacts of water use scenarios on the natural flow regime are then compared to the river flow objectives to assess the resource status of the catchment and to develop a final catchment scale abstraction licensing strategy.

The need to develop a rapid, nationally consistent approach to estimating natural and artificially-influenced flow regimes on ungauged rivers led to the development of the Micro LOW FLOWS software system in the early 1990s (Young *et al.*, 2000a, 2000b). This system enabled flow duration statistics to be estimated from catchment descriptors for a river reach selected from a digital river network. The system incorporated the regionalized low flow models developed by Gustard *et al.* (1992). A revised technical specification, together with a high level of end-user consultation led to the subsequent development of the Low Flows 2000 software, which was designed to provide an improved decision support system for the water resource managers in the UK and to implement new, advanced models for estimating resource availability. The system is underpinned by regionalized hydrological models which enable the natural, long term FDCs to be estimated for any UK river reach, mapped at a 1:50 000 scale. Both long term 'annual' statistics (considering variability within a year) and 'monthly' statistics (considering variability within a calendar month) are provided. The impact of artificial influences is simulated using a geographically referenced database that quantifies seasonal water use associated with individual features.

Region of Influence Model

The regionalized models employed within Low Flows 2000 are based on a Region Of Influence (ROI) approach, which removes the need for a priori identification of regions and instead develops a 'region' of catchments similar to the ungauged catchment (Sections 6.7.4 and 8.5.4). The dynamic construction of a region, based upon the similarity of the properties of the gauged catchments to those of an ungauged, was originally termed the Region of Influence (ROI) approach by Burn (1990). The ROI method has been used in flood estimation (e.g. Robson & Reed, 1999; Burn & Goel, 2000) and the adaptations for estimating 'annual' and 'monthly' flow duration statistics is described fully by Holmes *et al.* (2002b, 2002c). In summary, similarity between the ungauged catchment and other catchments is assessed based on the distribution of Hydrology Of Soil Types (HOST) classes, a hydrologically-based soil classification system (Boorman *et al.*, 1995). The HOST classification is used as a surrogate for a formal, hydrogeological classification. A region of the ten most similar catchments is then identified from a good quality data set of catchments with natural flow regimes. Estimates of the flow statistics for the ungauged catchment are then calculated as a weighted average of the observed (standardized) flow duration statistics for the ten catchments in the region.

The standardized annual FDC is re-scaled by multiplying by an estimate of annual mean flow from a national runoff grid derived from a daily soil moisture accounting model (Holmes *et al.*, 2002a). A similar approach is used to derive the flow duration curve for any month based on a distribution of annual runoff within the year (Holmes & Young, 2002).

Low Flows 2000

Low Flows 2000 incorporates the regionalized hydrological models within a PC-based software framework using contemporary programming tools. A GIS-based graphical interface provides access to spatial data sets of catchment characteristics and climatic variables required for the application of the regionalized models. These are defined for the entire UK at a 1×1 km resolution. A 1:50 000 scale vectored-digital river network and set of digitised catchment boundaries are used in conjunction with a digital terrain model (DTM) to define catchment boundaries.

In application, natural flow duration statistics are obtained by first selecting a point on the digital river network which defines the catchment outlet. A boundary is automatically generated by selecting one of two methods; the 'analogue' method (based on river network drainage-density, Sekulin *et al.*, 1992) or the 'digital' method (based on the DTM, Morris & Heerdegen, 1988; Morris & Flavin, 1990). This boundary is overlain onto the spatial data sets to obtain the catchment descriptors, such as the distributions of HOST classes within the catchment, and the climatic variables required by the underlying hydrological models described previously. The required natural flow duration statistics are returned at a 'monthly' and 'annual' resolution and are displayed in tabular and graphical form. Examples of the output from the software are shown on Figure 11.3.

Water use within the catchment is simulated by utilizing data stored in a flexible database system based on the CEH Water Information System (WIS) (Moore, 1997). Seasonal water use patterns associated with point influences including abstractions, discharges and impounding reservoirs are geographically referenced to enable catchment-based data retrievals. The net influences acting within a catchment are calculated by summing the individual influences, where discharges and releases from reservoirs are positive and abstractions are negative. The 'influenced' flow duration statistics are presented, together with the natural statistics, at a 'monthly' and 'annual' resolution. Hence, practitioners can make a comparison between the natural regime and associated river flow objectives, and the regime as modified by current water use within the

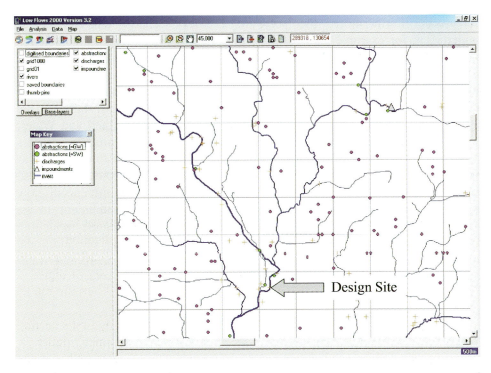

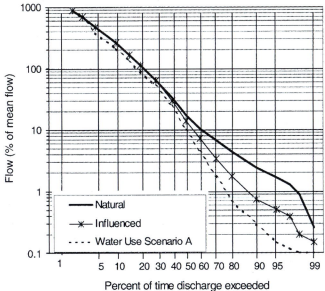

Figure 11.3 Flow duration curves for natural and influenced flow regimes obtained from the Low Flows 2000 software (published with the approval of CEH).

catchment. Furthermore, the software enables scenario analysis to be undertaken, for example to investigate the effect of increasing the abstraction rates across a particular catchment to simulate a future water use strategy.

Following a national implementation program in 2001 the Agency adopted the Low Flows 2000 system as the standard decision support tool for low flow estimation in ungauged catchments. In excess of 120 trained staff use the system on a weekly basis at over 90 installations across England and Wales. The system is routinely used as an operational tool in the determination process associated with granting new abstraction licences and consents to discharge, as well as reviewing applications for renewal of licences. Low Flows 2000 is also supporting the CAMS process by delivering artificially-influenced flow statistics enabling water-stressed river reaches within a catchment to be identified. The Low Flows 2000 is designed to provide a generic modelling framework and is therefore suited to the incorporation of additional functionality. For example, the software is being enhanced to incorporate water quality models for planning purposes (Feijtel *et al.*, 1997).

11.4.3   Small scale hydropower – India and Nepal

Small-scale hydropower schemes (Figure 11.4) are generally run-of-the-river, which means they have *no artificial storage* to provide a constant supply of water. Such schemes, therefore, rely entirely on the flow conditions of the river to generate electricity. There are many components to a run-of-the-river hydropower scheme and there are many issues for the prospective developer to consider. For example, the proximity of the scheme to the consumer and implications for transmission, the accessibility of the site, the extent of civil engineering works, the practicalities of acquiring and maintaining the electro-mechanical equipment, the mitigation of environmental impacts, the likely profitability of the scheme, and so on. However, there is one vital ingredient, without which none of the above can be contemplated seriously: a sufficient and reliable supply of water. In small-scale hydropower design, as with many other water resource projects, such as water supply, irrigation and water quality management (Chow, 1964; Warnick, 1984), the conventional method for describing the availability of water in a river is by using the flow duration curve. The FDC is an effective way of characterizing the temporal variation of flows as it identifies the proportion of time any given discharge (flow) is equalled or exceeded. It is a particularly useful measure for hydropower assessment as the area under the curve represents the volume of water that is potentially available for power generation (ESHA, 1994). Thus the full range of flows and not just 'low flows' are critical for small hydropower design.

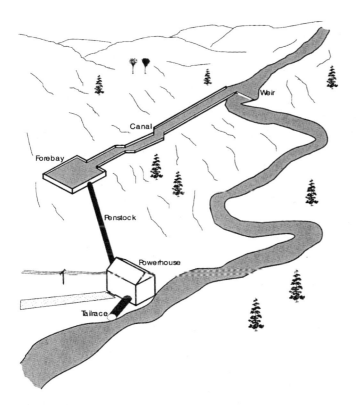

Figure 11.4 Typical layout for a run-of-the-river small-scale hydropower scheme (published with the approval of CEH).

At locations where there is good-quality long-term river flow data available, the FDC can be derived directly from the analysis of observed data (Section 5.2). However, many prospective sites are located in remote areas where the river has never been gauged and there is no historic flow data in the vicinity from which a reliable FDC could be obtained. It is not uncommon for schemes at such ungauged sites to be implemented solely on the basis of an FDC derived from a relatively short two-year campaign of field measurements (Chowdhary *et al.*, 1997). However, there is no guarantee that the flows encountered during this period are representative of conditions in the long term.

The Flow Duration Curve in hydropower design

Figure 11.5 illustrates that not all of the water that flows in the river will be available for hydropower generation. A certain amount of residual flow, $Q_{residual}$, should be left in the river to meet the needs of water users immediately downstream of the scheme and to preserve the ecology of the river (Chapter 10).

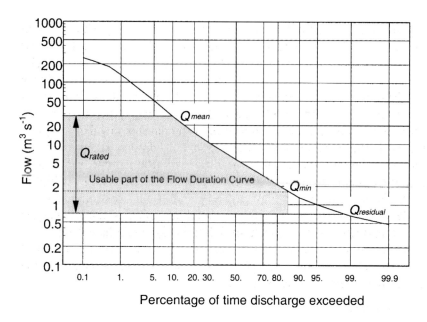

Percentage of time discharge exceeded

Figure 11.5 The flow duration curve in hydropower design (Q_{min} designates the minimum flow at which the turbine generates power), (published with the approval of CEH).

Neither is it possible to harness all of the flow; turbines that would operate at high flows could not function at the lowest flows. For initial design the maximum discharge is normally set to the mean flow, Q_{mean}. The maximum flow available for power production, Q_{rated}, is the difference between Q_{mean} and $Q_{residual}$. The minimum flow of the turbine, Q_{min}, is the flow below which power production stops. The residual flow, rated flow and minimum flow, together with the FDC for the site, thus determine the volume of water the scheme will use.

Study Area

The Himalayas are characterized by high mountains. Rapid changes in relief, a monsoon-influenced climate and a relatively sparse network of hydrometric and meteorological stations compound the difficulty of hydrological modelling. In 1997, a large-scale regional study of the Himalayan region of India was completed as part of the UNDP-GEF Hilly Hydro Project in India (AHEC, 1997). The study defined nine homogeneous regions, using 10-day average flow series to derive a single standardized flow duration curve for each region. The standardizing parameter was the catchment mean flow, which was described by a simple catchment-area-based regression relationship in each region (Singh *et al.*, 2001). More recently a regional model has been developed for Nepal and the state of Himachel Pradesh in northern India. The methodology for Himachel Pradesh is presented here which was confined to a relatively hydrologically homogeneous rain- and snow-fed region at an elevation between 2000 and 5000 m.a.s.l. Having favourable topography, relatively easy access and a reliable supply of water throughout the year, this part of the State is the most suitable for small-scale hydropower development. At lower altitudes (below 2000 m) there is little variation in topography and, with runoff derived solely from rainfall, rivers tend to be ephemeral. At higher altitudes (above 5000 m) the remoteness and harsh conditions rule-out extensive small hydro development. Although objective methods, such as cluster analysis and principal component analysis (Sections 5.5 and 8.3), could have been used, the homogeneous region was delineated according to the advice of local hydrologists, with reference to topographical maps and snowline data from the Himachel Pradesh Remote Sensing Agency.

Database Development

To develop a regional model requires good-quality hydrometric data sets from catchments that are representative of the study area, together with meteorological and other spatial data that adequately describe the region's climate and hypsographical features. The first step was to identify gauged catchments, having long time series of flows with limited artificial influences that were sufficiently widespread to be representative of the flow regimes in the study area. Data from 60 gauging stations in the rain-and snow-fed region were obtained from State authorities. Subsequent quality control checks to identify catchments suitable for the low-flow analysis reduced the final sample data set to 41. The catchment boundary for every catchment was drawn from 1:50 000 topographic map sheets, digitised and stored as polygons in the Arc/Info Geographical Information System (GIS). A variety of spatial data sets, describing the climate and physiographical nature of the study area, were also identified.

Model Description

The regional model for Himachal Pradesh (Rees *et al.*, 2002) is based on the premise that, as in the UK and Europe (Rees *et al.*, 2000), the shape of the flow duration curve at any point along a river is influenced by the natural storage characteristics of the upstream catchment. In a catchment where impermeable soils (or geology) predominate, there is very little natural storage and most of the rain drains rapidly to the river, hence a steep flow duration curve results. In contrast, for a catchment having natural storage, for example one with permeable geology, the slope of the FDC is much flatter, indicating less variability in flows (Figure 5.3). When the FDC is standardized by the mean flow, certain low flow statistics can be used to describe the whole flow duration curve. The Q_{95} flow is a particularly good indicator of hydrological response and a low Q_{95}, relative to the mean flow, generally indicates an impermeable catchment with very little storage, while a relatively high Q_{95} typifies a permeable catchment with storage (Section 5.3.1).

This relationship between Q_{95} and hydrogeology was used to develop a multivariate linear regression model (Section 8.3) relating the observed Q_{95} (expressed as a percentage of $\overline{Q}$) to the proportional extent of different soil (or geology) classes, including a class for snow and ice, within a catchment. The relationship yields an estimate of standardized Q_{95} for each soil (or geology type) encountered. A map of standardized Q_{95}, based on the distribution of soils (or geology), can thus be derived for the region. Overlaying the boundary of an ungauged catchment onto the map enables the Q_{95} of that catchment to be determined. The approach of Gustard *et al.* (1992) was adopted to extend the estimation from Q_{95} to a full FDC. A family of flow duration 'type' curves were derived for the region by averaging curves derived from observed flow data having similar values of Q_{95}. The shape of the estimated FDC at the ungauged site is then determined by the predicted value of Q_{95}. An estimate of the catchment mean flow is necessary to rescale the FDC in absolute units. A detailed analysis of the precipitation, flow and altitude revealed that the distribution of runoff in the rain- and snow-fed region of Himachal Pradesh was best described by means of a simple linear relationship between annual runoff depth and altitude.

Software Application of the Model

The model described aims to provide a means of estimating the water availability for small-scale hydropower schemes in Himachal Pradesh. A software package (HydraA-HP) has been developed, incorporating the Q_{95} model and the mean flow model as separate grids at a spatial resolution of 1 km. It allows even those with a minimal understanding of hydrology to rapidly estimate the flow duration curve and, hence, the hydropower potential at any prospective site. The user of the software is simply required to enter the geographical coordinates that delineate the catchment boundary upstream of the site of interest. The software identifies the constituent grid cells and derives a flow duration curve (FDC) specifically for the site. Referring to the FDC, the user is then required to enter the rated (or design) flow, the residual (or ecological) flow and head conditions at the site.

The software proceeds to calculate the likely hydropower potential (annual energy output, maximum power and rated capacity) for up to eight types of

Power Potential Report						⊠

Site: Demo_NP
Run Date / Time: 05 December 2003 at 11:24

					Show details for:
Mean Flow:	23.90 m³/s	Gross Hydraulic Head:	10.00 m		○ Individual Turbines
Provisional Rated Flow:	23.04 m³/s	Nett Hydraulic Head:	9.30 m		⊙ Full Site (3 turbines)
Residual Flow:	3.87 m³/s	Site Rated Flow:	19.17 m³/s		

Applicable Turbines	Gross Annual Average Output	Nett Annual Average Output	Maximum Power Output	Rated Capacity	Minimum Site Flow
Francis Open Flume	5352.5	5084.9	1547.8	1485.9	9.62
Francis Spiral Case	5581.8	5302.7	1605.5	1541.3	9.62
Propellor	4581.1	4352.0	1530.3	1469.1	16.33
Crossflow	5231.7	4970.1	1399.1	1309.6	6.75
Semi Kaplan	5314.6	5048.9	1539.1	1440.6	9.62
Kaplan	5661.2	5378.1	1561.8	1461.8	7.70
	MWhr	MWhr	kW	kW	m3/sec

Flow Regime Results File: c:\progra~1\hydra_np\data\tumling.frr

Plot Power/Flow	Plot Flow/Duration	Plot Power/Duration	Save	Print	Quit

Figure 11.6 Power potential report (published with the approval of CEH).

turbine. With the values of hydraulic head and rated flow entered by the user, the software refers to the operational envelopes of eight standard turbine types to identify which will operate under the stated conditions. For each selected turbine type, the software determines the usable part of the flow duration curve (Figure 11.5). Then, by referring to the relevant flow-efficiency curves it calculates the average annual energy potential (MWh) and the power generation capability (kW) of the site. The single page report that results from this procedure is shown in Figure 11.6.

Further Developments

This case study has shown it is possible to develop a regional multi-variate regression-based model for estimating percentiles of the FDC in the Himalayan region. Applied to the HydrA-HP software, the model will provide hydropower engineers and planners in Himachal Pradesh with a tool to provide reasonable estimates of the small-scale hydropower potential of any number of prospective sites. This enables more detailed flow gauging exercises to be focused effectively on a smaller subset of sites with the greatest hydropower potential.

As with any hydrological model, the reliability and accuracy of the results depends on the quality of the data from which it was derived. To significantly improve the model and enable a more accurate assessment of the water resources generally, it is clear that the spatial and temporal representativeness of the hydrometeorological networks of the region would need to be improved.

Having shown that the method can be successfully applied in the Himalayan region, there are many possibilities for it to be applied beyond Himachal Pradesh. Its application is not exclusive to hydropower and could have value for a host of water resource and environmental management purposes including the design of irrigation schemes, the planning of surface water abstractions for public water supply or determining acceptable levels of effluent discharge. With water stress increasingly becoming an issue of concern in the region, as the population and economy continues to grow, improved monitoring of the hydrological cycle together with the continued development of effective management tools, such as the one described, will be essential.

11.4.4   Design standards in Europe

In many countries national standards have been developed which describe procedures for low flow analysis. For example, in the *Czech Republic* hydrological analyses are based on the Czech Standard (ČSN) 75 1400 for surface water hydrological data adopted in 1997. The Standard defines hydrological procedures for deriving the flow duration curve, the return period of annual minima for different durations and the threshold method (Section 5.3). The investigations of low flow and droughts have been given a high priority in engineering hydrological practice and scientific research in *Russia*, where low flows determine the reliability of water-supply systems, power production and aquatic ecosystems protection. This is reflected in different national standards which regulate the estimation of the low flow volume and allowable water abstraction rates during dry periods. The main national standard is Building Standard (BS) 2-01-14-83 "Estimation of design hydrological parameters". This standard for estimating low flows recommends three approaches according to the availability of data:

a)   statistical analysis where observed data are available;

b)   the method of analogy or regional analysis at ungauged sites;

c)   using an analogue catchment to extend short flow records.

Low flow analysis in *Slovakia* typically includes annual minimum and flow duration curve analysis as well as the estimation of $Q_{97.3}$ and $Q_{99.7}$ as key low flow indices. There are a number of national, state and regional design standards relating to water resources. These are prepared, approved and published under the Slovak Institute for Standardization (SUTN). The most important low flow standard under preparation is STN 75 1420 "Hydrological data for surface water: Quantification of low flows".

In *Poland* the leading role in hydrological monitoring and establishing water standards is played by the Institute of Meteorology and Water Management (IMGW), which carries out management and engineering design. Considerable attention has recently been directed to estimating drought severity and drought extent and in estimating low flows at ungauged sites (Stachy *et al.*, 1991; Czamara *et al.*, 1997).

The index 'common low flow' used in *Norway* is based on a formulation in the Norwegian Water Resources Act, which stipulates that if there are abstractions in a basin, the residual flow should not be less than the 'common low flow'. The 'common low flow' may not fulfil the users needs, in which case the index will be used as a starting point for a subjective determination of the residual flow. The index is calculated by first removing the 15 smallest values every year in a daily streamflow record, then calculating the annual minimum series and finally removing one third of the lowest annual minimum flows. The smallest daily annual minimum streamflow in the remaining series is defined as the 'common low flow'. It has been shown (Erichsen & Tallaksen, 1995) that for Norwegian basins this index is highly correlated with the mean annual one day minimum flow, MAM(1-day), and Q_{95}.

11.4.5 Navigation and international treaties – The Meuse

The Meuse rises in France and flows through Belgium to the Netherlands. The length of the river is 874 km and the basin area is 33 000 km^2, approximately equal to the area of the Netherlands. The Border Meuse flows in a natural valley and forms the border between the Netherlands and Flanders. There are no weirs or navigational structures and the river has many meanders with a gravel bed at this point. These features are unique for the Netherlands and Belgium and as a result the river has a very high ecological and amenity value (Figure 11.7). In 1995 the Netherlands and Flanders signed the Meuse Treaty. The starting point was equal sharing of water for both partners and a common responsibility for the Border Meuse area. Very low discharges are known to be detrimental for ecosystem health (slow flow, large areas of exposed bed, low oxygen content) and a minimum discharge of $10 \, \mathrm{m^3 \, s^{-1}}$ was proposed. The most important function of the Meuse and the canals are:

Figure 11.7 The River Meuse at low flow (published with the approval of Rijkswaterstaat Directie Limburg, the Netherlands).

a) flood control;

b) cooling water and water for hydroelectric plants;

c) water demand for industry, agriculture, navigation, public water supply and environmental protection.

Some of these functions are consumptive in nature (industry and agriculture), however, the largest water users, the environment and navigation, are non-consumptive. Navigation uses water during the passage of ships through locks. In dry years, for two or three months a year there is not enough water to meet all demands and to avoid conflicts a national system of giving priorities to different functions has been developed. For navigation three operational measures are taken:

a) To pump water back from the low section of a lock to the high one. For example in the Juliana Canal there are 2 locks with a total difference of water levels of 23 m. Each lock has pumps, which pump water back for 3 months a year up to a maximum of 12 m^3 s^{-1} constantly day and night. This is very expensive. It also causes pollution and diminishes natural energy sources.

b) To decrease the use of water during locking. This can be achieved by the installation of reservoirs where some of the water from a lock chamber can be stored instead of being discharged at the lower section of the canal, or by siphoning lockage (water is exchanged between two parallel chambers). These measures are rather expensive: it costs money to install the necessary equipment and locking takes more time which is inconvenient for ships.

c) Although inconvenient for navigation, to change the frequency of locking: instead of locking every ship that arrives, to wait until the chamber is full of ships.

Future increases in economic activity in the region are likely to increase the level of shipping traffic and hence the issue of water allocation under drought will become more pressing. Additionally, the impact of climate change may put further pressure on water resources in the region.

11.4.6 Mitigating drought impacts in Europe

Water managers cannot prevent drought from occurring but they can mitigate the impacts of drought on water consumers and the environment. The most significant development in the last thirty years has been to recognize that water is an expensive commodity that must be managed wisely and that resources cannot be continuously developed to meet an ever-rising demand. Reduction in the per capita use of water will contribute to reducing the impacts of extreme drought and mitigate the impacts of abstractions on stream ecology. Improvements in water management have been brought about by significant reductions in leakage from water distribution systems, by introducing water meters and water pricing (to reduce demand), by improvements in the water efficiency of domestic appliances (washing machines, dishwaters), reducing the water requirements for flush toilets and raising public awareness of the need to conserve water, particularly during drought events. Large water efficiencies have also been introduced in industry and opportunities exist for improvements in irrigation efficiency, particularly in using tickle irrigation to replace spray irrigation.

In many countries these initiatives have led to reduced demand for water and hence a reduction in the frequency of adverse drought impacts. In addition, early warning of the onset of drought is used to request water users to use water wisely. As a drought progresses more formal steps can be taken. For example, in England and Wales drought contingency plans must be produced by Water Companies, which will identify how measures to make significant reductions in demand will be implemented. The regulatory authority – the Environment Agency or Water Companies – can apply for a Drought Order which can impose restrictions on the use of water and/or can vary the abstraction conditions during times of exceptional shortages of rainfall. Restrictions include

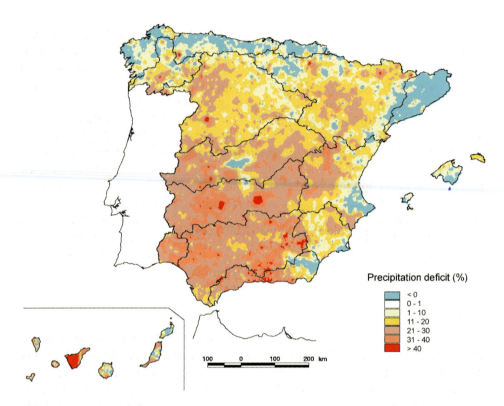

Precipitation deficit (%)

	< 0
	0 - 1
	1 - 10
	11 - 20
	21 - 30
	31 - 40
	> 40

100   0   100   200 km

Figure 11.8 Precipitation deficit in Spain during the 1990–1995 drought (Ministerio de Medio Ambiente, 2000).

the non-essential use of water, e.g. for washing cars or leisure facilities and variations to license conditions include the reduction in releases below reservoirs. There have also been advances in regional efficiencies in the use of water, notably in linking surface and groundwater sources, improved use of multi-reservoir systems and application of more advanced forecasting and control systems.

There have been major advances in the planning and operation of water resources in *Spain*. Much of the country has a Mediterranean climate with prolonged dry periods in summer. The increased demand for water for both the expansion of tourism and developments in irrigation for agricultural production has placed great stress on water resources during recent extreme drought. The response to this has been to develop proposals for further development of major water transfer schemes and to improve the management and operation of water resources during droughts. A key element of this has been the early identification of drought through continuous monitoring and analysis of rainfall, river flows and reservoir volumes. This has required key water resource areas or units to be identified in major catchments, to develop drought indicators for monitoring drought severity, to develop procedures for real time data collection and to carry out spatial analysis of drought response. This now provides an early warning of the onset of drought for different resources units, e.g. reservoirs, aquifers, direct river abstraction. As an example of this, Figure 11.8 shows the precipitation deficit for Spain for the period 1990 to 1995. It illustrates the area of Spain with the most extreme precipitation deficit in relation to the long-term average conditions. Production of these drought response maps on a routine operational basis provides water resource planners with early warning of drought, enabling them to take steps to mitigate the impact of drought. For example, the early introduction of small restrictions on water use may avoid major future reductions if the drought is prolonged.

11.5 Summary

This chapter has described the key drivers which determine the appropriate drought and low flow analysis method e.g. the scale of the catchment, the water sector and the organizational environment. For each different sector a different operational problem has to be resolved, ranging from detailed seasonal analysis for irrigation requirements, to the use of the flow duration curve for small hydropower to a wide range of simple indices and more complex instream flow models for environmental protection. The framework for the decision process is national and international water law and policy, implemented through formal or informal 'best practice'. Although a sectorial approach to water management continues, it is being replaced by resolving water resource issues and conflicts. This requires all demands on the river system to be identified, measured or estimated and used together with the natural hydrological system to make informed operational decisions. This process is being supported by major policy developments, for example the European Water Framework Directive, which provides a procedure for legal enforcement of integrated catchment management. In response to this, national research and development programmes are providing tools to assist the operational hydrologist in implementing a more integrated approach.

A review of operational analysis of low flows in the UK has found that the most common analysis method used is the Flow Duration Curve. Advances in estimation of low flow statistics at ungauged sites have been achieved through GIS development, improved regionalization techniques and by ensuring that local observations are made to complement long flow records. A key issue in ensuring operational uptake of new design techniques is to provide software implementation of research results, training in the use of these techniques and, most important, to ensure that there is a clear understanding by the researcher of the needs of the operational hydrologist. A key challenge for the research hydrologist is to ensure that *techniques are used operationally*. Despite major research advances in modelling and statistical analysis, many agencies are using very simple traditional techniques for drought and low flow design and operation. A second challenge for reducing uncertainty is to advance databases quantifying the spatial and temporal impacts of artificial influences and improving the techniques for incorporating their effects into water resource design procedures.

In addition to improving design, there is a need to improve the use of models and analysis techniques for real time operation and forecasting. Reductions in uncertainty in short- and medium-range weather forecasting will provide opportunities for improved management of water resource schemes during extreme drought events. Although some climate change scenarios indicate an increased frequency, duration and magnitude of drought (Section 2.6), there are a number of advances which can be made to mitigate drought impact including improved demand management, improved efficiency of domestic, industrial and agricultural water use and advanced integrated basin management. It is essential that these opportunities are fully utilized in the developed world, but of even greater importance that through *capacity building and investment* they are also established in developing countries.

12

Outlook

Alan Gustard, Henny A.J. van Lanen, Lena M. Tallaksen

12.1 Introduction

A key challenge for the twenty first century is to overcome the world water crises by providing access to freshwater, improved sanitation, secure food supply and meeting energy demands for an increasing global population. This will contribute to improved health, economic development and poverty reduction (HRH Prince of Orange, 2003). In both developed and developing countries there is increasing pressure to improve the reliability of water resource schemes and to enhance ecosystems degraded by reservoirs, diversions, over abstraction and pollution. Both surface and groundwater resources are under greatest pressure during drought and with population growth, human migration, increasing per capita demand, climate and land-use change the human pressures will increase. For these challenges to be met, it is essential that water resources are developed and operated so that the impacts of drought on the livelihoods of people and ecosystems are minimized. This can only be achieved by developing and disseminating best practice guidelines based on a thorough understanding of drought processes, good quality hydrological data and analytical techniques appropriate to a wide range of environments. This chapter presents a forward look to the potential advances in research and application relating to streamflow and groundwater drought. The conceptual diagram in Figure 12.1 illustrates how different aspects of drought dealt with in this chapter can be integrated to yield a holistic approach to drought management.

12.2 Data and monitoring

To advance our understanding of hydrological drought it is necessary to continuously obtain new information, and observations are the prime source of information. Estimation methods and tools cannot generate new knowledge, but

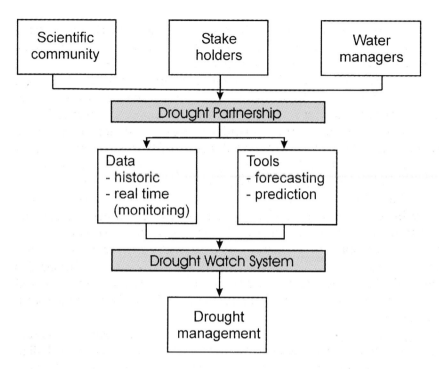

Figure 12.1 Components of an integrated drought management system.

are important means of extracting as much information as possible from the observations, i.e. the data. Uncertainty in both forecasting (Section 12.3) and predicting (Section 12.5) is reduced if more data are available and of good quality. Techniques for *monitoring* drought, including recording, processing and disseminating data (Chapter 4) are in some countries well established. In many parts of the world, however, the resources allocated to environmental monitoring are inadequate and both the volume and data quality are in decline (Deutsches IHP/OHP-Nationalkomitee, 2003). The highest priority is therefore to ensure that there is a long-term commitment *over several decades* to increase the resources allocated to data collection and to disseminate good practice particularly, but not only, in developing countries. Networks need to be expanded and designed in an optimal way. Statistical techniques for analysing spatial variability can assist in getting as much information as possible from the observations, evaluating the worth of additional measurements and designing monitoring networks. In addition to data collected operationally (e.g. streamflow, groundwater level, precipitation), there is a need to collect soil moisture and evaporation data, which are critical for both understanding drought processes and predicting drought response (Chapter 3). Data processing and quality control also need to be improved, and different environmental data sets integrated into GISs and freely disseminated.

It is anticipated that there will be continued improvements in sensors, data loggers and processing software. The spatial resolution, global coverage, frequency of measurement and record length will continue to advance by the measurement of land use, soil conditions, snow and ice from space (e.g. Wagner *et al.*, 2003). This will have direct benefits in forecasting and for estimating the impacts of e.g. land-use change and potential deglaciation on dry season flows. Advances in remote sensing for measurement of surface water levels should also improve the availability of data from large rivers and lakes, particularly in remote locations. The use of remotely sensed data depends on ground truth data for evaluation and calibration, and should be considered a supplement to, and not a replacement of, existing networks. There is further a special need to provide low cost sensors for at site measuring. In Europe this relates in particular to logging of groundwater and surface water levels as well as water quality to support the implementation of the Water Framework Directive (Chapter 11).

Hydrological design has a major influence on catchments in densely populated areas with great pressures from competing water users, including direct river abstractions and discharges, groundwater pumping and water transfer. It also affects pristine catchments, mainly due to the construction of large reservoirs and water transfer schemes (Chapters 9 and 11). There is an urgent need for improved *monitoring and modelling of artificial influences*. This includes access to information on the location, volume and timing of these influences using direct measurements when possible for the dominant impacts and estimation procedures for the large number of indirect and minor influences. Although some impacts are relatively simple (for instance, the effect of a sewage discharge on low flows) others are more complex, such as the impact of groundwater pumping on low flows or the impact of urbanization (Chapter 9). In Europe, the Water Framework Directive will give added impetus to the holistic approach of integrated catchment management, in contrast to the historical approach of considering issues separately for each water sector.

For an optimal management of water resources, for example the operation of a hydropower scheme or the control of abstractions, it is essential to monitor and have access to data in real time. Advances in sensor and telecommunication technology can improve the acquisition and processing of real time data, which will enable authorities to forecast and respond more rapidly to drought. Meanwhile, the widespread use of the web as a platform for dissemination of real time data means that water authorities and the general public will be better informed of the status of the environment. At the global scale there is a need to advance operational design and forecasting in developing countries.

12.3   Drought forecasting and early warning

Droughts are regional events that develop slowly. They are caused by persistent and large-scale atmospheric circulations, e.g. lasting anticyclone pressure fields (Chapter 2). Considerable progress has been made in recent years to understand the relationship between synoptic weather patterns and the occurrence of drought. In particular, links have been found between drought severity and synoptic scale anomalies of the atmospheric pressure, such as the El Niño Southern Oscillation (teleconnections). The link between synoptic weather conditions or other correlative variables such as sea-surface temperature and drought (e.g. Hoerling & Kumar, 2003), is a prime area of research that offers a major advance in medium and long term forecasting. This may ultimately allow an early warning of drought events even months before they occur.

A combination of remotely sensed data and land based observations, preferable combined in a GIS, offers potential for improved forecasting (Section 12.2). In particular there is a need to refine the spatial and temporal resolution of the observations, keeping in mind that droughts are not local, but regional events. It is anticipated that major advances in our understanding and forecasting of the synoptic behaviour of drought would result if real time data were analysed at regional centres as extreme hydrological events often occur on a scale that reaches beyond national boundaries (Section 12.8).

Drought forecasting should be seen as an integrated part of an early warning system for drought (Drought Watch System). Such a system should be tailor-made according to the requirements of the stakeholders and water managers and be based upon the most recent scientific developments in drought research, communication technology and GIS (Figure 12.1). It must be based on different types of data and tools to forecast and predict drought, and monitor the variables that control the development of drought in the region of interest. For example, in a semi-arid climate such as the Mediterranean region, the winter precipitation largely determines whether or not a drought will develop in the following dry summer (Chapter 3). In addition, quantitative information is needed about human pressures, such as groundwater abstractions and water withdrawal for irrigation. In time of drought, a limited number of indices that are able to summarize the status of the water resources are required (Chapter 5). Further research is needed to assess the applicability of different variables and indices in different regions of the world and thus allow guidelines for drought monitoring and early warning techniques to be formulated.

12.4 Drought processes and indices

A variety of hydrological drought indices and approaches for characterizing drought have been developed (Chapter 5). These are tailored to fit the data availability, climatic and catchment conditions, the particular problem under study and the needs of the respective region. A majority of the approaches have been developed for perennial rivers and are limited to at-site analysis. In semi-arid and arid regions where the rivers only flow occasionally, other types of data, such as reservoir and groundwater level, are likely to provide more information about the drought situation. In these regions precipitation is generally low whereas the inter- and intra-annual variability is high. Clearly, this affects overland flow, groundwater recharge and streamflow generation. There is a need to further test and develop suitable drought indices for these regions. It is anticipated that through the UNESCO-IHP programme, hydrological processes in semi-arid and arid zones will be given a higher priority in both basic and applied research. In mountain environments there is a need to predict the long-term impact of deglaciation on dry season flows (Dyurgerov & Meier, 1997). This is of particular importance in the Himalayas and Andes where low flows derive primarily from glacial melt water and are critical for the agriculture and drinking water of the region.

The complex processes involved in the propagation of a drought through the hydrological system, from a precipitation deficit to a drought in soil moisture, groundwater and streamflow, need to be better understood and studies are needed in different hydroclimatological regions. In particular, the understanding of the role of thick unsaturated zones needs to be developed. It is expected that knowledge on the stream-groundwater interaction will significantly increase over the coming decade. The spatial and temporal distribution of groundwater discharge into the stream is crucial to understand the development of streamflow drought. Key aspects include here the role of the valley-bottom hydrogeology and the recharge dynamics of aquifers.

Improvement is needed in the derivation of drought indices for a wide range of hydrological regimes to ensure more robust and flexible indices. One example is the threshold level method where techniques to deal with seasonality, minor droughts and dependent droughts are topics warranting further research. As drought commonly covers large, often diverse areas, drought studies are likely to benefit from a multivariable approach that allows the drought to be characterized using a combination of different hydrological variables, such as streamflow, groundwater and reservoir data. This would also allow the spatial structure of the drought to be considered in heterogeneous areas. Regional drought monitoring would further be advanced by the integration of hydrological, meteorological, soil moisture and socio-economic drought characteristics. In particular, complex indices that combine different aspects of a drought are increasingly requested by operational management.

12.5 Drought prediction

The present focus on issues related to environmental and climate change has increased the importance of the uncertainty aspect of forecasting and prediction applying stochastic as well as deterministic models. Prediction as opposed to forecasting deals with the estimation of future conditions without reference to a specific time. Predictive models are used in combination with historical data to extrapolate observations both in time and space. Sometimes data are absent, too short or of poor quality. Furthermore, there may be a requirement for answering 'what if' questions related to the potential impact of human activities in a catchment. The predictive models can be classified as stochastic or physically-based and further model development is elaborated below.

12.5.1 Statistical procedures

Traditional frequency analysis is based on the assumption of stationarity, i.e. there is no change over time in the observation series (Chapter 6). An important aspect of human pressures on water resources is the challenge of non-stationarity. The future will not be like the past, both climate and land-use changes will change the frequency and magnitude of extreme hydrological events. Recent climate research suggests that the observed increase in air temperature is not only a result of a shift of the statistical distribution towards warmer temperatures, but also caused by an increase in the annual variability (Schär *et al.*, 2004). An increase in variability would strongly affect the occurrence of heatwaves and droughts in the future. In general we need to cope with larger uncertainties in our estimates as the value of historical data reduces if the future changes, and this is an important topic for research.

Drought characteristics such as the distribution and moments of crossing properties for a natural or artificially influenced hydrologic system, can also be derived using random models and simulated series, e.g. the Monte-Carlo method (Chapter 7). The use of long (and many) simulated series permits the uncertainties in these characteristics to be quantified, those related to the natural variability of the processes studied as well as those related to a limited amount of observations. Random models and simulated series applied either to describe the process as such or the errors of a deterministic model, is one possible approach to the problem of uncertainty. Other more recent approaches that bring insight into the uncertainty aspect are so-called resampling methods, for example jackknifing and bootstrapping. These methods are applicable when large data sets are available and several new data sets are regenerated from the total set of data, from which the statistical analysis can be performed.

Frequency analysis further assumes the data to be homogeneous, and in regions with strong seasonality one extreme value series needs to be obtained for each season or generating process. Two important aspects are the choice of season and the selection of events within each season. Selection of events that best fulfil the requirement of independent, identically distributed data is an area that requires further research and should encompass time series from a range of hydrological regimes. Applications of the partial duration series approach have mainly focused on exceedances over an upper limit and have demonstrated the sensitivity of the results to the choice of upper limit. This choice should not be an arbitrary choice but be based on extreme value theory. A similar theoretical approach for minimum values ought to be further developed. In the latter case the presence of zero values in the time series raises several methodological questions, and more studies are needed to guide the choice of drought characteristics in perennial as well as in intermittent rivers.

Extreme value analysis of time series of hydrological drought characteristics should have a theoretical base and preferable be governed by knowledge on the physical processes operating. The derived distribution approach allows catchment properties such as antecedent moisture conditions to be included in the formulation of the extreme value model (Blöschl & Sivapalan, 1995b). It is thus less sensitive to sample variability and could allow the influence of environmental change on extreme events to be evaluated. Its potential for regional frequency analysis and studies of scaling in extreme events like flood and drought needs to be further investigated. Scaling invariance and the related topic of synchronicity in events, linked to e.g. the spatial structure of high pressure fields over a large territory, are complicated features that require further insight.

Regional frequency analysis combines information from several sites in a homogeneous region to permit estimation at the ungauged site and to reduce the uncertainty in at-site estimation. In the latter case the use of regional information means that less emphasis is put on single observations and is in general recommended. However, for both regional drought frequency analysis and analysis of the spatial behaviour of historical droughts only a few studies exist. Further development of regional procedures and more experience are strongly needed, including regional drought studies in different hydroclimatological regions.

Regional statistical methods for estimation of drought indices at the ungauged site (Chapter 8) range from simple empirical relationships to more complex multivariate models. A main limitation on their further development is the limited availability of good quality hydrological data for model development and calibration, and the absence of the most appropriate catchment descriptors. There is a need to ensure that all properties of a catchment are in digital form and where necessary, to develop new descriptors. For example, although it is recognized that hydrogeology is a key variable in controlling the low flow response of a catchment, it is rarely incorporated explicitly in regional estimation models. It is also necessary to improve estimates of *model uncertainty*, for instance, by testing the model on a subset of catchments not used in calibration This will lead to improvements in selecting the best model structure and parameter values, and enable different modelling approaches to be more rigorously tested.

Current regional models are generally steady-state models that lack a formal deterministic based representation of hydrological processes and hence regression-based models cannot be used for predictive modelling of scenarios that assume changes in land use or climate or for assessments that are essentially time-dependent problems; for example yield assessment of complex water resource schemes. As meteorological data availability improves, current regional models will be augmented by the development of regionalized process-based time series models. In these models the parameters and indeed the structure, will be defined explicitly or implicitly by the characteristics of a catchment.

12.5.2   Physically-based models

Physically-based models that are satisfactory developed, calibrated and validated can make a major contribution to understanding drought processes in a catchment. There have been considerable advances in hydrological modelling at a range of spatial scales based on advances in our knowledge of land surface processes and the movement of water in the subsurface (Chapter 3). Models range from simple, more conceptual based models to advanced spatially-distributed, process orientated models. The latter may play a key role in exploring the possible impacts of changing hydrological conditions, e.g. groundwater abstraction, river flow augmentation or land-use change (Chapter 9). So far, physically-based models lack the ability to account for feedback mechanisms in the physical system. Coupled climate and hydrological models can allow the sensitivity of hydrological variables to a changing climate or land use to be predicted. The performance of the models depends, however, largely on future improvements in the resolution of climate model scenarios and in techniques for downscaling. Today's resolution is too coarse to account for local and regional features in topography and models do not account for feedback in the land part of the hydrological cycle. Meanwhile impact assessment requires comprehensive models including as many domains as possible, i.e. integrated atmosphere-plant-soil, groundwater and surface water models.

Physically-based models in hydrology have traditionally focused on reducing the uncertainty in model prediction based on calibration of a single catchment or small sub-sets of catchments. Models have often been used in a deterministic way implying that model input data have no uncertainties. If the probability density functions of the input data are known, stochastic fields for distributed data can be generated and a Monte Carlo procedure adopted to simulate model results, including statistical moments. Furthermore, it is expected that data assimilation techniques will improve our knowledge about drought processes (e.g. Troch *et al.*, 2003). Data assimilation enables observations and model predictions to be merged to arrive at physically consistent and optimal predictions of state variables. It has been an active field of research in meteorology for more than a decade. The advanced technology that links remote sensing data, GIS and spatially-distributed hydrological models will facilitate further model development, improve model prediction and reduce the uncertainty in our modelling results (e.g. Schuurmans *et al.*, 2003).

One of the most important priorities for hydrological modelling is to build on existing expertise in single catchments to *regionalize daily continuous simulation models*. Driven by inputs of daily precipitation and evaporation and with parameters determined through regional calibration combined with local catchment information (e.g. Gottschalk *et al.*, 2001), it will be feasible to generate long periods of daily flows at ungauged sites, extend short records, and predict the impact of land-use change or the sensitivity of low flows to climate variability. The regionalization of these models will enable operational hydrologists to carry out simulation tasks for all river reaches within a region.

12.6 Advancing stream ecology and flow management

Although the role of river flow in structuring river ecosystems has been recognized for many years, the multidisciplinary science of linking river flow management to stream ecology is relatively new. Many countries are active in developing techniques for setting minimum flows for environmental protection. A high priority is to understand and model the interaction of stream biology with both river discharge and the secondary factors affected by discharge regulation, e.g. water velocity, depth, sediment, temperature and water quality. For such models to be predictive they must go beyond making simple predictions of structure (e.g. suitability for individual species) towards understanding the effects of artificial influences on ecological functioning (e.g. food webs, energy flow and production). During drought periods with reduced streamflow the wetted area and hence available habitat is reduced, stream temperature increases, dissolved oxygen falls and the dilution of effluents is reduced leading to a deterioration in water quality. By definition, extreme drought occurs infrequently and it is thus essential that very long term monitoring and process studies are carried out in order to measure the *resilience (or sensitivity) of river ecology to extreme drought* in a wide range of environments.

One of the most important applications of improving our understanding of river ecosystems is to develop Instream flow methods to enable rivers to be managed more effectively (Chapter 10), particularly to minimise the impact of river abstractions (and reservoir regulation) on the abundance and diversity of stream biota and to understand the interactions between physical and chemical habitat stress and the response of biota to extreme events. There continues to be a demand for very rapid assessment techniques where the ratio of an abstraction to the natural low flow is low. Simple statistics of low flow have traditionally been used, but the availability of national field programs and associated databases describing river biota, channel geometry and substrate combined with flow information should lead to significant advances in rapid assessment techniques. This will be given increased importance in Europe in response to the Water Framework Directive (Chapter 11). The WFD is giving a much higher priority to improving the ecology of European rivers and floodplains based on improved monitoring, decision making based on objective scientific analysis and public participation in the decision making process.

The increased international experience of using *habitat* based models for a wide range of species, life stages, river environments and operational problems will result in further improvements to these models and their application. Of particular importance is moving beyond selection of the most important species and life stage and moving towards more integrated ecological assessment based on communities. At whatever level, derivation of appropriate habitat suitability models based on field data continues to attract controversy and requires further understanding. Recent advances in quantifying fish habitat preferences in terms of energy budgets (e.g. Guensch *et al.*, 2001) should eventually lead to improvements to operational instream flow models. Most research and application of generic modelling techniques have been focused on ecosystems in temperate climates. However there is an urgent need to extend model development and application to developing counties particularly in the tropics where the pressure on water resources is highest. Finally there is a need to bridge the gap between biological sciences and hydrological sciences in order to make significant progress in unravelling the complex interaction between stream biota and their environment.

12.7 Communication and transfer of knowledge

Advances in drought research become beneficial to society when it is communicated to the scientific community, to stakeholders and the general public. *Scientific dissemination* should reach the academic, consultancy, water management and policy-making communities through activities such as workshops and training courses including both end-users and scientific experts. In order to advance the public understanding of drought there is a need to develop appropriate learning material to assist the teaching of hydrology at the undergraduate level. At the university level a greater priority should be given to teaching drought processes and analytical techniques for analysing drought response. It is further recommended to organise international study courses to impart basic knowledge and recent development in drought research and to promote harmonization of tools across larger regions.

The basic requirements for low flow and drought information for the design of river abstraction schemes, effluent dilution, hydropower, drought planning and management will continue. However, in response to changing policy objectives, priorities will change and advances in research will lead to improvements in both forecasting and prediction. There is considerable opportunity for improving the *transfer of existing knowledge* from the research to the user community including government departments, environmental protection agencies, water boards, water utilities and consultants. For example, by organizing short courses on drought hydrology including the most advanced analytical and modelling techniques and encouraging scientific seminars and workshops to include representatives of both the research and operational community. There is also a need to structure the knowledge to the specifics of the questions being asked by that community. New analysis requirements, and hence research requirements, will be identified through this process. For gauged locations it is essential that software is available to analyse the range of low flow indices with tabular and graphical outputs in a form appropriate for the decision-making and reporting of operational agencies. To estimate flows at ungauged sites there is a need to integrate existing regional models with digital databases of catchment descriptors. This may require major national digitizing programs to convert existing maps into digital form or the bringing together of disparate data bases from meteorology, hydrogeology, soils survey, river networks and topography.

Advances in operational hydrology will be accelerated if long-term partnerships can be established between the user and research community. This enables the operational requirements to be clearly specified before a research program is initiated. It is important that the final product is compatible with the available data, the skills of the organization and its policy objectives. Often research and development of new methodology is transferred through software. It is, however, essential that its implementation is supported over several years through training and assistance, enabling refinements to the decision support system to be made following advances in the underpinning research.

An essential part of *public communication* is to account for the needs and behaviour of the affected people. Drought should be perceived as a natural phenomenon that society must adapt to, and drought impact should be handled using a risk-based (proactive) approach rather than a crisis-based (reactive) approach as is the practice in most countries today (Panu & Sharma, 2002). In the latter case no plans are prepared in advance and responses are first initiated once the drought occurs and the impacts are perceived. Communication is most effective when it is passed both ways and risk communication should preferable be based on local knowledge, i.e. personal experience, tradition and public participation. This can subsequently be a valuable information source for the water managers and contribute to raise the general public awareness of drought.

12.8   Drought partnerships

Many of the concepts and proposals outlined in this final chapter are being developed at the national level. These include a more strategic approach to research planning, the development of partnerships between the user and the research community and the whole life support of models and software including software development, installation and training (Figure 12.1). Hydrological data are a precious resource and although limited and indeed an endangered species, it is increasingly available on the web (Section 12.2). The access to near real-time hydrological data has improved our ability to monitor drought development and take steps to mitigate the impact. There are considerable benefits of sharing data, analytical techniques, operational software and human resources between countries. In the United States a National Drought Mitigation Center has been established to advise all states on the compilation of mandatory drought mitigation plans and to give drought forecasts (Chapter 2). This has also facilitated the exchange of experiences in drought management and has led to the development of more efficient drought mitigation strategies. Box 12.1 outlines a number of activities which have been proposed for establishing a European Drought Partnership. These are not unique to Europe and if implemented could lead to considerable progress in understanding drought processes, monitoring drought development and mitigating drought impacts.

Box 12.1

European Drought Partnership

A coordinated pan-European Drought Partnership has been proposed within the framework of the ASTHyDA project (ASTHyDA, 2004) to improve the coordination of drought related activities in Europe. This is an idea originally put forward at an international workshop on drought and drought mitigation in Europe (Vogt & Somma, 2000). The partnership would act as a platform to initiate and discuss scientific progress on drought research within the academic society, but as important be a meeting place between multidisciplinary experts involved in drought research and management, including other technical and socio-economic disciplines. It would thus ensure a regular exchange of information and assist in the integration of new knowledge in multidisciplinary sectors. It would further act as a regional competence centre and be a forum for discussing policy issues related to sustainable water resources management in a pan-European context, e.g. implications of the Water Framework Directive on drought management.

Activities could include the following:

a) Develop a real time monitoring system to identify the growth and decay of droughts based on precipitation, river flows and groundwater data.

b) Identify gaps in knowledge and coordinate research activities.

c) Develop analytical tools for estimating drought severity, extent and frequency.

d) Support harmonization of methods for drought analysis.

e) Share experiences in drought planning and management.

f) Establish best practice guidelines for drought monitoring, forecasting, prediction and mitigation to support operational drought management.

g) Improve mobility of operational and research staff.

h) Develop links with international programmes and national drought activities outside Europe.

i) To carry out specific research projects, e.g. estimating event frequency from non-stationary data, impact of climate change on drought and medium-range drought forecasting.

j) Develop and implement training activities.

k) Communicate with stakeholders and the public.

12.9   Conclusions

Faced with the ever increasing demand for more water due to population increase and human development along with an increased uncertainty in our forecasts and predictions, the future provide many challenges for water management during periods of water scarcity. It is essential that we are able to assess the future need for water, the current status of the global water resources, the potential impact of climate and land-use change both on demand and supply, in particular how to cope during periods of severe drought. Good management practise cannot prevent a drought from happening, but it can assist in minimizing the impact of the drought. The drought hazard has major impacts on nature and society through large economic losses in agricultural, hydropower and industrial production, human suffering and environmental damage.

In response to a growing concern about the drought hazard this book has aimed to synthesize existing knowledge on processes and estimation methods for streamflow and groundwater drought. It can be concluded that there is a need to advance our understanding of how different processes control hydrological drought in different regions of the world and to further develop appropriate analysing tools, assessing several disciplinary perspectives and impact of human pressures. This requires more high quality data and a coordinated research effort, preferable through regional competence centres that are not limited to national boundaries. Such centres, virtual or real, would guarantee continuity in drought research, and if assigned operational and public responsibility, the important link between science, management and society would be ensured. An institutional framework for drought management and mitigation at the regional level would represent a major joint effort to reduce the risk associated with drought by developing a proactive approach to drought mitigation. It should be equally built on research, operational responsibilities and public communication. Along with an increased public awareness of the drought hazard, it could give drought the attention it deserves.

Appendix 2.1

Köppen's Climate Classification

The Köppen system defines five major climate groups according to monthly or annual values of temperature and precipitation. Subtypes within the major groups are designated by a second letter, and from the combination of the two letter groups, 12 distinct climates result. An overview of their main features is given in Table A2.1.1.

Table A2.1.1 Köppen's Climate Classification (adapted from FAO, 1997)

Climate Class	Subtype	Description
A: Tropical		Temperature of the coldest month: > 18°C. This is the climate where the most water- and heat-demanding crops (e.g. oilpalm and rubber) are grown. The climate is also ideal for yams, cassava, maize, rice, bananas and sugar cane.
	Af	No dry season: ≥ 60 mm of rainfall in the driest month
	Am	Monsoon type: short dry season but sufficient moisture to keep ground wet throughout the year
	Aw	Distinct dry season: one month with precipitation < 60 mm
B: Dry		Arid regions where annual evaporation exceeds annual precipitation. Even the wettest variants of this climate are characterized by a marked dry season. The climate is, therefore, mostly unsuitable for crops that require year-round moisture. The main crops are usually millet, sorghum and groundnuts. Sunshine is usually high, which leads to high productivity where a sufficiently long rainy season or irrigation ensure a sufficient water supply. Rice, sugar cane and maize are also common crops under this climate.
	BS	Steppe climate
	BW	Desert

(→ Table continued on next page)

Table A2.1.1 (continued)

Climate Class	Subtype	Description
C: Temperate		Average temperature of the coldest month: < 18°C and > −3°C, and average temperature of warmest month: > 10°C The main crops are temperate cereals such as wheat, barley and Irish (white) potatoes. An important variant of this climate is the Mediterranean climate, characterized by the olive tree, and also highly suitable for grapes.
	Cw	Winter dry season: at least ten times as much precipitation in wettest month of summer as in driest month of winter
	Cs	Summer dry season: at least three times as much rain in wettest month of winter as in driest month of summer, the latter having: < 30 mm precipitation
	Cf	No dry season: ≥ 30 mm precipitation in the driest month, difference between wettest month and driest month less than for Cw and Cs
D: Cold		Average temperature of the warmest month: > 10°C and that of coldest month: < −3°C. This climate grows essentially the same crops as the temperate climate, but seasons tend to be shorter and limited at the beginning and end by frost.
	Df	No dry season: ≥ 30 mm of rain in the driest month, difference between wettest month and driest month less than for Cw and Cs
	Dw	Winter dry season: at least ten times as much precipitation in wettest month of summer as in driest month of winter)
E: Polar		Average temperature of the warmest month: < 10°C. No crops are grown under this climate.
	ET	Tundra, average temperature of warmest month: > 0°C
	EF	No month with temperature: > 10°C

Additional qualifiers

To denote further variation in climate, a third letter can be added to the climate codes (Table A2.1.1) as described below.

A-climates

For the tropical climates an **i** can be added (isothermal subtype) if the annual range of temperature: $< 5°C$.

B-climates

The dry climates are further subdivided as **h** (subtropical desert with average temperature: $> 18°C$), **k** (cool dry climate of the middle latitude deserts), and **k'** (temperature of the warmest month: $< 18°C$).

C-climates

The temperate climates are further subdivided as: **a** (hot summer, average temperature of warmest month: $> 22°C$), **b** (cool summer, average temperature of warmest month: $< 22°C$), and **c** (cool and short summer, less than four months: $> 10°C$).

Note that the temperate climate as defined by C-only extends into what is actually B-climates, as the B-subtypes are superimposed on the other four major climate groups where only temperature is used to classify the climate.

D-climates

The cold climates are further subdivided as: **a** (hot summer, average temperature of warmest month: $> 22°C$), **b** (cool summer, average temperature of warmest month: $< 22°C$), **c** (cool and short summer, less than four months: $> 10°C$), and **d** (average temperature of coldest month: $< -38°C$).

E-climates

For the polar climates a **d** may be added to the classes if the average temperature of the coldest month: $< -38°C$.

Appendix 3.1

Water Balance of Catchment Domains

This appendix defines the water balance and its equations. These equations are used in Chapter 3 to explain the impact of lack of precipitation or high potential evapotranspiration on the development of hydrological drought (propagation of a pulse through the hydrological system).

The hydrological system of a catchment consists of three spatial domains, namely the soil, the groundwater and the surface water systems. The continuity equation applies to all three domains:

$$IN - OUT = \frac{\Delta S}{\Delta t} \qquad (A3.1.1)$$

where IN is inputs [LT^{-1}], OUT is outputs [LT^{-1}], and $\Delta S/\Delta t$ [LT^{-1}] is the change in stored water S [L] per unit of time t [T].

Soil domain

The water balance of the soil domain can be written as follows:

$$P - ET - q_{if} - I = \frac{\Delta S_{so}}{\Delta t} \qquad (A3.1.2)$$

where P is precipitation [LT^{-1}], ET is actual evapotranspiration [LT^{-1}], q_{if} is throughflow [LT^{-1}], I is groundwater recharge [LT^{-1}], and S_{so} [L] is stored soil water.

In areas with vegetation that intercepts significant amounts of precipitation (e.g. coniferous forest), the evaporation of intercepted water can be a major process (Figure 3.4). In this case Equation A3.1.2 can be extended to:

$$P_{gr} - E_i - P_n = \frac{\Delta S_{veg}}{\Delta t} \qquad (A3.1.2a)$$

$$P_n - ET - q_{if} - I = \frac{\Delta S_{so}}{\Delta t} \qquad (A3.1.2b)$$

where P_{gr} is gross precipitation at the top of the vegetation [LT^{-1}], P_n is net precipitation [LT^{-1}] reaching the soil surface, E_i is evaporation of interception water [LT^{-1}], and S_{veg} [L] is stored water on the leaves, trunks and stems.

In regions where overland flow and storage at the soil surface take place (small-depressions, snow cover; Figure 3.3), Equation A3.1.2 is replaced by the following two equations:

$$P - q_s - q_{of} = \frac{\Delta S_{su}}{\Delta t} \qquad\qquad\qquad (A3.1.2c)$$

$$q_s - q_{if} - I = \frac{\Delta S_{so}}{\Delta t} \qquad\qquad\qquad (A3.1.2d)$$

where q_s is infiltration across the soil surface $(P - ET)$ [LT^{-1}], q_{of} is overland flow [LT^{-1}], and S_{su} [L] is stored water at the surface (small-depressions, snow cover).

Actual evapotranspiration is equal to potential evapotranspiration, PE [LT^{-1}], when soil moisture content is sufficient. Figure 3.3 gives a schematic representation of the subsurface with relevant hydrological processes and state variables (e.g. water tables, stores). In areas with deep water tables the unsaturated zone is often subdivided into a root zone and thick unsaturated subsoil. Separate water balance equations are then solved for the root zone and the subsoil, with the flux across the bottom of the root zone q_r as output from the root zone and input to the subsoil (percolation). Stored soil water S_{so} is subdivided into water stored in the root zone S_{sor} and in the unsaturated subsoil S_{sos}. Throughflow q_{if} might take place if very fine-grained layers, e.g. clay lenses, occur in the unsaturated subsoil (Section 3.2). It takes place when the downward flux in the subsoil temporarily exceeds the saturated conductivity of the very fine-grained layer. A perched water table develops and the interflow begins. Another form of throughflow is concentrated saturated subsurface flow of water through natural pipes (e.g. mole tunnels) or macropores made by plant root systems, especially trees. Throughflow can directly feed a stream, but it can also break out at the foothill with rock debris considerably contributing to quick flow. A flow diagram (Figure 3.4) illustrates the water balance components and fluxes (transfer of water).

Groundwater domain

The groundwater domain of a catchment consists of one or more aquifers, which are separated by slowly permeable aquitards. If the groundwater divide coincides with the topographic catchment borders (surface water catchment), the water balance equation of the groundwater system has the following simple form:

$$I - q_{gr} = \frac{\Delta S_{gr}}{\Delta t} \tag{A3.1.3}$$

where q_{gr} is groundwater discharge to the surface water system [LT^{-1}], and S_{gr} [L] is stored groundwater.

In cases where the groundwater divide does not coincide with the surface water catchment (Figure 3.4), Equation A3.1.3 needs to be extended:

$$I + q_{gr,in} - q_{gr,out} = \frac{\Delta S_{gr}}{\Delta t} \tag{A3.1.3a}$$

where $q_{gr,in}$ is groundwater inflow [LT^{-1}], and $q_{gr,out}$ is groundwater outflow [LT^{-1}] across the border of the surface water catchment.

Thus, the groundwater discharge q_{gr} from the catchment is replaced by the difference between groundwater outflow and inflow. In most catchments with regional systems (Figure 3.1) the groundwater divide does not coincide with the surface water divide due to the hydrogeological setting.

In cases in a region where two or more aquifers occur that are separated by a semi-permeable layer (aquitard), Equation A3.1.3a needs to be written as follows:

$$I + q_{gr,in} - q_{gr,out} - \frac{(H_s - H_{de})}{c} = \frac{\Delta S_{gr}}{\Delta t} \tag{A3.1.3b}$$

where H_s and H_{de} are hydraulic heads [L] of the shallow and the deep aquifers, and c is the hydraulic resistance [T] of the aquitard. Thus, if $H_s - H_{de} > 0$ then downward leakage occurs and if $H_s - H_{de} < 0$ upward seepage is present.

Surface water domain

The balance of the surface water domain of a catchment has the following form:

$$Q = A_c(q_{of} + q_{if} + q_{gr}) + A_s(P_s - E_s) \tag{A3.1.4}$$

where Q is streamflow at the outlet [L^3T^{-1}], A_c is catchment area excluding large surface water bodies [L^2], A_s is surface water area [L^2], E_s is evaporation of open water surface [LT^{-1}], and P_s is precipitation directly falling on the surface water system [LT^{-1}].

In Equation A3.1.4 groundwater discharge directly into the lake is neglected. For normal surface water bodies (e.g. brooks, rivers) in river basins change in surface water storage (ΔS_s) is negligible for the catchment water balance. However, for large surface water bodies, such as lakes and reservoirs, a storage term needs to be introduced:

$$Q = A_c(q_{of} + q_{if} + q_{gr}) + A_s(P_s - E_s) + \frac{\Delta S_s}{\Delta t} \qquad\qquad \text{(A3.1.4a)}$$

Appendix 6.1

Probability Distributions
(Modified from DHI Water & Environment (2003))

A6.1.1 Generalized Extreme Value (GEV) distribution

Definition

Parameters: ξ (location), α (scale), κ (shape)

Range: $\alpha > 0$, $\xi + \alpha/\kappa \le x < \infty$ for $\kappa < 0$, $-\infty \le x \le \xi + \alpha/\kappa$ for $\kappa > 0$

Special case: Gumbel distribution for $\kappa = 0$

$$f(x) = \frac{1}{\alpha}\left[1 - \frac{\kappa(x-\xi)}{\alpha}\right]^{1/\kappa - 1} \exp\left(-\left[1 - \frac{\kappa(x-\xi)}{\alpha}\right]^{1/\kappa}\right) \tag{A6.1.1}$$

$$F(x) = \exp\left(-\left[1 - \frac{\kappa(x-\xi)}{\alpha}\right]^{1/\kappa}\right) \tag{A6.1.2}$$

$$x_p = \xi + \frac{\alpha}{\kappa}\left(1 - [-\ln(p)]^{\kappa}\right) \tag{A6.1.3}$$

Moments

$$\mu = \xi + \frac{\alpha}{\kappa}\left[1 - \Gamma(1+\kappa)\right] \tag{A6.1.4}$$

$$\sigma^2 = \left(\frac{\alpha}{\kappa}\right)^2\left[\Gamma(1+2\kappa) - [\Gamma(1+\kappa)]^2\right] \tag{A6.1.5}$$

$$\gamma_3 = \text{sgn}(\kappa)\frac{-\Gamma(1+3\kappa) + 3\Gamma(1+\kappa)\Gamma(1+2\kappa) - 2[\Gamma(1+\kappa)]^3}{\left[\Gamma(1+2\kappa) - [\Gamma(1+\kappa)]^2\right]^{3/2}} \tag{A6.1.6}$$

where $\text{sgn}(\kappa)$ is plus or minus 1 depending on the sign of κ, and $\Gamma(.)$ is the gamma function.

L-moments

$$\lambda_1 = \xi + \frac{\alpha}{\kappa}\left[1 - \Gamma(1+\kappa)\right] \tag{A6.1.7}$$

$$\lambda_2 = \frac{\alpha}{\kappa}\left(1 - 2^{-\kappa}\right)\Gamma(1+\kappa) \tag{A6.1.8}$$

$$\tau_3 = \frac{2\left(1 - 3^{-\kappa}\right)}{1 - 2^{-\kappa}} - 3 \tag{A6.1.9}$$

Moment estimates

The shape parameter κ can be estimated from the skewness estimator (Equation A6.1.6) using a Newton-Raphson iteration scheme. In this scheme, an analytic expression of the derivative of the gamma function based on Euler's psi function is used. Moment estimates of ξ and α are subsequently obtained from:

$$\hat{\alpha} = \frac{\hat{\sigma}|\hat{\kappa}|}{\sqrt{\Gamma(1+2\hat{\kappa}) - \left[\Gamma(1+\hat{\kappa})\right]^2}} \quad , \quad \hat{\xi} = \hat{\mu} - \frac{\hat{\alpha}}{\hat{\kappa}}\left[1 - \Gamma(1+\hat{\kappa})\right] \tag{A6.1.10}$$

L-moment estimates

For estimation of the shape parameter κ the approximation given by Hosking (1991) can be used:

$$\hat{\kappa} = 7.817740c + 2.930462c^2 + 13.641492c^3 + 17.206675c^4 \tag{A6.1.11}$$

where:

$$c = \frac{2}{3 + \hat{\tau}_3} - \frac{\ln 2}{\ln 3} \tag{A6.1.12}$$

If $\tau_3 < -0.1$ or $\tau_3 > 0.5$, the approximation is less accurate and Netwon-Raphson iteration can be applied for further refinement. L-moment estimates of ξ and α are subsequently obtained from:

$$\hat{\alpha} = \frac{\hat{\lambda}_2 \hat{\kappa}}{(1 - 2^{-\hat{\kappa}})\Gamma(1+\hat{\kappa})} \quad , \quad \hat{\xi} = \hat{\lambda}_1 - \frac{\hat{\alpha}}{\hat{\kappa}}\left[1 - \Gamma(1+\hat{\kappa})\right] \tag{A6.1.13}$$

Maximum likelihood estimates

Maximum likelihood estimates of the GEV parameters can be obtained using the modified Newton-Raphson algorithm presented by Hosking (1985).

A6.1.2 Gumbel (GUM) distribution

Definition

Parameters: ξ (location), α (scale)
Range: $\alpha > 0$, $-\infty < x < \infty$

$$f(x) = \frac{1}{\alpha} \exp\left[-\frac{x-\xi}{\alpha} - \exp\left(-\frac{x-\xi}{\alpha} \right) \right] \tag{A6.1.14}$$

$$F(x) = \exp\left[-\exp\left(-\frac{x-\xi}{\alpha} \right) \right] \tag{A6.1.15}$$

$$x_p = \xi - \alpha \ln\left[-\ln(p) \right] \tag{A6.1.16}$$

Moments

$$\mu = \xi + \alpha \gamma_E \tag{A6.1.17}$$

$$\sigma^2 = \frac{\pi^2 \alpha^2}{6} \tag{A6.1.18}$$

where $\gamma_E = 0.5772\ldots$ is Euler's constant.

L-moments

$$\lambda_1 = \xi + \alpha \gamma_E \tag{A6.1.19}$$

$$\lambda_2 = \alpha \ln 2 \tag{A6.1.20}$$

Moment estimates

Moment estimates of ξ and α are obtained from Equations A6.1.17 and A6.1.18:

$$\hat{\alpha} = \frac{\sqrt{6}\hat{\sigma}}{\pi} \quad , \quad \hat{\xi} = \hat{\mu} - \hat{\alpha}\gamma_E \tag{A6.1.21}$$

Gumbel (1954) proposed a finite sample size correction to the moment estimates.

L-moment estimates

L-moment estimates of ξ and α are obtained from Equations A6.1.19 and A6.1.20:

$$\hat{\alpha} = \frac{\hat{\lambda}_2}{\ln 2} \quad , \quad \hat{\xi} = \hat{\lambda}_1 - \hat{\alpha}\gamma_E \tag{A6.1.22}$$

Maximum likelihood estimates

The maximum likelihood estimate of α can be obtained by solving:

$$\frac{\sum_{i=1}^{n} x_i \exp\left(-\frac{x_i}{\alpha}\right)}{\sum_{i=1}^{n} \exp\left(-\frac{x_i}{\alpha}\right)} + \alpha = \frac{1}{n}\sum_{i=1}^{n} x_i \tag{A6.1.23}$$

using Newton-Raphson iteration. The estimate of ξ is subsequently obtained from:

$$\exp\left(\frac{\xi}{\alpha}\right)\sum_{i=1}^{n} \exp\left(-\frac{x_i}{\alpha}\right) = n \tag{A6.1.24}$$

A6.1.3 Frechét (FRE) distribution

Definition

Parameters: ξ (location), α (scale), κ (shape)
Range: $\alpha > 0$, $\kappa > 0$, $\xi < x < \infty$

$$f(x) = \frac{\kappa}{\alpha}\left(\frac{x-\xi}{\alpha}\right)^{-(\kappa+1)} \exp\left[-\left(\frac{x-\xi}{\alpha}\right)^{-\kappa}\right] \tag{A6.1.25}$$

$$F(x) = \exp\left[-\left(\frac{x-\xi}{\alpha}\right)^{-\kappa}\right] \tag{A6.1.26}$$

$$x_p = \xi + \alpha\left[-\ln(p)\right]^{-1/\kappa} \tag{A6.1.27}$$

Moments

$$\mu = \xi + \alpha\Gamma\left(1-\frac{1}{\kappa}\right) \tag{A6.1.28}$$

$$\sigma^2 = \alpha^2\left[\Gamma\left(1-\frac{2}{\kappa}\right) - \left[\Gamma\left(1-\frac{1}{\kappa}\right)\right]^2\right] \tag{A6.1.29}$$

$$\gamma_3 = \frac{\Gamma\left(1-\dfrac{3}{\kappa}\right) - 3\Gamma\left(1-\dfrac{1}{\kappa}\right)\Gamma\left(1-\dfrac{2}{\kappa}\right) + 2\left[\Gamma\left(1-\dfrac{1}{\kappa}\right)\right]^3}{\left[\Gamma\left(1-\dfrac{2}{\kappa}\right) - \left[\Gamma\left(1-\dfrac{1}{\kappa}\right)\right]^2\right]^{3/2}} \tag{A6.1.30}$$

where $\Gamma(.)$ is the gamma function. The Frechét distribution is defined only for skewness larger than the skewness of the Gumbel distribution, i.e. $\gamma_3 > 1.1396$.

Moment estimates

For estimation of κ the method proposed by Kadoya (1962) can be employed. The method is numerically complicated and not described here. Having estimated κ, moment estimates of ξ and α are subsequently obtained from:

$$\hat{\alpha} = \frac{\hat{\sigma}}{\sqrt{\Gamma\left(1-\dfrac{2}{\hat{\kappa}}\right) - \left[\Gamma\left(1-\dfrac{1}{\hat{\kappa}}\right)\right]^2}} \quad , \quad \hat{\xi} = \hat{\mu} - \hat{\alpha}\Gamma\left(1-\frac{1}{\hat{\kappa}}\right) \tag{A6.1.31}$$

A6.1.4 Weibull (WEI) distribution

Definition

Parameters: ξ (location), α (scale), κ (shape)
Range: $\alpha > 0$, $\kappa > 0$, $\xi < x < \infty$
Special case: Exponential distribution for $\kappa = 1$

$$f(x) = \frac{\kappa}{\alpha}\left(\frac{x-\xi}{\alpha}\right)^{\kappa-1} \exp\left[-\left(\frac{x-\xi}{\alpha}\right)^\kappa\right] \tag{A6.1.32}$$

$$F(x) = 1 - \exp\left[-\left(\frac{x-\xi}{\alpha}\right)^\kappa\right] \tag{A6.1.33}$$

$$x_p = \xi + \alpha\left[-\ln(1-p)\right]^{1/\kappa} \tag{A6.1.34}$$

The Weibull distribution is a reverse Generalized Extreme Value distribution with parameters:

$$\xi_{GEV} = \xi_{WEI} - \alpha_{WEI} \quad , \quad \alpha_{GEV} = \frac{\alpha_{WEI}}{\kappa_{WEI}} \quad , \quad \kappa_{GEV} = \frac{1}{\kappa_{WEI}} \tag{A6.1.35}$$

where subscripts *GEV* and *WEI* refer to Generalized Extreme Value and Weibull distributions, respectively.

Moments

$$\mu = \xi + \alpha \Gamma \left(1 + \frac{1}{\kappa} \right) \tag{A6.1.36}$$

$$\sigma^2 = \alpha^2 \left[\Gamma \left(1 + \frac{2}{\kappa} \right) - \left[\Gamma \left(1 + \frac{1}{\kappa} \right) \right]^2 \right] \tag{A6.1.37}$$

$$\gamma_3 = \frac{\Gamma \left(1 + \frac{3}{\kappa} \right) - 3\Gamma \left(1 + \frac{1}{\kappa} \right) \Gamma \left(1 + \frac{2}{\kappa} \right) + 2 \left[\Gamma \left(1 + \frac{1}{\kappa} \right) \right]^3}{\left[\Gamma \left(1 + \frac{2}{\kappa} \right) - \left[\Gamma \left(1 + \frac{1}{\kappa} \right) \right]^2 \right]^{3/2}} \tag{A6.1.38}$$

where $\Gamma(.)$ is the gamma function.

L-moments

$$\lambda_1 = \xi + \alpha \Gamma \left(1 + \frac{1}{\kappa} \right) \tag{A6.1.39}$$

$$\lambda_2 = \alpha \left(1 - 2^{-1/\kappa} \right) \Gamma \left(1 + \frac{1}{\kappa} \right) \tag{A6.1.40}$$

$$\tau_3 = 3 - \frac{2 \left(1 - 3^{-1/\kappa} \right)}{1 - 2^{-1/\kappa}} \tag{A6.1.41}$$

Moment estimates

If ξ is known, the moment estimate of κ can be obtained by combining Equations A6.1.36 and A6.1.37, which is solved using Newton-Raphson iteration. The moment estimate of α is then given by:

$$\hat{\alpha} = \frac{\hat{\mu} - \xi}{\Gamma \left(1 + \frac{1}{\hat{\kappa}} \right)} \tag{A6.1.42}$$

If ξ is unknown, the moment estimate of κ can be obtained from the skewness estimator (Equation A6.1.38) using Newton-Raphson iteration.

Moment estimates of ξ and α are given by:

$$\hat{\alpha} = \frac{\hat{\sigma}}{\sqrt{\Gamma\left(1+\frac{2}{\hat{\kappa}}\right)-\left[\Gamma\left(1+\frac{1}{\hat{\kappa}}\right)\right]^2}} \quad , \quad \hat{\xi} = \hat{\mu} - \hat{\alpha}\Gamma\left(1+\frac{1}{\hat{\kappa}}\right) \tag{A6.1.43}$$

L-moment estimates

If ξ is known, L-moment estimates of α and κ are given by:

$$\hat{\kappa} = -\frac{\ln 2}{\ln\left(1-\frac{\hat{\lambda}_2}{\hat{\lambda}_1}\right)} \quad , \quad \hat{\alpha} = \frac{\hat{\lambda}_1 - \xi}{\Gamma\left(1+\frac{1}{\hat{\kappa}}\right)} \tag{A6.1.44}$$

If ξ is unknown, the shape parameter can be estimated from the approximate formula (Equation A6.1.11) for estimation of the shape parameter of the GEV distribution using $-\tau_3$ and $\kappa_{GEV} = 1/\kappa$. L-moment estimates of ξ and α are then given by:

$$\hat{\alpha} = \frac{\hat{\lambda}_2}{\left(1-2^{-1/\hat{\kappa}}\right)\Gamma\left(1+\frac{1}{\hat{\kappa}}\right)} \quad , \quad \hat{\xi} = \hat{\lambda}_1 - \hat{\alpha}\Gamma\left(1+\frac{1}{\hat{\kappa}}\right) \tag{A6.1.45}$$

Maximum likelihood estimates

If ξ is known, the maximum likelihood estimate of κ can be obtained by solving:

$$\frac{1}{\kappa} = \frac{\displaystyle\sum_{i=1}^{n}(x_i-\xi)^\kappa \ln(x_i-\xi)}{\displaystyle\sum_{i=1}^{n}(x_i-\xi)^\kappa} - \frac{1}{n}\sum_{i=1}^{n}\ln(x_i-\xi) \tag{A6.1.46}$$

using Newton-Raphson iteration. The maximum likelihood estimate of α is subsequently obtained from:

$$\hat{\alpha} = \left[\frac{1}{n}\sum_{i=1}^{n}(x_i-\xi)^{\hat{\kappa}}\right]^{1/\hat{\kappa}} \tag{A6.1.47}$$

A6.1.5 Generalized Pareto (GP) distribution

Definition

Parameters: ξ (location), α (scale), κ (shape)
Range: $\alpha > 0$, $\xi \leq x < \infty$ for $\kappa < 0$, $\xi \leq x \leq \xi + \alpha/\kappa$ for $\kappa > 0$
Special case: Exponential distribution for $\kappa = 0$

$$f(x) = \frac{1}{\alpha}\left[1 - \kappa\frac{x-\xi}{\alpha}\right]^{1/\kappa - 1} \tag{A6.1.48}$$

$$F(x) = 1 - \left[1 - \kappa\frac{x-\xi}{\alpha}\right]^{1/\kappa} \tag{A6.1.49}$$

$$x_p = \xi + \frac{\alpha}{\kappa}\left[1 - (1-p)^\kappa\right] \tag{A6.1.50}$$

Moments

$$\mu = \xi + \frac{\alpha}{1+\kappa} \tag{A6.1.51}$$

$$\sigma^2 = \frac{\alpha^2}{(1+\kappa)^2(1+2\kappa)} \tag{A6.1.52}$$

$$\gamma_3 = \frac{2(1-\kappa)\sqrt{1+2\kappa}}{(1+3\kappa)} \tag{A6.1.53}$$

L-moments

$$\lambda_1 = \xi + \frac{\alpha}{1+\kappa} \tag{A6.1.54}$$

$$\lambda_2 = \frac{\alpha}{(1+\kappa)(2+\kappa)} \tag{A6.1.55}$$

$$\tau_3 = \frac{(1-\kappa)}{(3+\kappa)} \tag{A6.1.56}$$

Moment estimates

If ξ is known, moment estimates of α and κ are given by:

$$\hat{\kappa} = \frac{1}{2}\left[\left(\frac{\hat{\mu}-\xi}{\hat{\sigma}}\right)^2 - 1\right] \quad , \quad \hat{\alpha} = (\hat{\mu}-\xi)(1+\hat{\kappa}) \tag{A6.1.57}$$

If ξ is unknown, κ can be estimated from the skewness estimator (Equation A6.1.53) using a Newton-Raphson iteration scheme. Moment estimates of ξ and α are subsequently obtained from:

$$\hat{\alpha} = \hat{\sigma}(1+\hat{\kappa})\sqrt{1+2\hat{\kappa}} \quad , \quad \hat{\xi} = \hat{\mu} - \frac{\hat{\alpha}}{1+\hat{\kappa}} \tag{A6.1.58}$$

L-moment estimates

If ξ is known, L-moment estimates of α and κ are given by:

$$\hat{\kappa} = \frac{\hat{\lambda}_1 - \xi}{\hat{\lambda}_2} - 2 \quad , \quad \hat{\alpha} = (\hat{\lambda}_1 - \xi)(1+\hat{\kappa}) \tag{A6.1.59}$$

If ξ is unknown, L-moment estimates are given by:

$$\hat{\kappa} = \frac{1-3\hat{\tau}_3}{1+\hat{\tau}_3} \quad , \quad \hat{\alpha} = \hat{\lambda}_2(1+\hat{\kappa})(2+\hat{\kappa}) \quad , \quad \hat{\xi} = \hat{\lambda}_1 - \frac{\hat{\alpha}}{1+\hat{\kappa}} \tag{A6.1.60}$$

Maximum likelihood estimates

The log-likelihood function reads:

$$L = -n\ln\alpha + \frac{1-\kappa}{\kappa}\sum_{i-1}^{n}\ln\left[1 - \frac{\kappa}{\alpha}(x_i - \xi)\right] \tag{A6.1.61}$$

If ξ is known, the maximum likelihood estimates can be obtained by solving:

$$\frac{\partial L}{\partial \alpha} = 0 \quad , \quad \frac{\partial L}{\partial \kappa} = 0 \tag{A6.1.62}$$

using a modified Newton-Raphson iteration scheme (Hosking & Wallis, 1987).

A6.1.6 Exponential (EXP) distribution

Definition

Parameters: ξ (location), α (scale)
Range: $\alpha > 0$, $\xi \le x < \infty$

$$f(x) = \frac{1}{\alpha} \exp\left[-\frac{x-\xi}{\alpha}\right] \tag{A6.1.63}$$

$$F(x) = 1 - \exp\left[-\frac{x-\xi}{\alpha}\right] \tag{A6.1.64}$$

$$x_p = \xi - \alpha \ln(1-p) \tag{A6.1.65}$$

Moments

$$\mu = \xi + \alpha \tag{A6.1.66}$$

$$\sigma^2 = \alpha^2 \tag{A6.1.67}$$

L-moments

$$\lambda_1 = \xi + \alpha \tag{A6.1.68}$$

$$\lambda_2 = \frac{\alpha}{2} \tag{A6.1.69}$$

Moment estimates

If ξ is known, α can be estimated from the sample mean value:

$$\hat{\alpha} = \hat{\mu} - \xi \tag{A6.1.70}$$

If ξ is unknown, moment estimates are given by:

$$\hat{\alpha} = \hat{\sigma} \quad , \quad \hat{\xi} = \hat{\mu} - \hat{\alpha} \tag{A6.1.71}$$

L-moment estimates

If ξ is known, the L-moment estimate of α is identical to the moment estimate. If ξ is unknown, L-moment estimates are given by:

$$\hat{\alpha} = 2\hat{\lambda}_2 \quad , \quad \hat{\xi} = \hat{\lambda}_1 - \hat{\alpha} \tag{A6.1.72}$$

Maximum likelihood estimates

If ξ is known, the maximum likelihood estimate of α is identical to the moment and the L-moment estimate.

A6.1.7 Pearson Type 3/Gamma (P3/GAM) distribution

Definition

Parameters: ξ (location), α (scale), κ (shape)

Range: $\kappa > 0$, $\xi \le x < \infty$ for $\alpha > 0$, $-\infty \le x \le \xi$ for $\alpha < 0$
Special cases: Exponential distribution for $\kappa = 1$ and $\alpha > 0$
Normal distribution for $\gamma = 0$

$$f(x) = \frac{1}{|\alpha|\Gamma(\kappa)}\left(\frac{x-\xi}{\alpha}\right)^{\kappa-1} \exp\left(-\frac{x-\xi}{\alpha}\right) \tag{A6.1.73}$$

$$F(x) = \begin{cases} G\left(\kappa, \dfrac{x-\xi}{\alpha}\right) & , \alpha > 0 \\[3mm] 1 - G\left(\kappa, \dfrac{x-\xi}{\alpha}\right) & , \alpha < 0 \end{cases} \tag{A6.1.74}$$

$$x_p = \xi + \alpha u_p \tag{A6.1.75}$$

where $\Gamma(.)$ is the gamma function, and $G(.,.)$ is the incomplete gamma integral. No explicit expression of the quantile function is available. The standardized quantile u_p can be determined as the solution of $F(u) = p$ where $u = (x - \xi)/\alpha$ using Newton-Raphson iteration.

Moments

$$\mu = \xi + \alpha\kappa \tag{A6.1.76}$$

$$\sigma^2 = \alpha^2\kappa \tag{A6.1.77}$$

$$\gamma_3 = \begin{cases} \dfrac{2}{\sqrt{\kappa}} & , \alpha > 0 \\[3mm] -\dfrac{2}{\sqrt{\kappa}} & , \alpha < 0 \end{cases} \tag{A6.1.78}$$

L-moments

$$\lambda_1 = \xi + \alpha\kappa \tag{A6.1.79}$$

$$\lambda_2 = \frac{|\alpha|}{\sqrt{\pi}}\frac{\Gamma\left(\kappa + \dfrac{1}{2}\right)}{\Gamma(\kappa)} \tag{A6.1.80}$$

$$\tau_3 = \begin{cases} 6I_{1/3}(\kappa, 2\kappa) - 3 & , \alpha > 0 \\[2mm] -6I_{1/3}(\kappa, 2\kappa) + 3 & , \alpha < 0 \end{cases} \tag{A6.1.81}$$

where $I_x(.,.)$ is the incomplete beta function ratio. Rational-function approximations of τ_3 as a function of κ are given by Hosking & Wallis (1997).

Moment estimates

If ξ is known, moment estimates of α and κ can be obtained from Equations A6.1.76 and A6.1.77

$$\hat{\kappa} = \frac{(\hat{\mu} - \xi)^2}{\hat{\sigma}^2} \quad , \quad \hat{\alpha} = \frac{\hat{\sigma}^2}{\hat{\mu} - \xi} \tag{A6.1.82}$$

If ξ is unknown, the shape parameter κ can be estimated from the skewness estimator (Equation A6.1.78). The skewness estimator can be corrected according to the bias correction formula given by Bobée & Robitaille (1975). If γ_3 or n (sample size) fall outside the ranges of the Bobée-Robitaille formula, the skewness can be corrected using the following general bias correction:

$$\hat{\gamma}_3^* = \frac{\sqrt{n(n-1)}}{n-2} \hat{\gamma}_3 \tag{A6.1.83}$$

Moment estimates of ξ and α are obtained from Equations A6.1.76 and A6.1.77:

$$\hat{\alpha} = \text{sgn}(\hat{\gamma}_3^*) \frac{\hat{\sigma}}{\sqrt{\hat{\kappa}}} \quad , \quad \hat{\xi} = \hat{\mu} - \hat{\alpha}\hat{\kappa} \tag{A6.1.84}$$

where $\text{sgn}(\hat{\gamma}_3^*)$ is plus or minus 1, depending on the sign of $\hat{\gamma}_3^*$.

L-moment estimates

If ξ is known, L-moment estimates of α and κ can be obtained from Equations A6.1.79 and A6.1.80. For estimation of κ, rational-function approximations of κ as a function of the L-coefficient of variation τ_2, are given in Hosking (1991). For estimation of κ when ξ is unknown, rational-function approximations of κ as a function of the L-skewness can be applied (Hosking & Wallis, 1997).

Maximum likelihood estimates

If ξ is known, maximum likelihood estimates are obtained from the following set of equations:

$$\sum_{i=1}^{n} \ln(x_i - \xi) - n\ln\alpha - n\psi(\kappa) = 0 \quad , \quad \alpha = \frac{1}{\kappa} \frac{1}{n} \sum_{i=1}^{n} (x_i - \xi) \tag{A6.1.85}$$

where $\psi(.)$ is Euler's psi function.

A6.1.8 Log-Pearson Type 3 (LP3) distribution

Definition

Parameters:	ξ (location), α (scale), κ (shape)
Range:	$\kappa > 0$, $\exp(\xi) \leq x < \infty$ for $\alpha > 0$, $0 \leq x \leq \exp(\xi)$ for $\alpha < 0$
Special case:	2-parameter Log-Normal distribution for $\gamma_y = 0$

If X is distributed according to a Log-Pearson Type 3 distribution, then $Y = \ln(X)$ is Pearson Type 3 distributed. The parameters ξ, α and κ are, respectively, the location, scale and shape parameter of the corresponding Pearson Type 3 distribution.

$$f(x) = \frac{1}{x|\alpha|\Gamma(\kappa)}\left(\frac{\ln(x) - \xi}{\alpha}\right)^{\kappa-1} \exp\left(-\frac{\ln(x) - \xi}{\alpha}\right) \qquad (A6.1.86)$$

$$F(x) = \begin{cases} G\left(\kappa, \dfrac{\ln(x) - \xi}{\alpha}\right) & , \alpha > 0 \\ 1 - G\left(\kappa, \dfrac{\ln(x) - \xi}{\alpha}\right) & , \alpha < 0 \end{cases} \qquad (A6.1.87)$$

$$x_p = \exp(\xi + \alpha u_p) \qquad (A6.1.88)$$

where $\Gamma(.)$ is the gamma function, and $G(.,.)$ is the incomplete gamma integral. No explicit expression of the quantile function is available. The standardized quantile u_p can be determined as the solution of $F(u) = p$ where $u = (\ln(x) - \xi)/\alpha$ using Newton-Raphson iteration.

Moment estimates

Parameter estimates can be obtained from the sample moments of the logarithmic transformed data $\{y_i = \ln(x_i), i = 1, 2, ..., n\}$ following the same approach as for the P3 distribution for unknown ξ (Section A6.1.7). Alternatively, an estimation method based on the moments in real space can be applied (Bobée, 1975).

L-moment estimates

Parameter estimates can be obtained from the sample L-moments of the logarithmic transformed data $\{y_i = \ln(x_i), i = 1, 2, ..., n.\}$ as outlined in Hosking & Wallis (1997).

A6.1.9 Log-Normal (LN) distribution

Definition

Parameters: ξ (location), μ_y (mean), σ_y (standard deviation)
Range: $\sigma_y > 0$, $x > \xi$

If X is distributed according to a Log-Normal distribution, then $Y = \ln(X - \xi)$ is normally distributed. The parameters μ_y and σ_y^2 are the population mean and variance of Y.

$$f(x) = \frac{1}{(x-\xi)\sigma_y\sqrt{2\pi}}\exp\left[-\frac{1}{2}\left(\frac{\ln(x-\xi)-\mu_y}{\sigma_y}\right)^2\right] \tag{A6.1.89}$$

$$F(x) = \Phi\left(\frac{\ln(x-\xi)-\mu_y}{\sigma_y}\right) \tag{A6.1.90}$$

$$x_p = \xi + \exp\left(\mu_y + \sigma_y\Phi^{-1}(p)\right) \tag{A6.1.91}$$

where $\Phi(.)$ and $\Phi^{-1}(.)$ are, respectively, the cumulative distribution function and the quantile function of the standard Normal distribution.

Moments

$$\mu_x = \xi + \exp\left[\mu_y + \frac{1}{2}\sigma_y^2\right] \tag{A6.1.92}$$

$$\sigma_x^2 = \left(\exp\left[2\mu_y + \sigma_y^2\right]\right)\left(\exp\left(\sigma_y^2\right)-1\right) \tag{A6.1.93}$$

$$\gamma_{3x} = 3\phi + \phi^3 \quad , \quad \phi = \sqrt{\exp\left(\sigma_y^2\right)-1} \tag{A6.1.94}$$

L-moments

$$\lambda_{1,y} = \mu_y \tag{A6.1.95}$$

$$\lambda_{2,y} = \frac{\sigma_y}{\sqrt{\pi}} \tag{A6.1.96}$$

Moment estimates

If ξ is known, moment estimates of μ_y and σ_y are given by the sample mean and standard deviation of the logarithmic transformed data $\{y_i = \ln(x_i - \xi)$, $i = 1, 2, \ldots, n\}$.

If ξ is unknown, it can be estimated based on a lower bound quantile estimator of ξ (Iwai, 1947; Stedinger, 1980) or based on the sample moments in real space $\{x_i, i = 1, 2, ..., n\}$ where a bias correction of the sample skewness is adopted (Bobée & Robitaille, 1975; Ishihara & Takase, 1957).

L-moment estimates

If ξ is known, μ_y and σ_y can be estimated from the sample L-moments of the logarithmic transformed data $\{y_i = \ln(x_i - \xi), i = 1, 2, ..., n\}$:

$$\hat{\mu}_y = \hat{\lambda}_{1,y} \quad , \quad \hat{\sigma}_y = \sqrt{\pi}\hat{\lambda}_{2,y} \tag{A6.1.97}$$

Maximum likelihood estimates

If ξ is known, maximum likelihood estimates of μ_y and σ_y are given by:

$$\hat{\mu}_y = \frac{1}{n}\sum_{i=1}^{n}\ln(x_i - \xi) \quad , \quad \hat{\sigma}_y = \sqrt{\frac{1}{n}\sum_{i=1}^{n}\left[\ln(x_i - \xi) - \hat{\mu}_y\right]^2} \tag{A6.1.98}$$

If ξ is unknown, the maximum likelihood estimate of ξ can be obtained by solving:

$$\frac{\partial L}{\partial \xi} = \frac{\partial}{\partial \xi}\left(\sum_{i=1}^{n}\ln\left(\sqrt{2\pi}\sigma(x_i - \xi)\right) + \frac{1}{2}\sum_{i=1}^{n}\left(\frac{\ln(x_i - \xi) - \mu}{\sigma}\right)^2\right) = 0 \tag{A6.1.99}$$

using a bisection iteration scheme. The parameter estimates of μ_y and σ_y are subsequently obtained from Equation A6.1.98.

Appendix 6.2

L-Moment Estimators

Generally, the L-moments of X are defined to be the quantities:

$$\lambda_r \equiv r^{-1} \sum_{k=0}^{r-1} (-1)^k \binom{r-1}{k} E\{X_{(r-k:r)}\} \qquad (A6.2.1)$$

where r is the order of moment ($r = 1, 2, \ldots$), and

$$\binom{r-1}{k} = \frac{(r-1)!}{k!(r-1-k)!} \qquad (A6.2.2)$$

is the number of combinations of any k items from $(r-1)$ items.

L-moments are defined for a probability distribution (Appendix 6.1), but in practice estimation of L-moments must be made from a random sample drawn from an unknown distribution. Let $x_{(1:n)} \le x_{(2:n)} \le \ldots \le x_{(n:n)}$ be the ordered sample from the observed sample of size n. The L-moment of order r, λ_r, is a function of the expected order statistics of a sample of size r, and can be estimated using a *U-statistic* (Hoeffding, 1948). A U-statistic is the corresponding function of the sample order statistics averaged over all subsamples of size r which can be drawn from the observed sample of size n.

It is, however, not necessary to iterate over all subsamples of size r, which can be quite large even for a relatively small sample size n. The L-moment statistic can be expressed explicitly as a linear combination of order statistics of a sample of size n (Hosking, 1990). The order statistics, $X_{(i:n)}$, are commonly expressed in terms of *probability weighted moments* (PWMs), and procedures based on PWMs and L-moments are equivalent. Estimators of L-moments obtained using unbiased PWM estimators, b_r, are given by:

$$l_r = \sum_{k=0}^{r-1} p^*_{r-1,k} b_k \qquad (A6.2.3)$$

where:

$$b_r = n^{-1} \sum_{i=1}^{n} \frac{(i-1)(i-2)\ldots(i-r)}{(n-1)(n-2)\ldots(n-r)} x_{(i:n)} \quad , \qquad p^*_{r,k} = (-1)^{r-k} \binom{r}{k}\binom{r+k}{k}$$

The first four L-moments are calculated as:

$$\lambda_1 = \beta_0$$
$$\lambda_2 = 2\beta_1 - \beta_0$$
$$\lambda_3 = 6\beta_2 - 6\beta_1 + \beta_0 \quad \text{(A6.2.4)}$$
$$\lambda_4 = 20\beta_3 - 30\beta_2 + 12\beta_1 - \beta_0$$

Replacing sample estimators, b_r, into Equation A6.2.4 provides the corresponding estimates for the L-moments. A simple, but biased estimator of b_r can be obtained using a *plotting position estimator* of $p_{(i:n)}$ (Stedinger *et al.*, 1993):

$$b_r^* = n^{-1} \sum_{i=1}^{n} x_{(i:n)} \left[p_{(i:n)} \right]^r \quad \text{(A6.2.5)}$$

The choice $p_{(i:n)} = (i\text{-}0.35)/n$ has proved to give good results for the Generalized Pareto, GEV and Wakeby distributions (Hosking, 1990). A library of FORTRAN subroutines useful for L-moment analyses is provided by Hosking (1991), and Stedinger *et al.* (1993) describe how to obtain the software by e-mail.

Wang (1996) provides *direct estimators* of L-moments which eliminates the need for introducing PWMs. The estimation procedure follows closely the definition of L-moments by covering all possible combinations in a more efficient way. For the sample value $x_{(i:n)}$ there are $(i - 1)$ values $\leq x_{(i:n)}$ and $(n - i)$ values $\geq x_{(i:n)}$, and for each subsample of size r, the number of values drawn from each of these categories are considered. The first four direct estimators are given by:

$$l_1 = \binom{n}{1}^{-1} \sum_{i=1}^{n} x_{(i:n)}$$

$$l_2 = \frac{1}{2} \binom{n}{2}^{-1} \sum_{i=1}^{n} \left[\binom{i-1}{1} - \binom{n-i}{1} \right] x_{(i:n)}$$

$$l_3 = \frac{1}{3} \binom{n}{3}^{-1} \sum_{i=1}^{n} \left[\binom{i-1}{2} - 2\binom{i-1}{1}\binom{n-i}{1} + \binom{n-i}{2} \right] x_{(i:n)} \quad \text{(A6.2.6)}$$

$$l_4 = \frac{1}{4} \binom{n}{4}^{-1} \sum_{i=1}^{n} \left[\binom{i-1}{3} - 3\binom{i-1}{2}\binom{n-i}{1} + 3\binom{i-1}{1}\binom{n-i}{2} - \binom{n-i}{3} \right] x_{(i:n)}$$

A Fortran subroutine is included which allows the reader to calculate the L-moments using the derived direct estimators.

The numerical values of the unbiased sample estimators using U-statistics, PWMs and direct estimators are the same. Landwehr *et al.* (1979) recommend the use of biased estimates of PWMs and L-moments, since such estimators often produce quantile estimates with lower root-mean-square error than unbiased alternatives. Although there is no theoretical reason for preferring plotting-position estimators, experience has shown that they sometimes yield better results when a distribution is fitted to data at a single site (Hosking & Wallis, 1995). The unbiased estimators are recommended for calculating L-moment diagrams and for use with regionalization procedures where unbiasedness is important (Stedinger *et al.*, 1993). Unbiased estimators are preferred because they have less bias for estimating L-moment ratios (Vogel & Fennessey, 1993).

References

Acreman, M.C. & Sinclair, C.D. (1986) Classification of drainage basins according to their physical characteristics: An application for flood frequency analysis in Scotland. *J. Hydrol.* **84**, 365–380.

Adamowski, K. (1996) Nonparametric estimation of low-flow frequencies. *J. Hydraul. Eng.* **122**(1), 46–49.

Adler, M.-J., Busuioc, A. & Ghioca, M. (1999) Atmospheric processes leading to droughty periods in Romania. In: *Hydrological Extremes: Understanding, Predicting, Mitigating* (ed. by L. Gottschalk, J.-C. Olivry, D. Reed & D. Rosbjerg), IAHS Publ. no. 255, 29–35.

Ahmed, K.M., Ravenscroft, P., Hasan, M.K., Burgess, W.G., Dottridge, J. & Wonderen, J.J. van (1999) The Dupi Tila aquifer of Dhaka, Bangladesh: Hydraulic and hydrochemical response to intensive exploitation. In: *Groundwater in the Urban Environment. Selected City Profiles* (ed. by J. Chilton), Balkema Publishers, Rotterdam, the Netherlands, 19–30.

Ahn, C.-H. & Tateishi R. (1994) Development of a Global 30-minute grid Potential Evapotranspiration Data Set. *Journal of the Japan Soc. Photogrammetry and Remote Sensing*, **33**(2), 12–21.

Akaike, H. (1974) A new look at statistical model identification. *IEEE Trans. Automat. Contr.* **AC-19**(16), 716–722.

Aksoy, H., Bayazit, M. & Wittenberg, H. (2001) Probabilistic approach to modelling of recession curves. *Hydrolog. Sci. J.* **46**(2), 269–286.

Allan, J.D. (1995a) Section 3.1. In: *Stream Ecology: Structure and Function of Running Waters.* Kluwer Academic Publishers, Dordrecht, the Netherlands.

Allan, J.D. (1995b) Section 3.2 In: *Stream Ecology: Structure and Function of Running Waters.* Kluwer Academic Publishers, Dordrecht, the Netherlands.

Alternate Hydro Energy Centre (AHEC) (1997) *Development of Regional Flow Duration Models for preparation of Zonal Plan for Small Hydro Power Projects.* UNDP-GEF Hilly Hydro Project, National Institute of Hydrology India, Roorkee, India.

Amani, A. & Nguetora, M. (2002) Evidence d'une modification du régime hydrologique du fleuve Niger à Niamey. In: *FRIEND 2002 – Regional Hydrology: Bridging the Gap between Research and Practice* (ed. by H.A.J. van Lanen & S. Demuth), IAHS Publ. no. 274, 449–456.

Ambramowitz M. & Stegun, I.A. eds. (1965) Section 26.3 In: *Handbook of Mathamatical Functions with Formulas, Graphs, and Mathematical Tables.* Dover Publications, New York.

Anderson, M.G. & Burt, T.P. (1980) Interpretation of recession flow. *J. Hydrol.* **46**, 89–101.

Anderson, M.G., Walling, D.E. & Bates, P.D. (1996) *Floodplain Processes.* Wiley.

Andréassian, V. (2004) Waters and forests: from historical controversy to scientific debate (review). *J. Hydrol.* **291**, 1–27.

Andreini, M., Vlek, P. & Giesen, N. van de (2002) Water sharing in the Volta basin. In: *FRIEND 2002 – Regional Hydrology: Bridging the Gap between Research and Practice* (ed. by H.A.J. van Lanen & S. Demuth), IAHS Publ. no. 274, 329–335.

Annear, T.C. et al. (16 authors) (2002) *Instream Flows for Riverine Stewardship.* Instream flow council, USA.

Appelo, C.A.J. & Postma D. (1993) *Geochemistry, Groundwater and Pollution.* Balkema Publishers, Rotterdam, the Netherlands.

Appleyard, S.J., Davidson, W.A. & Commander, D.P. (1999) The effects of urban development in the utilisation of groundwater resources in Perth, Western Australia. In: *Groundwater in the Urban Environment, Selected City Profiles* (ed. by J. Chilton), Balkema Publishers, Rotterdam, the Netherlands, 97–104.

Arihood, L.D. & Glatfelter, D.R. (1986) Method for Estimating Low Flow Characteristics of Ungauged Streams in Indiana. *U.S. Geol. Survey Open-File Reports.*

Arihood, L.D. & Glatfelter, D.R. (1991) Method for Estimating Low-Flow Characteristics of Ungauged Streams in Indiana. *USGS Water Suppl. Pap.* no. 2372.

Armbruster, J.T. (1976) An infiltration index useful in estimating low-flow characteristics of drainage basins. *J. Res. US Geol. Survey* 4(5), 533–538.

Armbruster, V. (2002) Grundwasserneubildung in Baden-Württemberg (Groundwater recharge in Baden-Württemberg). *Freiburger Schriften zur Hydrologie*, Band 17, Freiburg, Germany.

Armitage, P.D. & Cannan, C.E. (2000) Annual changes in summer patterns of mesohabitat distribution and associated macroinvertebrate assemblages. *Hydrol. Process.* 14, 3161–3179.

Armitage P.D., Pardo I. & Brown A. (1995) Temporal constancy of faunal assemblages in 'mesohabitats' – application to management? *Arch. Hydrobiol.* 133, 367–387.

Armour, C.L. & Taylor, J.G. (1991) Evaluation of the instream flow incremental methodology by U.S. Fish and Wildlife Service field users. *Fisheries* 16(5), 36–43.

Arnell, N.W. (1994) Variations over time in European hydrological behaviour: A spatial perspective. In: *FRIEND – Flow Regimes from International Experimental and Network Data* (ed. by P. Seuna, A. Gustard, N.W. Arnell & G.A. Cole), IAHS Publ. no. 221, 179–184.

Arnell, N.W. (1995) Grid mapping of river discharge. *J. Hydrol.* 167, 39–56.

Arnell, N.W. (1999) Climate change and global water resources. *Global Environ. Chang.* 9, 31–49.

Aronica, G., Candela, A. & Santoro, M. (2002) Changes in the hydrological response of two Sicilian basins affected by fire. In: *FRIEND 2002 – Regional Hydrology: Bridging the Gap between Research and Practice* (ed. by H.A.J. van Lanen & S. Demuth), IAHS Publ. no. 274, 163–169.

Arthington, A.H., King, J.M., O'Keeffe, J.H., Bunn, S.E., Day, J.A., Pusey, B.J., Bluhdorn, B.R. & Tharme, R. (1992) Development of an holistic approach for assessing environmental flow requirements of riverine ecosystems. In: *Water Allocation for the Environment* (ed. by J.J. Pilgram & B.P. Hooper), The Centre for Water Policy Research, University of New England, Armidale, 69–76.

Ashkar, F., Jabi, N.El. & Issa, M. (1998) A bivariate analysis of the volume and duration of low-flow events. *Stoch. Hydrol. Hydraul.* 12, 97–116.

ASTHyDA (2004) *Analysis, Synthesis and Transfer of Knowledge and Tools on Hydrological Drought Assessment through a European Network.* Available from: <http://drought.uio.no>.

Bahrenberg, G. & Giese, E. (1975) *Statistische Methoden und ihre Anwendung in der Geographie* (Statistical methods and their application within Geography). Teubner Studienbücher der Geographie, Stuttgart.

Barnes, S.L. (1973) Mesoscale objective map analysis using weighting time series observations. *NOAA. Tech. Memo.* ERL. NSSL-62. U.S Department of Commerce.

Barry, R.G. & Chorley, R.J. (1992) *Atmosphere, Weather and Climate.* 6th edn, Routledge, London.

Bartholow, J.M. (1989) Stream Temperature Investigations: Field and Analytic Methods. U.S. Fish and Wildlife Service Biological Report 89(17). *Instream flow information paper* no. 13.

Bates, B.C. & Davies, P.K. (1988) Effect of baseflow separation procedures on surface runoff models. *J. Hydrol.* **103**, 309–322.

Bayazit, M. & Önöz, B. (2002) LL-moments for estimating low flow quantiles. *Hydrolog. Sci. J.* **47**(5), 707–720.

Beard, L.R. (1965) Use of interrelated records to simulate streamflow. *J. Hydr. Eng. Div.-ASCE* **87**(HY5).

Beecher, H.A. (1990) Standards for instream flows. *Rivers* **1**, 97–109.

Beersma, J.J., Hurk, B.J.J.M. van der & Können, G.P. (2001) Weer and water in de 21ᵉ eeuw (Weather and water in the 21ᵗʰ century). *Report Royal Netherlands Meteorological Institute*, De Bilt, the Netherlands.

Beisel, J-N., Usseglio-Polatera, P., Thomas, S. & Moreteau, J-C. (2000) The spatial heterogeneity of a river bottom: a key factor determining macroinvertebrate communities. *Hydrobiologia* **442/423**, 163–171.

Bell, M.C. (1986) *Fisheries Handbook of Engineering Requirements and Biological Criteria.* U.S. Corps of Engineers, Portland.

Bell, V.A., Elliott, J.M. & Moore, R.J. (2000) Modelling the effects of drought on the population of brown trout in Black Brows Beck. *Ecol. Model.* **127**, 141–159.

Belore, H.S., Ashfield, D.F. & Singh, S.P. (1990) Regional analysis of low flow characteristics for central and southeastern regions. In: *Proc. Envron. Res. Technol. Trans. Conf. Toronto, Ontario* 1, 258–269.

Beran, M. & Rodier, J.A. (1985) *Hydrological Aspects of Drought.* Studies and Reports in Hydrology no. 39, UNESCO-WMO, Paris.

Berdowski, J., Guicherit, R. & Heij, B. (2001) *The Climate System.* Balkema Publishers, Rotterdam, the Netherlands.

Berger, J.O. (1985) *Statistical Decision Theory and Bayesian Analysis.* 2ⁿᵈ edn, Springer.

Bernier, J. (1964) La prévision statistique des bas débits. In: *Symp. on Surface Waters* (General Assembly of Berkley), IAHS Publ. no. 63, 340–351.

Bernier, J. (1970) Inventaire des modèles de processus stochastiques applicables à la description des debits journaliers des rivières. *Rev. Inst. Int. Stat.* **38**(1), 49–61.

Biggs, B.J.F. & Close, M.E. (1989) Periphyton biomass dynamics in gravel bed rivers: the relative effects of flows and nutrients. *Freshwater Biol.* **22**, 209–231.

Biggs, B.J.F. & Thomsen, A. (1995) Disturbance of stream periphyton by perturbations in shear stress: time to structural failure and differences in community resistance. *J. Phycol.* **31**, 233–241.

Biggs, B.J.F. & Smith, R.A. (2002) Taxonomic richness of stream benthic algae: effects of flood disturbance and nutrients. *Limnol. Oceanogr.* **47**, 1175–1186.

Biggs, B.J.F., Goring, D.G. & Nikora, V.I. (1998a) Subsidy and stress responses of stream periphyton to gradients in water velocity as a function of community growth form. *J. Phycol.* **34**, 598–607.

Biggs, B.J.F., Stevenson, R.J. & Lowe, R.L. (1998b) A habitat matrix conceptual model for stream periphyton. *Arch. Hydrobiol.* **143**, 21–56.

Biggs, B.J.F., Smith, R.A. & Duncan, M.J. (1999) Velocity and sediment disturbance of periphyton in headwater streams: biomass and metabolism. *J. N. Am. Benthol. Soc.* **18**, 222–241.

Bishop, G.D. & Church, M.R. (1992) Automated approaches for regional runoff mapping in the northeastern United States. *J. Hydrol.* **138**, 361–383.

Blöschl G. & Sivapalan M. (1995a) Scale Issues in Hydrologic Modelling: A Review. *Hydrol. Process.* **9**, 251–290.

Blöschl, G. & Sivapalan, M. (1995b) Scale issues in hydrological modelling: a review. In: *Scale Issues in Hydrological Modelling* (ed. by J.D. Kalma & M. Sivapalan), Wiley, Chichester, 9–48.

Blyth, E.M., Dolman, A.J. & Noilhan, J. (1994) The effect of forest on mesoscale rainfall. An example from HAPEX-MOBILHY. *J. Appl. Meteorol.* **33**, 445–454.

Bobée, B. (1975) The log Pearson Type 3 distribution and its application in hydrology. *Water Resour. Res.* **11**(5), 681–689.

Bobée, B. & Robitaille, R. (1975) Correction of bias in the estimation of the coefficient of skewness. *Water Resour. Res.* **11**(6), 851–854.

Bolgov, M.V. (2002) Stochastic models with periodic-correlation of seasonal river runoff variations. In: *Hydrological Models for Environmental Management* (ed. by M.V. Bolgov et al.) Nato Science Series 2, Environmental Security, Vol.79, Kluwer, Dordrecht, the Netherlands, 51–66.

Boorman, D.B. & Hollis, J.M. (1990) *Hydrology of Soil Types: a Hydrologically-based Classification of the Soils of England and Wales*. MAFF conference of river and coastal engineering, Loughborough University, UK.

Boorman, D.B., Hollis, J.M. & Lilly, A. (1995) Hydrology of Soil Types: a Hydrologically-based Classification of the Soils of the United Kingdom. *Institute of Hydrology Report 126*, Institute of Hydrology, Wallingford, UK.

Bosch, J.M. & Hewlett, J.D. (1982) A review of catchment experiments to determine the effect of vegetation changes on water yield and evapotranspiration. *J. Hydrol.* **55**, 3–23.

Boussinesq, J. (1877) Essai sur la theorie des eaux courantes. *Les Mémoires de l'Académie des Sciences Institut Française* **23**(1), 252–260.

Bovee, K.D. (1978) Probability of Use Criteria for the Family Salmonidae. U.S. Fish and Wildlife Service Biological Services Program FWS/OBS-78/07, *Instream Flow Information Paper* no. 4.

Bovee, K.D. (1982) A Guide to Stream Habitat Analysis Using the Instream Flow Incremental Methodology. U.S. Fish and Wildlife Service Biological Services Program FWS/OBS-82/26, *Instream Flow Information Paper* no. 12.

Box, G.E.P. & Cox, D.R. (1964) An analysis of transformations. *J.R. Statist. Soc. B* **26**(211).

Box, G.E.P. & Jenkins, G.M. (1970) *Time Series Analysis: Forecasting and Control*. Holden-Day, San Francisco.

Bradford, R.B. (2000) Drought events in Europe. In: *Drought and Drought Mitigation in Europe* (ed. by J. Vogt & F. Somma), Advances in Natural and Technological Hazards Research, Vol. 14. Kluwer Academic Publisher, Dordrecht, the Netherlands, 7–20.

Bradford, R.B. & Marsh, T.J. (2003) Defining a network of benchmark catchments for the UK. *P. I. Civil Eng. Water & Maritime Engineering* **156**, Issue WM2, 106–116.

Bradley, J.V. (1968) *Distribution-Free Statistical Tests*. Prentice-Hall, New York.

Brecht, R. & Harper, D.M. (2002) Towards an understanding of human impact upon hydrology of Lake Naivasha, Kenya. *Hydrobiologia* **488**, 1–11.

Browne, T.J. (1980) The problem of the spatially representing low stream flow. *Water Services* **84**(1010), 223–226.

Brownlee, K.A. (1960) *Statistical Theory and Methodology in Science and Engineering*. Wiley, New York.

Bruin, H.A.R. de (1987) From Penman to Makkink. In: *Evaporation and Weather* (ed. by J.C. Hooghart), Proc. and Information no. 39, TNO Committee on Hydrological Research, The Hague, the Netherlands, 5–31.

Brutsaert, W. & Nieber, J.L. (1977) Regionalized drought flow hydrographs from a mature glaciated plateau. *Water Resour. Res.* **13**(3), 637–643.

Burn, D.H. (1989) Cluster analysis as applied to regional flood frequency. *J. Water Res. Pl.-ASCE* **115**(5), 567–582.

Burn, D.H. (1990) An appraisal of the "region of influence" approach to flood frequency analysis. *Hydrol. Sci. J.* **35**(24), 149–165.

Burn, D.H. & Goel, N.K. (2000) The formation of groups for regional flood frequency estimation. *Hydrol. Sci. J.* **45**(1), 97–112.

Calder, I.R. (1992) Hydrologic effects of land-use change. In: *Handbook of Hydrology* (ed. by D.R. Maidment), McGraw-Hill, New York, 13.1–13.50.

Calow, P. & Petts, G.E. (1992) *The Rivers Handbook, Volume 1*. Blackwell Science.

Calow, P. & Petts, G.E. (1994) *The Rivers Handbook, Volume 2*. Blackwell Science.

Calow, R., Robins, N., Macdonald, A. & Nicol, A. (1999) Planning for groundwater drought in Africa. In: Proceedings of the International Conference on Integrated Drought Management: Lessons for Sub-Saharan Africa. 20–22 September 1999, Pretoria, South-Africa. IHP-V, *Technical Documents in Hydrology* no. 35, 255–270.

Carter, R.F. (1983) Effects of the Drought of 1980–81 on Stream Flow and on Groundwater Levels in Georgia. *U.S. Geological Survey Water Resources Investigations Report* 83-4158, Renton, Virginia, USA.

Carter, R.F., Hopkins, E.H. & Perlman, H.A. (1988) Low-Flow Profiles of the Tennessee River Tributaries in Georgia. *USGS Water Resources Investigations Report* 88-4049.

Caruso, B.S. (2000) Evaluation of low flow frequency analysis methods. *J. Hydrol. NZ* **39**(1), 19–47.

Champion, R., Lenard, C.T & Mill, T.M. (1996) An introduction to abstract splines. *Mathematical Scientist* **21**, 8–26.

Chang, T.J. & Stenson, J.R. (1990) Is it realistic to define a 100-year drought for water management? *Water Resour. Bull.* **26**(5), 823–829.

Chang, T.J. & Teoh, C.B. (1995) Use of the kriging method for studying characteristics of ground water droughts. *J. Am. Water Res. As.* **31**(6), 1001–1007.

Chapman, D. (1992) *Water Quality Assessments, A Guide to the Use of Biota, Sediments and Water Environmental Monitoring*. Chapman & Hall, London.

Chapman, T.G. (1999) A comparison of algorithms for stream flow recession and base flow separation. *Hydrol. Process.* **13**(5), 701–714.

Chiew, F.H.S. & McMahon, T.A. (2002) Global ENSO-streamflow teleconnection, streamflow forecasting and interannual variability. *Hydrolog. Sci. J.* **47**, 505–522.

Chow, V.T. (1964) *Handbook of Applied Hydrology*. McGraw-Hill, New York.

Chow, V.T., Maidment, D.R. & Mays, L.W. (1988) *Applied Hydrology*. McGraw-Hill, New York.

Chowdhary, H., Singh, R.D. & Kumar, A. (1997) Regional flow duration model for Himalayan region of the Indian state of Uttar Pradesh. In: *Hidroenergia 97, Proc. Fifth Int. Conf. Sept 29–Oct 1 1997*, Dublin, Ireland.

Chung, C. & Salas, J.D. (2000) Drought occurrence probabilities and risks of dependent hydrological processes. *J. Hydrol. Eng.* **5**(3), 259–268.

Clausen, B. & Pearson, C.P. (1995) Regional frequency analysis of annual maximum streamflow drought. *J. Hydrol.* **173**, 111–130.

Clausen, B. & Biggs, B.J.F. (1997) Relationships between benthic biota and hydrological indices in New Zealand streams. *Freshwater Biol.* **38**, 327–342.

Clausen, B. & Biggs, B.J.F. (2000) Flow variables for ecological studies in temperate streams: groupings based on covariance. *J. Hydrol.* **237**, 184–197.

Clausen, B., Young, A.R. & Gustard, A. (1994) Modelling the impact of groundwater abstraction on low river flows. In: *FRIEND – Flow Regimes from International Experimental and*

Network Data (ed. by P. Seuna, A. Gustard, N.W. Arnell & G.A. Cole), IAHS Publ. no. 221, 77–86.

Closs, G.P. & Lake, P.S. (1996) Drought, differential mortality and the co-existence of a native and an introduced fish species in a southeast Australian intermittent stream. *Freshwater Biol.* **14**, 311–316.

Coles, S. (2001) *An Introduction to Statistical Modeling of Extreme Values.* Springer Series in Statistics. Springer-Verlag, London.

Collier, C.G. (1998) Climate and climate change. In: *Encyclopaedia of Hydrology and Water Resources* (ed. by R. W. Hersey & W.F. Rhodes), Kluwer, Dordrecht, the Netherlands, 122–130.

Collier, K.J. & Wakelin, M.D. (1995) *Instream Habitat Use by Blue Duck on Tongariro River.* NIWA Science and Technology series 28, National Institute of Water and Atmospheric Research, Hamilton, New Zealand.

Connell, J.H. (1978) Diversity in tropical rain forests and coral reefs. *Science* **199**, 1302–1309.

COPA-COGECA (2003) Assessment of the Impact of Heat Wave and Drought of the Summer 2003 on Agriculture and Forestry. *Report POCC7616E, CEEC,* Brussels.

Council of the European Communities (2000) Directive 2000/60/EC of the European Parliament and of the Council of 23 October 2000 establishing a framework for Community action in the field of water policy. *Official Journal OJL* 327.

CPC (2002) *Typical Impacts of Warm (El Nino/Southern Oscillation – ENSO) and Cold Episodes.* Available from: <http://www.cpc.ncep.noaa.gov/products/analysis_monitoring/impacts/enso.html> [last visited 2002].

CPC (2003) *Famines Early Warning Systems Network.* Available from: <http://www.fews.net/> [last visited 2003].

Cressie, N.A.C. (1993) *Statistics for Spatial Data.* Wiley, New York.

Cressman G.P. (1959) An operational objective analysis system. *Mon. Weather Rev.* **87**, 367–374.

Croker, K.M., Young, A.R., Zaidman, M.D. & Rees, G. (2003) Flow duration curve estimation in ephemeral catchments in Portugal. *Hydrolog. Sci. J.* **48** (3), 427–439.

CRU (2002) *Climate Monitor.* NAOI, available from: <http://www.cru.uea.ac.uk/cru/climon/data/nao/> [last visited: 2002].

Cruces, J., Bradford, R., Bromley, Estrela, T., Crooks, S. & Martinez Cortina, L. (2000) Contribution of models to the assessment of alternative catchment plans. In: *Groundwater and River Resources Action Programme on a European Scale (GRAPES)* (ed. by M. Acreman), Technical Report, Institute of Hydrology, Wallingford, UK, 215–245.

Cunnane, C. (1978) Unbiased plotting positions – a review. *J. Hydrol.* **37**, 205–222.

Cunnane, C. (1979) A note on the Poisson Assumption in Partial Duration Series Models. *Water Resour. Res.* **15**(2), 489–494.

Cunnane, C. (1985) Factors affecting choice of distribution for flood series. *Hydrolog. Sci. J.* **30**(1), 25–36.

Cunnane, C. (1988) Methods and merits of regional flood frequency analysis. *J. Hydrol.* **100**, 269–290.

Custodio, E. (1997) Groundwater quantity and quality changes related to land and water management around urban areas: Blessings and misfortunes. In: *Groundwater in the Urban Environment. Problems, Processes and Management* (ed. by J. Chilton), Vol. 1, Balkema Publishers, Rotterdam, the Netherlands, 11–22.

Cutforth, H.W., McConkey, B.G., Woodvine, R.J., Smith, D.G., Jefferson, P.G. & Akinremi, O.O. (1999) Climate change in the semiarid prairie of southwestern Saskatchewan: late winter-early spring. *Can. J. Plant Sci.* **79**, 343–350.

Czamara, A. (1998) *Influence of Chosen Drainage Systems on Groundwater Resources.* Zeszyty Naukowe Akademii Rolniczej w Wroclaw, no. 340, Poland.

Czamara, W., Jakubowski, W. & Radczuk, L. (1997) Probabilistic Analysis of extreme low flows in selected catchments in Poland. In: *FRIEND'97 – Regional Hydrology: Concepts and Models for Sustainable Water Resource Management* (ed. by A. Gustard, S. Blazkova, M. Brilly, S. Demuth, J. Dixon, H. van Lanen, C. Llasat, S. Mkhandi & E. Servat), IAHS Publ. no. 246, 159–168.

Dalrymple, T. (1960) Flood frequency analyses. *USGS Water Supply Paper* 1543A, 11–51.

Davie, T. (2003) *Fundamentals of Hydrology*. Routledge Fundamentals of Physical Geography, London, New York.

Davis, J.C. (1973) *Statistics and Data Analysis in Geology*. Wiley, New York.

Davis, S.N. & Wiest, R.J. de (1966) *Hydrogeology*. Wiley.

Day, J.B.W. & Rodda, J.C. (1978) The effects of the 1975–76 drought on groundwater and aquifers. *Proc. R. Soc. London Ser.-A* 363, 55–68.

deMenocal, P.B. (2001) Cultural response to climate change during the late Holocene. *Science*, 292, 667–673.

Demuth, S. (1993) Untersuchungen zum Niedrigwasser in West-Europa (European low flow study). *Freiburger Schriften zur Hydrologie*, Band 1, Freiburg, Germany.

Demuth, S. & Hagemann, I. (1993) Case Study of regionalizing base flow in SW Germany applying a hydrogeological index. In: *Flow Regimes from International Experimental and Network Data (FRIEND), Vol. I Hydrological Studies* (ed. by A. Gustard), Institute of Hydrology, Wallingford, UK, 86–98.

Demuth, S. & Heinrich, B. (1997) Temporal and spatial behavior of drought in south Germany. In: *FRIEND'97 – Regional Hydrology: Concepts and Models for Sustainable Water Resource Management* (ed. by A. Gustard, S. Blazkova, M. Brilly, S. Demuth, J. Dixon, H. van Lanen, C. Llasat, S. Mkhandi & E. Servat), IAHS Publ. no. 246, 151–157.

Demuth, S. & Stahl, K. (2001) *ARIDE – Assessment of the Regional Impact of Droughts in Europe*. Final Report, EU contract ENV4-CT-97-0553, Institute of Hydrology, University of Freiburg, Freiburg, Germany.

Demuth, S., Lehner, B. & Stahl, K. (2000) Assessment of the vulnerability of a river system to drought. In: *Drought and Drought Mitigation in Europe* (ed. by J.V. Vogt & F. Somma), Natural and Technological Hazards Series, Kluwer Academic Publishers, Dordrecht, the Netherlands, 209–219.

Dettinger, M.D. & Diaz, H.F. (2000) Global characteristics of stream flow seasonality and variability. *J. Hydrometeorol.* 1(3), 289–310.

Deutsches IHP/OHP-Nationalkomittee (2003) *Hydrological Networks for Integrated and Sustainable Water Resources Management*. International Workshop, Koblenz, Germany, 22–23 October 2003.

DHI Water & Environment (2003) *EVA – Extreme Value Analysis*. Reference manual, DHI Software 2003, Danmark.

Dingman, S.L. (2002) *Physical Hydrology*. 2nd edn, Prentice Hall, Upper Saddle River, NJ, USA.

Dingman S.L, Seely-Reynolds, D.M & Reynolds, R.C. (1988) Application of kriging to estimate mean annual precipitation in a region of orographic influence. *Water Resour. Bull.* 24, 329–339.

Dolman, A.J., Verhagen, A. & Rovers, C.A. eds (2003) *Global Environmental Change and Land Use*. Kluwer Academic Publishers, Dordrecht, the Netherlands.

Domokos, M. & Sass, J. (1990) Long-term water balance for sub-catchments and partial national areas in the Danube basin. *J. Hydrol.* 112, 267–292.

Downing, R.A., Price, M. & Jones, G.P. (1993) *The Hydrogeology of the Chalk of North-West Europe*. Clarendon Press, Oxford, UK.

Dracup, J.A., Lee, K.S. & Paulson, E.G. (1980) On the definition of droughts. *Water Resour. Res.* 16(2), 297–302.

Draper, N.R. & Smith, H. (1998) *Applied Regression Analysis.* 2nd edn, Wiley, New York.

Drayton, R.S., Kidd, C.H.R., Mandeville, A.N. & Miller, J.B. (1980) A regional analysis of floods and low flows in Malawi. *Institute of Hydrology Report* no. 72, Wallingford, UK.

Dregne, H.E. (2000) Drought and desertification. In: *DROUGHT, A Global Assessment,* Vol II (ed. by D.A. Wilhite), Routledge Hazards and Disasters Series, Routledge, London, 231–240.

Dreher, J.E., Pramberger, F. & Rezabek, H. (1985) Faktorenanalyse – eine Möglichkeit zur Ermittlung hydrographisch ähnlicher Bereiche in einem Grundwassergebiet (Factor analysis – a possibility for the derivation of hydrographically similar regions within a groundwater catchment). *Mitteilungsblatt des Hydrographischen Dienstes in Österreich* **54**, 1–12.

Dunbar, M.J., Gustard, A., Acreman, M.C. & Elliot, C.R.N. (1998) Overseas approaches to setting river flow objectives. *R&D Technical Report* W6-161. Environment Agency and Institute of Hydrology, Wallingford, UK.

Duncan, M.J., Suren, A.M. & Brown, S.L.R. (1999) Assessment of streambed stability in steep, bouldery streams: development of a new analytical technique. *J. N. Am. Benthol. Soc.* **18**, 445–456.

Dvorak, V., Hladny, J. &. Kašpárek, L. (1997) Climate change hydrology and water resources impact and adaptation for selected river Basins in the Czech Republic. *Climatic Change* **36**, 93–106.

DVWK (1983) Niedrigwasseranalyse, Teil 1: Statistische Analyse des Niedrigwasserabflusses (Low flow analysis, part 1: statistical analysis of low flows). *DVWK – Regeln zur Wasserwirtschaft,* Heft 120, Germany.

DVWK (1996) Ermittlung der Verdunstung von Land- und Wasserflächen (Derivation of evaporation from land- and water surfaces). *DVWK – Regeln zur Wasserwirtschaft,* Heft 238, Germany.

Dyck, S. (1976) *Angewandte Hydrologie Teil I. Berechnung und Regelung des Durchflusses der Flüsse* (Applied hydrology, part I. Calculation and regulation of the discharge of streams). VEB Verlag für Bauwesen, Berlin.

Dyurgerov, M.B. & Meier, M.F. (1997) Mass balance of mountain and sub-polar glaciers: A new global assessment for 1961–1990. *Arctic Alpine Res.* **29**(4), 379–391.

Ebraheem, A.M., Garamoon, H.K., Riad, S., Wycisk, P. & Seif el Nasr, A.M. (2003) Numerical modelling of groundwater resource management in the East Oweinat area, southwest Egypt. In: *Hydrology of Mediterranean and Semiarid Regions* (ed. by E. Servat, W. Najem, C. Leduc & A. Shakeel), IAHS Publ. no. 278, 15–23.

Eckhardt, K. & Ulbrich, U. (2003) Potential impacts of climate change on groundwater recharge and streamflow in a central European low mountain range. *J. Hydrol.* **284**, 244–252.

EEA (2001) *Sustainable Water Use in Europe. Part 3: Extreme Hydrological Events: Floods and Droughts.* European Environment Agency, Copenhagen.

Eltahir, E.A.B. & Yeh, P. J.-F. (1999) On the asymmetric response of aquifer water level to floods and droughts in Illinois. *Water Resour. Res.* **35**(4), 1199–1217.

Embrechts, P., Klüppelberg, C. & Mikosch, T. (1997) *Modelling Extremal Events for Insurance and Finance.* Springer-Verlag Berlin.

Environment Agency (2002a) Technical Sheet 8 – Flow Accretion Curves. In: A review of techniques for applied hydrology in low flow investigations, *Technical Report* W6-057/TR, Bristol, UK.

Environment Agency (2002b) *Managing Water Abstraction: the Catchment Abstraction Management Strategy Process.* Environment Agency, Bristol, UK.

Erichsen, B. & Tallaksen, L.M. (1995) Sammenlikning av ulike lavvannsmål i 47 norske nedbørfelt (Comparison of different low flow indices in 47 Norwegian catchments). *Hydro Nova Report,* Norway.

Ernst, L.F. (1978) Drainage of undulating sandy soils with high groundwater tables. *J. Hydrol.* **39**, 1–50.

ESHA (1994) *Layman's Guidebook on how to Develop a Small Hydro Site*. Commission of the European Communities, Directorate General for Energy (DG-XVII), Brussels.

EU-MEDIN (2003) *Summary of Disaster News*. Available from: <http://www.eu-medin.org> [last visited August 2003].

FAO (1997) *SD: Environment: Global Climate Maps*. Available from: <http://www.fao.org/waicent/faoinfo/sustdev/EIdirect/climate/EIsp0002.htm> [last visited 2003].

Favreau, G., Leduc, C., Marlin, C., Dray, M., Taupin, J.D., Massault, M., Le Gal la Salle, C. & Babic, M. (2002) Estimate of recharge of a rising water table in semi-arid Niger from ^{3}H and ^{14}C modeling. *Ground Water* **40**, 144–151.

Feddes, R.A. (1987) Crop factors in relation to Makkink reference-crop evapotranspiration. In: *Evaporation and Weather* (ed. by J.C. Hooghart), Proc. and Information no. 39, TNO Committee on Hydrological Research, The Hague, the Netherlands, 33–45.

Federer, C.A. (1973) Forest transpiration greatly speeds streamflow recession. *Water Resour. Res.* **9**(6), 1599–1604.

Feijtel, T.C.J., Boeije, G., Matthies, M., Young, A.R., Morris, G., Gandolfi, C., Hansen, B., Fox, K., Holt, M., Koch, V., Schroeder, R., Cassani, G., Schowanek, D. & Rosenblom, J. (1997) Development of a Geography-referenced Regional Exposure Assessment Tool for European Rivers – GREAT-ER. Contribution to GREAT-ER, 1. *Chemosphere* **34**, 2351–2373.

Fiering, M.B. (1961) Queuing theory and simulation in reservoir design. *J. Hydr. Eng. Div.-ASCE* **87**(HY6).

Finke, W. & Dornblut, I. (1998) Statistische Niedrigwasseranalysen augewählter Pegel der Elbe (Statistical low-flow analyses of selected station data from the River Elbe). *Deutsche Gewässerkundliche Mitteilungen* **42**(5), 186–195.

Fisher, R.A. & Tippett, L.H.C. (1928) Limiting forms of the frequency distribution of the largest or smallest member of a sample. *Proc. Cambridge Phil. Soc.* **24**(2), 180–190.

Fisher, S.G., Gray, L.J., Grimm, N.B. & Bush, D.E. (1982) Temporal succession in a desert stream ecosystem following flash flooding. *Ecol. Monogr.* **52**, 93–110.

Foster, I., Gurnell, A. & Webb, B. (1995) *Sediment and Water Quality in River Catchments*. Wiley.

Foster, S., Morris, B., Lawrence, A. & Chilton, J. (1999) Groundwater impacts and issues in developing cities – An introductory review. In: *Groundwater in the Urban Environment. Selected City Profiles* (ed. by J. Chilton), Balkema, Rotterdam, the Netherlands, 3–16.

Fowler, K.K. (1992) Description and effects of the 1988 drought on groundwater levels, stream flow, and reservoir levels in Indiana. *U.S. Geological Survey Water Resources Investigations Report* 91–4100, Renton, Virginia, USA.

Fréchet, M. (1927) Sur la loi de probabilité de l'écart maximum. *Ann. Soc. Polonaise de Mathématique (Cracow)* **6**, 93–117.

Freeze, R.A. & Cherry, J.A. (1979) *Groundwater*. Prentice Hall, London.

Friel, E.A., Embree, W.N., Jack, A.R. & Atkins, J.T. (1989) Low flow characteristics of streams in West Virginia. *U.S. Geological Survey Water Resources Investigation Report* 88–4072, Renton, Virginia, USA.

Fuchs, L. & Rubach, H. (1983) Niedrigwasseranalyse unter besonderer Berücksichtigung einer regionalen Aussage (Low flow analysis with special emphasis on regional results). *Wasser & Boden* **1**, 13–17.

Furey, P.R. & Gupta, V.K. (2001) A physically based filter for seperating base flow from streamflow time series. *Water Resour. Res.* **37**(11), 2709–2722.

Gellens, D. & Roulin, E. (1998) Streamflow response of Belgian catchments to IPCC climate change scenarios. *J. Hydrol.* **210**, 242–258.

Geoghegan, I. (2003) *Hungary's Shrinking Balaton Fuels Climate Change Fears.* Reuter, available from: http://www.enn.com/news/2003-09-03/s_7986.asp [last visited November 2003].

Giller, P.S. & Malmqvist, B. (1998) *The Biology of Streams and Rivers.* Oxford University Press.

Glos, E. & Lauterbach, D. (1972) Regionale Verallgemeinerung von Niedrigwasserdurchflüssen mit Wahrscheinlichkeitsaussage (Regional generalisation of low flows with declarations of probability). *Mitteilungen des Institutes für Wassserwirtschaft,* Heft 37, VEB Verlag für Bauwesen, Berlin.

Goddard, G.C. (1963) Water supply characteristics of North-Carolina streams. *Geological Survey Water Supply Paper* 1761.

Good, P. (1994) *Permutation Tests, a Practical Guide to Resampling Methods for Testing Hypothesis.* Springer-Verlag, New York.

Gordon, N.D., McMahon, T.A. & Finlayson, B.L. (1992) *Stream hydrology (an Introduction for Ecologists).* Wiley.

Gottschalk, L. (1975) Stochastic modelling of monthly river runoff. *Bulletin Series A* no. 45, Department of Water Resources Engineering, Lund Institute of Technology, Lund, Sweden.

Gottschalk, L. (1976) Frequency of dry years. In: *Proceedings Nordic Hydrological Conference, 1976, Reykjavik, part II,* 75–86.

Gottschalk, L. (1985) Hydrological regionalization of Sweden. *Hydrolog. Sci. J.* **30**(1), 65–83.

Gottschalk, L. & Perzyna, G. (1993) Low flow distribution along a river. In: *Extreme Hydrological Events: Precipitation, Floods and Droughts* (ed. by Z.W. Kundzewicz, D. Rosbjerg, S.P. Simonovic & K. Takeuchi) (Proc. Yokohama Symp. July, 1993), IAHS Publ. no. 213, 33–41.

Gottschalk, L. & Krasovskaia, I. (1998) Grid Estimation of Runoff Data. Report of the WCP-Water Project B.3: Development of Grid-related Estimates of Hydrological Variables, WCASP-46, *WMO-TD* no. 870, WMO, Geneva, Switzerland.

Gottschalk, L., Tallaksen, L.M. & Perzyna, G. (1997) Derivation of low flow distribution functions using recession curves. *J. Hydrol.* **194**, 239–262.

Gottschalk, L., Beldring, S., Engeland, K., Tallaksen, L.M., Sælthun, N.R. & Kolberg, S. (2001) Regional/macroscale hydrological modelling: a Scandinavian experience. *Hydrolog. Sci. J.* **46**(6), 963–982.

Gould, J.L. & Keeton, W.T. (1996) *Biological Science.* 6th edn, Norton & Company.

Gray, L.J. (1981) Species composition and life histories of aquatic insects in a lowland Sonoran Desert stream. *Am. Midl. Nat.* **106**, 229–242.

Greatbach, R.J. (2000) The North Atlantic Oscillation. *Stoch. Env. Res. Risk A.* **14**, 213–242.

Greenwood, J.A., Landwehr, J.M., Matalas, N.C. & Wallis, J.R. (1979) Probability weighted moments: Definition and relation to parameters of several distributions expressible in inverse form. *Water Resour. Res.* **15**, 1049–54.

Gregory, J.M., Mitchell, J.F.B. & Brady, A.J. (1997) Summer drought in northern midlatitudes in a time dependent CO_2 climate experiment. *J. Climate* **10**, 662–686.

Greis, N.P. & Wood, E.F. (1981) Regional flood frequency estimation and network design. *Water Resour. Res.* **17**(4), 1167–1177. Correction, *Water Resour. Res.* **19**(2), 589–590, 1983.

Grove, A. (1998) Variability of African River Discharges and Lake Levels. In: *Tropical Climatology, Meteorology and Hydrology* (ed. by G. Demarée, J. Alexandre & M. de Dapper), Royal Meteorological Institute of Belgium, Brussels.

Guensch, G.R., Hardy, T.B. & Addley, R.C. (2001) Examining feeding strategies and position choice of drift-feeding salmonids using an individual-based, mechanistic foraging model. *Can. J. Fish. Aquat. Sci.* **58**, 446–457.

Gumbel, E.J. (1954) Statistical theory of droughts. *J. Hydraul. Eng.-ASCE* **439**(HY), 1–19.

Gumbel, E.J. (1958) *Statistics of Extremes.* Columbia Univ. Press, New York.

Gustard, A. ed. (1993) *Flow Regimes from International Experimental and Network Data (FRIEND)*, 3 volumes. Institute of Hydrology, Wallingford, UK.

Gustard, A. & Gross, R. (1989) Low Flow Regimes of Northern and Western Europe. In: *FRIENDS in Hydrology* (ed. by L. Roald, K. Nordseth & K.A. Hassel), IAHS Publ. no. 187, 205–213.

Gustard, A., Bullock, A. & Dixon, J.M. (1992) Low flow estimation in the United Kingdom. *Institute of Hydrology Report* no. 108, Wallingford, UK.

Gustard, A. & Cole, G. eds. (2002) *FRIEND – A Global Perspective 1998–2002*. Centre for Ecology and Hydrology (CEH), Wallingford, UK.

Gustard, A., Roald, L.A., Demuth, S., Lumadjeng, H.S. & Gross, R. (1989) *Flow Regimes From Experimental and Network Data (FREND), 2 volumes, Vol. I Hydrological Studies*. UNESCO, IHP III, Project 6.1, Institute of Hydrology, Wallingford, UK.

Haan, C.T. (1977) *Statistical Methods in Hydrology*. The Iowa State University Press, Ames, USA.

HAD (Hydrologischer Atlas von Deutschland) (2001) Atlastafel 3.5 Mittlere jährliche Abflusshöhe (Map 3.5: mean annual total runoff). Bundesanstalt für Umwelt, Naturschutz und Reaktorsicherheit, Germany.

Halford, K.J. & Mayer, G.C. (2000) Problems associated with estimating ground water discharge and recharge from stream-discharge records. *Ground Water*, 38(3), 331–342.

Hall, F.R. (1968) Base flow recessions – a review. *Water Resour. Res.* 4(5), 973–983.

Hansen, E. (1971) *Analyse af hydrologiske tidsserier* (Analysis of Hydrological Time Series). Laboratoriet for Hydraulik, Danmarks Tekniske Høyskole, Polyteknisk Forlag, Lyngby, Danmark.

Harper, D.M., Boar, R., Everard, M. & Hickley P. (2003) *Lake Naivasha, Kenya*. Kluwer Academic Publishers, Dordrecht, the Netherlands.

Harper, D.M & Ferguson, A.J.D. (1995) *The Ecological Basis for River Management*. Wiley.

Hasnain, S.I. (2002) Himalayan glaciers meltdown: impact on south Asian rivers. In: *FRIEND 2002 – Regional Hydrology: Bridging the Gap between Research and Practice* (ed. by H.A.J. van Lanen & S. Demuth), IAHS Publ. no. 274, 417–423.

Hayes, D.C. (1990) Low flow characteristics of streams in Virginia. *USGS Open-File Report* 89–586.

Hayes, J.W. (1995) Spatial and temporal variation in the relative density and size of juvenile brown trout in the Kakanui River, North Otago, New Zealand. *New Zeal. J. Mar. Fresh.* 29, 393–408.

Hayes, J.W. & Jowett, I.G. (1994) Microhabitat models of large drift-feeding brown trout in three New Zealand rivers. *N. Am. J. Fish. Manage.* 14, 710–725.

Hayes, M.J. (1999) *What is Drought? Drought Indices*. National Drought Migation Center (NDMC), available from: <http://www.drought.unl.edu/whatis/indices.htm> [last visited March 2004].

Hayes, M.J. (2002) *What is Drought? Drought and Climate Change*. National Drought Migation Center (NDMC), available from: <http://www.drought.unl.edu/whatis/cchange.htm> [last visited 2002].

Heddinghaus, T.R. & Sabol, P. (1991) A review of the Palmer Drought Severity Index and where do we go from here? In: *Proc. 7th Conf. on Applied Climatology*, September 10–13, 1991, American Meteorological Society, Boston, 242–246.

Heggenes, J. (1988) Physical habitat selection by brown trout (Salmo trutta) in riverine systems. *Nordic Journal of Freshwater Research* 64, 74–90.

Heggenes, J. (1996) Habitat selection by brown trout (Salmo trutta) and young atlantic salmon (S. salar) in streams: static and dynamic hydraulic modelling. *Regul. River.* 12, 155–169.

Helsel, D.R. & Hirsch, R.M. (1992) *Statistical Methods in Water Resources*. Studies in Environmental Sciences 49, Elsevier.

Henriques, A.G. & Santos M.J.J. (1999) Regional drought distribution model. *Phys. Chem. Earth Pt. B* **24**(1/2), 19–22.

Herschy, R.W. (1995) *Streamflow Measurement*. 2nd edn, Chapman & Hall, London.

Herschy, R.W. & Fairbridge, R.W. (1998) *Encyclopedia of Hydrology and Water Resources*. Encyclopedia of Earth Sciences Series, Kluwer Academic Publishers, Dordrecht, the Netherlands.

Hewlett, J.D. & Hibbert, A.R. (1967) Factors affecting the response of small watersheds to precipitation in humid areas. In: *Int. Symp. on Forest Hydrology* (ed. by W.E. Sopper & H.W. Lull), Pergamon, Oxford, UK, 275–290.

Hibbert, A.R. (1967) Forest treatment effects on water yield. In: *Forest Hydrology* (ed. by W.E. Sopper & H.W. Lull), Pergamon Press, Oxford, UK, 536–538.

Hicks, D.M. & Davies, T. (1997) Erosion and sedimentation in extreme events. (ed. by M.P. Mosley), New Zealand Hydrological Society, Wellington, New Zealand, 117–141.

Hipel, K.W. & McLeod, A.I. (1994) *Time Series Modelling of Water Resources and Environmental Systems*. Elsevier, Amsterdam.

Hirschboeck, K.K. (1988) Flood Hydroclimatology. In: *Flood Geomorphology* (ed. by V.R. Baker, R.C. Kochel, & P.C. Patton), Wiley, 27–49.

Hisdal, H. (2002) Regional aspects of drought. PhD Thesis Faculty of Mathematics and Natural Scienes, University of Oslo, No. 221, Oslo.

Hisdal, H. & Tallaksen, L.M. (2002) Handling non-extreme events in extreme value modelling of streamflow drought. In: *FRIEND 2002 – Regional Hydrology: Bridging the Gap between Research and Practice* (ed. by H.A.J. van Lanen & S. Demuth), IAHS Publ. no. 274, 281–288.

Hisdal, H. & Tallaksen, L.M. (2003) Estimation of regional meteorological and hydrological drought characteristics. *J. Hydrol.* **281**(3), 230–247.

Hisdal, H., Erup, J., Gudmundsson, K., Hiltunen, T., Jutman, T., Ovesen, N.B. & Roald, L.A. (1995) Historical runoff variations in the nordic countries. *NHP Report* 37, Norwegian Hydrological Council, Oslo.

Hisdal, H., Stahl, K., Tallaksen, L.M. & Demuth, S. (2001) Have streamflow droughts in Europe become more severe or frequent? *Int. J. Climatol.* **21**, 317–333.

Hoeffding, W. (1948) A class of statistics with asymptotically Normal distribution. *Ann. Math. Stat.* **19**, 293–325.

Hoerlin, M. & Kumar, A. (2003) The perfect ocean for drought. *Science* **299**, 691–694.

Holder, R.L. (1985) *Multiple Regression in Hydrology*. Institute of Hydrology, Wallingford, UK.

Holko, L., Herrmann, A., Uhlenbrook, S., Pfister, L. & Querner, E.P. (2002) Groundwater runoff separation – test of applicability of a simple separation method under varying natural conditions. In: *FRIEND 2002 – Regional Hydrology: Bridging the Gap between Research and Practice* (ed. by H.A.J. van Lanen & S. Demuth), IAHS Publ. no. 274, 101–108.

Holmes, M.G.R. & Young, A.R. (2002) Estimating seasonal low flow statistics in ungauged catchments. In: *Proc. 8th National Hydrology Symposium – Birmingham*, British Hydrological Society, 97–102.

Holmes, M.G.R., Young, A.R., Gustard, A. & Grew, R. (2002a) A new approach to estimating mean flow in the United Kingdom. *Hydrol. Earth Syst. Sc.* **6**(4), 709–720.

Holmes, M.G.R., Young, A.R., Gustard, A. & Grew, R. (2002b) A region of influence approach to predicting Flow Duration Curves within ungauged catchments. *Hydrol. Earth Syst. Sc.* **6**(4), 721–731.

Hoopes, R.L. (1974) Flooding, as the result of Hurricane Agnes, and its effect on a macrobenthic community in an infertile headwater stream in central Pennsylvania. *Limnol. Oceanogr.* **19**, 853–857.

Hornberger, G.M., Ebert, J. & Remson, I. (1970) Numerical solution of the Boussinesq equation for aquifer-stream interaction. *Water Resour. Res.* **6**(2), 601–608.

Hosking, J.R.M (1985) Algorithm AS215: Maximum-likelihood estimation of the parameters of the generalized extreme-value distribution. *Appl. Stat.-J. Roy. St. C* **34**, 301–310.

Hosking, J.R.M. (1990) L-moments: Analysis and estimation of distributions using linear combinations of order statistics. *J. Roy. Stat. Soc. B* **52**(1), 105–124.

Hosking, J.R.M. (1991) Fortran routines for use with the method of L-moments. *Research Report* RC17097, IBM Research Division, Yorktown Heights, New York.

Hosking, J.R.M. & Wallis, J.R. (1987) Parameter and quantile estimation for the generalized Pareto distribution. *Technometrics* **29**(3), 339–349.

Hosking, J.R.M. & Wallis, J.R. (1988) The effect of intersite dependence on regional flood frequency analysis. *Water Resour. Res.* **24**(4), 588–600.

Hosking, J.R.M. & Wallis, J.R. (1993) Some statistics useful in regional frequency analysis. *Water Resour. Res.* **29**(2), 271–281. Correction, *Water Resour. Res.* **31**(1), 251, 1995.

Hosking, J.R.M. & Wallis, J.R. (1995) A comparison of unbiased and plotting-position estimators of L moments. *Water Resour. Res.* **31**(8), 2019–2025.

Hosking, J.R.M. & Wallis, J.R. (1997) *Regional Frequency Analysis, an Approach Based on L-moments.* Cambridge University Press, UK.

Hosking, J.R.M., Wallis, J.R. & Wood, E.F. (1985) An appraisal of the regional flood frequency procedure in the UK Flood Studies Report. *Hydrolog. Sci. J.* **30**(1), 85–109.

HRH Prince of Orange (2003) International Year of Freshwater: Water is everybody's business. *Nat. Resour. Forum* **27**, 87–88.

Hubbard, E.F., Landwehr, J.M. & Barker, A.R. (1997) Temporal variability in the hydrologic regimes of the United States. In: *FRIEND'97 – Regional Hydrology: Concepts and Models for Sustainable Water Resource Management* (ed. by A. Gustard, S. Blazkova, M. Brilly, S. Demuth, J. Dixon, H. van Lanen, C. Llasat, S. Mkhandi & E. Servat), IAHS Publ. no. 246, 97–103.

Hughes, M.K. & Brown, P.M. (1992) Drought frequency in central California since 101 B.C. recorded in giant sequoia tree rings. *Clim. Dynam.* **6**, 161–167.

Hurst, H.E. (1951) Long term storage capacity of reservoirs. *T. Am. Soc. Civ. Eng.* **116**.

Hutchinson, P.D. (1993) Calculation of a base flow index for New Zealand catchments. *Report* WS818, Ministry of Works and Development (now NIWA), Christchurch, New Zealand.

Hutjes, R.W.A., Kabat, P. & Dolman, A.J. (2003) Land cover and the climate system. In: *Global Environmental Change and Land Use* (ed. by A.J. Dolman, A. Verhagen & C.A. Rovers), Kluwer Academic Publishers, Dordrecht, the Netherlands, 73–110.

Hynes, H.B.N. (1970) *The Ecology of Running Waters.* University of Toronto Press.

Iman, R.L. & Conover, W.L. (1983) *A Modern Approach to Statistics.* Wiley, New York

Institute of Hydrology (1980) *Low Flows Studies Report,* 3 volumes. Institute of Hydrology, Wallingford, UK.

IPCC (2001a) *Climate Change 2001: The Scientific Basis.* IPCC Third Assessment Report, Intergovernmental Panel of Climate Change (IPCC), UNEP, WMO, Cambridge University Press, Cambridge, UK.

IPCC (2001b) *Climate Change 2001: Impacts, Adaptation and Vulnerability.* IPCC Third Assessment Report, Intergovernmental Panel of Climate Change (IPCC), UNEP, WMO, Cambridge University Press, Cambridge, UK.

Ishihara, T. & Takase, N. (1957) The logarithmic normal distribution and its solution based on moment method (orig. in Japanese). *Trans. JSCE* **47**, 18–23.

Iwai, S. (1947) On the asymmetric distribution in hydrology. Collection of Treaties. *J. Civil Eng. Soc.* **2**, 93–116 (orig. in Japanese).

James, L.D. & Thompson, W.O. (1970) Least squares estimation of constants in a linear recession model. *Water Resour. Res.* **6**(4), 1062–1069.

Jenkinson, A.F. (1955) The frequency distribution of the annual maximum (or minimum) values of meteorological elements. *Quat. J. Roy. Meteor. Soc.* **81**, 158–171.

Johns, T.C. (1996) A description of the second Hadley centre coupled model (HadCM2). *Climate Research Technical Note* 71, Hadley Centre, United Kingdom Meteorological Office, Bracknell, UK.

Jones, P.D., Jonsson, T. & Wheeler, D. (1997) Extension to the North Atlantic Oscillation using early instrumental pressure observations from Gibraltar and SW Iceland. *Int. J. Climatol.* **17**, 1433–1450.

Jorde, K. (1997) Ökologisch begründete, dynamische Mindestwasserregelungen bei Ausleitungs-kraftwerken. (Ecologically based, dynamic regulations for minimum flows at water power stations). *Mitteilungen des Instituts für Wasserbau*, Heft 90, Universität Stuttgart, Stuttgart, Germany.

Jousma, G. & Roelofsen, F.J. (2003) Inventory of existing guidelines and protocols for groundwater assessment and monitoring. *IGRAC report*, International Groundwater Resources Assessment Centre, Utrecht, the Netherlands.

Jowett, I.G. (1982) The incremental approach to studying stream flows: New Zealand case studies. In: *River Low Flows: Conflicts of Water Use* (ed. by R.H.S. McColl). Ministry of Works and Development, Water and Soil Miscellaneous Publication 47, New Zealand, 9–15.

Jowett, I.G. (1989) River hydraulic and habitat simulation, RHYHABSIM computer manual. *New Zealand Fisheries Miscellaneous Report* 49, Ministry of Agriculture and Fisheries, Christchurch, New Zealand.

Jowett, I.G. (1992) Models of the abundance of large brown trout in New Zealand rivers. *N. Am. J. Fish. Manage.* **12**, 417–432.

Jowett, I.G. (1993) Minimum flow requirements for instream habitat in the Waiau River, Southland, from the Mararoa Weir to the Borland Burn. Report to the Waiau Working Party, *NIWA miscellaneous report* 46.

Jowett, I.G. (1995) Spatial and temporal variability in brown trout abundance: a test of regression models. *Rivers* **5**, 1–12.

Jowett, I.G. (1997a) Environmental effects of extreme flows. In: *Floods and Droughts: the New Zealand Experience* (ed. by M.P. Mosley & Ch.P. Pearson), New Zealand Hydrological Society Inc., Wellington North, New Zealand, 103–116.

Jowett, I.G. (1997b) Instream flow methods: a comparison of approaches. *Regul. River.* **13**, 115–127.

Jowett, I.G. (1999) *WAIORA – Interactive Technical Support for Water Allocation Decisions.* Proceedings of the 3rd International Ecohydraulics Conference, Salt Lake City, Utah, IAHR.

Jowett, I.G. (2000) Flow management. In: *New Zealand stream invertebrates: Ecology and Implications for Management* (ed. by K.J. Collier & K.J. Winterbourn), New Zealand Limnological Society, Hamilton, New Zealand, 289–312.

Jowett, I.G. & Richardson, J. (1995) Habitat preferences of common, riverine New Zealand native fishes and implications for flow management. *New Zeal. J. Mar. Fresh.* **29**, 13–23.

Jowett, I.G., Richardson, J. & McDowall, R.M. (1996) Relative effects of in-stream habitat and land use on fish distribution and abundance in tributaries of the Grey River, New Zealand. *New Zeal. J. Mar. Fresh* **30**, 463–475.

Kachroo, R.K. (1992) Storage required to augment low flows: a regional study. *Hydrol. Sci. J.* **37**, 247–261.

Kadoya (1962) On the applicable ranges and parameters of logarithmic normal distributions of the Slade type. Nougyou Doboku Kenkyuu. *Extra Publication* **3**, 12–27 (orig. in Japanese).

Kamp, G. van der, Hayashi, M. & Gallén, D. (2003) Comparing the hydrology of grassed and cultivated catchments in the semi-arid Canadian prairies. *Hydrol. Process.* **17**, 559–575.

Kartvelishvili, N.A. (1981) *Stochasticheskaga Gidrologiya* (Stochastic Hydrology). Gidrometeoizdat, Leningrad, USSR.

Kartvelishvili, N.A. & Gottschalk, L. (1976) Multivariate distributions in hydrology and river regulation. *Nord. Hydrol.* **7**, 265–280.

Kašpárek, L. (1998) Regional study of the impacts of climate change on hydrological conditions in the Czech Republic. *Report T.G. Masaryk Water Research Institute*, Prague.

Kašpárek, L. & Novický, O. (1997) Application of a physically-based model to identify factors causing hydrological droughts in western and central European basins. In: *FRIEND'97 – Regional Hydrology: Concepts and Models for Sustainable Water Resource Management* (ed. by A. Gustard, S. Blazkova, M. Brilly, S. Demuth, J. Dixon, H. van Lanen, C. Llasat, S. Mkhandi & E. Servat), IAHS Publ. no. 246, 197–204.

Kendall, M.G. (1975) *Rank Correlation Methods*. 4th edn, Charles Griffin, London.

Kendall, C. & McDonell, J.J. (1998) *Isotope Tracers in Catchment Hydrology*. Elsevier,

Khristoforov, A.V. & Samborski, T.V. (2001) A stochastic model of flood flow fluctuations. In: *Hydrological Models for Environmental Management* (ed. by M. Bolgov, L. Gottschalk, I. Krasovskaia & R.J. Moore), NATO Science Series. Kluwer Academic Publishers, Dordrecht, the Netherlands.

Killingtveit, Å. & Harby, A. (1994) Multi-purpose planning with the river system simulator – a decision support system for water resources planning and operation. In: *Proceedings of the first international symposium on habitat hydraulics,* Norwegian Institute of Technology, Trondheim, Norway.

Kim, T.-W., Valdés, J.B. & Yoo, C. (2003) Nonparametric approach for estimating return periods of droughts in arid regions. *J. Hydrolog. Eng.* **8**(5), 237–246.

King, J.M., Tharme, R.E. & deVilliers, M.S. (2001) *Environmental Flow Assessments for Rivers: Manual for the Building Block Methodology*. Freshwater Institute, University of Cape Town, South Africa.

Kirkhusmo, L.A. (1986) The use of groundwater monitoring data from the nordic countries. Norwegian Hydrological Programme, *NHP report* no. 19, Norway.

Kjeldsen, T.R., Lundorf, A. & Rosbjerg, D. (2000) Use of a two-component exponential distribution in partial duration modelling of hydrologicaldroughts in Zimbabwean rivers. *Hydrolog. Sci. J.* **45**(2), 285–298.

Klaassen, B. & Pilgrim, D.H. (1975) Hydrograph recession constants for New South Wales streams. *Australian Civil Engineer Transaction* CE17, **1**, 43–49.

Klein Tank, A., Wijngaard, J. & Engelen, A. van (2002) *Climate of Europe – The assessment of Observed Daily Temperature and Precipitation Extremes*. European Climate Assessment (ECA), KNMI, De Bilt, the Netherlands.

Klemeš, V. (1993) Probability of extreme hydrometeorological events – a different approach. In: *Extreme Hydrological Events: Precipitation, Floods and Droughts* (ed. by Z.W. Kundzewicz, D. Rosbjerg, S.P. Simonovic & K. Takeuchi) (Proc. Yokohama Symp. July, 1993), IAHS Publ. no. 213, 167–176.

Klemeš, V. (2000a) Tall tales about tails of hydrological distributions. I. *J. Hydrol. Eng.* **5**(3), 227–231.

Klemeš, V. (2000b) Tall tales about tails of hydrological distributions. II. *J. Hydrol. Eng.* **5**(3), 232–239.

Kliner, K. & Kněžek, M. (1974) Metoda separace podzemniho odtoku pri vyuziti pozorovani hladiny podzemni vody (The underground runoff separation method making use of the observation of groundwater table). *Journal of Hydrology and Hydrodynamics* **XXII**(5), 547–466.

Kogan, F. N. (1995) Application of vegetation index and brightness temperature for drought detection. *Adv. Space Res.* **15**, 91–100.

Köppen, W. (1918) Klassifikation der Klimate nach Temperatur, Niederschlag und Jahresablauf (Classification of climates according to temperature, precipitation and annual cycle). *Petermann. Geogr. Mitt.* **64**, 193–203 and 243–248.

Kottegoda, N.T. & Rosso, R. (1997) *Statistics, Probability, and Reliability for Civil and Environmental Engineers.* McGraw-Hill International Editions.

Krasovskaia, I. (1996) Sensitivity of the stability of river flow regimes to small fluctuations in temperature. *Hydrolog. Sci. J.* **41**(2), 251–264.

Krasovskaia, I. & Gottschalk, L. (1992) Stability of River Flow Regimes. *Nord. Hydrol.* **23**, 137–154.

Krasovskaia, I. & Gottschalk, L. (1995) Analysis of regional drought characteristics with empirical orthogonal functions. In: *New Uncertainty Concepts in Hydrology and Water Resources* (ed. by Z.W. Kundezewicz), International hydrology series, Cambridge University Press, Cambridge, UK, 163–167.

Krug, W.R., Gebert, W.A. & Graczyk, D.J. (1990) Map of mean annual runoff for the northeastern and mid-atlantic United States, Water Years 1951–1980. *US Geological Survey Water Resources Investigation Rep.* 88-4904.

Kruseman, G. (1997) Recharge from intermittent flow. In: *Recharge of Phreatic Aquifers in (Semi-)Arid Areas* (ed. by I. Simmers). Balkema Publishers, Rotterdam, the Netherlands, 145–200.

Kundzewicz, Z.W. (2002) Water and climate – The IPCC TAR perspective. In: XXII Nordic Hydrological Conference (ed. by Å. Killingtveit), *NHP Report* no. 47, 535–544.

Kundzewicz, Z.W. & Robson, A. eds. (2000) Detecting trend and other changes in hydrological data. World climate Programme – Water, World Climate Programme Data and Monitoring, WCDMP-45, *WMO-TD* no. 1013, Geneva.

Kundzewicz, Z.W., Rosbjerg, D., Simonovic, S.P. & Takeuchi, K. (1993) Extreme hydrological events in perspective. In: *Extreme Hydrological Events: Precipitation, Floods and Droughts* (ed. by Z.W. Kundzewicz, D. Rosbjerg, S.P. Simonovic & K. Takeuchi) (Proc. Yokohama Symp. July, 1993), IAHS Publ. no. 213, 1–7.

Laaha, G. (2002) Modelling summer and winter droughts as a basis for estimating river low flows. In: *FRIEND 2002 – Regional Hydrology: Bridging the Gap between Research and Practice* (ed. by H.A.J. van Lanen & S. Demuth), IAHS Publ. no. 274, 289–295.

Laaha, G. (2003) Process based regionalisation of low flows. PhD Thesis Faculty of Civil Engineering, Vienna University of Technology, Vienna, Austria.

Laaha, G. & Blöschl, G. (2003a) Saisonalität von Niederwasserspenden in Österreich (Seasonality of specific low flow discharges in Austria). *Mitteilungsblatt des Hydrographischen Dienstes in Österreich* 82, Austria.

Laaha, G. & Blöschl, G. (2003b) Niederwasseranalysen für österreichische Einzugsgebiete (Low flow analysis for catchments in Austria). *ÖWAV Seminar Wasserhaushalt und Wasserbewirtschaftung in niederschlagsarmen Gebieten,* 4.–5. Dezember 2003, Technische Universität Wien, Vienna, Austria.

Laird, K.R., Fritz, S.C. & Cumming, B.F. (1998) A diom-based reconstruction of drought intensity, duration and frequency from Moon Lake, North Dakota: a sub-decadal record of the last 2300 years. *J. Paleolimnol.* **19**, 161–179.

Lake, P.S. (2000) Disturbance, patchiness, and diversity in streams. *J. N. Am. Benthol. Soc.* **19**, 573–592.

Lall, U. (1995) Recent advances in non-paramteric function estimation: hydrological applications. *Rev. Geophys.* **33**(S), 1093–1102.

Lamouroux, N. & Capra, H. (2002) Simple predictions of instream habitat model outputs for target fish populations. *Freshwater Biol.* **47**, 1543–1556.

Landwehr, J.M, Matalas, N.C. & Wallis, J.R. (1979) Probability weighted moments compared with some traditional techniques in estimating Gumbel parameters and quantiles. *Water Resour. Res.* **15**(5), 1055–1064.

Lanen, H.A.J. van (2003) Groundwater networks and observation methods. In: *Groundwater Studies* (ed. by G. Kruseman & K. Rushton). Studies and Reports in Hydrology, UNESCO, Paris.

Lanen, H.A.J. van & Peters, E. (2000) Definition, effects and assessment of groundwater droughts. In: *Drought and Drought Mitigation in Europe* (ed. by J. Vogt & F. Somma), Advances in Natural and Technological Hazards Research, Vol. 14. Kluwer Academic Publisher, Dordrecht, the Netherlands, 49–61.

Lanen, H.A.J. van & Peters, E. (2002) Temporal variability of recharge as indicator for natural groundwater droughts in two climatically contrasting basins. In: *FRIEND 2002 – Regional Hydrology: Bridging the Gap between Research and Practice* (ed. by H A J. van Lanen & S. Demuth), IAHS Publ. no. 274, 101–108.

Lanen, H.A.J. van, Tallaksen, L.M., Kašpárek, L. & Querner, E.P. (1997) Hydrological drought analysis in the Hupsel basin using different physically-based models. In: *FRIEND'97 – Regional Hydrology: Concepts and Models for Sustainable Water Resource Management* (ed. by A. Gustard, S. Blazkova, M. Brilly, S. Demuth, J. Dixon, H. van Lanen, C. Llasat, S. Mkhandi & E. Servat), IAHS Publ. no. 246, 189–196.

Langbein, W.B. (1938) Some channel storage studies and their application to the determination of infiltration. *Transactions, American GeophysicalUnion* **19**, 435–445.

Laraque, A., Olivry, J.-C, Orange, D. & Marieu, B. (1997) Variations spatio-temporelles des régimes pluviométriques et hydrologiques en Afrique Central du début du siècle à nos jours. In: *FRIEND'97 – Regional Hydrology: Concepts and Models for Sustainable Water Resource Management* (ed. by A. Gustard, S. Blazkova, M. Brilly, S. Demuth, J. Dixon, H. van Lanen, C. Llasat, S. Mkhandi & E. Servat), IAHS Publ. no. 246, 257–263.

Leadbetter, M.R & Rootzén, H. (1988) Extremal theory for stochastic processes. *Ann. Probab.* **16**, 431–478.

Leduc, C., Favreau, G. & Schroeter, P. (2001) Long-term rise in Sahelian water-table: the Continental Terminal in South West Niger. *J. Hydrol.* **243**, 43–54.

Lehner, B. (1998a) Automated generation of a topologic river network system within a geographic enformation system (GIS) as a basis for catchment planning and river management. *Gesellschaft für technische Zusammenarbeit (G.T.Z.) Report,* Germany.

Lehner, B. (1998b) Automated generation of a topologic river network system within a GIS. In: *Angewandte Geographische Informationsverarbeitung* (Applied geographical information processing) (ed. by J. Strobel & F. Dollinger), Beiträge zum AGIT-Symposium Salzburg 98, 195–201.

Leonard, R. (1999) Climate change and groundwater, predicting how changes in the hydrological cycle affect water resources. *The Aquifer* **14**(2).

Lettenmaier, D.P., Wallis, J.R. & Wood, E.F. (1987) Effect of regional heterogeneity on flood frequency estimation. *Water Resour. Res.* **23**(2), 313–323.

Lewis-Beck, M.S. (1986) *Applied Regression – an Introduction.* Sage series: Quantitative applications in the social sciences no. 22.

L'Hôte, Y., Mahé, G., Somé, B. & Triboulet, J.P. (2002) Analysis of a Sahelian annual rainfall index from 1896 to 2000; the drought continues. *Hydrolog. Sci. J.* **47**(4), 563–572.

Lins, H.F. & Slack, J.R. (1999) Streamflow trends in the United States. *Geophys. Res. Lett.* **26**(2), 227–230.

Littlewood, I.G. Croke, B.F.W., Jakeman, A.J. & Sivapalan, M. (2003) The role of 'top-down' modeling for prediction in ungauged basins (PUB). *Hydrol. Process.* **17**, 1673–1679.

Lohman, K., Jones, J.R. & Perkins, B.D. (1992) Effects of nutrient enrichment and flood frequency on periphyton biomass in northern Ozark streams. *Can. J. Fish. Aquat. Sci.* **49**, 1198–1205.

Lu, L.-H. & Stedinger, J.R. (1992) Sampling variance of normalized GEV/PWM quantile estimators and a regional homogeneity test. *J. Hydrol.* **138**, 223–245.

Ludwig, A.H. & Tasker, G.D. (1993) Regionalizing of low-flow characteristics of Arkansas streams. *USGS Water Resources Investigations Report* 93–4013.

Madsen, H. & Rosbjerg, D. (1997a) The partial duration series method in regional index-flood modelling. *Water Resour. Res.* **33**(4), 737–746.

Madsen, H. & Rosbjerg, D. (1997b) Generalized least squares and empirical Bayes estimation in regional partial duration series index-flood modelling. *Water Resour. Res.* **33**(4), 771–781.

Madsen, H. & Rosbjerg, D. (1998) A regional Bayesian method for estimation of extreme streamflow droughts. In: *Statistical and Bayesian Methods in Hydrological Sciences* (ed. by E. Parent, B. Bobee, P. Hubert & J. Miquel), UNESCO, PHI Series, 327–340.

Madsen, H., Rosbjerg, D. & Harremoës, P. (1994) PDS-modelling and regional Bayesian estimation of extreme rainfalls. *Nord. Hydrol.* **25**(4), 279–300.

Madsen, H., Pearson, C.P. & Rosbjerg, D. (1997a) Comparison of annual maximum series and partial duration series for modeling extreme hydrologic events, 2. Regional modelling. *Water Resour. Res.* **33**(4), 759–769.

Madsen, H., Rasmussen, P.F. & Rosbjerg, D. (1997b) Comparison of annual maximum series and partial duration series methods for modeling extreme hydrologic events, 1. At-site modeling. *Wat. Resour. Res.* **33**(4), 747–757.

Madsen, H., Mikkelsen, P.S, Rosbjerg, D. & Harremoës, P. (2002) Regional Estimation of Rainfall Intensity-Duration-Frequency Curves Using Generalised Least Squares Regression of Partial Duration Series Statistics. *Water Resour. Res.* **38**(11), 1239.

Mahé, G., Dray, A., Paturel, E.J., Cres, A., Kone, F., Manga, M., Cres, F.N., Djoukam, J., Maiga, A., Ouedrago, M., Conway, D. & Servat, E. (2002) Climatic and anthropogenic impacts on the flow regime of the Nakambe River in Burkina Faso. In: *FRIEND 2002 – Regional Hydrology: Bridging the Gap between Research and Practice* (ed. by H.A.J. van Lanen & S. Demuth), IAHS Publ. no. 274, 61–68.

Maheras, P. (2000) Synoptic situations causing drought in the mediterranean basin. In: *Drought and Drought Mitigation in Europe* (ed. by J.V. Vogt & F. Somma), Kluwer Academic Publishers, Dordrecht, the Netherlands.

Majerčáková, O., Fendeková, M. & Lešková, D. (1997) The variability of hydrological series due to extreme climatic conditions and the possible change of the hydrological characteristics with respect to potential climate change. In: *FRIEND'97 – Regional Hydrology: Concepts and Models for Sustainable Water Resource Management* (ed. by A. Gustard, S. Blazkova, M. Brilly, S. Demuth, J. Dixon, H. van Lanen, C. Llasat, S. Mkhandi & E. Servat), IAHS Publ. no. 246, 59–66.

Mandelbrot, B.B. & Wallis, J.R. (1968) Noah, Joseph and operational hydrology. *Water Resour. Res.* **4**(5), 909–918.

Mandelbrot, B.B. & Wallis, J.R. (1969a) Computer experiments with fractional Gaussian noises, Part I. Averages and Variances. *Water Resour. Res.* **5**(1).

Mandelbrot, B.B. & Wallis, J.R. (1969b) Computer experiments with fractional Gaussian noises, Part II. Rescaled ranges and spectra. *Water Resour. Res.* **5**(1).

Mandelbrot, B.B. & Wallis, J.R. (1969c) Computer experiments with fractional Gaussian noises, Part III. Mathematical appendix. *Water Resour. Res.* **5**(1).

Marsh, T.J. (1999) Maximising the Utility of River Flow Data. In: *Hydrometry: Principles and Practices* (Herschy, R.W.), Wiley.

Marsh, T.J. (2002) Capitalising on river flow data to meet changing national needs – a UK perspective. *Flow Meas. Instrum.* **13**, 291–298.

Marsh, T.J., Monkhouse, R.A., Arnell, N.W., Lees, M.L. & Reynard, N.S. (1994) *The 1988–92 drought*. Institute of Hydrology, Wallingford, UK.

Martins, E.S. & Stedinger, J.R. (2000) Generalized maximum likelihood generalized extreme value quantile estimators for hydrologic data. *Water Resour. Res.* **36**(3), 737–744.

Martins, E.S. & Stedinger, J.R. (2001) Generalized maximum likelihood Pareto-Poisson estimators for partial duration series. *Water Resour. Res.* **37**(10), 2551–2557.

Matalas, N.C. (1963) Probability distribution of low flows. *U.S.G.S. Prof. Pap.* 434-A, Washington, D.C.

Matalas, N.C. (1991) Drought description. *Stoch. Hydrol. Hydraul.* **5**, 255–260.

Matalas, N.C. & Reiher, B.J. (1967) Some comments on the use of factor analysis. *Water Resour. Res.* **3**(1), 213–224.

Mathier, L., Perreault, L. & Bobeé, B. (1992) The use of geometric and gamma related distributions for frequency analysis of water deficit. *Stoch. Hydrol. Hydraul.* **6**, 239–254.

McClave, J.T., Dietrich, F.H. & Sincich, T. (1997) *Statistics.* 7th edn, Prentice-Hall International.

McConchie, J.A. (1992) Urban Hydrology. In: *Waters of New Zealand* (ed. by M.P. Mosley), New Zealand Hydrological Society, Wellington, New Zealand, 335–363.

McDonald, M.G. & Harbaugh, H.W. (1988) A Modular three-dimensional finite-difference groundwater flow model. *U.S. Geological Survey Techniques of Water-Resources Investigations*, Book 6.

McKay, G.A. (1976) Hydrological Mapping. In: *Facets of Hydrology* (ed. by J.C. Rodda), Wiley, London, 1–36.

McMahon, T.A. (1976) Low flow analyses of streams: details of computational procedures and annotated bibliography. *Research Report* 5, Monash University, Dep. of Civil Eng., Clayton, Australia.

McMahon, T.A. & Mein, R.G. (1978) *Reservoir Capacity and Yield.* Developments in Water Science 9, Elsevier, Amsterdam, the Netherlands.

McMahon, T.A. & Arenas, A.D. (1982) Methods of computation of low streamflow. *Studies and Reports in Hydrology* no. 36, UNESCO, Paris.

McMahon, T.A. & Mein R.G. (1986) *River and Reservoir Yield.* Water Resources Publications, Littleton, Colorado, USA.

Meddi, M. & Hubert, P. (2003) Impact de la modification du régime pluviométrique sur les ressources en eau du nord-ouest de l'Algérie. In: *Hydrology of Mediterranean and Semiarid Regions* (ed. by E. Servat, W. Najem, C. Leduc & A. Shakeel), IAHS Publ. no. 278, 229–235.

Meffe, G.K. (1984) Effects of abiotic disturbance on coexistence of predatory and prey fish species. *Ecology* **65**, 1525–1534.

Meigh, J., Tate, E. & McCartney, M. (2002) Methods for identifying and monitoring river flow drought in southern Africa. In: *FRIEND 2002 – Regional Hydrology: Bridging the Gap between Research and Practice* (ed. by H.A.J. van Lanen & S. Demuth), IAHS Publ. no. 274, 181–188.

Melin, R. (1954) *Vattenföringen i Sveriges floder* (Discharge data of Swedish rivers). Sveriges Meteorologiska och Hydrologiska Institut, Meddelande Serie D Nr 6, Kungl. boktryckerite P.A. Norstedt & Söner, Stockholm.

Mikkelsen, P.S., Madsen, H., Rosbjerg, D. & Harremoës, P. (1996) Properties of extreme point rainfall III: Identification of spatial inter-site correlation structure. *Atmos. Res.* **40**, 77–98.

Miles, J. C. (1985) The representation of flows to partially penetrating rivers using groundwater flow models. *J. Hydrol.* **82**, 341–355.

Milhous, R.T., Updike, M.A. & Schneider, D.M. (1989) Physical Habitat Simulation System Reference Manual – Version II. United States Fish and Wildlife Service, Fort Collins, Colorado, *Instream flow information paper* no. 26.

Ministerio de Medio Ambiente (2000) *Libro Blanco del Agua en Espana* (White Paper on Water in Spain). Centro de Publicaciones del Ministerio de Medio Ambiente, Madrid.

Minshall, G.W. (1968) Community dynamics of the benthic fauna in a woodland springbrook. *Hydrobiologia* **32**, 305–339.

Minshall, G.W. (1984) Aquatic insect-substratum relationships. In: *Ecology of Aquatic Insects* (ed. by V.H. Resh & D.M. Rosenberg), Praeger Scientific, New York, 358–400.

Mioduszewski, W., Kowalewski, Z., Ślesicka, A. & Querner, E.P. (1997) Groundwater management in the Jegrznia River valley. *In: FRIEND'97 – Regional Hydrology: Concepts and Models for Sustainable Water Resource Management* (ed. by A. Gustard, S. Blazkova, M. Brilly, S. Demuth, J. Dixon, H. van Lanen, C. Llasat, S. Mkhandi & E. Servat), IAHS Publ. no. 246, 181–187.

Mises, R. von (1936) La distribution de la plus grande de *n* valeurs. In: *Selected papers*, Vol II (ed. by P. Frank, S. Goldstein, M. Kac, W. Prager, G. Szegö & G. Birkhoff, 1964), Amer. Math. Soc., Providence, RI, USA, 271–294.

Moore, R.V. (1997) The Logical and Physical Design of the LOIS Database. LOIS Special Volume, *Sci. Total Environ.* **194/195**, 137–146.

Moors, E.J. & Dolman, A.J. (2003) Land-use change, climate and hydrology. In: *Global Environmental Change and Land Use* (ed. by A.J. Dolman, A. Verhagen & C.A. Rovers), Kluwer Academic Publishers, Dordrecht, the Netherlands, 139–166.

Moran, P.A. (1969) Statistical inference with bivariate gamma-distribution. *Biometrica* **56**(3), 123–136.

Morris, B.L., Seddique, A.A. & Ahmed, K.M. (2003) Response of the Dupi Tila aquifer to intensive pumping in Dhaka, Bangladesh. *Hydrogeol. J.* **11**, 496–503.

Morris, D. & Heerdegen, R. (1988) Automatically derived catchment boundaries and channel networks and their hydrological applications. *Geomorphology* **1**, 131–141.

Morris, D. & Flavin, R. (1990) A digital terrain model for hydrology. In: *Proc. 4th International Symposium on Spatial Data Handling, Zurich* no. 1, 250–262.

Mosley, M.P. (1983) Flow requirements for recreation and wildlife in New Zealand rivers – a review. *J. Hydrol. NZ* **22**, 152–174.

Mosley, M.P. & Pearson, C.P. eds. (1997) *Floods and Droughts: the New Zealand Experience.* New Zealand Hydrological Society Inc., Wellington North, New Zealand.

Moss, Z. (2001) Monitoring trout populations in the Waiau river (Southland) downstream of the Mararoa Weir. Report to Meridian Energy, Southland Fish and Game, Invercargill, New Zealand.

Moyle, P.B & Vondracek, B. (1985) Persistence and structure of the fish assemblage in a small California stream. *Ecology* **66**, 1–13.

Murphy, J. (2000) Predictions of climate change over Europe using statistical and dynamical downscaling techniques. *Int. J. Climatol.* **20**, 489–501.

Mutua, F.M. (1994) The use of the Akaike Information Criterion in the identification of an optimum flood frequency model. *Hydrolog. Sci. J.* **39**(3), 235–244.

Nakicenovic, N., Alcamo, J., Davis, G., Vries, B. de, Fenhann, J., Gaffin, S., Gregory, K., Grübler, A., Jung, T.Y., Kram, T., La Rovere, E.L., Michaelis, L., Mori, S., Morita, T., Pepper, W., Pitcher, H., Price, L., Raihi, K., Roehrl, A., Rogner, H.H., Sankovski, A., Schlesinger, M., Shukla, P., Smith, S., Swart, R., Rooijen, S. van, Victor, N. & Dadi, Z.

(2000) *IPCC Special Report on Emissions Scenarios.* Cambridge University Press, Cambridge, UK.

Nathan, R.J. & McMahon, T.A. (1990a) Evaluation of automated techniques for base flow and recession analysis. *Water Resour. Res.* **26**(7), 1465–1473.

Nathan, R.J. & McMahon, T.A. (1990b) Identification of homogeneous regions for the purposes of regionalisation. *J. Hydrol.* **121**, 217–238.

Nathan, R.J. & McMahon, T.A. (1990c) Practical aspect of low-flow frequency analysis. *Water Resour. Res.* **26**(9), 2135–2141.

Nathan, R.J. & McMahon, T.A. (1992) Estimating low flow characteristics in ungauged catchments. *Water Resour. Manag.* **6**, 85–100.

NDMC (2003) *U.S. Drought Monitor.* Available from: <http://www.drought.unl.edu/dm/> [last visited 2003].

Nehring, R.B. & Anderson, R.M. (1993) Determination of population-limiting critical salmonid habitats in Colorado streams using the physical habitat simulation system. *Rivers* **4**, 1–19.

NERC (1975) *Flood Studies Report,* 5 volumes. Natural Environment Research Council, UK.

Newcombe, C.P. & MacDonald, D.D. (1991) Effects of suspended sediments on aquatic ecosystems. *N. Am. J. Fish. Manage.* **11**, 72–82.

Nicholson, S.E. (2000) The nature of rainfall variability over Africa on time scales of decades to millennia. *Global Planet. Change* **26**, 137–158.

Nordin, C.F. & Rosbjerg, D.M (1970) Application of crossing theory in hydrology. *Bulletin of the International Association of Scientific Hydrology* **3**, 27–43.

Oberlin, G. & Desbos, E. (1997) *Flow Regimes from International Experimental and Network Data (FRIEND).* Third Report: 1994–1997, Cemagref Éditions, France.

Oerlemans J., Anderson, B. Hubbard, A., Huybrechts, Ph., Johannesson, T., Knap, W.H., Schmeits, M., Stroeven, A.P., Wal, R.S.W. van de, Wallinga, J. & Zuo Z. (1998) Modelling the response of glaciers to climate warming. *Clim. Dynam.* **14**, 267–274.

OFDA/CRED (2002) *EM-DAT: The OFDA/CRED International Disaster Database.* Université Catholique de Louvain, Brussels, available from: <http://www.cred.be/emdat> [last visited October 2002].

Oki, T. & Xue, J. (1998) *Investigation of River Discharge Variability in Sahel Desertification Experiment.* Reprint of 9th Symposium on Global Change Studies, 259–260.

Önöz, B. & Bayazit, M. (2002) Troughs under threshold modeling of minimum flows in perennial streams. *J. Hydrol.* **258**, 187–197.

Orth, D.J. & Maughan, O.E. (1982) Evaluation of the incremental methodology for recommending instream flows for fishes. *T. Am. Fish. Soc.* **111**, 413–445.

Otto, C.J. (1998) Monitoring tools and type of recording. In: *Monitoring for Groundwater Management in (Semi-)Arid Regions.* (ed. by H.A.J. van Lanen), Studies and Reports in Hydrology no. 57, UNESCO, Paris, 65–90.

Pálfai, I. (1990) Suša u 1990 godini (The drought of the year 1990). *Vode Vojvodine* **19**, 185–192. Novi Sad, Yugoslavia.

Palmer, W.C. (1965) Meteorological drought. *Research Paper* no. 45, U.S. Weather Bureau, Washington, D.C.

Palmer, W.C. (1968) Keeping track of crop moisture conditions, nationwide: The new crop moisture index. *Weatherwise* **21**(4), 156–161.

Panu, U.S. & Sharma, T.C. (2002) Challenges in drought research: some perspectives and future directions. *Hydrol. Sci. J.* **47**(S), 19–30.

Parzen, E. (1979) Nonparametric statistical data modelling. *J. Am. Stat. Assoc* **74**(365), 105–122.

Pearson, C.P. (1991) New Zealand regional flood frequency analysis using L-moments. *J. Hydrol. NZ,* **30**(2), 53–64.

Pearson, C.P. (1995) Regional frequency analysis of low flows in New Zealand rivers. *J. Hydrol. NZ* **33**, 94–122.

Peral Garcia, C., Mestre Barceló, A. & Garcia Merayo, J.L. (2000) The drought of 1991–95 in southern Spain. Analysis, repercussions, and response measures. In: *DROUGHT, A Global Assessment*, Vol I (ed. by D.A. Wilhite), Routledge Hazards and Disasters Series, Routledge, London, 267–380.

Pereira, L.S. & Keller, H.M. (1982) Factors affecting recession parameters and flow components in eleven small Pre-Alp basins. In: *Hydrological Aspects of Alpine and High Mountain Areas*. IAHS Publ. no. 138, 233–242.

Peters, E. (2003) Propagation of drought through groundwater systems. Illustrated in the Pang (UK) and Upper-Guadiana (ES) catchments. PhD Thesis Wageningen University, Wageningen, the Netherlands.

Peters E. & Lanen, H.A.J. van (2001) Environmental impact. In: *Assessment of the Regional Impact of Droughts in Europe (ARIDE)* (ed. by S. Demuth & K. Stahl), Final Report, EU contract ENV4-CT-97-0553, Institute of Hydrology, University of Freiburg, Freiburg, Germany, 40–45.

Peters, E. & Lanen, H.A.J van. (2004) Separation of the base flow component from streamflow using observed heads – illustrated for the Pang catchment (UK). *Hydrol. Process.* (Accepted).

Peters E., Lanen, H.A.J. van, Alvarez J. & Bradford, R.B.B. (2001) Groundwater droughts: evaluation of temporal variability of recharge in three European groundwater catchments. *ARIDE Technical Report* no.11, Wageningen, the Netherlands.

Peters, E., Torfs, P.J.J.F., Lanen, H.A.J. van & Bier, G. (2003) Propagation of drought through groundwater – a new approach using linear reservoir theory. *Hydrol. Process.* **17**(15), 3023–3040.

Pickands, J. (1975) Statistical inference using extreme order statistics. *Ann. Stat.* **3**, 119–131.

Poff, N.L. (1992) Why disturbances can be predictable: a perspective on the definition of disturbance in streams. *J. N. Am. Benthol. Soc.* **11**, 86–92.

Poff, N.L. & Ward, J.V. (1989) Implications of streamflow variability and predictability for lotic community structure: a regional analysis of streamflow patterns. *Can. J. Fish.Aquat. Sci.* **46**, 1805–1817.

Pokrovsky, D.S., Rogov, G.M. & Kuzevanov, K.I. (1999) The impact of urbanization on the hydrogeological conditions of Tomsk, Russia. In: *Groundwater in the Urban Environment. Selected City Profiles* (ed. by J. Chilton), Balkema Publishers, Rotterdam, the Netherlands, 217–224.

Pouilly M., Valentin S., Capra H., Ginot V. & Souchon Y. (1995) Note technique: Méthode des microhabitats: principes et protocoles d'application. *B. Fr. Pêche Piscic.* **336**, 41–54.

Price, M., Low, R.G. & McCann, C. (2000) Mechanisms of water storage and flow in the unsaturated zone of the chalk aquifer. *J. Hydrol.* **233**, 54–71.

Putarić, V. (2001) Floods and droughts as natural hazards in Vojvodina. In: *Environmental Management of the Rural Landscape in Central and Eastern Europe.* (ed. by H. van Es & D. Húska), Slovak Agricultural University, Nitra, Slovakia, 24–27.

Querner, E.P. (1988) Description of a regional groundwater flow model SIMGRO and some applications. *Agr. Water Manage.* **14**, 209–218

Querner, E.P. (1997) Description and application of the combined surface and groundwater model MOGROW. *J. Hydrol.* **192**, 158–188.

Querner E.P. & Lanen, H.A.J. van (2001) Impact assessment of drought mitigation measures in two adjacent Dutch basins using simulation modelling. *J. Hydrol.* **252,** 51–64.

Querner, E.P., Tallaksen, L.M., Kašpárek, L. & Lanen, H.A.J. van (1997) Impact of land-use, climate change and groundwater abstraction on streamflow droughts using physically-based models. In: *FRIEND'97 – Regional Hydrology: Concepts and Models for Sustainable Water*

Resource Management (ed. by A. Gustard, S. Blazkova, M. Brilly, S. Demuth, J. Dixon, H. van Lanen, C. Llasat, S. Mkhandi & E. Servat), IAHS Publ. no. 246, 171–179.

Quimpo, R.G. (1967) *Stochastic Models of Daily River Flow Sequences.* Colorado State University Paper No 18, Forth Collins, Colorado, USA.

Ramnarong, V. (1999) Evaluation of groundwater management in Bangkok: Positive and negative. In: *Groundwater in the Urban Environment. Selected City Profiles* (ed. by J. Chilton), Balkema Publishers, Rotterdam, the Netherlands, 51–62.

Rasmussen, P.F. & Rosbjerg, D. (1991) Prediction uncertainty in seasonal partial duration series. *Water Resour. Res.* **27**(11), 2875–2883.

Ravera, O. ed. (1991) *Terrestrial and Aquatic Ecosystems: Perturbation and Recovery.* Ellis Horwood Ltd., Chichester, UK.

Rees, G. & Demuth S. (2000) The application of modern information system technology in the European FRIEND project. *Wasser & Boden* **52**(3), 9–13.

Rees, H.G., Cole, G.A. & Gustard, A. (1996) Surface water quantity monitoring in Europe. European Environment Agency, *Topic Report* no. 3, Copenhagen.

Rees, H.G., Croker, K.M., Prudhomme, C. & Gustard, A. (2000) The European atlas of small-scale hydropower resources. In: *Conf. Proc. of Small Scale Hydro 2000.* Lisbon 8–12 May 2000, 191–202.

Rees, H.G., Croker, K.M., Singhal, M.K., Saraf, A.K., Kumar, A., Zaidman, M.D. & Gustard, A. (2002) Flow regime estimation for small-scale hydropower development in Himachal Pradesh. *Journal of Applied Hydrology* **XV**, (2/3), 77–90.

Reeve, C. & Watts, H. eds. (1994) *Drought, Pollution and Management.* Balkema Publishers, Rotterdam, the Netherlands.

Resh, V.H., Brown, A.V., Covich, A.P., Gurtz, M.E., Li, H.W., Minshall, G.W., Reice, S.R., Sheldon, A.L., Wallace, J.B. & Vissmar, R. (1988) The role of disturbance in stream ecology. *J. N. Am. Benthol. Soc.* **7**, 433–455.

Reznikovskij, A.Sh. (1969) *Vognoenergeticheskie rachety metodom Monte Carlo* (Water-Energetical Calculations with the Monte-Carlo Method). Energya, Moskva.

Richter, B.D., Baumgartner, J.V., Wigington, R. & Braun, D.P. (1997) How much water does a river need? *Freshwater Biol.* **37**, 231–249.

Riebsame, W.E., Changnon, S.A. Jr. & Carl, T.R. (1991) *Drought and Natural Resources Management in the United States: Impacts and Implications of the 1987–89 Drought.* Westview Press, Boulder, Colorado, USA.

Ries, K.G. & Friesz, P.J. (2000) Methods for estimating low-flow statistics for Massachusetts streams. *USGS Water-Resources Investigations Report* 00-4135.

Riggs, H.C. (1972) Regional Analysis of Stream-Flow Techniques. In: *Techniques of Water Research Investigations, Book 4: Hydrologica Analysis and Interpretation,* USGS Washington, D.C.

Riggs, H.C. (1985) *Stream Flow Characteristics.* Elsevier, Amsterdam.

Riggs, H.C., Caffey, J.E., Orsborn, J.F., Schaake, J.C. Jr., Singh, K.P. & Wallace, J.R. (1980) Characteristics of low flows. Report of an ASCE task commitee. *J. Hydraul. Eng.-ASCE* **106**(5), 717–731.

Riis, T. & Biggs, B.J.F. (2003) Hydrologic and hydraulic control of macrophyte establishment and performance in streams. *Limnol. Oceanogr.* **48**, 1488–1497.

Risle, J.C. (1994) Estimating the magnitude and frequency of low flows of streams in Massachusetts. *USGS Water Resources Investigations Report* 94-4100.

Robins, N.S., Calow, R.C., MacDonald, A.M., Macdonald, D.M.J., Gibbs, B.R., Orpen, W.R.G., Mtembezeka, P., Andrews, A.J., Appiah, S.O. & Banda, K. (1997) Final report – groundwater management in drought-prone areas of Africa. *British Geological Survey Report* WC/97/57, UK.

Robinson, P.J. & Henderson-Sellers, A. (1999) *Contemporary Climatology*. 2nd edn, Pearson Education Limited, Longman, Singapore.

Robson, A. & Reed, D.W. (1999) *Statistical Procedures for Flow Frequency Estimation. Flood Estimation Handbook Vol.3*. Institute of Hydrology, Wallingford, UK.

Rodda, J.C. (1998) Hydrological networks need improving! In: Proc. Int. conf. on world water resources at the beginning of the 21st century, *IHP-V Technical Documents in Hydrology* no. 18, UNSECO, Paris, 91–102.

Rodda, J.C. (2000) Drought and water resources. In: *DROUGHT, A Global Assessment*, Vol II (ed. by D.A. Wilhite), Routledge Hazards and Disasters Series, Routledge, London, 241–263.

Roeckner, E., Oberhuber, J.M., Bacher, A., Christoph, M. & Kirchner, I. (1996) ENSO variability and atmospheric response in a global coupled atmosphere-ocean GCM. *Clim. Dynam.* **12**, 737–754.

Roesner, L.A. & Yevjevich, V. (1966) Mathematical models for time series of monthly precipitation and monthly runoff. *Colorado State University Paper* no. 15, Forth Collins, Colorado, USA.

Rolf, H.L.M. (1989) *Verlaging von de Grondwaterstanda in Nederland* (Fall of the groundwater levels in the Netherlands). Ministry of transport, Public Works and Water Management, Lelystad, the Netherlands.

Ropelewski, C.F. & Halpert, M.S. (1987) Global and Regional Scale Precipitation patterns associated with the El Nino/Southern Oscillation. *Mon. Weather Rev.* **115**, 1606–1626.

Ropelewski, C.F. & Folland, C.K. (2001) Prospects for the prediction of meteorological drought. In: *DROUGHT, A Global Assessment*, Vol I (ed. by D.A. Wilhite), Routledge Hazards and Disasters Series, Routledge, London, 21–40.

Rosbjerg, D. (1977) Crossing and extremes in dependent annual series. *Nord. Hydrol.* **8**, 257–266.

Rosbjerg, D., Madsen, H. & Rasmussen, P.F. (1992) Prediction in partial duration series with generalized Pareto-distributed exceedances. *Water Resour. Res.* **28**(11), 3001–3010.

Rossi, F., Fiorentino, M. & Versace, P. (1984) Two-component extreme value distribution for flood frequency analysis. *Water Resour. Res.* **20**(7), 847–856.

Rossi, G., Benedini, M., Tsakiris, G. & Giakoumakis, S. (1992) On regional drought estimation and analysis. *Water Resour. Manag.* **6**, 249–277.

Rowe, L., Fahey, B., Jackson, R. & Duncan, M. (1997) Effects of land use on floods and low flows. In: *Floods and Droughts: the New Zealand Experience* (ed. by M.P. Mosley & Ch.P. Pearson), New Zealand Hydrological Society Inc., Wellington North, New Zealand, 89–102.

Salas, J.D. (1993) Ananlysis and modelling of hydrologic time series. In: *Handbook of Hydrology* (ed. by D.R. Maidment), McGraw-Hill, New York, 19.1–19.72.

Salas, J.D., Delleur, J.W., Yevjevich V. & Lane, W.L. (1980) *Applied Modelling of Hydrological Time Series*. Book Crafters, Chelsea.

Sankarasubramanian, A. & Srinivasan, K. (1999) Investigation and comparison of sampling properties of L-moments and conventional moments. *J. Hydrol.* **218**, 13–34.

Santos, M.A. (1983) Regional droughts: a stochastic characterization. *J. Hydrol.* **66**, 183–211.

Santos, M.J., Veríssimo, R., Fernandes, S., Orlando, M. & Rodrigues, R. (2000) Overview of regional meteorological drought analysis in western Europe. *ARIDE Technical Report* no. 10, INAG, Water Institute, Lisbon.

Santos, M.J., Veríssimo, R., Fernandes, S., Orlando, M. & Rodrigues, R. (2002) Meteorological droughts focused on a pan-European context. In: *FRIEND 2002 – Regional Hydrology: Bridging the Gap between Research and Practice* (ed. by H.A.J. van Lanen & S. Demuth), IAHS Publ. no. 274, 273–280.

Sauquet, E., Gottschalk, L. & Leblois, E. (2000) Mapping average annual runoff: a hierarchical approach applying a stochastic interpolation scheme. *Hydrolog. Sci. J.* **45**(6), 799–815.

Savenije, H.H.G. (1995) New definitions for moisture recycling and the relationship with land-use changes in the Sahel. *J. Hydrol.* **167**, 57–78.

Scarsbrook, M.R. & Townsend, C.R. (1993) Stream community structure in relation to spatial and temporal variation: a habitat templet study of two contrasting New Zealand streams. *Freshwater Biol.* **29**, 395–410.

Schär, C., Vidale, P.L., Lüthi, D., Frei, C., Häberli, C., Liniger, M.A. & Appenzeller, C. (2004) The role of increasing temperature variability in European summer heatwaves. *Nature* **427**, 332–336.

Schreiber, P. (1996) Regionalisierung des Niedrigwassers mit statistischen Verfahren unter Verwendung eines Geographischen Informationssystems (Regionalisation of low flow with statistical methods applying geographical information systems), *Freiburger Schriften zur Hydrologie*, Band 4, Freiburg, Germany.

Schreiber, P. & Demuth, S. (1997) Regionalization of low flows in Southwest Germany. *Hydrolog. Sci. J.* **42**(6), 845–858.

Schuurmans, J.M, Troch, P.A., Veldhuizen, A.A., Bastiaanssen, W.G.M. & Bierkens, M.F.P. (2003) Assimilation of remotely sensed latent heat flux in a distributed hydrological model. *Adv. Water Resour.* **26**(2), 151–159.

Schwartz, F.W. & Zhang, H. (2003) *Fundamentals of groundwater.* Wiley.

Scott, D. & Poynter, M. (1991) Upper temperature limits for trout in New Zealand and climate change. *Hydrobiologia* **222**, 147–151.

Scrimgeour, G.J. & Winterbourn, M.J. (1989) Effects of floods on epilithon and benthic macroinvertebrate populations in an unstable New Zealand river. *Hydrobiologia* **171**, 33–44.

Seegrist, D.W. & Gard, R. (1972) Effects of floods on trout in Sagehen Creek, California. *T. Am. Fish. Soc.* **101**, 478–482.

Sekulin, A.E., Bullock, A. & Gustard, A. (1992) Rapid calculation of catchment boundaries using an automated river network overlay technique. *Water Resour. Res.* **28**(8), 2101–2109.

Sellers, P.J. & Lockwood, J.G. (1981) A numerical simulation of the effects of changing vegetation type on surface hydroclimatology. *Climate Change* **3**, 121–136.

Sen, Z. (1977) Run-sums of annual flow series. *J. Hydrol.* **35**, 311–324.

Sen, Z. (1980) Regional drought and flood frequency analysis: theoretical consideration. *J. Hydrol.* **46**, 265–279.

Sen, Z. (1998) Probabilistic formulation of spatio-temporal drought pattern. *Theor. Appl. Climatol.* **61**, 197–206.

Shafer, B.A. & Dezman, L.E. (1982) Development of a Surface Water Supply Index (SWSI) to assess the severity of drought conditions in snowpack runoff areas. In: *Proceedings of the Western Snow Conference*, 164–175.

Shenkman, B.M., Pinneker, E.V. & Shenkman, I.B. (1999) Groundwater of the Irkutsk agglomeration. In: *Groundwater in the urban environment. Selected City Profiles* (ed. by J. Chilton), Balkema Publishers, Rotterdam, the Netherlands, 239–245.

Sheppard, D. (1968) A two-dimensional interpolation function for irregularly spaced data. In: *Proc. 1968 ACM National Conference*, 517–524.

Simmers, I. (2003) *Understanding Water in a Dry Environment – Hydrological Processes in Arid and Semi-arid Zones.* IAH International Contribution to Hydrogeology no. 23, Balkema Publishers, Rotterdam, the Netherlands.

Singh, K.P. (1969) Theoretical baseflow curves. *J. Hydr. Eng. Div.-ASCE* **6**, 2029–2048.

Singh, K.P. & Stahl, J.B. (1971) Derivation of base flow recession curves and parameters. *Water Resour. Res.* **7**(2), 292–303.

Singh, R.D., Mishra, S.K. & Chowdhary, H. (2001) Regional Flow-Duration Models for Large Number of Ungauged Himalayan Catchments for Planning Microhydro Projects. *J. Hydrol. Eng.* **6**(4), 310–316.

Singh, V.P. (1995) *Computer Models of Watershed Hydrology.* Water Resource Publications, LLC, Highlands Ranch, USA.

Smakhtin, V.Y. (2001) Low flow hydrology: a review. *J. Hydrol.* **240**, 147–186.

Smakhtin, V.Y. (2002) Regional low-flow studies in South Africa. In: *FRIEND'97 – Regional Hydrology: Concepts and Models for Sustainable Water Resource Management* (ed. by A. Gustard, S. Blazkova, M. Brilly, S. Demuth, J. Dixon, H. van Lanen, C. Llasat, S. Mkhandi & E. Servat), IAHS Publ. no. 246, 125–132.

Smakhtin, V.Y. & Toulouse, M. (1998) Relationships between low-flow characteristics of South African streams. *Water SA* **24**(2), 107–112.

Smakhtin, V.Y., Watkins, D.A. & Hughes, D.A. (1995) Preliminary analysis of low flow characteristics of South African rivers. *Water SA* **21**(3), 201–210.

Smakhtin, V.Y., Watkins, D.A., Hughes, D.A., Sami, K. & Smakhtina, O.Y. (1998) Methods of catchment-wide assessment of daily low-flow regimes in South Africa. *Water SA* **24**(3), 173–185.

Smith, R.L. (1989) Extreme value analysis of environmental time series: an application to trend detection in ground-level ozone. *Stat. Sci.* **4**(4), 367–393.

Smith, R.L. (2002) *Environmental Statistics, Lecture Notes for CBMS Course in Environmental Statistics.* Available from: <http://www.stat.unc.edu/postscript/rs/envnotes.ps> [last visited July 2001].

Soetrisno, S. (1999) Groundwater management problems: Comparative city case studies of Jakarta and Bandung, Indonesia. In: *Groundwater in the Urban Environment. Selected City Profiles* (ed. by J. Chilton), Balkema Publishers, Rotterdam, the Netherlands, 63–68.

Solow, A.R. & Huppert, A. (2003) On non-stationarity of ENSO. *Geophys. Res. Lett.* **30**(17), 1910.

Soulé, P.T. (1992) Spatial patterns of drought frequency and duration in the contiguous USA based onmultiple drought event definitions. *Int. J. Climatol.* **12**, 11–24.

Speer, P.R., Golden, H.C. & Patterson, J.F. (1964) Low flow characteristics of streams in the Mississippi embayment in Mississippi and Alabama. *U.S. Geological Survey Professional Paper* 448-I.

Stachy, J., Biernat, B., Bondarczuk, Z., Czarnecka, H., Dobrzynska, I. & Fal, B. (1991) *The Rules of the Computation of the Average Low Flow for the Polish Rivers.* Guide and Manual Publ. IMGW. Warszawa, Poland.

Stahl, K. (2001) Hydrological Drought – a Study across Europe. PhD Thesis Albert-Ludwigs-Universität Freiburg, *Freiburger Schriften zur Hydrologie* no. 15, Freiburg, Germany, also available from: <http://www.freidok.uni-freiburg.de/volltexte/202>.

Stall, J.B. (1962) Reservoir mass analysis by a low flow series. *J. Sanit. Eng. Div. ASCE.* **88**(SA5).

Stalnaker, C.B., Lamb, L., Henriksen, J., Bovee, K. & Bartholow, J. (1995) The instream flow incremental methodology: a primer for IFIM. National Biological Service, Fort Collins, *Biological Report* 29.

Stedinger, J.R. (1980) Fitting log normal distributions to hydrologic data. *Water Resour. Res.* **16**(3), 481–490.

Stedinger, J.R. (1983) Estimating a regional flood frequency distribution. *Water Resour. Res.* **19**(2), 503–510.

Stedinger, J.R. (1997) Expected probability and annual damage estimation. *J. Water Resour. Pl.-ASCE* **123**(2), 125–135.

Stedinger, J.R. & Tasker, G.D. (1985) Regional hydrologic analysis, 1. Ordinary, weighted and generalized least squares compared. *Water Resour. Res.* **21**(9), 1421–1432. Correction, *Water Resour. Res.* **22**(5), 844, 1986.

Stedinger, J.R., Vogel, R.M. & Foufoula-Georgiou, E. (1993) Frequency analysis of extreme events. In: *Handbook of Hydrology* (ed. by D.R. Maidment), McGraw-Hill, New York, 18.1–18.66.

Stein, A. (1998) Geostatistical procedures for analysing spatial variability and optimizing collection of monitoring data. In: *Monitoring for Groundwater Management in (Semi-)Arid Regions* (ed. by H.A.J. van Lanen), Studies and Reports in Hydrology no. 57, UNESCO, Paris, 91–106.

Stiebing, W.H. Jr. (1989) *Out of the desert? Archaeology and the Exodus/Conquest Narratives.* Prometheus Books, New York.

Strahler, A.N. (1957) Quantitative analysis of watershed geomorphology. *Eos, Transactions, American Geophysical Union* **38**, 913–920.

Strupczewski, W.G., Singh, V.P. & Feluch, W. (2001) Non-stationary approach to at-site frequency modelling I. Maximum likelihood estimation. *J. Hydrol.* **248**, 123–142.

Subrahmanyan, K., Prakash, B.A. & Shakeel, A. (2003) The impact of anthropogenic factors on groundwater regime in crystalline hard rock aquifers, in Andhra Pradesh, India. In: *Hydrology of Mediterranean and Semiarid Regions* (ed. by E. Servat, W. Najem, C. Leduc & A. Shakeel), IAHS Publ. no. 278, 396–402.

Suren, A.M. (1991) Brophytes as invertebrate habitat in two New Zealand alpine streams. *Freshwater Biol.* **26**, 399–418.

Suren, A.M., Biggs, B.J.F. & Weatherhead, M. (2002) Moawhango River: Suggested biological metric to meet consent requirements, and quantification of the effect of increased residual flows. *NIWA Client Report* no. CHC02/50, NIWA, Christchurch, New Zealand.

Suren, A.M., Biggs, B.J.F., Duncan, M.J. & Bergey, L. (2003) Benthic community dynamics during summer low-flows in two rivers of contrasting enrichment. 2. Invertebrates. *New Zeal. J. Mar. Fresh.* **37**(1), 71–84.

Svanidze, G.G. (1977) *Matematicheskae modelirovanie gidrologicheskykh ryadov* (Mathematical Modelling of Hydrological Series). Gidrometeoizdat, Leningrad, UdSSR.

Svoboda, M. (2002) *What is Drought? Drought Indices.* National Drought Migation Center (NDMC), available from: <http://www.drought.unl.edu/whatis/dustbowl.htm> [last visited 2002].

Szilagyi, J. & Parlange, M.B. (1998) Baseflow separation based on analytical solutions of the Boussinesq equation. *J. Hydrol.* **204**, 251–260.

Takasao, T., Takara, K. & Shimizu, A. (1986) A basic study on frequency analysis of hydrological data in the Lake Biwa basin (orig in Japanese). *Annuals, Disas. Prev. Res. Inst.*, Kyoto Univeristy, 29B-2, 157–171.

Tallaksen, L.M. (1987) An evaluation of the Base Flow Index (BFI). *Report Series in Hydrology* no. 16, University of Oslo, Norway.

Tallaksen, L.M. (1989) Analysis of time variability in recessions. In: *FRIENDS in Hydrology* (ed. by L. Roald, K. Nordseth & K.A. Hassel), IAHS Publ. no. 187, 85–96.

Tallaksen, L.M. (1995) A review of baseflow recession analysis. *J. Hydrol.* **165**, 349–370.

Tallaksen, L.M. (2000) Streamflow drought frequency analysis. In: *Drought and Drought Mitigation in Europe* (ed. by J.V. Vogt & F. Somma), Advances in Natural and Technological Hazards Research, Vol. 14, Kluwer Academic Publishers, Dordrecht, the Netherlands, 103–117.

Tallaksen, L.M. & Hisdal, H. (1997) Regional analysis of extreme streamflow drought duration and deficit volume. In: *FRIEND'97 – Regional Hydrology: Concepts and Models for Sustainable Water Resource Management* (ed. by A. Gustard, S. Blazkova, M. Brilly, S. Demuth, J. Dixon, H. van Lanen, C. Llasat, S. Mkhandi & E. Servat), IAHS Publ. no. 246, 141–150.

Tallaksen, L.M., Madsen, H. & Clausen, B. (1997) On the definition and modelling of streamflow drought duration and deficit volume. *Hydrolog. Sci. J.* **42**(1), 15–33.

Tase, N. (1976) Area-deficit-intensity characteristics of droughts. *Hydrology Papers* no. 87, Colorado State University, Fort Collins, USA.

Tase, N. & Yevjevich, V. (1978) Effects of size and shape of a region on drought coverage. *Hydrol. Sci. B.* **23**, 2, 6, 203–213.

Tate, E.L. & Freeman, S. (2000) Three modelling approaches for seasonal streamflow droughts in southern Africa: the use of censored data. *Hydrolog. Sci. J.* **45**(1), 27–42.

Tate, E.L. & Gustard, A. (2000) Drought definition: a hydrological perspective. In: *Drought and Drought Mitigation in Europe* (ed. by J.V. Vogt & F. Somma), Advances in Natural and Technological Hazards Research, Vol. 14, Kluwer Academic Publishers, Dordrecht, the Netherlands, 23–48.

Tate, E.L., Meigh, J., Prudhomme C. & McCartney, M. (2000) *Drought Assessment in Southern Africa Using River Flow Data*. DFID Report 00/4, Institute of Hydrology, Wallingford and Department for International Development (DFID), UK.

Teirney, L.D. & Jowett, I.G. (1990) Trout abundance in New Zealand rivers: an assessment by drift diving. *Freshwater Fisheries Report* 118. Freshwater Fisheries Centre, Ministry of Agriculture and Fisheries, Christchurch, New Zealand.

Telis, P.A. (1992) Techniques for estimating 7-day 10-year low-flow characteristics for ungauged sites on streams in Mississippi. *USGS Water Resources Investigations Report* 91-4130.

Tennant, D.L. (1976) Instream flow regimens for fish, wildlife, recreation, and related environmental resources. In: *Proceedings of the Symposium and Speciality Conference on Instream Flow Needs II* (ed. by J.F. Orsborn & C.H. Allman), American Fisheries Society, Bethesda, Maryland, USA, 359–373.

Tharme, R.E. (1996) Review of international methodologies for the quantification of the instream flow requirements of rivers. *Water Law Review Report for Policy Development*, Freshwater Research Unit, Zoology Department, University of Cape Town, South Africa.

Thiessen, A.H. (1911) Precipitation for large areas. *Mon. Weather Rev.* **39**, 1082–1084.

Thomas, D.M. & Benson, M.A. (1970) Generalisation of stream-rlow characteristics from drainage basin characteristics. *U.S. Geol. Survey, Open File Report*.

Thomsen, R. (1993) Future droughts, water shortages in parts of Western Europe. *Eos, Transactions, American Geophysical Union* **74**(14), 161, 164–165.

Todorovic, P. & Zelenhasic, E. (1970) A stochastic model for flood analysis. *Water Resour. Res.* **6**(6), 1641–1648.

Toebes, C. & Strang, D.D. (1964) On recession curves, 1-Recession equations. *J. Hydrol. NZ* **3**(2), 2–15.

Townsend, C.R., Scarsbrook, M.R & Doledec, S. (1997) The intermediate disturbance hypothesis, refugia, and biodiversity in streams. *Limnol. Oceanogr.* **42**, 938–949.

Troch, P.A., Troch, F.P. de & Brutsaert, W. (1993) Effective water table depth to describe initial conditions prior to storm rainfall in humid regions. *Water Resour. Res.* **29**(2), 427–434.

Troch, P.A., Paniconi, C. & McLaughlin, D. (2003) Catchment-scale hydrological modeling and data assimilation. *Adv. Water Resour.* **26**(2), 131–135.

Überla, K. (1968) *Faktorenanalyse* (factor analysis). Berlin.

Uijlenhoet, R., Wit, M.J.M. de, Warmerdam, P.M.M. & Torfs, P.J.J.F. (2001) Statistical analysis of daily discharge data of the River Meuse and its tributaries (1968–1998): Assessment of drought sensitivity. *Sub-department of Water Resources Report* no. 100, Wageningen University, Wageningen, the Netherlands.

UNDP/MARD (2003) *Report on Water Shortage Situation in Central Province*. Prepared by: UNDP/MARD – DISASTER MANAGEMENT UNIT, VIE/97/002. According to Report No.39/QLN dated 01 August 2003 issued by the Department for Water Resources and

Hydraulic Structures Management, Ministry of Agriculture and Rural Development, Hanoi, Vietnam.

UNEP & Center for Clouds, Chemistry and Climate (C^4) (2002) *The Asian Brown Cloud: Climate and other Environmental Impacts*. United Nations Environment Programme (UNEP), Nairobi.

UNESCO (1997) Southern African FRIEND. *Technical Documents in Hydrology* no. 15, UNESCO, Paris.

UNESCO/WMO (2003) *International Glossary of Hydrology*. Available from: <http://www.unesco.org/water/> [last visited December 2003].

UNFCCC (1992) *United Nations Framework Convention on Climate Change*. Available from: <http://unfccc.int/resource/ccsites/senegal/conven.htm> [last visited 2003].

US National Academy of Sciences (1975) *Understanding Climate Change*. NAS, Washington DC.

Valencia, D.R. & Schaake, J.C. (1973) Disaggregation processes in stochastic hydrology. *Water Resour. Res.* **9**(3).

Vannote, R.L., Minshall, G.W., Cummins, K.W., Sedell, J.R. & Cushing, C.E. (1980) The river continuum concept. *Can. J. Fish. Aquat. Sci.* **37**, 130–137.

Verschuren, D., Laird, K.R. & Cumming, B.F. (2000) Rainfall and drought in equatorial east Africa during the past 1100 years. *Nature* **403**, 410–414.

Vogel, R.M. (1986) The probability plot correlation coefficient test for the normal, lognormal and Gumbel distributional hypotheses. *Water Resour. Res.* **22**(4), 587–590. Correction, *Water Resour. Res.* **23**(10), 2013.

Vogel, R.M. & Stedinger, J.R. (1987) Generalized storage-reliability-yield relationships. *J. Hydrol.* **89**, 303–327.

Vogel, R.M. & Kroll, C.N. (1992) Regional geohydrologic- geomorphic relationships for the estimation of low-flow statistics. *Water Resour. Res.* **28**(9), 2451–2458.

Vogel, R.M. & Fennessey, N.M. (1993) L-moment diagrams should replace product moment diagrams. *Water Resour. Res.* **29**(6), 1745–1752.

Vogel, R.M. & Fennessey, N.M. (1994) Flow Duration Curves. I: New Interpretation and Confidence Intervals. *J. Water Res. Pl.-ASCE* **120**(4), 485–504.

Vogel, R.M. & Wilson, I. (1996) The probability distribution of flood, drought and average streamflow in the United States. *J. Hydrol. Eng.* **1**(2), 69–76.

Vogt, J.V. & Somma, F. eds. (2000) *Drought and Drought Mitigation in Europe*. Advances in Natural and Technological Hazards Research, Vol. 14, Kluwer Academic Publishers, Dordrecht, the Netherlands.

Vörösmarty, C.J., Green, P, Salisbury, J. & Lammers, R.B. (2000) Global Water Resources: Vulnerability from Global Climate Change and Population Growth. *Science* **289**, 284–288.

Vose, D. (2000) *Risk Analysis: a Quantitative Guide*. 2nd edn, Wiley, Chichester.

Wagner, W., Scipal, K. & Pathe, C. (2003) Evaluation of the agreement between the first global remotely sensed soil moisture data with model and precipitation data. *J. Geophys. Res.* **108**, 1–10.

Walker, K.F., Byrne, M., Hickey, C.W. & Roper, D.S. (2001) Freshwater Mussels (Hyriidae) of Australasia. In: *Ecology and Evolution of the Freshwater Mussels Unionoida* (ed. by G. Bauer & K. Wächtler), Springer-Verlag, Berlin, 5–31.

Wallis, J.R. (1980) Risk and uncertainties in the evaluation of flood events for the design of hydraulic structures. In: *Piene e Siccità* (ed. by E. Guggino, G. Rossi, & E. Todini), Fondazione Politecnica del Mediterraneo, Catania, Italy, 3–36.

Wallis, J.R. & Wood, E.F. (1985) Relative accuracy of log Pearson III procedures. *J. Hydraul. Eng.-ASCE* **111**(7), 1043–1056.

Wang, H.F. & Anderson, M.P. (1982) *Introduction to Groundwater Modelling. Finite Difference and Finite Element Methods.* Academic Press, San Diego, USA.

Wang, Q.J. (1996) Direct sample estimators of L moments. *Water Resour. Res.* **32**(12), 3617–3619.

Wang, Q.J. (1997) LH moments for statistical analysis of extreme events. *Water Resour. Res.* **33**(12), 2841–2848.

Ward, R.C. & Robinson, M. (2000) *Principles of Hydrology.* McGraw Hill.

Warnick, C.C. (1984) *Hydropower Engineering.* Prentice-Hall, New Jersey, USA.

Warrick, R.A. (1980) Drought in the Great Plains: A Case Study of Research on Climate and Society in the USA. In: *Climatic Constraints and Human Activities* (ed. by J. Ausubel & A.K. Biswas), IIASA Proceedings Series, Vol. 10. Pergamon Press, New York, 93–123.

Waters T.F. (1995) *Sediment in Streams. Sources, Biological Effects, and Control.* Bethesda, Maryland, American Fisheries Society.

Waugh, J., Freestone, H. & Lew, D. (1997) Historic floods and droughts in New Zealand. In: *Floods and Droughts: the New Zealand Experience* (ed. by M.P. Mosley & Ch.P. Pearson), New Zealand Hydrological Society Inc., Wellington North, New Zealand, 29–50.

Weibull, W. (1961) *Fatigue Testing and Analysis of Results.* Pergamon Press.

Weisberg, S. (1985) *Applied Linear Regression.* Wiley series in probability and mathematical statistics, New York.

Weiss, G. (1977) Shot noise models for the generation of synthetic streamflow data. *Water Resour. Res.* **13**(1), 101–108.

Werner, P.W. & Sundquist, K.J. (1951) On the groundwater recession curve for large watersheds. IAHS Publ. no. 33, 202–212.

Wesselink, A., Hagemann, I., Demuth, S. & Gustard, A. (1994) Computer application of regional low flow study in Baden-Württemberg. In: *FRIEND – Flow Regimes from International Experimental and Network Data* (ed. by P. Seuna, A. Gustard, N.W. Arnell & G.A. Cole), IAHS Publ. no. 221, 141–150.

Wetzel, K.L. & Bettandorff, J.M. (1986) Techniques for estimating streamflow characteristics in the eastern and interior coal provinces of the United States. *U.S. Geol. Survey Water Supply Paper* 2276, Alexandria, Virginia, USA.

White, I., Falkland, T. & Scott, D. (1999) Droughts in small coral islands: case study, South Tarawa, Kiribati. *IHP-V, Technical Documents in Hydrology* no 26.

White, P. (1997) Hydrological extremes and the groundwater system. In: *Floods and Droughts: the New Zealand Experience* (ed. by M.P. Mosley & Ch.P. Pearson), New Zealand Hydrological Society Inc., Wellington North, New Zealand, 143–158.

White, P.S. & Pickett, S.T.A. (1985) Natural disturbance and patch dynamics: an introduction. In: *The Ecology of Natural Disturbance and Patch Dynamics* (ed. by S.T.A. Pickett & P.S. White), Academic Press, San Diego, California, 3–13.

Wilding, T.K. (2003) Minimum flow report for the Tauranga area. *NIWA Client Report* HAM2003-043, National Institute of Water and Atmospheric Research, PO Box 115, Hamilton, New Zealand.

Wilhite, D.A. (2000a) Droughts as a natural hazard: concepts and definitions. In: *DROUGHT, A Global Assessment,* Vol I (ed. by D.A. Wilhite), Routledge Hazards and Disasters Series, Routledge, London, 3–18.

Wilhite, D.A. ed. (2000b) *DROUGHT, A Global Assessment,* Vol I & II. Routledge Hazards and Disasters Series, Routledge, London.

Wilhite, D.A. & Glantz, M.H. (1985) Understanding the drought phenomenon: The role of definitions. *Water Int.* **10**(3), 111–120.

Wilhite, D.A., Hayes, M.J., Knutson, C. & Smith, K.H. (2000a) Planning for Drought: Moving from crisis to risk management. *J. Am. Water Resour. A. 36,* 697–710.

Wilhite, D.A., Hayes, M.J. & Svoboda, M.D. (2000b) Drought monitoring and assessment: Status and trends in the United States. In: *Drought and Drought Mitigation in Europe* (ed. by J.V. Vogt & F. Somma), Advances in Natural and Technological Hazards Research, Vol. 14, Kluwer Academic Publishers, Dordrecht, the Netherlands, 149–160.

Willeke, G., Hosking, J.R.M., Wallis, J.R. & Guttman, N.B. (1994) The national drought atlas. *Institute for Water Resources Report* 94-NDS-4, U.S. Army Corps of Engineers, USA.

Williams, D.D. & Hynes, H.B.N. (1977) The ecology of temporary streams II. General remarks on temporary streams. *Int. Rev. Ges. Hydrobio.* **62**, 53–61.

Wilson, E.B. & Hilferty, M.M. (1931) The distribution of chi-square. *Proc. Natl. Acad. Sci. USA* **17**, 684.

Wilson, E.M. (1990) *Engineering Hydrology.* 4th edn, The Macmillan Press Ltd., UK.

Wiltshire, S.E. (1985) Grouping basins for regional flood frequency analysis. *Hydrol. Sci. J.* **30**(1), 151–159.

Wiltshire, S.E. (1986a) Identification of homogeneous regions for flood frequency analysis. *J. Hydrol.* **84**, 287–302.

Wiltshire, S.E. (1986b) Regional flood frequency analysis I: Homogeneity statistics. *Hydrolog. Sci. J.* **31**(3), 321–333.

Wit, M. de, Warmerdam, P., Torfs, P., Uijlenhoet, R., Roulin, E., Cheymol, A., Deursen, W. van, Walsum, P. van, Kwadijk, J., Ververs, M. & Buitenveld, H. (2001) Effect of climate change on the hydrology of the River Meuse. *Report Wageningen University*, Wageningen, the Netherlands.

Wittenberg, H. & Sivapalan, M. (1999) Watershed groundwater balance estimation using streamflow recession analysis and baseflow seperation. *J. Hydrol.* **219**, 20–33.

WMO (1994) *Guide to Hydrological Practices, Data Acquisition and Processing, Analysis, Forecasting and other Applications.* 5th edn, WMO-No.168, Geneva, Switzerland.

Woo, M.-K. & Tarhule, A. (1994) Streamflow droughts of northern Nigerian rivers. *Hydrolog. Sci. J.* **39**(1), 19–34.

Wood P.J. & Armitage P.D. (1997) Biological effects of fine sediment in the lotic environment. *Environ. Manage.* **21**, 203–217.

Wright, C.E. (1970) Catchment characteristics influencing low flows. *Water Water Eng.* **74**, 468–471.

Wright, C.E. (1974) The influence of catchment characteristics upon low flows in South East England. *Water Services* **78**, 227–230.

Yevjevich, V. (1964) Fluctuation of wet and dry years, 2. analysis by serial correlation. *Colorado State University Paper* no. 4, Forth Collins, Colorado, USA.

Yevjevich, V. (1967) An objective approach to definition and investigations of continental hydrologic droughts. *Hydrology Papers* 23, Colorado State University, Fort Collins, USA.

Yevjevich, V., Cunha, L. da & Vlachos, E. eds. (1983) *Coping with Droughts.* Water Resources Publications, Colorado, USA.

Young, A.R., Croker, K.M. & Sekulin, A.E. (2000a) Novel techniques for characterizing complex water use patterns within a network based statistical model. LOIS Special Volume, *Sci. Total Environ.* **251/252**, 277–291.

Young, A.R., Gustard, A., Bullock, A., Sekulin, A.E. & Croker, K.M. (2000b) A river network based hydrological model for predicting natural and influenced flow statistics at ungauged sites. LOIS Special Volume, *Sci. Total Environ.* **251/252**, 293–304.

Young, A.R., Round, C.E. & Gustard, A. (2000c) Spatial and temporal variations in the occurrence of low flow events in the UK. *Hydrol. Earth Syst. Sc.* **4**(1), 35–45.

Yu, Z. & Ito, E. (1999) Possible solar forcing of century-scale drought frequency in the northern Great Plains. *Geology* **27**(3), 263–266.

Yue, S., Hashino, M., Bobée, B., Rasmussen, P.F. & Ouarda, T.B.M.J. (1999) Derivation of streamflow statistics based on a filtered point process. *Stoch. Env. Res. Risk A.* **13**, 317–326.

Zaidman, M., Rees, G. & Gustard, A. (2001) Drought visualisation. In: *Assessment of the Regional Impact of Droughts in Europe* (ed. by S. Demuth & K. Stahl), Final Report, EU contract ENV4-CT-97-0553, Institute of Hydrology, University of Freiburg, Freiburg, Germany.

Zaidman, M., Keller, V., Young, A.R. & Cadman, D. (2003) Flow-duration-frequency behaviour of British rivers based on annual minima data. *J. Hydrol.* **277**, 195–213.

Zar, J.H. (1999) *Biostatistical analysis.* 4th edn, New Jersey, Prentice-Hall.

Zelenhasic, E. & Salvai, A. (1987) A method of streamflow drought analysis. *Water Resour. Res.* **23**(1), 156–168.

Ziegler, A.D., Sheffield, J., Wood, E.F., Nijssen, B., Maurer, E.P. & Lettenmaier, D.P. (2002) Detection of intensification of the global water cycle: the potential role of FRIEND. In: *FRIEND 2002 – Regional Hydrology: Bridging the Gap between Research and Practice* (ed. by H.A.J. van Lanen & S. Demuth), IAHS Publ. no. 274, 51–57.

Index